Engineering Materials

More information about this series at http://www.springer.com/series/4288

Jacek J. Skrzypek · Artur W. Ganczarski
Editors

Mechanics of Anisotropic Materials

Springer

Editors
Jacek J. Skrzypek
Solid Mechanics Division, Institute of Applied Mechanics
Cracow University of Technology
Krakow
Poland

Artur W. Ganczarski
Solid Mechanics Division, Institute of Applied Mechanics
Cracow University of Technology
Krakow
Poland

ISSN 1612-1317 ISSN 1868-1212 (electronic)
Engineering Materials
ISBN 978-3-319-17159-3 ISBN 978-3-319-17160-9 (eBook)
DOI 10.1007/978-3-319-17160-9

Library of Congress Control Number: 2015937362

Springer Cham Heidelberg New York Dordrecht London

Printed on acid-free paper

Springer International Publishing AG Switzerland is part of Springer Science+Business Media (www.springer.com)

To our Families
for their assistance and patience

Foreword

This book which was prepared by a team of Polish scientists from the University of Technology in Cracow and edited by Jacek J. Skrzypek and Artur W. Ganczarski is devoted to the *Mechanics of Anisotropic Materials*. This topic became more and more important during the past years. Anisotropy with respect to material behavior should be taken into account—otherwise recent design requirements like reduction of mass, precise shape prediction after mechanical treatment, etc., cannot be fulfilled. Including anisotropy in simulations, for example, based on commercial finite element codes is not a trivial task since the origins of anisotropy are related to the microstructure of materials as usual disregarded in phenomenological modeling. In addition, one has to distinguish initially anisotropic materials and materials that are initially isotropic, but under loading the evolution of anisotropy can be established. This makes the constitutive description not easier—instead of "pure" constitutive equations one needs additionally evolution equations describing the development of the anisotropy. Recently various monographs in this field were published. But there is a need for more because up to now no general accepted theory exists.

This monograph contains 7 chapters prepared by Halina and Władysław Egner, Artur W. Ganczarski, Szymon Hernik, and Jacek J. Skrzypek. The chapters are related to the theory and numerical simulation of anisotropic materials.

Chapter 1 is an introduction to the mechanics of anisotropic materials. This is at first a summary of the state of the art in this field. In addition, different representations of classical equations are given. It is obvious that the mathematical background is worked out and there are no open questions. Looking on the number of unknown material parameters the question arises of how to identify these parameters. The last part of Chap. 1 is devoted to the classification and description of anisotropy. This part is unique since a lot of special cases are presented and analogy between two fourth order tensors (the Hookean tensor and the tensor related to the von Mises yield and failure) is demonstrated.

Chapter 2 is focused on isotropic and anisotropic linear viscoelastic materials. The first part is a nice summary of the basics of rheological models. The extension of the correspondence principle to anisotropic materials is given.

Chapter 3 is devoted to the anisotropic behavior of composite materials. Anisotropy is the usual property of any composite (even in the case of short fibers with chaotic distribution, a significant anisotropy as a result of the injection molding process can be established. This anisotropy results in approximately 10 % error in displacement predictions, but 30 % in stress estimates. At first, the analogy between crystal lattice unit cell and composite representative element is discussed. This allows using the methods for modeling the symmetries in both cases. Second, the bounds and micromechanically based homogenization methods are presented. The concept of limit surfaces, which is usually not related to the micro-scale observation is an engineering tool to model the limit or failure behavior. It can be used for various limit states (for example, the beginning of plastic flow or damage processes). The mathematics (convexity and normality rules, for example) is presented in Chap. 4 and the authors demonstrate how to use the mathematical tools in phenomenological modeling with respect to engineering applications.

Chapters 5 and 6 present two topics which are at the moment under discussion in the scientific literature: the termination of the elastic range of pressure insensitive and sensitive materials. Both items are important for any isotropic and anisotropic initial yield or failure criteria. In addition, starting with Bridgeman's famous publications at the end of the 1940s/beginning of the 1950s of the past century, which discussed for the first time the influence of hydrostatic stress states on yielding and failure, more and more publications are related to this topic. Jacek J. Skrzypek and Artur W. Ganczarski present their own points of view. At the same time they compare their results with the others. It is obvious that a hierarchical embedding of various approaches can be stated.

Finally in Chap. 7 a short classification of constitutive equations for dissipative materials is given. This is an important task, especially if one deals with damage-induced anisotropy.

The present book is a monograph, but at the same time it can be recommended for a first reading. It can be used also as a textbook for students of Master's or PhD studies because of the excellent didactic representation.

Magdeburg
April 8, 2015

Prof. Dr.-Ing.habil. Dr.h.c.mult. Holm Altenbach
Lehrstuhl für Technische Mechanik
Institut für Mechanik
Fakultät für Maschinenbau
Otto-von-Guericke-Universität Magdeburg

Preface

This monograph is focused on constitutive description of mechanical behavior of engineering materials: both conventional (e.g., polycrystalline homogeneous isotropic or anisotropic metallic materials) and nonconventional ones (e.g., heterogeneous multicomponent, usually anisotropic composite materials) fabricated by modern material engineering. Effective material properties at the macrolevel depend on both the material microstructure (isotropic or originally anisotropic in general case) as well as dissipative phenomena occurred on fabrication and consecutive loading phase, resulting in irreversible microstructure changes. The material symmetry is a background and anisotropy is a core around which the book is formed. Revision of classical rules of enhanced constitutive description of materials, capable of capturing virgin or acquired anisotropy, hydrostatic pressure dependence, distortion of initial and subsequent yield/failure surfaces, as well as coupled several dissipative phenomena, such as (thermo)elastic, viscoelastic, elastic-plastic-damage, is necessary.

In the past decade new developed technologies for manufacturing of advanced engineering materials have stimulated numerous original papers addressed to more enhanced and rigorous constitutive description and its experimental verification. Among the recent books attempting to combine a progress in constitutive description of complex materials with modern engineering expectations, some can be mentioned. These are: *The Mechanics of Constitutive Modeling* by Ottosen and Ristinmaa, Elsevier 2005; *Advanced Materials and Structures for Extreme Operating Conditions* by Skrzypek, Ganczarski, Rustichelli and Egner, Springer 2008; *Innovative Technological Materials*, Eds. Rustichelli and Skrzypek, Springer 2010; *Continuum Damage Mechanics* by Murakami, Springer 2012; *Damage Mechanics in Metal Forming* by Saanouni, Wiley 2012; *Micromechanics of Composite Materials* by Aboudi, Arnold and Bednarcyk, Elsevier 2013; *Plasticity of Pressure Sensitive Materials*, Eds. Altenbach and Öchsner, Springer 2014, to mention only a few of them. A variety of pioneering original papers given by, e.g., Chaboche, Voyiadjis, Aboudi, Barlat, Khan, and many others, need to be summarized and presented in a comparative way, to emphasize their significance in a growth of knowledge in the field addressed. The present monograph is an attempt to build a bridge between a large number of the new technology inspired and well-experimented established research papers on one hand, and the systematic and comparative study from the viewpoint of rigorous classical thermodynamics-based constitutive descriptions of anisotropic materials on the other hand.

A concise classification of anisotropic materials with respect to symmetry of elastic matrices referred to as crystal lattice symmetry, and the extended analogy between symmetries of constitutive material matrices (elastic and yield/failure), are discussed in Chap. 1. This chapter provides necessary tools for enhanced constitutive description of materials which exhibit the virgin anisotropy or the damage or phase change acquired anisotropy, following microstructural changes. Apart from classical definitions of single tensor invariants, the choice of state variables necessary to describe irreversible microstructural changes accompanying coupled dissipative phenomena, as well as basic definitions of common invariants of either two second-order tensors (e.g., stress/strain and damage tensors) or two different-order tensors (e.g., stress/strain and fourth-order structural tensors), are given.

The aim of Chap. 2 is to show useful enhancement of the Alfrey–Hoff analogy to a broader class of material anisotropy, for which separation of the volumetric and shape change effects from total viscoelastic deformation does not occur. Such extension requires use of the vector–matrix notation for description of the general constitutive response of anisotropic linear viscoelastic material. When implemented to anisotropic composite materials, which exhibit linear viscoelastic response, the classically used homogenization techniques for averaged elastic matrix can be implemented to viscoelastic work-regime for associated fictitious elastic Representative Unit Cell of composite material. Next, subsequent application of the inverse Laplace transformation (cf. Haasemann and Ulbricht) is applied. Similarly, the well-established Hill upper and lower bounds for effective elastic matrices can be extended to anisotropic linear viscoelastic composite materials. In the space of transformed variable, instead of time space, the classical homogenization rules for fictitious elastic composite materials can be adopted.

Mechanics of composite materials in the past decade was one of the most rapidly explored and developed engineering areas, basically due to huge progress in composite fabrication and engineering use. The main problem related to Chap. 3 is focused on and how to correctly predict averaged effective properties by implementation of numerous homogenization techniques. Useful classification of composites with respect to the format of effective stiffness matrix, based on analogy between the crystal lattice symmetry and respective configuration of reinforcement in the RUC, is given. The conventionally used Hill theorem on upper and lower bounds by Voigt and Reuss isotropic estimates, for approximate determination of stiffness and compliance matrices of anisotropic composite, is studied. A consistent application of the Hill theorem to the elements of elastic stiffness or compliance matrices enables to rule out some peculiarities of the Poisson ratio diagrams met in the respective bibliography. The new effective approximation of the mechanical modules of unidirectionally reinforced composites by use of weighted average between the Voigt and Reuss upper and lower estimates is also proposed.

The general nature of yield or failure criteria terminating elastic range of isotropic or anisotropic materials is discussed in Chap. 4. The hydrostatic pressure sensitivity of anisotropic materials can be captured either by the first stress and second common deviatoric invariants' direct use, or by the second common stress invariant use in an indirect fashion. Tension/compression asymmetry in anisotropic materials is accounted for either by presence of the first common invariant (translation only) or third common invariant (distortion). Comparison of two ways to capture anisotropic response: more rigorous explicit common invariant formulations, or implicit approaches based on extension of traditional isotropic criteria in terms of transformed invariants capable of capturing a complete distortion (Barlat, Khan, etc.), is shown. Convexity requirement of limit surfaces is discussed and compared for two material behaviors by use of: Drucker's material stability postulate extended to multi-dissipative response, or Sylvester's stability condition based on positive definiteness of the tangent stiffness or compliance matrices of hyperelastic material. Generalized Drucker's postulate based on elastic-plastic stiffness matrix is also shown.

Basic features of isotropic or anisotropic initial yield criteria are discussed in Chap. 5 following explicit versus implicit formulations. The explicit description of anisotropy is rigorously based on theory of common invariants. The implicit approach involves linear transformation tensor of the Cauchy stress to enhance the classical isotropic criteria for capturing anisotropy, hydrostatic pressure sensitivity, and asymmetry of yield surface. The advantages and differences of both formulations are critically presented. Incidental convexity loss of the classical Hill'48 yield surface in the case of strong orthotropy is examined and highlighted in contrast to unconditionally stable von Mises–Hu–Marin criterion. Different transversely isotropic yield criteria are distinguished in light of irreducibility or reducibility to the isotropic Huber–von Mises criterion in the transverse isotropy plane, and the appropriate class of tetragonal symmetry (classical Hill's formulation) or hexagonal symmetry (hexagonal Hill or von Mises–Hu–Marin criteria) are considered. The new hybrid formulation, applicable for some engineering materials based on additional bulge test, is also proposed.

Chapter 6 comprises yield/failure initiation criteria, discussed in detail with respect to the three following effects: the hydrostatic pressure dependence, tension/compression asymmetry, isotropic or anisotropic response. In case of anisotropic materials the explicit formulation, based on either all three common invariants or first and second common invariants, are addressed especially to case of transverse isotropy where difference between tetragonal versus hexagonal symmetry is highlighted. A mixed approach to formulate the pressure sensitive and tension/compression asymmetric initial failure criteria, capable to describe fully distorted limit surfaces, which are based on both all stress invariants and the second common invariant, are proposed. It is particularly addressed to orthotropic materials where fourth-order linear transformation tensors are used to achieve extension of the respective isotropic criteria.

Chapter 7 presents the general features of thermodynamically based constitutive modeling. The type of constitutive modeling, based on a hypothesis that the state of a material is entirely determined by certain values of some variables of state, is well adapted to the formulation of constitutive equations for deformable solids with several dissipative phenomena. The classification of constitutive equations is presented for the following materials: elastic-damage, elastic-plastic, thermo-elastic-(visco)plastic, and elastic-plastic-damage, in a critical and comparative way. Damage acquired anisotropy and unilateral damage effect are accounted for. When plasticity is considered, an alternative multiscale approach, based on polycrystalline calculations for the description of yield anisotropy and its evolution with accumulated deformation, is also discussed. As an example of thermo-plastic coupling, the fatigue behavior in nonisothermal conditions is analyzed. Numerical simulations which indicate the significant influence of temperature rate on the response of constitutive model when cyclic thermo-mechanical loading is considered, are performed.

Finally, all recent trends to account for modeling material anisotropy and coupling of dissipative phenomena have been highlighted and compared. The advantages and difficulties of both a traditional explicit concept of consistent common invariant-based polynomial formulation versus recently dynamically developed implicit approach by extension of isotropic criteria with use of linear transformation of the Cauchy stress tensor, are critically reviewed. Formal and complete analysis of couplings between several dissipative phenomena (e.g., thermo-plastic coupling, damage-plasticity coupling, nonisothermal thermo-damage-plasticity coupling, etc.) are systematically analyzed in frame of irreversible thermodynamics including internal variables.

Jacek J. Skrzypek
Artur W. Ganczarski

Acknowledgments

A scope of this monograph can be considered as enrichment and essential extension of the author's two earlier books: *Mechanics of Novel Materials* by Artur W. Ganczarski and Jacek J. Skrzypek, and *Constitutive Modeling of Coupled Dissipative Phenomena in Engineering Materials* by Halina Egner, both published (in Polish) by Cracow University of Technology Publishing House, in 2013.

The editors appreciate the thorough work and valuable remarks of Profs. Zbigniew Kowalewski, Andrzej Seweryn, and Andrzej Litewka, who reviewed the above books. The editors do appreciate also informal remarks by Profs. Akhtar Khan, Holm Altenbach, and others who brought about important impacts on the final shape of this monograph.

The Grant UMO-2011/03/B/ST8/05132 from the National Science Center NCN, Poland is gratefully acknowledged.

Contents

Contributors

H. Egner Solid Mechanics Division, Institute of Applied Mechanics, Cracow University of Technology, Kraków, Poland

W. Egner Division of Technical Mechanics, Institute of Applied Mechanics, Cracow University of Technology, Kraków, Poland

Artur W. Ganczarski Solid Mechanics Division, Institute of Applied Mechanics, Cracow University of Technology, Kraków, Poland

S. Hernik Solid Mechanics Division, Institute of Applied Mechanics, Cracow University of Technology, Kraków, Poland

Jacek J. Skrzypek Solid Mechanics Division, Institute of Applied Mechanics, Cracow University of Technology, Kraków, Poland

Notations

General Rules

- superscripts ^{e},^{d} ,^{p} ,^{ph} ,^{v} ,^{T} refer to elastic, damage, plastic, phase change, viscous, and thermal quantities, respectively
- superscripts ^{m},^{r} refer to matrix and reinforcement of composite, respectively
- superscript ^{res} refers to residual quantity
- superscripts ^{V},^{R} refer to Voigt or Reuss estimate of quantity
- left subscripts preceding matrices $_{sec}\mathbb{X}$ or $_{tan}\mathbb{X}$ refer to tangent or secant matrices, respectively
- tilde over symbol $\tilde{}$ denotes effective quantity including effect of damage
- line over symbol $\bar{}$ denotes quantity averaged over representative element
- double line over symbol $=$ denotes quantity after weighting homogenization
- roof over symbol $\hat{}$ denotes Lapace's transform of quantity

Operators

x	Scalar
$x_i, \boldsymbol{x}$	Vector
$x_{ij}, \boldsymbol{x}$	Second-rank tensor
$x_{ijkl}, \mathbb{X}$	Fourth-rank tensor
$[\boldsymbol{X}]$	Matrix
$[\boldsymbol{X}]^{\mathrm{T}}$	Matrix transposed
det	Determinant
$\|x\|$	Absolute value
$\|\|x\|\|$	Norm
$\langle x \rangle$	Macaulay bracket $= \begin{cases} x & \text{if } x \geq 0 \\ 0 & \text{if } x < 0 \end{cases}$

$\frac{\mathrm{d}}{\mathrm{d}x}$ Operator of derivative

$\dot{x}$ Time derivative $= \frac{\mathrm{d}x}{\mathrm{d}t}$

$\frac{\mathrm{D}}{\mathrm{D}t}$ Absolute derivative

$\frac{\partial}{\partial x}$ Partial derivative

$\mathcal{D}$ Operator of time derivative $= \frac{\partial}{\partial t}$

∇ Nabla $= \frac{\partial}{\partial x} + \frac{\partial}{\partial y} + \frac{\partial}{\partial z}$

Δ Increment

$\boldsymbol{x} \otimes \boldsymbol{y}$ Diadic product of vectors

$\boldsymbol{x} \cdot \boldsymbol{y}$ Contracted product of tensors

$\mathrm{tr}(\boldsymbol{x})$ Trace operator or first invariant of tensor

$J_2(\boldsymbol{x})$ Second invariant of tensor

$\mathrm{H}(t)$ Heaviside's function

δ_{ij} Kronecker's symbol

$\mathcal{L}, \mathcal{L}_\mathrm{C}$ Laplace and Laplace–Carson integral transforms

$\exp(x)$ Exponential function

$x * y$ Convolution operator

Latin and Greek Letters Used as Symbols

a	c	e	g	i	k	m	p	s	u	w	y	α	ε	ϑ	ν	ρ	ϕ	ω	Λ
A	C	E	G	I	K	M	P	S	U	W	Y	β	ζ	κ	ξ	σ	φ	Γ	Φ
b	d	f	h	j	l	n	r	t	v	x	z	γ	η	λ	π	τ	χ	Δ	Ψ
B	D	F	H	J	L	N	R	T	V	X	Z	δ	θ	μ	ρ	υ	ψ	Θ	Ω

Symbols

a, b Inner and outer radius of cylinder, respectively

a, b, c Lattice edges

a_{ij} Direction cosines

A Area of surface

$\boldsymbol{A}$ Strain concentration tensor

A, B, C Material coefficients in Burzyński yield/failure criterion

b_i Vector of body forces

b, h, l Dimensions of representative unit cell

$\boldsymbol{B}$ Stress concentration tensor

c	Coefficient of cohesion, material constant in Drucker's criterion
C	Complementary energy per unit volume
$D, D_i, \boldsymbol{D}, \mathbb{D}$	Damage parameter, vector, second-rank and fourth-rank tensors
$\mathbf{e}, e_{ij}$	Strain deviator
E	Young's modulus
E_{ii}	Axial modules (generalized Young's modulus)
$E_{ijkl}, E_{ijkl}^{-1}, \mathbb{E}, \mathbb{E}^{-1}$	Hooke's stiffness or compliance tensors
$E(t)$	Relaxation function
${}^{\mathrm{ve}}E_{ijkl}(t), {}^{\mathrm{ve}}\mathbb{E}(t)$	Tensor of relaxation functions
f	Yield function, dissipation potential
F	Force
g	Plastic potential function
G	Kirchhoff's modulus
$\mathcal{G}$	Gibbs' potential function
G_{ij}	Shear modules (generalized Kirchhoff's modules)
$h_{ij}^{(1)}, h_{ijkl}^{(2)}, h_{ijklmn}^{(3)}$	Structural anisotropy tensors after Kowalsky
$[\mathcal{H}_{ij}]$	Hessian's matrix
I	Sectional moment of inertia
$I_{ijkl}, \mathbb{I}$	Four-rank unit tensor
$I_{i\varepsilon}$	Principal invariants of strain tensor
$I_{i\sigma}$	Principal invariants of stress tensor
J_{ie}	Basic invariants of strain deviator
$J_{i\varepsilon}$	Basic invariants of strain tensor
J_{is}	Basic invariants of stress deviator
$J_{i\sigma}$	Basic invariants of stress tensor
$\{J^{\mathrm{k}}\}$	Vector of thermodynamic conjugate forces
$J(t)$	Creep compliance function
${}^{\mathrm{ve}}J_{ijkl}(t), {}^{\mathrm{ve}}\mathbb{J}(t)$	Tensor of creep functions
k	Strength, yield point stress
$k_{\mathrm{c}}, k_{\mathrm{t}}, k_{\mathrm{s}}$	Yield point stress under compression, tension, and pure shear, respectively
k_x, k_y, k_z	Plastic tension limits along axes x, y, z, respectively
k_{yz}, k_{zx}, k_{xy}	Plastic shear limits in planes $(yz), (zx), (xy)$
K	Bulk modulus
l	Length
m	Compressive to tensile strength ratio
M	Bending moment
$\boldsymbol{M}^{(i)}$	Second-rank structural tensors
$\mathbb{M}$	Fourth-rank damage effect tensor
n_{ij}	Direction cosines
$\boldsymbol{n}$	Unit vector

N	Axial force
p	Cumulative plastic stain, pressure
p_i	Traction vector
$\mathcal{P}, \mathcal{Q}$	Linear differential operators
q_{ij}	Scalar products of corresponding direction cosines
q_i	Outward heat flux
$[\mathbb{Q}]$	Transformation matrix
r	Heat source intesity
R	Hosford–Backhofen parameter
$\mathbb{R}$	Fourth-rank continuity tensor
s	Entropy per unit volume
$\boldsymbol{s}, s_{ij}$	Stress deviator
$\boldsymbol{S}', \boldsymbol{S}''$	Transformed Cauchy stress deviators
t	Time
T	Absolute temperature
u	Internal energy per unit volume
u_i	Displacement vector
v_i	Material velocity
V	Volume
V_{f}	Volume fraction
$\{V^{\mathrm{k}}\}$	Vector of internal structural variables
w	Deflection of beam
$\mathcal{W}$	Strain energy per unit volume
x_i, x	Euler's (space) coordinates
X_i	Lagrange's (material) coordinates
$\mathbf{1}$	Second-rank unit tensor
α	Material constant in Drucker–Prager's criterion
α, β, γ	Lattice angles
γ_{ij}	Shear strain
$\varepsilon_{ij}, \boldsymbol{\varepsilon}$	Strain tensor
ε_{ij}^*	Strain tensor including crack/opening effect
ε_{m}	Mean strain
$\varepsilon_{\mathrm{off}}$	Plastic offset
$\epsilon_{ij}, \in_{ij}$	Green's and Almansi's strain tensors, respectively
η	Coefficient of dynamic viscosity
η_i	Coefficients of free energy function in Murakami–Kamiya's model
$\eta_{i(jk)}$	Rabinovich's modules
ϑ_i	Coefficients of Gibbs' complementary function in Hayakawa–Murakami's model
κ	Curvature
λ	Extension, plastic multiplier
π_{ij}	Tensor of plastic anisotropy of second-rank
μ, λ	Lamé's coefficients

$\mu_{ij(kl)}$	Chencov's modules
ν	Poisson's ratio
ν_{ij}	Generalized Poisson's coefficients
ξ, ρ, θ	Haigh–Westergaard's coordinates
ϕ	Angle of internal friction
ρ	Mass density
$\sigma_{ij}, \boldsymbol{\sigma}$	Stress tensor
σ_{eq}	Effective stress
σ_{h}	Hydrostatic stress
σ_{m}	Mean stress
ζ	Material constant
τ_{ij}	Components of shear stress
ψ	Helmholtz's free energy
Γ	Domain
Σ_{ij}	Second Piola–Kirchhoff's stress tensor
$\boldsymbol{\Sigma}$	Transformed Cauchy stress tensor
Θ	Volume change, dilatation
$\Pi_{ij}, \Pi_{ijkl}, \pi, \mathbb{III}$	Structural tensors of plastic/failure anisotropy
Ω_{ij}	Skew-symmetric spin tensor

Chapter 1
Introduction to Mechanics of Anisotropic Materials

Artur W. Ganczarski, H. Egner and Jacek J. Skrzypek

Abstract This book is focused on constitutive description of mechanical behavior of engineering materials: both conventional (e.g., polycrystalline homogeneous isotropic or anisotropic metallic materials) and nonconventional ones (e.g., heterogeneous multicomponent usually anisotropic composite materials) fabricated by modern material engineering. Effective material properties at the macrolevel depend on both the material microstructure (isotropic or originally anisotropic in general case) and on dissipative phenomena occurred on fabrication and consecutive loading phase resulting in irreversible microstructure changes (acquired anisotropy). The material symmetry is a background and anisotropy is a core around which the book is formed. In this way a revision of classical rules of enhanced constitutive description of materials is required. The aim of this introductory chapter lies in providing, apart from classical definitions of tensor single invariants, also the choice of state variables necessary to describe irreversible microstructure changes accompanying coupled dissipative phenomena, and basic definitions of common invariants of either two second-order tensors (e.g., stress/strain and damage tensors) or two different-order tensors (e.g., stress/strain and fourth-order structural tensors). Concise classification of anisotropic materials with respect to symmetry of elastic matrices as referred to the crystal lattice symmetry is given, and extended analogy between symmetries of constitutive material matrices (elastic and yield/failure) is also discussed. Next, strain and complementary energy as function of either stress/strain invariants (initial elastic isotropy) or common stress/strain—damage invariants (damage acquired anisotropy) are mentioned. Constitutive equation of linear elasticity in terms of common invariants of strain and structural orthotropic tensors is given. The scope of

A.W. Ganczarski (✉) · H. Egner · J.J. Skrzypek
Solid Mechanics Division, Institute of Applied Mechanics,
Cracow University of Technology, al. Jana Pawła II 37, 31-864 Kraków, Poland
e-mail: Artur.Ganczarski@pk.edu.pl

H. Egner
e-mail: Halina.Egner@pk.edu.pl

J.J. Skrzypek
e-mail: Jacek.Skrzypek@pk.edu.pl

J.J. Skrzypek and A.W. Ganczarski (eds.), *Mechanics of Anisotropic Materials*,
Engineering Materials, DOI 10.1007/978-3-319-17160-9_1

this chapter provides necessary tools for more extended constitutive description of materials which exhibit either virgin anisotropy or damage or phase change acquired anisotropy following microstructure changes.

Keywords Single or common tensor invariants · Material symmetry and constitutive matrices · Virgin or acquired anisotropy · Shear and volumetric change coupling · Strain energy of anisotropic materials · Damage and phase change state variables · Constitutive tensors analogy

1.1 Second-Order Tensors

1.1.1 Stress Tensor and Stress Tensor Invariants

Stress tensor $\boldsymbol{\sigma}$, when mathematical σ_{ij} $i, j = 1, 2, 3$, or $i, j = x, y, z$, and engineering notations are used is furnished as

$$[\sigma_{ij}] = \begin{bmatrix} \sigma_{11} & \sigma_{12} & \sigma_{13} \\ \sigma_{21} & \sigma_{22} & \sigma_{23} \\ \sigma_{31} & \sigma_{32} & \sigma_{33} \end{bmatrix} = \begin{bmatrix} \sigma_{xx} & \sigma_{xy} & \sigma_{xz} \\ \sigma_{yx} & \sigma_{yy} & \sigma_{yz} \\ \sigma_{zx} & \sigma_{zy} & \sigma_{zz} \end{bmatrix} = \begin{bmatrix} \sigma_{x} & \tau_{xy} & \tau_{xz} \\ \tau_{yx} & \sigma_{y} & \tau_{yz} \\ \tau_{zx} & \tau_{zy} & \sigma_{z} \end{bmatrix} \tag{1.1}$$

where x, y, z denote cartesian coordinate system.

When symmetry of the stress tensor $\sigma_{ij} = \sigma_{ji}$ is assumed, the stress tensor can be represented as *columnar stress vector* as follows:

$$\{\boldsymbol{\sigma}\} = \{\sigma_{11}, \sigma_{22}, \sigma_{33}, \sigma_{23}, \sigma_{13}, \sigma_{12}\}^{\mathrm{T}} = \begin{Bmatrix} \sigma_{11} \\ \sigma_{22} \\ \sigma_{33} \\ \sigma_{23} \\ \sigma_{13} \\ \sigma_{12} \end{Bmatrix} \tag{1.2}$$

When the definition of *stress deviator* is assumed as

$$s_{ij} = \sigma_{ij} - \frac{1}{3}\sigma_{kk}\delta_{ij} = \sigma_{ij} - \sigma_{\mathrm{h}}\delta_{ij} = \sigma_{ij} - \frac{1}{3}\mathrm{tr}\,(\boldsymbol{\sigma})\,\delta_{ij} \tag{1.3}$$

where $\sigma_{\mathrm{h}} = \frac{1}{3}\sigma_{kk}$ denotes either hydrostatic or mean stress, while $\delta_{ij} = \begin{cases} 1 & i = j \\ 0 & i \neq j \end{cases}$ denotes Kronecker's symbol, decomposition of the stress tensor into the *stress axiator* and the *stress deviator* takes the following form:

$$\boldsymbol{\sigma} = \sigma_{\mathrm{h}}\mathbf{1} + \boldsymbol{s} \tag{1.4}$$

where absolute notation $\sigma_{\mathrm{h}}\mathbf{1}$ and $\boldsymbol{s}$ are used for the stress axiator and the stress deviator, respectively

$$[\sigma_{\mathrm{h}}\mathbf{1}] = \begin{bmatrix} \sigma_{\mathrm{h}} & 0 & 0 \\ 0 & \sigma_{\mathrm{h}} & 0 \\ 0 & 0 & \sigma_{\mathrm{h}} \end{bmatrix}$$
$$[\boldsymbol{s}] = \begin{bmatrix} \sigma_x - \sigma_{\mathrm{h}} & \tau_{xy} & \tau_{xz} \\ \tau_{yx} & \sigma_y - \sigma_{\mathrm{h}} & \tau_{yz} \\ \tau_{zx} & \tau_{zy} & \sigma_z - \sigma_{\mathrm{h}} \end{bmatrix} = \begin{bmatrix} s_{xx} & s_{xy} & s_{xz} \\ s_{yx} & s_{yy} & s_{yz} \\ s_{zx} & s_{zy} & s_{zz} \end{bmatrix} \tag{1.5}$$

Classical *stress transformation* rule from i, j to k, l directions is

$$\sigma_{kl} = a_{ki} a_{lj} \sigma_{ij} \tag{1.6}$$

where second-order tensor transformation rule is applied and a_{ki}, a_{lj} denote direction cosines of the transformation from the original frame $i, j = x, y, z$ in the new reference frame $k, l = \xi, \eta, \zeta$. It is possible to distinguish the specific transformation into eigendirections (principal directions) for which the corresponding stress tensor takes the diagonal representation

$$\begin{bmatrix} \sigma_{xx} & \sigma_{xy} & \sigma_{xz} \\ \sigma_{yx} & \sigma_{yy} & \sigma_{yz} \\ \sigma_{zx} & \sigma_{zy} & \sigma_{zz} \end{bmatrix} \xrightarrow{\text{transformation}} \begin{bmatrix} \sigma_1 & 0 & 0 \\ 0 & \sigma_2 & 0 \\ 0 & 0 & \sigma_3 \end{bmatrix} \tag{1.7}$$

Three *principal stresses* are determined as real roots of the cubic equation, being solution of eigenproblem for the stress tensor $\boldsymbol{\sigma}$

$$\boldsymbol{\sigma} = \lambda \mathbf{1} \tag{1.8}$$

where $\lambda_i = \sigma_1, \sigma_2, \sigma_3$ stand for eigenvalues. These principal stresses are real roots of the characteristic equation of stress tensor $\lambda_i = \sigma_i$

$$\det(\boldsymbol{\sigma} - \lambda \mathbf{1}) = 0 \tag{1.9}$$

which can be rewritten in the equivalent fashion

$$\sigma^3 - I_{1\sigma}\sigma^2 + I_{2\sigma}\sigma - I_{3\sigma} = 0 \tag{1.10}$$

Three coefficients of the characteristic equation (1.10) $I_{1\sigma}, I_{2\sigma}, I_{3\sigma}$ are called the *principal invariants of* the *stress tensor* and may be defined in terms of stress components

$$
\begin{aligned}
I_{1\sigma} &= \mathrm{tr}(\boldsymbol{\sigma}) = \sigma_{ii} = \sigma_{xx} + \sigma_{yy} + \sigma_{zz} && [\mathrm{MPa}] \\
I_{2\sigma} &= \begin{vmatrix} \sigma_{xx} & \sigma_{xy} \\ \sigma_{yx} & \sigma_{yy} \end{vmatrix} + \begin{vmatrix} \sigma_{yy} & \sigma_{yz} \\ \sigma_{zy} & \sigma_{zz} \end{vmatrix} + \begin{vmatrix} \sigma_{zz} & \sigma_{zx} \\ \sigma_{xz} & \sigma_{xx} \end{vmatrix} \\
&= \sigma_{xx}\sigma_{yy} + \sigma_{yy}\sigma_{zz} + \sigma_{zz}\sigma_{xx} - (\sigma_{xy}^2 + \sigma_{yz}^2 + \sigma_{zx}^2) && [\mathrm{MPa}^2] \\
I_{3\sigma} &= \det\boldsymbol{\sigma} = \begin{vmatrix} \sigma_{xx} & \sigma_{xy} & \sigma_{xz} \\ \sigma_{yx} & \sigma_{yy} & \sigma_{yz} \\ \sigma_{zx} & \sigma_{zy} & \sigma_{zz} \end{vmatrix} = \sigma_{xx}\sigma_{yy}\sigma_{zz} \\
&+ 2\sigma_{xy}\sigma_{yz}\sigma_{zx} - (\sigma_{xx}\sigma_{yz}^2 + \sigma_{yy}\sigma_{xz}^2 + \sigma_{zz}\sigma_{xy}^2) && [\mathrm{MPa}^3]
\end{aligned}
\tag{1.11}
$$

Apart from the principal invariants, the *basic stress invariants* also called the *generic stress invariants* are of particular importance, namely

$$
\begin{aligned}
J_{1\sigma} &= \sigma_{ii} = \mathrm{tr}\,(\boldsymbol{\sigma}) && [\mathrm{MPa}] \\
J_{2\sigma} &= \tfrac{1}{2}\sigma_{ij}\sigma_{ji} = \tfrac{1}{2}\mathrm{tr}\left(\boldsymbol{\sigma}^2\right) && [\mathrm{MPa}^2] \\
J_{3\sigma} &= \tfrac{1}{3}\sigma_{ij}\sigma_{jk}\sigma_{ki} = \tfrac{1}{3}\mathrm{tr}\left(\boldsymbol{\sigma}^3\right) && [\mathrm{MPa}^3]
\end{aligned}
\tag{1.12}
$$

It is seen that the basic stress invariants can be interpreted as traces of subsequent powers of stress tensor $\boldsymbol{\sigma}$, $\boldsymbol{\sigma}^2 = \boldsymbol{\sigma}\cdot\boldsymbol{\sigma}$, $\boldsymbol{\sigma}^3 = \boldsymbol{\sigma}\cdot\boldsymbol{\sigma}\cdot\boldsymbol{\sigma}$, if appropriate coefficients 1, 1/2, 1/3 are used. Note that the basic invariants differ from the principal invariants, which are coefficients of the characteristic equation (1.10).

The basic stress invariants $J_{1\sigma}$, $J_{2\sigma}$, $J_{3\sigma}$ are expressed in terms of the principal stress invariants $I_{1\sigma}$, $I_{2\sigma}$, $I_{3\sigma}$ as follows:

$$
\begin{aligned}
J_{1\sigma} &= I_{1\sigma} \\
J_{2\sigma} &= \tfrac{1}{2}I_{1\sigma}^2 - I_{2\sigma} \\
J_{3\sigma} &= \tfrac{1}{3}I_{1\sigma}^3 - I_{1\sigma}^2 I_{2\sigma} + I_{3\sigma}
\end{aligned}
\tag{1.13}
$$

The reciprocal relations are

$$
\begin{aligned}
I_{1\sigma} &= J_{1\sigma} \\
I_{2\sigma} &= \tfrac{1}{2}J_{1\sigma}^2 - J_{2\sigma} \\
I_{3\sigma} &= \tfrac{1}{6}J_{1\sigma}^3 - J_{1\sigma}^2 J_{2\sigma} + J_{3\sigma}
\end{aligned}
\tag{1.14}
$$

Decomposition of the stress tensor into the stress axiator (spherical tensor) and the stress deviator (1.3–1.5) leads to the following system of *principal* or *generic invariants of* the *stress deviator*

$$
\begin{aligned}
J_{1s} &= s_{ii} = \mathrm{tr}(\boldsymbol{s}) = 0 && [\mathrm{MPa}] \\
J_{2s} &= \tfrac{1}{2}s_{ij}s_{ji} = \tfrac{1}{2}\mathrm{tr}\left(\boldsymbol{s}^2\right) && [\mathrm{MPa}^2] \\
J_{3s} &= \tfrac{1}{3}s_{ij}s_{jk}s_{ki} = \tfrac{1}{3}\mathrm{tr}\left(\boldsymbol{s}^3\right) && [\mathrm{MPa}^3]
\end{aligned}
\tag{1.15}
$$

where, in similar fashion as in Eq. (1.12), subsequent powers of the stress deviator $\mathbf{s}, \mathbf{s}^2 = \mathbf{s} \cdot \mathbf{s}, \mathbf{s}^3 = \mathbf{s} \cdot \mathbf{s} \cdot \mathbf{s}$ are used. Note that the first basic deviatoric stress invariant J_{1s} is equal to zero according to definition (1.3).

Additionally, some *engineering tensor stress invariants* characterized by the stress dimension homogeneity [MPa], by contrast to the above defined basic invariants of different dimensions [MPa], [MPa^2], [MPa^3] are frequently used as

$$\begin{aligned} \sigma_{\mathrm{h}} &= \tfrac{1}{3}J_{1\sigma} = \tfrac{1}{3}\mathrm{tr}\,(\boldsymbol{\sigma}) = \tfrac{1}{3}\boldsymbol{\sigma} : \mathbf{1} = \tfrac{1}{3}\sigma_{kk} \quad [\mathrm{MPa}] \\ \sigma_{\mathrm{eq}} &= \sqrt{3J_{2s}} = \sqrt{\tfrac{3}{2}s_{ij}s_{ji}} \quad [\mathrm{MPa}] \end{aligned} \tag{1.16}$$

The first of them σ_{h} is easily recognized as the *mean stress* and the second σ_{eq} represents the commonly used stress intensity also called the *effective stress*.

1.1.2 Strain Tensor and Strain Tensor Invariants

Strain tensor $\boldsymbol{\varepsilon} = \varepsilon_{ij}$ when uniform mathematical notation $i, j = 1, 2, 3$ or $i, j = x, y, z$, and the engineering notation are used, is furnished as

$$[\varepsilon_{ij}] = \begin{bmatrix} \varepsilon_{11} & \varepsilon_{12} & \varepsilon_{13} \\ \varepsilon_{21} & \varepsilon_{22} & \varepsilon_{23} \\ \varepsilon_{31} & \varepsilon_{32} & \varepsilon_{33} \end{bmatrix} = \begin{bmatrix} \varepsilon_{xx} & \varepsilon_{xy} & \varepsilon_{xz} \\ \varepsilon_{yx} & \varepsilon_{yy} & \varepsilon_{yz} \\ \varepsilon_{zx} & \varepsilon_{zy} & \varepsilon_{zz} \end{bmatrix} = \begin{bmatrix} \varepsilon_{x} & \frac{1}{2}\gamma_{xy} & \frac{1}{2}\gamma_{xz} \\ \frac{1}{2}\gamma_{yx} & \varepsilon_{y} & \frac{1}{2}\gamma_{yz} \\ \frac{1}{2}\gamma_{zx} & \frac{1}{2}\gamma_{zy} & \varepsilon_{z} \end{bmatrix} \tag{1.17}$$

where x, y, z denote cartesian coordinate frame.

Transformation of the strain tensor is described in a similar fashion as the stress tensor transformation (1.6), namely

$$\varepsilon_{kl} = a_{ki}a_{lj}\varepsilon_{ij} \tag{1.18}$$

Similarly, the *principal strains* can be obtained by solution of the eigenproblem of the tensor $\boldsymbol{\varepsilon}$

$$\boldsymbol{\varepsilon} = \lambda\mathbf{1} \tag{1.19}$$

or equivalently as solution of characteristic equation of strain tensor

$$\varepsilon^3 - I_{1\varepsilon}\varepsilon^2 + I_{2\varepsilon}\varepsilon - I_{3\varepsilon} = 0 \tag{1.20}$$

Coefficients of the above equation $I_{1\varepsilon}, I_{2\varepsilon}, I_{3\varepsilon}$ denote the *principal invariants of* the small (linearized) *strain tensor* and are defined as the homogeneous, scalar functions of the strain components

$$\begin{aligned}
I_{1\varepsilon} &= \mathrm{tr}(\boldsymbol{\varepsilon}) = \varepsilon_{ii} = \varepsilon_{xx} + \varepsilon_{yy} + \varepsilon_{zz} \\
I_{2\varepsilon} &= \begin{vmatrix} \varepsilon_{xx} & \varepsilon_{xy} \\ \varepsilon_{yx} & \varepsilon_{yy} \end{vmatrix} + \begin{vmatrix} \varepsilon_{yy} & \varepsilon_{yz} \\ \varepsilon_{zy} & \varepsilon_{zz} \end{vmatrix} + \begin{vmatrix} \varepsilon_{zz} & \varepsilon_{zx} \\ \varepsilon_{xz} & \varepsilon_{xx} \end{vmatrix} \\
&= \varepsilon_{xx}\varepsilon_{yy} + \varepsilon_{yy}\varepsilon_{zz} + \varepsilon_{zz}\varepsilon_{xx} - (\varepsilon_{xy}^2 + \varepsilon_{yz}^2 + \varepsilon_{zx}^2) \\
I_{3\varepsilon} &= \begin{vmatrix} \varepsilon_{xx} & \varepsilon_{xy} & \varepsilon_{xz} \\ \varepsilon_{yx} & \varepsilon_{yy} & \varepsilon_{yz} \\ \varepsilon_{zx} & \varepsilon_{zy} & \varepsilon_{zz} \end{vmatrix} = \varepsilon_{xx}\varepsilon_{yy}\varepsilon_{zz} + 2\varepsilon_{xy}\varepsilon_{yz}\varepsilon_{zx} \\
&\quad - (\varepsilon_{xx}\varepsilon_{yz}^2 + \varepsilon_{yy}\varepsilon_{xz}^2 + \varepsilon_{zz}\varepsilon_{xy}^2)
\end{aligned} \tag{1.21}$$

If symmetry of the strain tensor is assumed equivalent representation of the strain tensor in the form of *columnar strain vector* may be applied as

$$\{\varepsilon\} = \{\varepsilon_{11}, \varepsilon_{22}, \varepsilon_{33}, \varepsilon_{23}, \varepsilon_{13}, \varepsilon_{12}\}^{\mathrm{T}} = \begin{Bmatrix} \varepsilon_{11} \\ \varepsilon_{22} \\ \varepsilon_{33} \\ \varepsilon_{23} \\ \varepsilon_{13} \\ \varepsilon_{12} \end{Bmatrix} \tag{1.22}$$

When the definition of the *strain deviator*, analogous to the stress deviator (1.3), is used, we arrive at

$$e_{ij} = \varepsilon_{ij} - \frac{1}{3}\varepsilon_{kk}\delta_{ij} = \varepsilon_{ij} - \varepsilon_{\mathrm{m}}\delta_{ij} = \varepsilon_{ij} - \frac{1}{3}\mathrm{tr}\,(\varepsilon)\,\delta_{ij} \tag{1.23}$$

where ε_{m} denotes *mean* (volumetric) *strain*. Decomposition of the strain tensor into the *strain axiator* and the *strain deviator* is given according to the scheme

$$\boldsymbol{\varepsilon} = \varepsilon_{\mathrm{m}}\mathbf{1} + \boldsymbol{e} \tag{1.24}$$

when the absolute notation was used, where $\varepsilon_{\mathrm{m}}\mathbf{1}$ and $\boldsymbol{e}$ denote the strain axiator and the strain deviator, respectively

$$\begin{aligned}
[\varepsilon_{\mathrm{m}}\mathbf{1}] &= \begin{bmatrix} \varepsilon_{\mathrm{m}} & 0 & 0 \\ 0 & \varepsilon_{\mathrm{m}} & 0 \\ 0 & 0 & \varepsilon_{\mathrm{m}} \end{bmatrix} \\
[\boldsymbol{e}] &= \begin{bmatrix} \varepsilon_x - \varepsilon_{\mathrm{m}} & \frac{1}{2}\gamma_{xy} & \frac{1}{2}\gamma_{xz} \\ \frac{1}{2}\gamma_{yx} & \varepsilon_y - \varepsilon_{\mathrm{m}} & \frac{1}{2}\gamma_{yz} \\ \frac{1}{2}\gamma_{zx} & \frac{1}{2}\gamma_{zy} & \varepsilon_z - \varepsilon_{\mathrm{m}} \end{bmatrix} = \begin{bmatrix} e_{xx} & e_{xy} & e_{xz} \\ e_{yx} & e_{yy} & e_{yz} \\ e_{zx} & e_{zy} & e_{zz} \end{bmatrix}
\end{aligned} \tag{1.25}$$

The *basic* or the *generic strain tensor invariants* are defined in analogous fashion as in Eq. (1.12)

$$\begin{aligned} J_{1\varepsilon} &= \varepsilon_{ii} = \mathrm{tr}\,(\boldsymbol{\varepsilon}) \\ J_{2\varepsilon} &= \tfrac{1}{2}\varepsilon_{ij}\varepsilon_{ji} = \tfrac{1}{2}\mathrm{tr}\left(\boldsymbol{\varepsilon}^2\right) \\ J_{3\varepsilon} &= \tfrac{1}{3}\varepsilon_{ij}\varepsilon_{jk}\varepsilon_{ki} = \tfrac{1}{3}\mathrm{tr}\left(\boldsymbol{\varepsilon}^3\right) \end{aligned} \tag{1.26}$$

The basic strain tensor invariants $J_{1\varepsilon}$, $J_{2\varepsilon}$, $J_{3\varepsilon}$ are expressed in terms of the principal strain invariants $I_{1\varepsilon}$, $I_{2\varepsilon}$, $I_{3\varepsilon}$ as

$$\begin{aligned} J_{1\varepsilon} &= I_{1\varepsilon} \\ J_{2\varepsilon} &= \tfrac{1}{2}I_{1\varepsilon}^2 - I_{2\varepsilon} \\ J_{3\varepsilon} &= \tfrac{1}{3}I_{1\varepsilon}^3 - I_{1\varepsilon}^2 I_{2\varepsilon} + I_{3\varepsilon} \end{aligned} \tag{1.27}$$

The reciprocal relationships are

$$\begin{aligned} I_{1\varepsilon} &= J_{1\varepsilon} \\ I_{2\varepsilon} &= \tfrac{1}{2}J_{1\varepsilon}^2 - J_{2\varepsilon} \\ I_{3\varepsilon} &= \tfrac{1}{6}J_{1\varepsilon}^3 - J_{1\varepsilon}^2 J_{2\varepsilon} + J_{3\varepsilon} \end{aligned} \tag{1.28}$$

The *principal invariants of the strain deviator* may be determined in an analogous way as the principal invariants of the stress deviator (1.15), namely

$$\begin{aligned} J_{1e} &= e_{ii} = \mathrm{tr}(\boldsymbol{e}) = 0 \\ J_{2e} &= \tfrac{1}{2}e_{ij}e_{ji} = \tfrac{1}{2}\mathrm{tr}\left(\boldsymbol{e}^2\right) \\ J_{3e} &= \tfrac{1}{3}e_{ij}e_{jk}e_{ki} = \tfrac{1}{3}\mathrm{tr}\left(\boldsymbol{e}^3\right) \end{aligned} \tag{1.29}$$

1.1.3 Matrix Representation of Stress and Strain Tensors

Stress σ_{ij} and strain ε_{ij} are the second-rank tensors having in general $3^2 = 9$ components, since each of indices i, j runs from 1 to 3. Each of them can be interpreted as linear transformation of a certain vector to another vector. In case of the stress tensor, linear transformation of direction cosines n_j into a traction vector p_i according to rule

$$p_i = \sigma_{ij} n_j \tag{1.30}$$

is written down or

$$\begin{aligned} p_1 &= \sigma_{11}n_1 + \sigma_{12}n_2 + \sigma_{13}n_3 \\ p_2 &= \sigma_{21}n_1 + \sigma_{22}n_2 + \sigma_{23}n_3 \\ p_3 &= \sigma_{31}n_1 + \sigma_{32}n_2 + \sigma_{33}n_3 \end{aligned} \tag{1.31}$$

when the expanded form is used. Applying the *matrix-vector notation* to the above formulae the equivalent form is reached

$$\begin{Bmatrix} p_1 \\ p_2 \\ p_3 \end{Bmatrix} = \begin{bmatrix} \sigma_{11} & \sigma_{12} & \sigma_{13} \\ \sigma_{21} & \sigma_{22} & \sigma_{23} \\ \sigma_{31} & \sigma_{32} & \sigma_{33} \end{bmatrix} \begin{Bmatrix} n_1 \\ n_2 \\ n_3 \end{Bmatrix} \tag{1.32}$$

In Eq. (1.32) the second-rank *stress tensor* is represented by the 3×3 tensor *representation matrix* and analogously, the strain tensor representation matrix

$$[\boldsymbol{\sigma}] = \begin{bmatrix} \sigma_{11} & \sigma_{12} & \sigma_{13} \\ \sigma_{21} & \sigma_{22} & \sigma_{23} \\ \sigma_{31} & \sigma_{32} & \sigma_{33} \end{bmatrix} \qquad [\boldsymbol{\varepsilon}] = \begin{bmatrix} \varepsilon_{11} & \varepsilon_{12} & \varepsilon_{13} \\ \varepsilon_{21} & \varepsilon_{22} & \varepsilon_{23} \\ \varepsilon_{31} & \varepsilon_{32} & \varepsilon_{33} \end{bmatrix} \tag{1.33}$$

Due to symmetry conditions of both the stress $\sigma_{ij} = \sigma_{ji}$ and the strain $\varepsilon_{ij} = \varepsilon_{ji}$ tensors, both representation matrices are symmetric, comprising 6 independent components each. When engineering notation is used, replacing $1, 2, 3$ frame by x, y, z cartesian coordinate frame, and introducing appropriate notation $\sigma_{ij} = \tau_{ij}$ and $\varepsilon_{ij} = \frac{1}{2}\gamma_{ij}$ for $i \neq j$, we arrive at

$$[\boldsymbol{\sigma}] = \begin{bmatrix} \sigma_{xx} & \tau_{xy} & \tau_{xz} \\ & \sigma_{yy} & \tau_{yz} \\ & & \sigma_{zz} \end{bmatrix} \qquad [\boldsymbol{\varepsilon}] = \begin{bmatrix} \varepsilon_{xx} & \frac{1}{2}\gamma_{xy} & \frac{1}{2}\gamma_{xz} \\ & \varepsilon_{yy} & \frac{1}{2}\gamma_{yz} \\ & & \varepsilon_{zz} \end{bmatrix} \tag{1.34}$$

1.1.4 Decomposition of Strains

In the case of infinitesimal deformation the total strain ε_{ij} can be expressed as the sum of the *elastic (reversible) strain* $\varepsilon^{\mathrm{e}}_{ij}$, *inelastic (irreversible) strain* $\varepsilon^{\mathrm{I}}_{ij}$, and *thermal strain* $\varepsilon^{\mathrm{T}}_{ij}$:

$$\varepsilon_{rs} = \varepsilon^{\mathrm{e}}_{rs} + \varepsilon^{\mathrm{I}}_{rs} + \varepsilon^{\mathrm{T}}_{rs} \tag{1.35}$$

In the process of deformation, various *microstructural rearrangements* of material structure may take place, for example, the changes in density and configuration of dislocations, the development of microscopic cavities, changes from primary to secondary phase, etc. All these rearrangements may contribute to both reversible and irreversible strains (cf. Abu Al-Rub and Voyiadjis [1]), therefore:

$$\begin{aligned} \varepsilon^{\mathrm{e}}_{rs} &= \varepsilon^{\mathrm{E}}_{rs} + \varepsilon^{\mathrm{ed}}_{rs} + \varepsilon^{\mathrm{eph}}_{rs} + \cdots \\ \varepsilon^{\mathrm{I}}_{rs} &= \varepsilon^{\mathrm{p}}_{rs} + \varepsilon^{\mathrm{d}}_{rs} + \varepsilon^{\mathrm{ph}}_{rs} + \cdots \end{aligned} \tag{1.36}$$

where $\varepsilon^{\mathrm{E}}_{rs}$ is a "pure" elastic strain, and $\varepsilon^{\mathrm{ed}}_{rs}, \ldots, \varepsilon^{\mathrm{p}}_{rs}, \ldots$ are respectively the reversible and irreversible components of the total strain induced by *dissipative phenomenon*

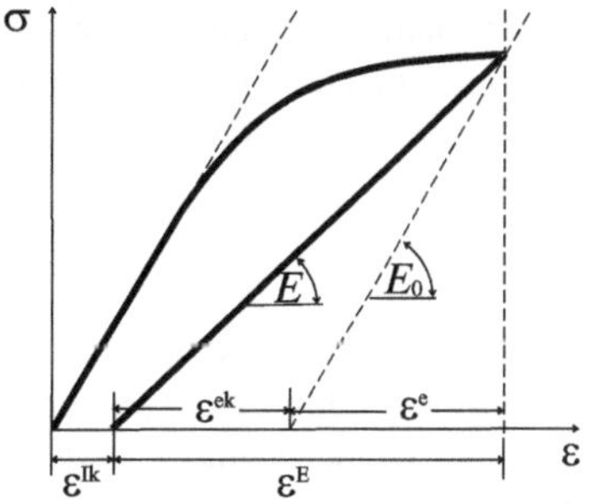

Fig. 1.1 Components of the strain tensor ε^k_{ij} induced by kth dissipative phenomenon

(see Fig. 1.1), e.g., *plastic flow*, *damage*, *phase transformation*, etc. For example, in the case of *thermo-elastic-plastic-damage material* the total strain tensor ε_{ij} is expressed as

$$\varepsilon_{rs} = \underbrace{\varepsilon^{\mathrm{E}}_{rs} + \varepsilon^{\mathrm{ed}}_{rs}}_{\varepsilon^{\mathrm{e}}_{rs}} + \underbrace{\varepsilon^{\mathrm{Id}}_{rs} + \varepsilon^{\mathrm{p}}_{rs}}_{\varepsilon^{\mathrm{I}}_{rs}} + \varepsilon^{\mathrm{T}}_{rs} \tag{1.37}$$

while its damage-induced component, $\varepsilon^{\mathrm{d}}_{rs}$, consists of both *reversible* (ed) and *irreversible* (Id) *damage strain* terms:

$$\varepsilon^{\mathrm{d}}_{rs} = \varepsilon^{\mathrm{ed}}_{rs} + \varepsilon^{\mathrm{Id}}_{rs} \tag{1.38}$$

1.2 Fourth-Order Tensors and Matrix Representation

1.2.1 Stiffness and Compliance Matrices—Voigt's Notation

General *linear elasticity equation* for anisotropic material, frequently called the *generalized Hooke law*, takes the forms

$$\varepsilon_{ij} = E^{-1}_{ijkl}\sigma_{kl} \qquad \sigma_{ij} = E_{ijkl}\varepsilon_{kl} \tag{1.39}$$

where the *fourth-rank elasticity tensors*, *stiffness* E_{ijkl} or *compliance* E^{-1}_{ijkl}, are defined, in general by $3^4 = 81$ components, since each of indices i, j, k, l runs through 1, 2, 3. Because of the symmetry of the stress $\sigma_{kl} = \sigma_{lk}$ and the strain $\varepsilon_{ij} = \varepsilon_{ji}$ tensors, both the stiffness and compliance tensors are symmetric with respect to change of indices in pairs $i \leftrightarrow j$ and $k \leftrightarrow l$

$$E_{ijkl} = E_{jikl} = E_{ijlk} \qquad E^{-1}_{ijkl} = E^{-1}_{jikl} = E^{-1}_{ijlk} \tag{1.40}$$

Additionally, because of property of *positive definiteness* of *strain energy* or *complementary energy* the symmetry with respect to change of indices between pairs $ij \leftrightarrow kl$ must also hold

$$E_{ijkl} = E_{klij} \qquad E^{-1}_{ijkl} = E^{-1}_{klij} \tag{1.41}$$

Because of symmetry conditions (1.40) and (1.41) from among 81 components of stiffness or compliance tensors, only 21 are independent. In order to describe the *generalized Hooke's law* (1.39) by use of *vector-matrix Voigt's notation*, stress and strain tensors are written as columnar stress and strain vectors, if the following scheme of change between tensor $i, j = 1, 2, 3$ and vectors $k = 1, 2, \ldots, 6$ indices holds:

$$\begin{array}{ccccccc} ij & 11 & 22 & 33 & 23, 32 & 31, 13 & 12, 21 \\ & \downarrow & \downarrow & \downarrow & \downarrow & \downarrow & \downarrow \\ k & 1 & 2 & 3 & 4 & 5 & 6 \end{array} \tag{1.42}$$

From the above scheme we obtain the following *representations of stress and strain tensors*:

$$\begin{array}{l} [\sigma_{ij}] = \begin{bmatrix} \sigma_{11} & \sigma_{12} & \sigma_{13} \\ & \sigma_{22} & \sigma_{23} \\ & & \sigma_{33} \end{bmatrix} \rightarrow \begin{bmatrix} \sigma_1 & \sigma_6 & \sigma_5 \\ & \sigma_2 & \sigma_4 \\ & & \sigma_3 \end{bmatrix} \rightarrow \begin{Bmatrix} \sigma_1 \\ \sigma_2 \\ \sigma_3 \\ \sigma_4 \\ \sigma_5 \\ \sigma_6 \end{Bmatrix} \\ [\varepsilon_{ij}] = \begin{bmatrix} \varepsilon_{11} & \varepsilon_{12} & \varepsilon_{13} \\ & \varepsilon_{22} & \varepsilon_{23} \\ & & \varepsilon_{33} \end{bmatrix} \rightarrow \begin{bmatrix} \varepsilon_1 & \varepsilon_6 & \varepsilon_5 \\ & \varepsilon_2 & \varepsilon_4 \\ & & \varepsilon_3 \end{bmatrix} \rightarrow \begin{Bmatrix} \varepsilon_1 \\ \varepsilon_2 \\ \varepsilon_3 \\ \varepsilon_4 \\ \varepsilon_5 \\ \varepsilon_6 \end{Bmatrix} \end{array} \tag{1.43}$$

Analogous scheme is applied to the first and second pairs of indices of stiffness and compliance tensors

$$\begin{array}{ll} E_{ijkl} = E_{mn}, \; E^{-1}_{ijkl} = E^{-1}_{mn} & \text{if } m \text{ or } n \text{ go through } 1, 2, 3 \\ 2E_{ijkl} = E_{mn}, \; 2E^{-1}_{ijkl} = E^{-1}_{mn} & \text{if } m \text{ or } n \text{ go through } 4, 5, 6 \\ 4E_{ijkl} = E_{mn}, \; 4E^{-1}_{ijkl} = E^{-1}_{mn} & \text{if both } m \text{ and } n \text{ go through } 4, 5, 6 \end{array} \tag{1.44}$$

where appropriate factors 2 or 4 are applied.

For instance, if the axial strain ε_{11} is considered the transformation scheme is as follows:

$$\begin{array}{ccccccccccccc}
\varepsilon_{11} & = & E^{-1}_{1111}\sigma_{11} & + & E^{-1}_{1122}\sigma_{22} & + & E^{-1}_{1133}\sigma_{33} & + & 2E^{-1}_{1123}\sigma_{23} & + & 2E^{-1}_{1113}\sigma_{13} & + & 2E^{-1}_{1112}\sigma_{12} \\
\downarrow & & \downarrow & & \downarrow & & \downarrow & & \downarrow & & \downarrow & & \downarrow \\
\varepsilon_{1} & = & E^{-1}_{11}\sigma_{1} & + & E^{-1}_{12}\sigma_{2} & + & E^{-1}_{13}\sigma_{3} & + & E^{-1}_{14}\sigma_{4} & + & E^{-1}_{15}\sigma_{5} & + & E^{-1}_{16}\sigma_{6}
\end{array} \tag{1.45}$$

In case the shear strain ε_{23} is considered, the following are furnished:

$$\begin{array}{ccccccccccccc}
\varepsilon_{23} & = & E^{-1}_{2311}\sigma_{11} & + & E^{-1}_{2322}\sigma_{22} & + & E^{-1}_{2333}\sigma_{33} & + & 2E^{-1}_{2323}\sigma_{23} & + & 2E^{-1}_{2313}\sigma_{13} & + & 2E^{-1}_{2312}\sigma_{12} \\
\downarrow & & \downarrow & & \downarrow & & \downarrow & & \downarrow & & \downarrow & & \downarrow \\
2\varepsilon_{4} & = & 2E^{-1}_{2311}\sigma_{1} & + & 2E^{-1}_{2322}\sigma_{2} & + & 2E^{-1}_{2333}\sigma_{3} & + & 4E^{-1}_{2323}\sigma_{4} & + & 4E^{-1}_{2313}\sigma_{5} & + & 4E^{-1}_{2312}\sigma_{6} \\
\downarrow & & \downarrow & & \downarrow & & \downarrow & & \downarrow & & \downarrow & & \downarrow \\
\gamma_{4} & = & E^{-1}_{41}\sigma_{1} & + & E^{-1}_{42}\sigma_{2} & + & E^{-1}_{43}\sigma_{3} & + & E^{-1}_{44}\sigma_{4} & + & E^{-1}_{45}\sigma_{5} & + & E^{-1}_{46}\sigma_{6}
\end{array} \tag{1.46}$$

Finally, the *generalized Hooke's law* (1.39) is represented in *vector-matrix notation* as follows:

$$\begin{array}{ll}
\varepsilon_i = E^{-1}_{ij}\sigma_j & (i = 1, 2, 3,\ j = 1, 2, \ldots, 6) \\
\gamma_i = E^{-1}_{ij}\sigma_j & (i = 4, 5, 6,\ j = 1, 2, \ldots, 6)
\end{array} \tag{1.47}$$

or

$$\{\varepsilon\} = [\mathbb{E}^{-1}]\,\{\boldsymbol{\sigma}\} \tag{1.48}$$

or equivalently

$$\{\boldsymbol{\sigma}\} = [\mathbb{E}]\,\{\varepsilon\} \tag{1.49}$$

where $[\mathbb{E}]$ or $[\mathbb{E}^{-1}]$ denote representation matrices of elastic stiffness or compliance tensors, whereas $\{\varepsilon\}$ and $\{\boldsymbol{\sigma}\}$ denote the columnar vectors of strain and stress, respectively. When *columnar vectors of stress* and *strain* are used as well as elasticity matrices are explicitly written down, Hooke's law is furnished as

$$\begin{Bmatrix} \varepsilon_1 \\ \varepsilon_2 \\ \varepsilon_3 \\ \gamma_4 \\ \gamma_5 \\ \gamma_6 \end{Bmatrix} = \left[\begin{array}{ccc|ccc}
E^{-1}_{11} & E^{-1}_{12} & E^{-1}_{13} & E^{-1}_{14} & E^{-1}_{15} & E^{-1}_{16} \\
E^{-1}_{21} & E^{-1}_{22} & E^{-1}_{23} & E^{-1}_{24} & E^{-1}_{25} & E^{-1}_{26} \\
E^{-1}_{31} & E^{-1}_{32} & E^{-1}_{33} & E^{-1}_{34} & E^{-1}_{35} & E^{-1}_{36} \\
\hline
E^{-1}_{41} & E^{-1}_{42} & E^{-1}_{43} & E^{-1}_{44} & E^{-1}_{45} & E^{-1}_{46} \\
E^{-1}_{51} & E^{-1}_{52} & E^{-1}_{53} & E^{-1}_{54} & E^{-1}_{55} & E^{-1}_{56} \\
E^{-1}_{61} & E^{-1}_{62} & E^{-1}_{63} & E^{-1}_{64} & E^{-1}_{65} & E^{-1}_{66}
\end{array}\right] \begin{Bmatrix} \sigma_1 \\ \sigma_2 \\ \sigma_3 \\ \tau_4 \\ \tau_5 \\ \tau_6 \end{Bmatrix} \tag{1.50}$$

or

$$\begin{Bmatrix} \sigma_1 \\ \sigma_2 \\ \sigma_3 \\ \tau_4 \\ \tau_5 \\ \tau_6 \end{Bmatrix} = \left[\begin{array}{ccc|ccc} E_{11} & E_{12} & E_{13} & E_{14} & E_{15} & E_{16} \\ E_{21} & E_{22} & E_{23} & E_{24} & E_{25} & E_{26} \\ E_{31} & E_{32} & E_{33} & E_{34} & E_{35} & E_{36} \\ \hline E_{41} & E_{42} & E_{43} & E_{44} & E_{45} & E_{46} \\ E_{51} & E_{52} & E_{53} & E_{54} & E_{55} & E_{56} \\ E_{61} & E_{62} & E_{63} & E_{64} & E_{65} & E_{66} \end{array}\right] \begin{Bmatrix} \varepsilon_1 \\ \varepsilon_2 \\ \varepsilon_3 \\ \gamma_4 \\ \gamma_5 \\ \gamma_6 \end{Bmatrix} \tag{1.51}$$

where engineering notation for shear stress $\tau_j = \sigma_j$ $(j = 4, 5, 6)$ is used.

It should be mentioned that symmetric *stiffness* $[E_{ij}] = [E_{ji}]$ and symmetric *compliance* $[E_{ij}^{-1}] = [E_{ji}^{-1}]$ *matrices*, both having dimension 6×6, are representation matrices of fourth-rank elasticity tensors E_{ijkl} or compliance E_{ijkl}^{-1}. Transformation of each matrix to another coordinate frame can be performed if the matrix notation (1.51) is replaced by the tensor notation (1.39), or by use of the appropriate transformation matrix $[\mathbb{Q}]$

$$[\mathbb{E}'] = [\mathbb{Q}]^{\mathrm{T}}[\mathbb{E}][\mathbb{Q}] \tag{1.52}$$

For instance, if the stiffness matrix is considered the transformation matrix takes the form

$$[\mathbb{Q}] = \begin{bmatrix} q_{11}q_{11} & q_{12}q_{12} & q_{13}q_{13} & q_{12}q_{13} & q_{13}q_{11} & q_{12}q_{11} \\ q_{21}q_{21} & q_{22}q_{22} & q_{23}q_{23} & q_{23}q_{22} & q_{23}q_{21} & q_{22}q_{21} \\ q_{31}q_{31} & q_{32}q_{32} & q_{33}q_{33} & q_{33}q_{32} & q_{33}q_{31} & q_{32}q_{31} \\ 2q_{31}q_{21} & 2q_{32}q_{22} & 2q_{33}q_{23} & q_{33}q_{22}+q_{32}q_{23} & q_{33}q_{21}+q_{31}q_{23} & q_{31}q_{22}+q_{32}q_{21} \\ 2q_{31}q_{11} & 2q_{32}q_{12} & 2q_{33}q_{13} & q_{33}q_{12}+q_{32}q_{13} & q_{33}q_{11}+q_{31}q_{13} & q_{31}q_{12}+q_{32}q_{11} \\ 2q_{21}q_{11} & 2q_{12}q_{22} & 2q_{13}q_{23} & q_{13}q_{22}+q_{12}q_{23} & q_{13}q_{21}+q_{11}q_{23} & q_{11}q_{22}+q_{12}q_{21} \end{bmatrix} \tag{1.53}$$

the coefficients of which are scalar products of corresponding direction cosines $q_{ij} = n_i n_j$ between both coordinate frames.

1.2.2 The Choice of State Variables

The irreversible rearrangements of the internal structure can be represented by a group of variables describing the current state of *material microstructure*:

$$\{V^k\} = \{V^{\mathrm{p}}, V^{\mathrm{d}}, V^{\mathrm{ph}}, \ldots\} \tag{1.54}$$

where V^k may be scalars, vectors, or even rank tensors. For damage description, in the case where the damaged material remains isotropic, the current state of damage is often represented by the scalar variable $V^{\mathrm{d}} = D$ denoting the volume fraction of cracks and voids $\mathrm{d}V^{\mathrm{d}}$ in the total volume $\mathrm{d}V^0$. *Damage acquired orthotropy* requires a second-order tensor, for example, the classical *Murakami–Ohno* [38] *tensor* $V^{\mathrm{d}}_{ij} = D_{ij}$, see Eq. (1.61). In the most general case of anisotropy the description of *damage* needs to be embodied in an *eight-order tensor* (cf. Cauvin and Testa [6]), while the principle of strain equivalence allows using *fourth-order tensors*, see Sect. 1.2.3. For *phase transformation* analysis the *scalar variable* $V^{\mathrm{ph}} = \xi$ is commonly adopted (cf. Egner and Skoczeń [14]), which denotes the *volume fraction of the* secondary *phase* in the total volume of the two-phase Representative Volume Element. However, a scalar variable is not capable of describing the *acquired anisotropy* due to partially directional nature of the *secondary inclusions* in the *primary matrix*. Therefore, instead of scalar variable a *second-order phase change tensor* can be defined in analogy to the damage tensor:

$$\boldsymbol{V}^{\mathrm{ph}} = \boldsymbol{\xi} = \sum_{i=1}^{3} \xi_i \boldsymbol{n}_i \otimes \boldsymbol{n}_i \tag{1.55}$$

where ξ_i describes the ratio of the secondary phase area $\mathrm{d}A_i^{\mathrm{ph}}$ to the total area $\mathrm{d}A_i^0$ on the principal plane of normal unit vector $\boldsymbol{n}_i$ (cf. Egner [13]). Another group of state variables consists of *internal (hidden) variables* corresponding to the *modifications of loading surfaces*:

$$\begin{aligned}\{h^k\} = \Big\{& r^{\mathrm{p}}, \alpha^{\mathrm{p}}_{ij}, l^{\mathrm{p}}_{ijkl}, g^{\mathrm{p}}_{ijklmn}, \dots \\ & r^{\mathrm{d}}, \alpha^{\mathrm{d}}_{ij}, l^{\mathrm{d}}_{ijkl}, g^{\mathrm{d}}_{ijklmn}, \dots \Big\}\end{aligned} \tag{1.56}$$

where r^{p}, r^{d} correspond to *isotropic expansion* of the *loading surface*, α^{p}_{ij}, α^{d}_{ij} affect *loading surface translatoric displacements*, l^{p}_{ijkl}, l^{d}_{ijkl} are hardening tensors of the fourth order which includes varying lengths of axes and *rotation of the loading surface*, and g^{p}_{ijklmn}, g^{d}_{ijklmn} describe changes of the curvature of the *loading surface (distortion)* related to appropriate dissipative phenomenon (cf. Kowalsky et al. [27], see Fig. 1.2). The complete set of *state variables* $\{V_{\mathrm{st}}\}$ reflecting the current state of the thermodynamic system consists of *observable variables*: elastic (or total) strain tensor $\varepsilon^{\mathrm{e}}_{ij}$ and absolute temperature T, and two groups of *microstructural* $\{V^k\}$ and *hardening* $\{h^k\}$ *state variables*:

$$\{V_{\mathrm{st}}\} = \{\varepsilon^{\mathrm{e}}_{ij}, T;\, V^{\mathrm{p}}, V^{\mathrm{d}}, V^{\mathrm{ph}}, \dots, h^{\mathrm{p}}, h^{\mathrm{d}}, h^{\mathrm{ph}}, \dots\} \tag{1.57}$$

When *thermo-elastic-plastic-damage* two-phase material is considered, the exemplary set of state variables for a general case of hardening/softening effects induced by different dissipative phenomena is further listed in Table 7.1.

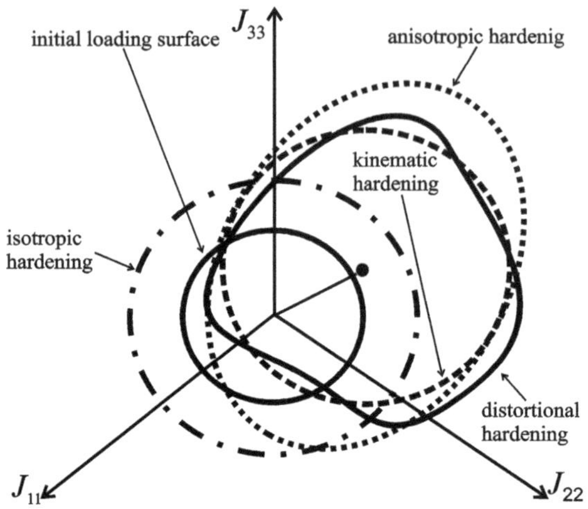

Fig. 1.2 Modifications of the loading surface in the space of thermodynamic conjugate forces $\{J^k\}$ (after [13])

When the material is subjected to *reverse tension-compression cycles*, the unsymmetrical behavior in tension and compression is observed as the *unilateral response* due to partial *crack closure effect*. To describe the phenomenon of the *unilateral damage*, also called the *damage deactivation* or the *crack closure/opening effect*, a decomposition of the stress or strain tensors into the positive or negative projection is usually introduced using the fourth-rank *projection operators* (cf. Krajcinovic [30]; Bielski et al. [4]):

$$\varepsilon^*_{ij} = \sum_{I=1}^{3} \kappa(\varepsilon_I) n^{(\varepsilon)}_{iI} n^{(\varepsilon)}_{jI} n^{(\varepsilon)}_{Ik} n^{(\varepsilon)}_{Il} \varepsilon_{kl} = B^{(\varepsilon)}_{ijkl} \varepsilon_{kl} \tag{1.58}$$

where the fourth-rank tensor $B^{(\varepsilon)}_{ijkl}$ is built of directional cosines between the principal and the current spatial systems, $n^{(\varepsilon)}_{iI}$ and $\kappa(\varepsilon_I) = \mathrm{H}(a) + \zeta \mathrm{H}(-a)$, H is a Heaviside function and ζ is a material constant.

1.2.3 Damage and Damage Effect Tensors

So far constitutive description of material has not accounted for influence of damage. Damage means existence of microvoids and microcracks in the material that result in essential deterioration of mechanical properties at the macroscale, such as strength and stiffness or compliance.

In the simplest case when microvoids are spherical and homogeneously distributed in material, damage is described by the scalar damage variable D, usually called the damage parameter, Fig. 1.3

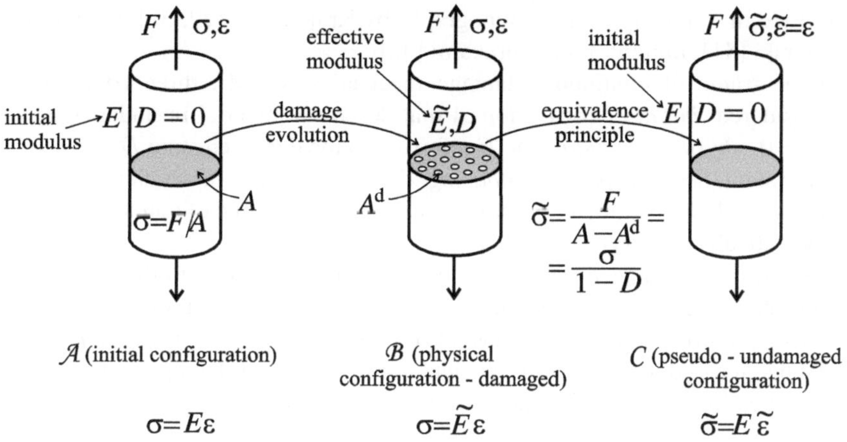

Fig. 1.3 Three configurations used in CDM: **a** initial, **b** damaged, **c** effective pseudo-undamaged

$$D = \frac{A^{\mathrm{d}}}{A} \tag{1.59}$$

Scalar damage variable D, introduced by Kachanov [24] and Rabotnov [46], represents the loss of effective area from the initial A to the damaged A^{d} states. In order to generalize the scalar damage variable to the case when microvoids exhibit clearly directional nature, the *vector damage variable* D_i, is proposed by Davison and Stevens [11], Kachanov [25], Krajcinovic and Fonseka [28]

$$D_i = \frac{A_i^{\mathrm{d}}}{A_i} \quad i = 1, 2, 3 \tag{1.60}$$

Murakami and Ohno [39] introduced more general damage variable defined by the symmetric *second-rank damage tensor* $\boldsymbol{D}$, capable of capturing an *orthotropic damage* nature

$$\boldsymbol{D} = \begin{bmatrix} D_{11} & D_{12} & D_{13} \\ & D_{22} & D_{23} \\ & & D_{33} \end{bmatrix} \tag{1.61}$$

Recently, researches aimed towards correct description of damage mechanism in elastic-brittle *rock-like materials*, *ceramics* or *concrete* led to definition of the *fourth-rank damage tensors*, e.g., Chaboche [8], Krajcinovic [29] or Lubarda and Krajcinovic [35]. Apart from the above-mentioned damage variables possessing clear geometric interpretation other damage variables referring to physical planes, described in details e.g., by Gambarotta and Lagomarsino [15], Seweryn and Mróz [48] should also be mentioned. More general *classifications of damage variables* were listed in following subject monographs by Krajcinovic [29, 30], Skrzypek and Ganczarski [51], Betten [3] or Murakami [40].

In the frame of continuum damage mechanics (CDM), three configurations are considered: *initial configuration* $\mathcal{A}$ that describes material in undamaged state $D(\mathcal{A}) = 0$, *physical configuration* $\mathcal{B}$ referring to the damaged state $D(\mathcal{B}) \neq 0$, and the equivalent, *fictitious pseudo-undamaged configuration* $\mathcal{C}$ in which real heterogeneous material is substituted by a homogeneous material, free of damage $D(\mathcal{C}) = 0$, as schematically is shown in Fig. 1.3.

The physical (damaged) configuration $\mathcal{B}$ is equivalent to the effective (pseudo-undamaged) configuration $\mathcal{C}$ in a certain sense, for instance, of *strain equivalence* Chaboche [7], *stress equivalence* Taher et al. [54], or *elastic strain energy equivalence* Cordebois and Sidoroff [10]. In physical configuration $\mathcal{B}$ damage state manifests through the effective elasticity modulus $\widetilde{E}$, for instance,

$$\widetilde{E} = E(1 - D) \qquad \text{or} \qquad \widetilde{E} = E(1 - D)^2 \tag{1.62}$$

where the hypotheses of strain or stress equivalence (first formula) or elastic energy equivalence are used. Contrarily, in the effective configuration $\mathcal{C}$ damage state manifests by the definition of the *effective variables* $\widetilde{\sigma}$, $\widetilde{\varepsilon}$, respectively

$$\widetilde{\sigma} = \sigma \frac{E}{\widetilde{E}}, \quad \widetilde{\varepsilon} = \varepsilon \qquad \text{or} \qquad \widetilde{\sigma} = \sigma \sqrt{\frac{E}{\widetilde{E}}}, \quad \widetilde{\varepsilon} = \varepsilon \sqrt{\frac{\widetilde{E}}{E}} \tag{1.63}$$

or equivalently

$$\widetilde{\sigma} = \frac{\sigma}{1 - D}, \quad \widetilde{\varepsilon} = \varepsilon \qquad \text{or} \qquad \widetilde{\sigma} = \frac{\sigma}{1 - D}, \quad \widetilde{\varepsilon} = \varepsilon(1 - D) \tag{1.64}$$

The *damage effect matrix*, being matrix representation of the *damage effect tensor*

$$[\mathbb{M}] = \left[\mathrm{diag}\,\{M_{11}, M_{22}, M_{33}, M_{44}, M_{55}, M_{66}\}\right] \tag{1.65}$$

is expressed in terms of the *damage parameter D* as follows:

$$[\mathbb{M}] = \frac{1}{1 - D} \left[\mathrm{diag}\,\{1, 1, 1, 1, 1, 1\}\right] \tag{1.66}$$

where the diagonal form is applicable.

Damage effect matrix plays an essential role in definitions of the *damage effective stress tensor* $\widetilde{\boldsymbol{\sigma}}$

$$\{\widetilde{\sigma}\} = [\mathbb{M}]\,\{\sigma\} = \left\{\frac{\sigma_x}{1 - D}, \frac{\sigma_y}{1 - D}, \frac{\sigma_z}{1 - D}, \frac{\tau_{yz}}{1 - D}, \frac{\tau_{xz}}{1 - D}, \frac{\tau_{xy}}{1 - D}\right\}^{\mathrm{T}} \tag{1.67}$$

and the *damage effective compliance (stiffness) matrix* $[\widetilde{\mathbb{E}}^{-1}]$

$$[\widetilde{\mathbb{E}}^{-1}] = [\mathbb{M}]^{\mathrm{T}}[\mathbb{E}^{-1}][\mathbb{M}] = \frac{1}{E(1-D)^2}\left[\begin{array}{ccc|ccc} 1 & -\nu & -\nu & & & \\ & 1 & -\nu & & & \\ & & 1 & & & \\ \hline & & & 1+\nu & & \\ & & & & 1+\nu & \\ & & & & & 1+\nu \end{array}\right] \quad (1.68)$$

For brevity, in all the above equations (1.62–1.68) the assumption of material isotropy in undamaged state ($\mathcal{A}$) is applied.

Assumption of the isotropic damage nature is too strong a simplification since usually microvoids or microcracks are of oval or directional shapes. A proper damage description requires application of orthotropic damage representation (1.61), which under the assumption of the principal damage frame reduces to the diagonal form, where D_1, D_2, D_3 components may be interpreted by reduction of effective areas 1, 2, 3 (1.60), hence

$$\boldsymbol{D} = \begin{bmatrix} D_1 & & \\ & D_2 & \\ & & D_3 \end{bmatrix} \qquad D_i = \frac{A_i^{\mathrm{d}}}{A_i} \qquad i = 1, 2, 3 \quad (1.69)$$

Chosen *representations* of the *damage effect matrix* based on various hypotheses, after Chen and Chow [9], Skrzypek [49], Murakami [40], can be defined as follows:

$$[\mathbb{M}_1] = \begin{bmatrix} \frac{1}{1-D_1} & & & & & \\ & \frac{1}{1-D_2} & & & & \\ & & \frac{1}{1-D_3} & & & \\ & & & \frac{1}{\sqrt{(1-D_2)(1-D_3)}} & & \\ & & & & \frac{1}{\sqrt{(1-D_3)(1-D_1)}} & \\ & & & & & \frac{1}{\sqrt{(1-D_1)(1-D_2)}} \end{bmatrix} \quad (1.70)$$

or

$$[\mathbb{M}_2] = \begin{bmatrix} \frac{1}{1-D_1} & & & & & \\ & \frac{1}{1-D_2} & & & & \\ & & \frac{1}{1-D_3} & & & \\ & & & \frac{1}{1-0.5(D_2+D_3)} & & \\ & & & & \frac{1}{1-0.5(D_3+D_1)} & \\ & & & & & \frac{1}{1-0.5(D_1+D_2)} \end{bmatrix} \quad (1.71)$$

or

$$[\mathbb{M}_3] = \begin{bmatrix} \frac{1}{1-D_1} & & & & & \\ & \frac{1}{1-D_2} & & & & \\ & & \frac{1}{1-D_3} & & & \\ & & & \frac{1}{2}\left(\frac{1}{1-D_2}+\frac{1}{1-D_3}\right) & & \\ & & & & \frac{1}{2}\left(\frac{1}{1-D_3}+\frac{1}{1-D_1}\right) & \\ & & & & & \frac{1}{2}\left(\frac{1}{1-D_1}+\frac{1}{1-D_2}\right) \end{bmatrix} \tag{1.72}$$

The damage effective stress can be defined, for instance in the following two ways, both satisfying symmetry of the effective stress $\tilde{\boldsymbol{\sigma}}(\boldsymbol{D})$:

$$\{\tilde{\sigma}\} = [\mathbb{M}_1^{1/2}]^{\mathrm{T}}\{\sigma\}[\mathbb{M}_1^{1/2}] \longrightarrow [\tilde{\sigma}] = \begin{bmatrix} \frac{\sigma_x}{1-D_1} & \frac{\tau_{xy}}{\sqrt{(1-D_1)(1-D_2)}} & \frac{\tau_{xz}}{\sqrt{(1-D_1)(1-D_3)}} \\ & \frac{\sigma_y}{1-D_2} & \frac{\tau_{yz}}{\sqrt{(1-D_2)(1-D_3)}} \\ & & \frac{\sigma_z}{1-D_3} \end{bmatrix} \tag{1.73}$$

or

$$\{\tilde{\sigma}\} = [\mathbb{M}_1]^{\mathrm{T}}\{\sigma\}[\mathbb{M}_1] \longrightarrow [\tilde{\sigma}] = \begin{bmatrix} \frac{\sigma_x}{(1-D_1)^2} & \frac{\tau_{xy}}{(1-D_1)(1-D_2)} & \frac{\tau_{xz}}{(1-D_1)(1-D_3)} \\ & \frac{\sigma_y}{(1-D_2)^2} & \frac{\tau_{yz}}{(1-D_2)(1-D_3)} \\ & & \frac{\sigma_z}{(1-D_3)^2} \end{bmatrix} \tag{1.74}$$

Exemplary effective compliance matrices take the following representations, Skrzypek and Ganczarski [51]:

$$[\tilde{\mathbb{E}}^{-1}] = [\mathbb{M}_1]^{\mathrm{T}}[\mathbb{E}^{-1}][\mathbb{M}_1] =$$

$$\frac{1}{E}\left[\begin{array}{ccc|ccc} \frac{1}{(1-D_1)^2} & \frac{-\nu}{(1-D_1)(1-D_2)} & \frac{-\nu}{(1-D_1)(1-D_3)} & & & \\ \frac{-\nu}{(1-D_2)(1-D_1)} & \frac{1}{(1-D_2)^2} & \frac{-\nu}{(1-D_2)(1-D_3)} & & & \\ \frac{-\nu}{(1-D_3)(1-D_1)} & \frac{-\nu}{(1-D_3)(1-D_2)} & \frac{1}{(1-D_3)^2} & & & \\ \hline & & & \frac{1+\nu}{(1-D_2)(1-D_3)} & & \\ & & & & \frac{1+\nu}{(1-D_3)(1-D_1)} & \\ & & & & & \frac{1+\nu}{(1-D_1)(1-D_2)} \end{array}\right] \tag{1.75}$$

or

$$[\widetilde{\mathbb{E}}^{-1}] = \frac{1}{2}\left([\mathbb{M}_2][\mathbb{E}^{-1}] + [\mathbb{E}^{-1}][\mathbb{M}_2]\right) =$$
$$\frac{1}{E}\left[\begin{array}{ccc|ccc} \frac{1}{1-D_1} & \frac{-\nu}{1-0.5(D_1+D_2)} & \frac{-\nu}{1-0.5(D_1+D_3)} & & & \\ \frac{-\nu}{1-0.5(D_2+D_1)} & \frac{1}{1-D_2} & \frac{-\nu}{1-0.5(D_2+D_3)} & & & \\ \frac{-\nu}{1-0.5(D_3+D_1)} & \frac{-\nu}{1-0.5(D_3+D_2)} & \frac{1}{1-D_3} & & & \\ \hline & & & \frac{1+\nu}{1-0.5(D_2+D_3)} & & \\ & & & & \frac{1+\nu}{1-0.5(D_1+D_3)} & \\ & & & & & \frac{1+\nu}{1-0.5(D_1+D_2)} \end{array}\right] \tag{1.76}$$

In both cases, for the sake of brevity, material isotropy at the undamaged state was assumed.

In a more general case of full damage anisotropy the *fourth-rank damage tensor* D_{ijkl}, built of 21 independent components, should be used.

Following Cauvin and Testa [6] the effective stiffness tensor is defined as

$$\widetilde{\mathbb{E}} = (\mathbb{I} - \mathbb{D}) : \mathbb{E} = \mathbb{R} : \mathbb{E} \tag{1.77}$$

where fourth-rank tensors $\mathbb{R}$ and $\mathbb{D}$ stand for damage effect and damage tensors, respectively. In general case of full *damage anisotropy* the 6×6 *matrix representation* of the *fourth-rank damage tensor* is as follows:

$$[\mathbb{D}] = \begin{bmatrix} D_{11} & D_{12} & D_{13} & D_{14} & D_{15} & D_{16} \\ & D_{22} & D_{23} & D_{24} & D_{25} & D_{26} \\ & & D_{33} & D_{34} & D_{35} & D_{36} \\ & & & D_{44} & D_{45} & D_{46} \\ & & & & D_{55} & D_{56} \\ & & & & & D_{66} \end{bmatrix} \tag{1.78}$$

As a particular case the *orthotropic damage* is considered as example for which the unsymmetric orthotropic damage matrix reduces to

$$[\mathbb{D}] = \begin{bmatrix} D_{11} & D_{12} & D_{13} & & & \\ D_{21} & D_{22} & D_{23} & & & \\ D_{31} & D_{32} & D_{33} & & & \\ & & & D_{44} & & \\ & & & & D_{55} & \\ & & & & & D_{66} \end{bmatrix} \tag{1.79}$$

In the particular case when the orthotropic symmetry of damaged material is considered, the damage tensor takes the following *matrix representation*, after Cauvin and Testa [6], also Ganczarski [17]:

$$[\mathbb{D}] = \begin{bmatrix} D_{1111} & D_{1122} & D_{1133} & & & \\ D_{2211} & D_{2222} & D_{2233} & & & \\ D_{3311} & D_{3322} & D_{3333} & & & \\ & & & 2D_{2323} & & \\ & & & & 2D_{1313} & \\ & & & & & 2D_{1212} \end{bmatrix} \tag{1.80}$$

defined by 12 independent elements, in general nonsymmetric because three elements under diagonal D_{2211}, D_{3311}, D_{3322} are truly independent.

In the narrower case of *transverse isotropy* (in the plane 2, 3), number of independent elements of the tensor D_{ijkl} reduces to 5, namely D_1, D_2, D_3, D_4, D_5

$$[\mathbb{D}] = \begin{bmatrix} D_1 & D_2 & D_2 & & & \\ D_{2'} & D_3 & D_4 & & & \\ D_{2'} & D_4 & D_3 & & & \\ & & & D_3 - D_4 & & \\ & & & & D_5 & \\ & & & & & D_5 \end{bmatrix} \tag{1.81}$$

Two components $D_{2211} = D_{3311} = D_{2'}$ are dependent, and expressed as

$$D_{2'} = \frac{1}{1-\nu}\left[D_2 + \nu\left(D_1 - D_3\right) - \nu D_4\right] \tag{1.82}$$

This kind of transverse isotropy will further be classified as *transverse isotropy* case *of hexagonal symmetry* (5 independent components in contrast to another *transverse isotropy of tetragonal symmetry* where all 6 components are truly independent, see Table 1.1).

The 6×6 *transversely isotropic compliance matrix* is of the following form:

$$[\tilde{\mathbb{E}}^{-1}] = \begin{bmatrix} \frac{1}{E_1} & -\frac{\nu_{12}}{E_1} & -\frac{\nu_{12}}{E_1} & & & \\ -\frac{\nu_{12}}{E_1} & \frac{1}{E_2} & -\frac{\nu_{23}}{E_2} & & & \\ -\frac{\nu_{12}}{E_1} & -\frac{\nu_{23}}{E_2} & \frac{1}{E_2} & & & \\ & & & \frac{1}{G_{23}} & & \\ & & & & \frac{1}{G_{12}} & \\ & & & & & \frac{1}{G_{12}} \end{bmatrix} \tag{1.83}$$

in which damage affected modules expressed in terms of damage variables are

Table 1.1 *Classification of anisotropic elastic materials* with respect to *stiffness matrix symmetry* referring to crystal lattice cf. Nye [42]

Conventional unit cells of space lattices	Stiffness matrix $[\mathbb{C}]$	Conventional unit cells of space lattices	Stiffness matrix $[\mathbb{C}]$
1. 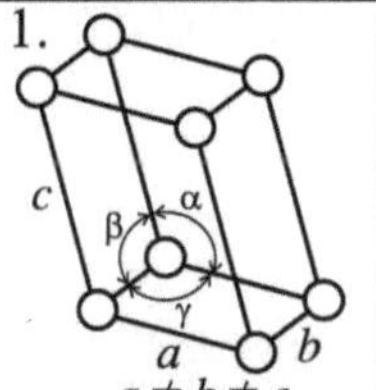 $a \neq b \neq c$ $\alpha \neq \beta \neq \gamma \neq 90^\circ$ triclinic lattice	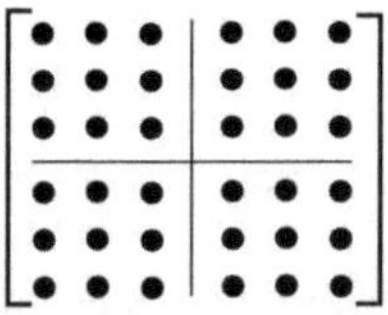 anisotropic Hooke's (21 constants)	5. 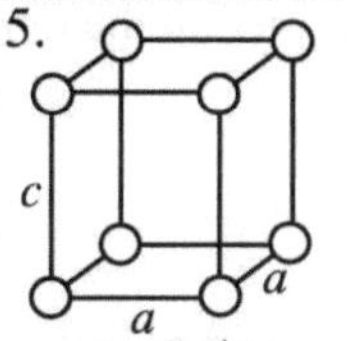 $a = b \neq c$ $\alpha = \beta = \gamma = 90^\circ$ tetragonal lattice	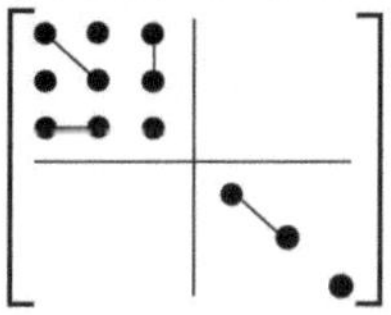 transversely isotropic tetragonal Hooke's (6 constants)
2. 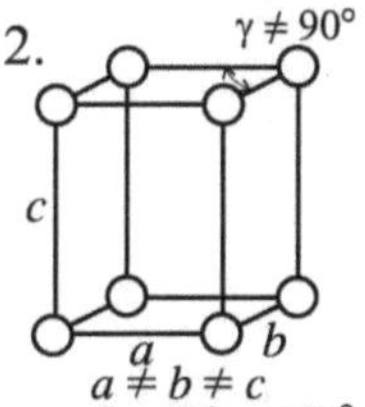 $a \neq b \neq c$ $\alpha = \beta = 90^\circ, \gamma \neq 90^\circ$ monoclinic lattice	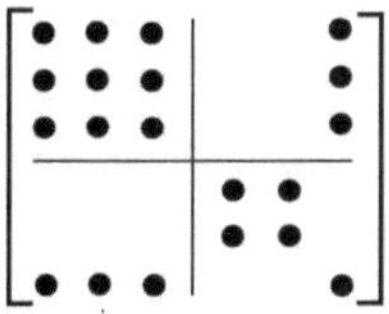 monoclinic or oblique Hooke's anisotropy	6. 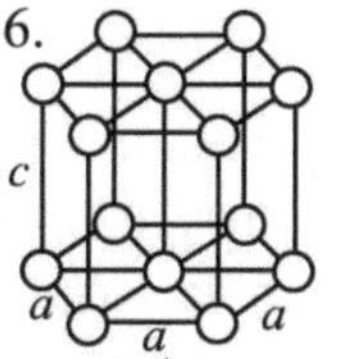 $a \neq c$ $\alpha = 90^\circ, \gamma = 120^\circ$ hexagonal lattice	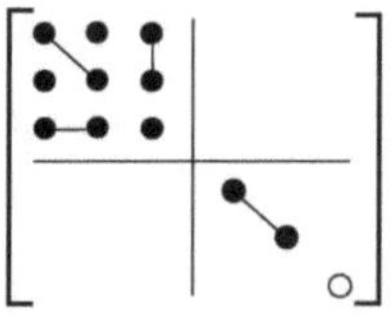 hexagonal Hooke's (5 constants)
3. 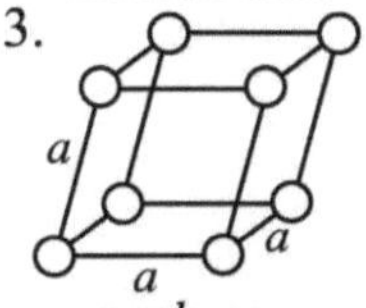 $a = b = c$ $\alpha = \beta = \gamma \neq 90^\circ$ rhombohedral lattice	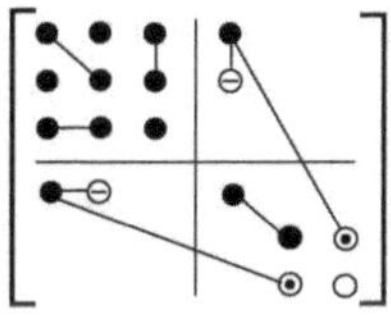 trigonal anisotropy (6 constants)	7. 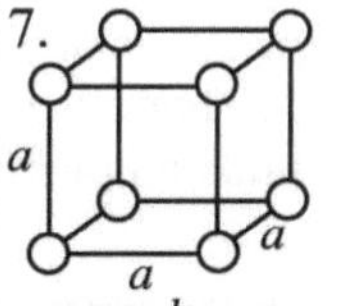 $a = b = c$ $\alpha = \beta = \gamma = 90^\circ$ cubic (regular) lattice	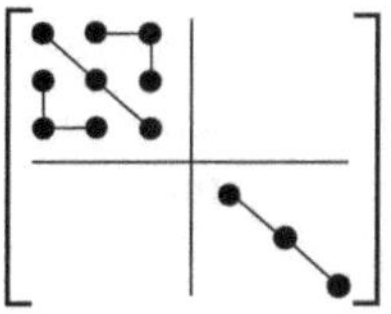 Cubic Hooke's (3 constants)
4. 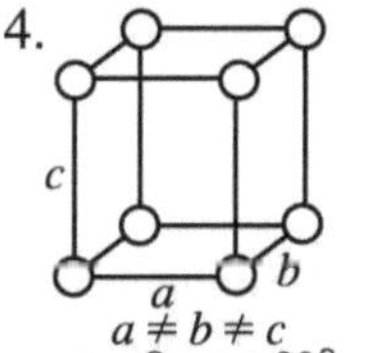 $a \neq b \neq c$ $\alpha = \beta = \gamma = 90^\circ$ orthorhombic lattice	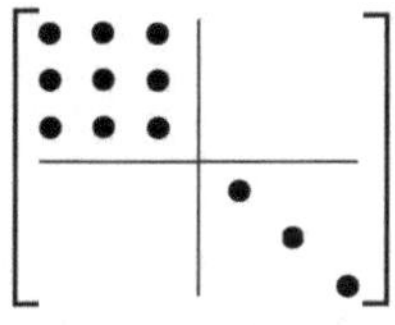 orthotropic Hooke's (9 constants)	8. 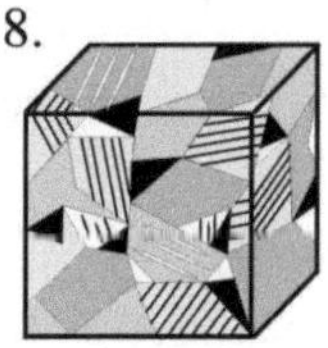 polycrystal	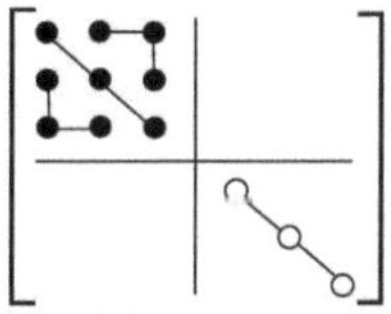 isotropy (2 constants)

$$
\begin{aligned}
E_1 &= E\frac{(1-D_1)(1-D_3-D_4)-2D_2D_{2'}}{1-D_3-D_4-2\nu D_{2'}}\\
\nu_{12} &= \frac{\nu(1-D_3-D_4)-(1-\nu)D_{2'}}{1-D_3-D_4-2\nu D_{2'}}\\
E_2 &= E\frac{(1-D_3+D_4)[(1-D_1)(1-D_3-D_4)-2D_2D_{2'}]}{(1-D_1)(1-D_3-\nu D_4)-\nu D_2(1-D_3+D_4)-(1+\nu)D_2D_{2'}}\\
\nu_{23} &= \frac{(1-D_1)(\nu-\nu D_3-D_4)+\nu D_2(1-D_3+D_4)-(1+\nu)D_2D_{2'}}{(1-D_1)(1-D_3-\nu D_4)-\nu D_2(1-D_3+D_4)-(1+\nu)D_2D_{2'}}\\
G_{23} &= \frac{E}{2(1+\nu)}(1-D_3-D_4)\\
G_{12} &= \frac{E}{2(1+\nu)}(1-D_5)
\end{aligned}
\tag{1.84}
$$

More accurate description of *anisotropic damage* may be provided by use of *fabric tensors*, see Murakami [40], Voyiadjis and Kattan [55], Yun-bing and Xing-fu [56], Lubarda and Krajcinovic [35]. For this reason a unit spherical surface around a given point $P(x)$ in the RVE is considered (see Fig. 1.4), and the *directional distribution* $\xi(\boldsymbol{n})$ of the *microvoid density* on the unit sphere is defined as a *polynomial function* of the *direction vector* $\boldsymbol{n}$

$$\xi(\boldsymbol{n}) = D_0 + D_{ij} f_{ij}(\boldsymbol{n}) + D_{ijkl} f_{ijkl}(\boldsymbol{n}) + \cdots \tag{1.85}$$

Expression (1.85) is a *generalized Fourier series* with respect to the *irreducible tensor bases* $f_{ij}(\boldsymbol{n})$, $f_{ijkl}(\boldsymbol{n})$, ...

$$\begin{aligned} f_{ij}(\boldsymbol{n}) &= n_i n_j - \tfrac{1}{3}\delta_{ij} \\ f_{ijkl}(\boldsymbol{n}) &= n_i n_j n_k n_l - \tfrac{1}{7}(\delta_{ij} n_k n_l + \delta_{ik} n_j n_l + \delta_{il} n_j n_k \\ &\quad + \delta_{jk} n_i n_l + \delta_{jl} n_i n_k + \delta_{kl} n_i n_j) + \tfrac{1}{5\times 7}(\delta_{ij} n_k n_l \\ &\quad + \delta_{ik} n_j n_l + \delta_{il} n_j n_k) \end{aligned} \tag{1.86}$$

The *tensor bases* $f_{ij}(\boldsymbol{n})$, $f_{ijkl}(\boldsymbol{n})$, ... are symmetric with respect to the indices, consist of even-order tensor components, and have vanishing trace.

The tensors $D_0, D_{ij}, D_{ijkl}, \ldots$ characterize the *directional distribution of damage*, and are called *fabric tensors*. For given $\xi(\boldsymbol{n})$ they can be derived by calculating the following integrals (cf. Murakami [40]):

$$\begin{aligned} D_0 &= \tfrac{1}{4\pi} \int_{S^2} \xi(\boldsymbol{n}) \mathrm{d}\Omega \\ D_{ij} &= \tfrac{1}{4\pi} \tfrac{3\times 5}{2} \int_{S^2} \xi(\boldsymbol{n}) f_{ij}(\boldsymbol{n}) \mathrm{d}\Omega \\ D_{ijkl} &= \tfrac{1}{4\pi} \tfrac{3\times 5\times 7\times 9}{2\times 3\times 4} \int_{S^2} \xi(\boldsymbol{n}) f_{ijkl}(\boldsymbol{n}) \mathrm{d}\Omega \end{aligned} \tag{1.87}$$

The even-order tensors D_0, D_{ij}, D_{ijkl}, ... represent completely the damage state of the materials, and have been used as the internal state variables in thermodynamic modeling of creep and brittle damage, see Onat and Leckie [43], Lacy et al. [32].

Concluding, it is worth to mention that *virgin material anisotropy* may either manifest from the very beginning of the elastic response when appropriate anisotropic

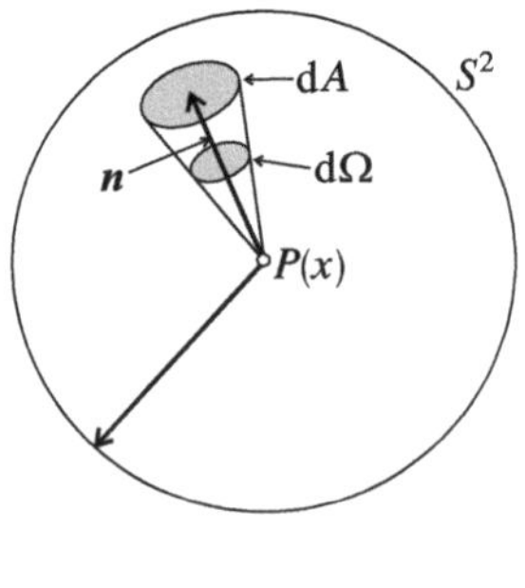

Fig. 1.4 Unit spherical surface to represent directional void distribution

formulation of Hooke's law is required or at damage initiation phase when *damage acquired anisotropy* appears as shown above. In the last case the elasticity matrix at the virgin state may have isotropic nature, whereas after some dissipative process initiates it changes to anisotropic form.

1.3 Common Invariants of the Second-Order and Fourth-Order Tensors

1.3.1 Common Invariants of Two Second-Order Tensors: The Stress/Strain and the Damage Tensors

A fundamental (irreducible) set of common invariants of two second-order tensors comprises 10 invariants. In a particular case when the common *strain-damage space* $(\boldsymbol{\varepsilon}, \boldsymbol{D})$ is considered they are furnished as

$$\begin{aligned}
J_{1\varepsilon} &= \mathrm{tr}(\boldsymbol{\varepsilon}) = \varepsilon_{ii} \\
J_{2\varepsilon} &= \tfrac{1}{2}\mathrm{tr}(\boldsymbol{\varepsilon}\cdot\boldsymbol{\varepsilon}) = \tfrac{1}{2}\varepsilon_{ij}\varepsilon_{ji} \\
J_{3\varepsilon} &= \tfrac{1}{3}\mathrm{tr}(\boldsymbol{\varepsilon}\cdot\boldsymbol{\varepsilon}\cdot\boldsymbol{\varepsilon}) = \tfrac{1}{3}\varepsilon_{ij}\varepsilon_{jk}\varepsilon_{ki} \\
J_{1D} &= \mathrm{tr}(\boldsymbol{D}) = D_{ii} \\
J_{2D} &= \tfrac{1}{2}\mathrm{tr}(\boldsymbol{D}\cdot\boldsymbol{D}) = \tfrac{1}{2}D_{ij}D_{ji} \\
J_{3D} &= \tfrac{1}{3}\mathrm{tr}(\boldsymbol{D}\cdot\boldsymbol{D}\cdot\boldsymbol{D}) = \tfrac{1}{3}D_{ij}D_{jk}D_{ki} \\
J_{1\varepsilon D} &= \mathrm{tr}(\boldsymbol{\varepsilon}\cdot\boldsymbol{D}) = \varepsilon_{ij}D_{ji} \\
J_{2\varepsilon D} &= \mathrm{tr}(\boldsymbol{\varepsilon}\cdot\boldsymbol{\varepsilon}\cdot\boldsymbol{D}) = \varepsilon_{ij}\varepsilon_{jk}D_{ki} \\
J_{3\varepsilon D} &= \mathrm{tr}(\boldsymbol{\varepsilon}\cdot\boldsymbol{D}\cdot\boldsymbol{D}) = \varepsilon_{ij}D_{jk}D_{ki} \\
J_{4\varepsilon D} &= \mathrm{tr}(\boldsymbol{\varepsilon}\cdot\boldsymbol{\varepsilon}\cdot\boldsymbol{D}\cdot\boldsymbol{D}) = \varepsilon_{ij}\varepsilon_{jk}D_{kl}D_{li}
\end{aligned} \tag{1.88}$$

When another *stress-damage* commonly used *space* $(\boldsymbol{\sigma}, \boldsymbol{D})$ is considered the following holds:

$$\begin{aligned}
J_{1\sigma} &= \mathrm{tr}(\boldsymbol{\sigma}) = \sigma_{ii} \\
J_{2\sigma} &= \tfrac{1}{2}\mathrm{tr}(\boldsymbol{\sigma}\cdot\boldsymbol{\sigma}) = \tfrac{1}{2}\sigma_{ij}\sigma_{ji} \\
J_{3\sigma} &= \tfrac{1}{3}\mathrm{tr}(\boldsymbol{\sigma}\cdot\boldsymbol{\sigma}\cdot\boldsymbol{\sigma}) = \tfrac{1}{3}\sigma_{ij}\sigma_{jk}\sigma_{ki} \\
J_{1D} &= \mathrm{tr}(\boldsymbol{D}) = D_{ii} \\
J_{2D} &= \tfrac{1}{2}\mathrm{tr}(\boldsymbol{D}\cdot\boldsymbol{D}) = \tfrac{1}{2}D_{ij}D_{ji} \\
J_{3D} &= \tfrac{1}{3}\mathrm{tr}(\boldsymbol{D}\cdot\boldsymbol{D}\cdot\boldsymbol{D}) = \tfrac{1}{3}D_{ij}D_{jk}D_{ki} \\
J_{1\sigma D} &= \mathrm{tr}(\boldsymbol{\sigma}\cdot\boldsymbol{D}) = \sigma_{ij}D_{ji} \\
J_{2\sigma D} &= \mathrm{tr}(\boldsymbol{\sigma}\cdot\boldsymbol{\sigma}\cdot\boldsymbol{D}) = \sigma_{ij}\sigma_{jk}D_{ki} \\
J_{3\sigma D} &= \mathrm{tr}(\boldsymbol{\sigma}\cdot\boldsymbol{D}\cdot\boldsymbol{D}) = \sigma_{ij}D_{jk}D_{ki} \\
J_{4\sigma D} &= \mathrm{tr}(\boldsymbol{\sigma}\cdot\boldsymbol{\sigma}\cdot\boldsymbol{D}\cdot\boldsymbol{D}) = \sigma_{ij}\sigma_{jk}D_{kl}D_{li}
\end{aligned} \tag{1.89}$$

1.3.2 Common Invariants of Two Different-Order Tensors: The Second Stress/Strain and the Fourth-Order Structural Tensors

The orthotropic material is characterized by three mutually perpendicular symmetry planes determined by three second-rank tensors called the *structural tensors* in terms of which the elastic strain energy $\mathcal{W}$ can be represented as

$$\mathcal{W} = \mathcal{W}(\varepsilon, \boldsymbol{M}^{(1)}, \boldsymbol{M}^{(2)}, \boldsymbol{M}^{(3)}) \tag{1.90}$$

When axes of material orthotropy coincide with axes of reference frame the structural tensors take the simplified forms

$$\boldsymbol{M}^{(1)} = \begin{bmatrix} 1 & 0 & 0 \\ & 0 & 0 \\ & & 0 \end{bmatrix} \quad \boldsymbol{M}^{(2)} = \begin{bmatrix} 0 & 0 & 0 \\ & 1 & 0 \\ & & 0 \end{bmatrix} \quad \boldsymbol{M}^{(3)} = \begin{bmatrix} 0 & 0 & 0 \\ & 0 & 0 \\ & & 1 \end{bmatrix} \tag{1.91}$$

for which the following holds:

$$\mathbf{1} = \boldsymbol{M}^{(1)} + \boldsymbol{M}^{(2)} + \boldsymbol{M}^{(3)} \tag{1.92}$$

Condition (1.92) means that the structural tensors are mutually dependent. Hence, *elastic strain energy* (1.90) can be represented *in terms of* two *structural tensors* chosen as independent, e.g., $\boldsymbol{M}^{(1)}$ and $\boldsymbol{M}^{(2)}$

$$\mathcal{W} = \mathcal{W}(\varepsilon, \boldsymbol{M}^{(1)}, \boldsymbol{M}^{(2)}) \tag{1.93}$$

Analogously, strain tensor can be written as $\varepsilon = \mathbf{1} \cdot \varepsilon = \varepsilon \cdot \mathbf{1}$, which finally leads to

$$\begin{aligned} \varepsilon &= \varepsilon \cdot \boldsymbol{M}^{(1)} + \varepsilon \cdot \boldsymbol{M}^{(2)} + \varepsilon \cdot \boldsymbol{M}^{(3)} = \varepsilon \cdot \mathbf{1} \\ \varepsilon &= \boldsymbol{M}^{(1)} \cdot \varepsilon + \boldsymbol{M}^{(2)} \cdot \varepsilon + \boldsymbol{M}^{(3)} \cdot \varepsilon = \mathbf{1} \cdot \varepsilon \end{aligned} \tag{1.94}$$

Summing up, the above equations assure symmetry of the strain tensor ε

$$\varepsilon = \frac{1}{2}(\varepsilon \cdot \boldsymbol{M}^{(1)} + \boldsymbol{M}^{(1)} \cdot \varepsilon) + \frac{1}{2}(\varepsilon \cdot \boldsymbol{M}^{(2)} + \boldsymbol{M}^{(2)} \cdot \varepsilon) + \frac{1}{2}(\varepsilon \cdot \boldsymbol{M}^{(3)} + \boldsymbol{M}^{(3)} \cdot \varepsilon) \tag{1.95}$$

The following *representation of elastic strain energy in terms of* 7 *invariants* can be obtained:

$$\begin{aligned} \mathcal{W} = \mathcal{W}\big[&\mathrm{tr}(\varepsilon), \tfrac{1}{2}\mathrm{tr}(\varepsilon \cdot \varepsilon), \tfrac{1}{3}\mathrm{tr}(\varepsilon \cdot \varepsilon \cdot \varepsilon), \\ &\mathrm{tr}(\varepsilon \cdot \boldsymbol{M}^{(1)}), \mathrm{tr}(\varepsilon \cdot \boldsymbol{M}^{(2)}), \mathrm{tr}(\varepsilon \cdot \boldsymbol{M}^{(3)}), \mathrm{tr}(\varepsilon \cdot \varepsilon \cdot \boldsymbol{M}^{(2)})\big] \end{aligned} \tag{1.96}$$

comprising both 3 *single strain invariants* and 4 *common strain and structural tensor invariants*. However, based on (1.95) two first-strain invariants can be represented as

$$\begin{aligned} \mathrm{tr}(\boldsymbol{\varepsilon}) &= \mathrm{tr}(\boldsymbol{\varepsilon}\cdot\boldsymbol{M}^{(1)}) + \mathrm{tr}(\boldsymbol{\varepsilon}\cdot\boldsymbol{M}^{(2)}) + \mathrm{tr}(\boldsymbol{\varepsilon}\cdot\boldsymbol{M}^{(3)}) \\ \mathrm{tr}(\boldsymbol{\varepsilon}\cdot\boldsymbol{\varepsilon}) &= \mathrm{tr}(\boldsymbol{\varepsilon}\cdot\boldsymbol{\varepsilon}\cdot\boldsymbol{M}^{(1)}) + \mathrm{tr}(\boldsymbol{\varepsilon}\cdot\boldsymbol{\varepsilon}\cdot\boldsymbol{M}^{(2)}) + \mathrm{tr}(\boldsymbol{\varepsilon}\cdot\boldsymbol{\varepsilon}\cdot\boldsymbol{M}^{(3)}) \end{aligned} \tag{1.97}$$

whereas the third strain invariant $\frac{1}{3}\mathrm{tr}(\boldsymbol{\varepsilon}\cdot\boldsymbol{\varepsilon}\cdot\boldsymbol{\varepsilon})$ is ignored because strain energy must be a quadratic function of strain $\boldsymbol{\varepsilon}$. For further details see Sect. 1.7.3.

1.4 Classification of Elastic Materials with Respect to Symmetry Groups and Classes

For further considerations, analogy between the *crystal lattice symmetry* groups and classes and corresponding *symmetry* of the *stiffness matrices* defined for crystalline materials might be useful (cf. e.g. Nye [42]). Unit cells of the eight conventional crystal lattices are demonstrated based on Love [34] and Jastrzebski [23], whereas corresponding *constitutive elasticity matrices* are schematically sketched applying Nye's graphics (symbol • refers to independent element, symbol ∘ refers to dependent element, symbols •—• or ∘—∘ represent pairs of identical matrix elements, symbols •—⊙ stand for pairs of elements in which one is doubled (effect of engineering notation applied to shear strain $\gamma_{ij} = 2\varepsilon_{ij}$), whereas symbols •—⊖ denote pairs of elements of the same absolute value but opposite signs, respectively.

1.4.1 Triclinic Hooke's Anisotropy (21 Constants)

Deformation of representative cube taken of the generally anisotropic material of *triclinic symmetry* subjected to exemplary axial tension along three axes is fully anisotropic. This means that it comprises both anisotropic axial strains (transformation of the cube to a rectangular prism) and anisotropic shear strains (transformation of the rectangular prism to a parallelepiped), as schematically sketched in Fig. 1.5. In such a case of general deformation the *elastic compliance matrix* is fully populated. In other words, all components of the *columnar stress vector* depend on all six components of the *columnar strain vector* (36 combinations). Final representation of compliance matrix for fully *anisotropic (triclinic) material* is as follows:

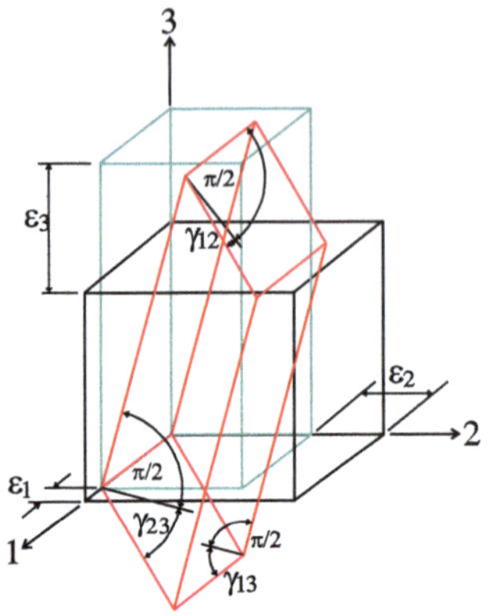

Fig. 1.5 Schematic deformation of representative cube of anisotropic triclinic material under uniaxial tension along three axes

$$
\left[\mathbb{E}^{-1}\right] =
\left[\begin{array}{ccc|ccc}
\frac{1}{E_{11}} & -\frac{\nu_{21}}{E_{11}} & -\frac{\nu_{31}}{E_{11}} & \frac{\eta_{23(1)}}{E_{11}} & \frac{\eta_{31(1)}}{E_{11}} & \frac{\eta_{12(1)}}{E_{11}} \\
-\frac{\nu_{12}}{E_{22}} & \frac{1}{E_{22}} & -\frac{\nu_{32}}{E_{22}} & \frac{\eta_{23(2)}}{E_{22}} & \frac{\eta_{31(2)}}{E_{22}} & \frac{\eta_{12(2)}}{E_{22}} \\
-\frac{\nu_{13}}{E_{33}} & -\frac{\nu_{23}}{E_{33}} & \frac{1}{E_{33}} & \frac{\eta_{23(3)}}{E_{33}} & \frac{\eta_{31(3)}}{E_{33}} & \frac{\eta_{12(3)}}{E_{33}} \\
\hline
\frac{\eta_{(1)23}}{G_{23}} & \frac{\eta_{(2)23}}{G_{23}} & \frac{\eta_{(3)23}}{G_{23}} & \frac{1}{G_{23}} & \frac{\mu_{31(23)}}{G_{23}} & \frac{\mu_{12(23)}}{G_{23}} \\
\frac{\eta_{(1)31}}{G_{31}} & \frac{\eta_{(2)31}}{G_{31}} & \frac{\eta_{(3)31}}{G_{31}} & \frac{\mu_{(23)31}}{G_{31}} & \frac{1}{G_{31}} & \frac{\mu_{12(31)}}{G_{31}} \\
\frac{\eta_{(1)12}}{G_{12}} & \frac{\eta_{(2)12}}{G_{12}} & \frac{\eta_{(3)12}}{G_{12}} & \frac{\mu_{(23)12}}{G_{12}} & \frac{\mu_{(31)12}}{G_{12}} & \frac{1}{G_{12}}
\end{array}\right]
\left[\begin{array}{ccc|ccc}
\bullet & \bullet & \bullet & \bullet & \bullet & \bullet \\
 & \bullet & \bullet & \bullet & \bullet & \bullet \\
 & & \bullet & \bullet & \bullet & \bullet \\
\hline
 & & & \bullet & \bullet & \bullet \\
 & & & & \bullet & \bullet \\
 & & & & & \bullet
\end{array}\right]
\tag{1.98}
$$

Symmetry of the *elastic compliance matrix* (1.98) results from symmetry of both stress and strain tensors, namely

$$
\begin{array}{lll}
\frac{\nu_{ij}}{E_{jj}} = \frac{\nu_{ji}}{E_{ii}} & \longrightarrow & \nu_{ij} E_{ii} = \nu_{ji} E_{jj} \\
\frac{\eta_{ij(k)}}{E_{kk}} = \frac{\eta_{(k)ij}}{G_{ij}} & \longrightarrow & \eta_{ij(k)} G_{ij} = \eta_{(k)ij} E_{kk} \\
\frac{\mu_{ij(ki)}}{G_{ki}} = \frac{\mu_{(ki)ij}}{G_{ji}} & \longrightarrow & \mu_{ij(ki)} G_{ji} = \mu_{(ki)ij} G_{ki}
\end{array}
\tag{1.99}
$$

In should be pointed out that the symmetry $\mathbb{E}^{-1}_{ij} = \mathbb{E}^{-1}_{ji}$ holds for elements of compliance matrix but not for corresponding *engineering material constants* $E_{ii}, \nu_{ij}, G_{ij}, \eta_{(i)jk}, \mu_{ij(ki)}$ as shown in (1.100) versus (1.98) (Table 1.2)

$$
\left[\mathbb{E}^{-1}\right] =
\left[\begin{array}{ccc|ccc}
E^{-1}_{11} & E^{-1}_{12} & E^{-1}_{13} & E^{-1}_{14} & E^{-1}_{15} & E^{-1}_{16} \\
E^{-1}_{21} & E^{-1}_{22} & E^{-1}_{23} & E^{-1}_{24} & E^{-1}_{25} & E^{-1}_{26} \\
E^{-1}_{31} & E^{-1}_{32} & E^{-1}_{33} & E^{-1}_{34} & E^{-1}_{35} & E^{-1}_{36} \\
\hline
E^{-1}_{41} & E^{-1}_{42} & E^{-1}_{43} & E^{-1}_{44} & E^{-1}_{45} & E^{-1}_{46} \\
E^{-1}_{51} & E^{-1}_{52} & E^{-1}_{53} & E^{-1}_{54} & E^{-1}_{55} & E^{-1}_{56} \\
E^{-1}_{61} & E^{-1}_{62} & E^{-1}_{63} & E^{-1}_{64} & E^{-1}_{65} & E^{-1}_{66}
\end{array}\right]
\tag{1.100}
$$

Table 1.2 Superposition of the strain tensor components of anisotropic material corresponding to subsequent stress tensor components

State	Strains: axial ε_1	ε_2	ε_3	Strains: shear γ_{23}	γ_{31}	γ_{12}
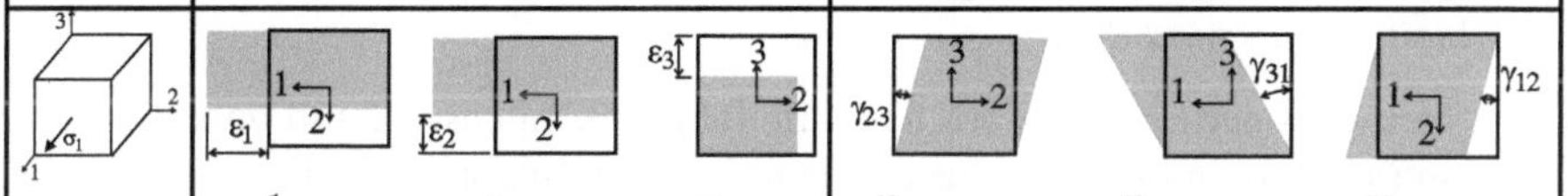						
	$\frac{1}{E_{11}}\sigma_1$	$-\frac{\nu_{21}}{E_{11}}\sigma_1$	$-\frac{\nu_{31}}{E_{11}}\sigma_1$	$\frac{\eta_{23(1)}}{E_{11}}\sigma_1$	$\frac{\eta_{31(1)}}{E_{11}}\sigma_1$	$\frac{\eta_{12(1)}}{E_{11}}\sigma_1$
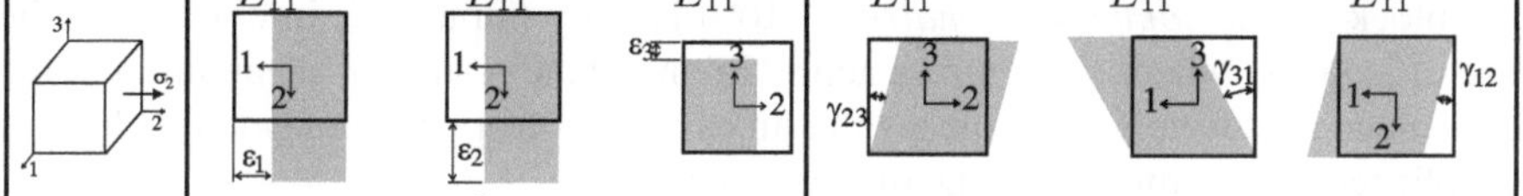						
	$-\frac{\nu_{12}}{E_{22}}\sigma_2$	$\frac{1}{E_{22}}\sigma_2$	$-\frac{\nu_{32}}{E_{22}}\sigma_2$	$\frac{\eta_{23(2)}}{E_{22}}\sigma_2$	$\frac{\eta_{31(2)}}{E_{22}}\sigma_2$	$\frac{\eta_{12(2)}}{E_{22}}\sigma_2$
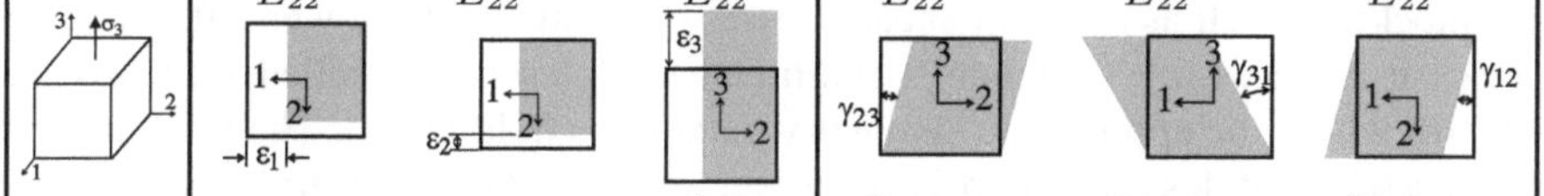						
	$-\frac{\nu_{13}}{E_{33}}\sigma_3$	$-\frac{\nu_{23}}{E_{33}}\sigma_3$	$\frac{1}{E_{33}}\sigma_3$	$\frac{\eta_{23(3)}}{E_{33}}\sigma_3$	$\frac{\eta_{31(3)}}{E_{33}}\sigma_3$	$\frac{\eta_{12(3)}}{E_{33}}\sigma_3$
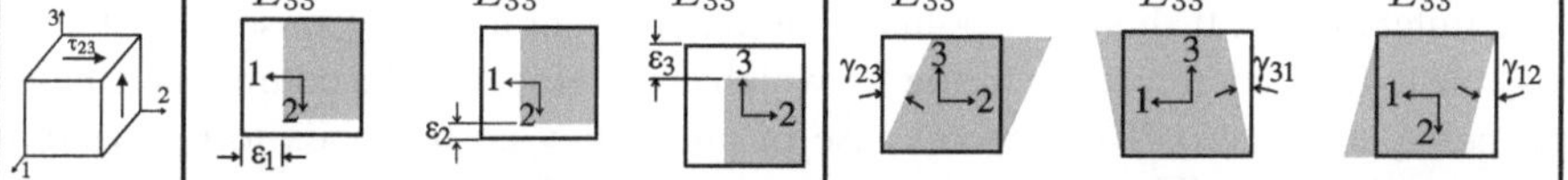						
	$\frac{\eta_{(1)23}}{G_{23}}\tau_{23}$	$\frac{\eta_{(2)23}}{G_{23}}\tau_{23}$	$\frac{\eta_{(3)23}}{G_{23}}\tau_{23}$	$\frac{1}{G_{23}}\tau_{23}$	$\frac{\mu_{(31)23}}{G_{23}}\tau_{23}$	$\frac{\mu_{(12)23}}{G_{23}}\tau_{23}$
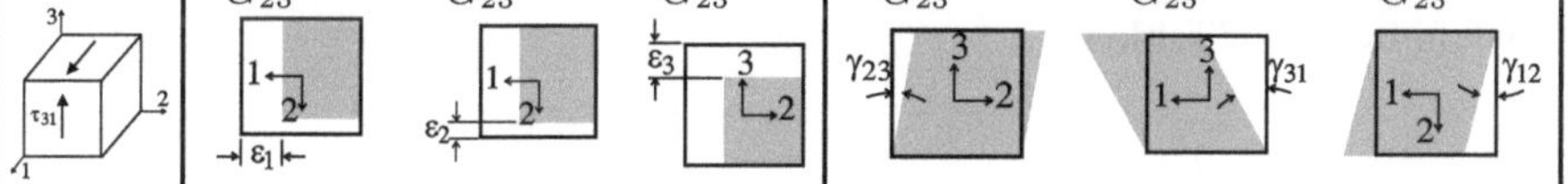						
	$\frac{\eta_{(1)31}}{G_{31}}\tau_{31}$	$\frac{\eta_{(2)31}}{G_{31}}\tau_{31}$	$\frac{\eta_{(3)31}}{G_{31}}\tau_{31}$	$\frac{\mu_{(23)31}}{G_{31}}\tau_{31}$	$\frac{1}{G_{31}}\tau_{31}$	$\frac{\mu_{(12)31}}{G_{31}}\tau_{31}$
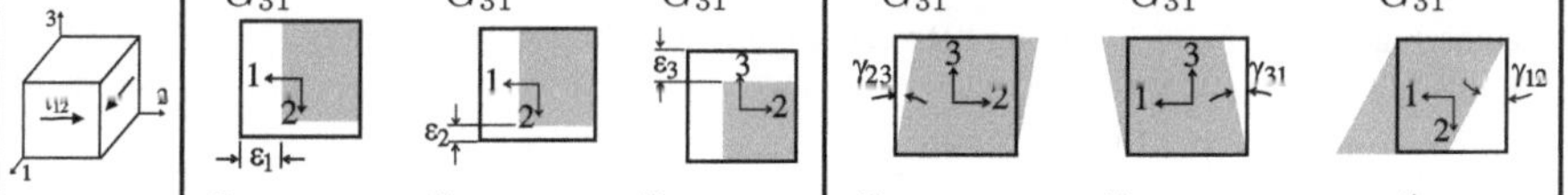						
	$\frac{\eta_{(1)12}}{G_{12}}\tau_{12}$	$\frac{\eta_{(2)12}}{G_{12}}\tau_{12}$	$\frac{\eta_{(3)12}}{G_{12}}\tau_{12}$	$\frac{\mu_{(23)12}}{G_{12}}\tau_{12}$	$\frac{\mu_{(31)11}}{G_{12}}\tau_{12}$	$\frac{1}{G_{12}}\tau_{12}$

Elastic engineering modules of five types can be sorted in the following way, after Lekhnitskii [33]:

- E_{ii}—axial modules (3 *generalized Young's modules*)
- G_{ij}—shear modules for planes parallel to the coordinate planes (3 *generalized Kirchhoff's modules*)
- ν_{ij}—Poisson's ratios characterizing the contraction in the direction of one axis when tension is applied in the direction of another axis (3 *generalized Poisson's coefficients*)

- $\mu_{ij(kl)}$—coefficients characterizing shears in planes parallel to the coordinate planes resulting from shear stresses acting in other planes parallel to the coordinate planes (3 *Chencov's modules*)
- $\eta_{i(jk)}$—mutual influence coefficients characterizing extensions in the directions of the coordinate axes resulting from shear stresses acting in the coordinate planes (9 *Rabinovich's modules*)

The aforementioned modules are listed in Table 1.3. In case of full anisotropy the shear stress acting in one plane results in a shear strain appearing in another plane. This effect is described by the three *Chencov modules*. Hence, the bottom right-hand side block of the *compliance matrix* (1.100) is fully populated, in contrast to the case of isotropy where shear stress acting in one plane results in shear strain in the same plane exclusively. This means that in case of isotropy the considered block of compliance matrix must have the diagonal form.

In order to describe effect of axial stresses on shear strains (upper right-hand side block), as well as effect of shear stresses on axial strains (lower left-hand side block), it is necessary to define 9 additional modules $\eta_{(i)jk}$, called *Rabinovich's modules* where the appropriate symmetry conditions hold (1.99). The total number of discussed modules is equal to 21. However, only 18 of them are truly independent because the compliance matrix $[\mathbb{E}^{-1}]$ has to obey transformation with respect to three Euler angles. It should be pointed out that in general case of anisotropy it is not possible to find any reference frame for which any element of the compliance matrix can be equal to zero. The general case of anisotropy corresponds to the *triclinic symmetry lattice cell* in which all three edges differ from each other and all three angles between them differ from each other and none of them is equal to 90°, as shown in item 1 of Table 1.1.

Table 1.3 *Engineering modules* defining elements of *elastic compliance matrix* (1.98) of fully anisotropic material

Engineering elastic modules	Coupling between		Corresponding axes or planes	Number of coefficients
	Stress	Strain		
E_{11}, E_{22}, E_{33}	Axial	Extension	The same axes 1 → 1, etc.	3
G_{12}, G_{32}, G_{31}	Shear	Shear strain	The same planes 12 → 12, etc.	3
$\nu_{21}, \nu_{31}, \nu_{32}$	Axial	Extension	Different exes 1 → 2, etc.	3
$\mu_{31(23)}, \mu_{12(23)}, \mu_{12(31)}$	Shear	Shear strain	Different planes 13 → 23, etc.	3
$\eta_{23(1)}, \ldots, \eta_{12(3)}$	Shear	Extension	Normal to shear plane 23 → 1, etc.	9

1.4.2 Monoclinic Hooke's Anisotropy (13 Constants)

Among anisotropic materials the narrower group called *monoclinic symmetry* can be distinguished. *Monoclinic or oblique symmetry* corresponds to *monoclinic space lattice cell* symmetry in which all three edges differ from each other, whereas two angles are equal to 90° and one is different, as shown in item 2 of Table 1.1. The corresponding stiffness matrix symmetry characterizes through incomplete population in which only 13 elements are not equal to zero, as shown below.

$$
\left[\mathbb{E}^{-1}\right] =
\left[\begin{array}{ccc|ccc}
\frac{1}{E_{11}} & -\frac{\nu_{21}}{E_{11}} & -\frac{\nu_{31}}{E_{11}} & & & \frac{\eta_{12(1)}}{E_{11}} \\
-\frac{\nu_{12}}{E_{22}} & \frac{1}{E_{22}} & -\frac{\nu_{32}}{E_{22}} & & & \frac{\eta_{12(2)}}{E_{22}} \\
-\frac{\nu_{13}}{E_{33}} & -\frac{\nu_{23}}{E_{33}} & \frac{1}{E_{33}} & & & \frac{\eta_{12(3)}}{E_{33}} \\
\hline
 & & & \frac{1}{G_{23}} & \frac{\mu_{31(23)}}{G_{23}} & \\
 & & & \frac{\mu_{(23)31}}{G_{31}} & \frac{1}{G_{31}} & \\
\frac{\eta_{(1)12}}{G_{12}} & \frac{\eta_{(2)12}}{G_{12}} & \frac{\eta_{(3)12}}{G_{12}} & & & \frac{1}{G_{12}}
\end{array}\right]
\tag{1.101}
$$

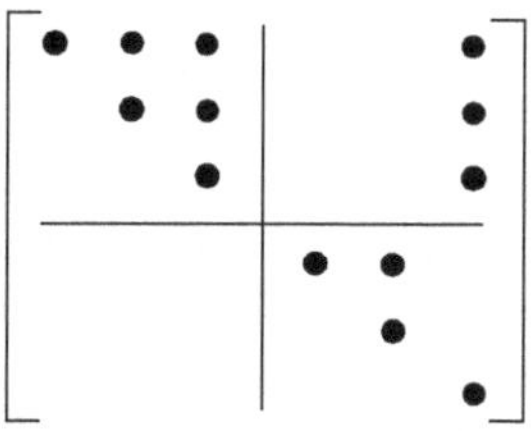

In other words, in case of monoclinic symmetry only three of the Rabinovich modules and only one of the Chencov modules are different from zero.

1.4.3 Trigonal/Rhombohedral Hooke's Anisotropy (6 Constants)

Another important narrower case of material anisotropy called trigonal anisotropy can be distinguished. The trigonal anisotropy corresponds to the *rhombohedral cell lattice* in which all three edges are equal to each other and all three angles are equal but different from 90°, as shown in item 3 of Table 1.1. The corresponding *compliance matrix* takes the following representation:

$$
\left[\mathbb{E}^{-1}\right] =
\left[\begin{array}{ccc|ccc}
\frac{1}{E_{11}} & -\frac{\nu_{21}}{E_{11}} & -\frac{\nu_{31}}{E_{11}} & \frac{\eta_{23(1)}}{E_{11}} & & \\
-\frac{\nu_{12}}{E_{22}} & \frac{1}{E_{11}} & -\frac{\nu_{31}}{E_{11}} & -\frac{\eta_{23(1)}}{E_{11}} & & \\
-\frac{\nu_{13}}{E_{33}} & -\frac{\nu_{13}}{E_{33}} & \frac{1}{E_{33}} & & & \\
\hline
\frac{\eta_{(1)23}}{G_{23}} & -\frac{\eta_{(1)23}}{G_{23}} & & \frac{1}{G_{23}} & & \\
 & & & & \frac{1}{G_{31}} & \frac{2\mu_{12(31)}}{G_{31}} \\
 & & & & \frac{2\mu_{(31)12}}{G_{12}} & \frac{2(1+\nu_{12})}{E_{11}}
\end{array}\right]
\tag{1.102}
$$

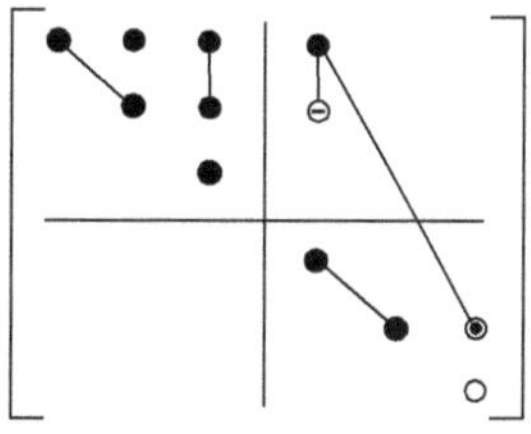

It is seen that in case of *trigonal symmetry* among *Rabinovich's modules* only two are nonzeroth but in fact only one of them is independent because they only differ in sign. Additionally, only one *Chencov's modulus* is different from zero but in fact it is the dependent modulus due to the specific coupling between components $2E_{14}^{-1} = E_{56}^{-1}$ and $E_{24}^{-1} = -E_{14}^{-1}$ as well as $E_{11}^{-1} = E_{22}^{-1}$, $E_{44}^{-1} = E_{55}^{-1}$, $E_{13}^{-1} = E_{23}^{-1}$ whereas $E_{66}^{-1} = (E_{11}^{-1} - E_{12}^{-1})/2$ must hold. Finally for trigonal symmetry only 6 elements of the compliance matrix are independent, see Berryman [2].

1.4.4 Orthorhombic Hooke's Orthotropy (9 Constants)

The majority of engineering materials exhibit a specific symmetry property, which may result in reduction of the number of nonzeroth elastic modules. It can be done when, for chosen symmetry group or class, some particular material directions are defined in such a way that transformation of the compliance matrix from an arbitrary coordinate frame to the given *structural symmetry frame* leads to the zeroth population of the top right-hand side and the bottom left-hand side blocks of the *compliance matrix* (1.98), and additionally the bottom right-hand side block possesses a diagonal form. In such practically important cases both the nine Rabinovich $\eta_{(i)jk}$ and the three Chencov $\mu_{ij(kl)}$ modules are equal to zero, and consequently, coupling between the shear stresses and elongations does not exist such that shear strains are produced exclusively by the action of stresses at the same planes. In this particular symmetry, called orthotropy, there exist three mutually perpendicular axes (1, 2, 3) that determine the three *material orthotropy planes*. The *orthotropy symmetry* case corresponds to the *orthorhombic lattice* in which all three edges differ each from other but all angles are equal to 90°, as presented in item 4 of Table 1.1.

$$
\left[\mathbb{E}^{-1}\right] =
\left[\begin{array}{ccc|ccc}
\frac{1}{E_{11}} & -\frac{\nu_{21}}{E_{11}} & -\frac{\nu_{31}}{E_{11}} & & & \\
-\frac{\nu_{12}}{E_{22}} & \frac{1}{E_{22}} & -\frac{\nu_{32}}{E_{22}} & & & \\
-\frac{\nu_{13}}{E_{33}} & -\frac{\nu_{23}}{E_{33}} & \frac{1}{E_{33}} & & & \\
\hline
 & & & \frac{1}{G_{23}} & 0 & 0 \\
 & & & & \frac{1}{G_{13}} & 0 \\
 & & & & & \frac{1}{G_{12}}
\end{array}\right]
\left[\begin{array}{ccc|ccc}
\bullet & \bullet & \bullet & & & \\
 & \bullet & \bullet & & & \\
 & & \bullet & & & \\
\hline
 & & & \bullet & & \\
 & & & & \bullet & \\
 & & & & & \bullet
\end{array}\right]
\tag{1.103}
$$

The following conditions must hold to assure matrix symmetry:

$$
\frac{\nu_{21}}{E_{11}} = \frac{\nu_{12}}{E_{22}} \qquad \frac{\nu_{13}}{E_{33}} = \frac{\nu_{31}}{E_{11}} \qquad \frac{\nu_{23}}{E_{33}} = \frac{\nu_{32}}{E_{22}} \tag{1.104}
$$

Finally, in case of orthotropy the number of independent material constants is nine, that is, three *generalized Hooke's modules* E_{11}, E_{22}, E_{33}, 3 *generalized Kirchhoff's modules* G_{12}, G_{23}, G_{31} and three *generalized Poisson's ratios* $\nu_{21}, \nu_{23}, \nu_{31}$.

1.4.5 Tetragonal Hooke's Transverse Isotropy (6 Constants)

For several engineering applications the general orthotropic symmetry model seems too complicated, since additional symmetry conditions frequently appear. Particularly, when conditions of isotropy hold in selected orthotropy plane the so-called transverse isotropy obeys.

In case of so-called *tetragonal symmetry* material properties in the plane (1, 2) satisfy condition of cubic symmetry, see item 5 of Table 1.1

$$E_{11} = E_{22}, \quad G_{13} = G_{23}, \quad \nu_{31} = \nu_{32} \tag{1.105}$$

Hence, in case of transverse isotropy of tetragonal symmetry the number of independent material constants is equal to 6: E_{11}, E_{33}, G_{23}, G_{12}, ν_{21}, ν_{31}. Corresponding crystal lattice is sketched in item 5 of Table 1.1, where *tetragonal lattice* being special case of the *orthorhombic lattice* with $a = b \neq c$ obeys.

When the constraints (1.105) are applied to compliance matrix (1.103) the transverse isotropy tetragonal symmetry case yields

$$\left[\mathbb{E}^{-1}\right] = \left[\begin{array}{ccc|ccc} \frac{1}{E_{11}} & -\frac{\nu_{21}}{E_{11}} & -\frac{\nu_{31}}{E_{11}} & & & \\ & \frac{1}{E_{11}} & -\frac{\nu_{31}}{E_{11}} & & & \\ & & \frac{1}{E_{33}} & & & \\ \hline & & & \frac{1}{G_{23}} & & \\ & & & & \frac{1}{G_{23}} & \\ & & & & & \frac{1}{G_{12}} \end{array}\right]$$

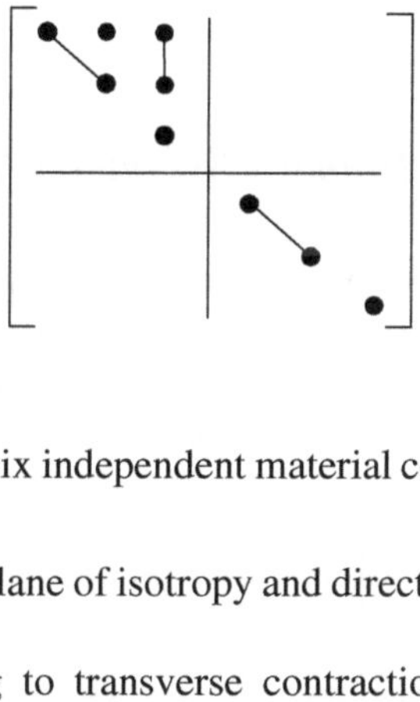

(1.106)

It follows from the constraints (1.105) that six independent material constants define the *tetragonal symmetry matrix*:

- E_{11}, E_{33}—two Young's modulus in the plane of isotropy and direction perpendicular to this plane,
- ν_{21}, ν_{31}—two Poisson's ratios referring to transverse contraction or swelling caused by tension or compression in direction perpendicular to isotropy plane,
- G_{12}, G_{23}—two different Kirchhoff's modules in the isotropy or orthotropy planes.

1.4.6 Hexagonal Hooke's Transverse Isotropy (5 Constants)

In special case of the transverse isotropy called *hexagonal symmetry* the additional constraint must obey for the shear modulus in the isotropy plane

$$G_{12} = \frac{E_{11}}{2(1+\nu_{21})} \quad \text{or} \quad E_{66}^{-1} = 2\left(E_{11}^{-1} - E_{12}^{-1}\right) \tag{1.107}$$

where modulus G_{12} is expressed in terms of the *transverse Young modulus* E_{11} and *transverse Poisson's ratio* ν_{21}. Hence, in case of the transverse isotropy of hexagonal symmetry the number of independent constants is equal to 5: E_{11}, E_{33}, G_{23}, ν_{21}, ν_{31}. A choice of the five independent material constants from among six can be performed in an optional way, for instance

$$
[\mathbb{E}^{-1}] = \left[\begin{array}{ccc|ccc}
\frac{1}{E_{11}} & -\frac{\nu_{21}}{E_{11}} & -\frac{\nu_{31}}{E_{11}} & & & \\
 & \frac{1}{E_{11}} & -\frac{\nu_{31}}{E_{11}} & & & \\
 & & \frac{1}{E_{33}} & & & \\
\hline
 & & & \frac{1}{G_{23}} & & \\
 & & & & \frac{1}{G_{23}} & \\
 & & & & & \frac{2(1+\nu_{21})}{E_{11}}
\end{array}\right] \tag{1.108}
$$

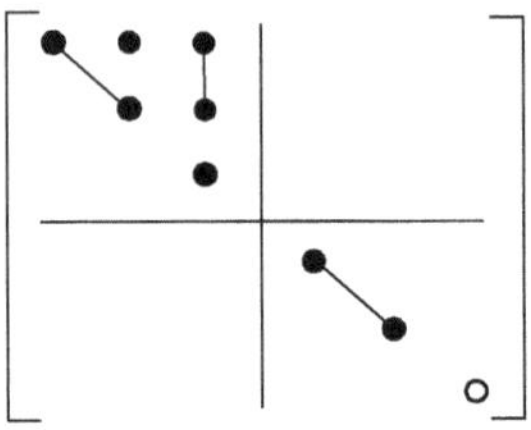

Rolled metals, some multi-phase composite materials, basalt, or columnar ice are examples of *transversely isotropic materials*, however, precise distinction between the *tetragonal* or *hexagonal symmetry classes* is often difficult (see for example Gan et al. [16]).

1.4.7 Cubic Hooke's Symmetry (3 Constants)

Further reduction in the number of independent constants leads to cubic symmetry for which the compliance matrix is characterized by three independent material constants $E_{11} = E_{22} = E_{33} = E$, $G_{23} = G_{31} = G_{12} = G$ and $\nu_{21} = \nu_{31} = \nu_{32} = \nu$. Hence, the following form of the compliance matrix is furnished:

$$
[\mathbb{E}^{-1}] = \left[\begin{array}{ccc|ccc}
\frac{1}{E} & -\frac{\nu}{E} & -\frac{\nu}{E} & & & \\
 & \frac{1}{E} & -\frac{\nu}{E} & & & \\
 & & \frac{1}{E} & & & \\
\hline
 & & & \frac{1}{G} & & \\
 & & & & \frac{1}{G} & \\
 & & & & & \frac{1}{G}
\end{array}\right] \tag{1.109}
$$

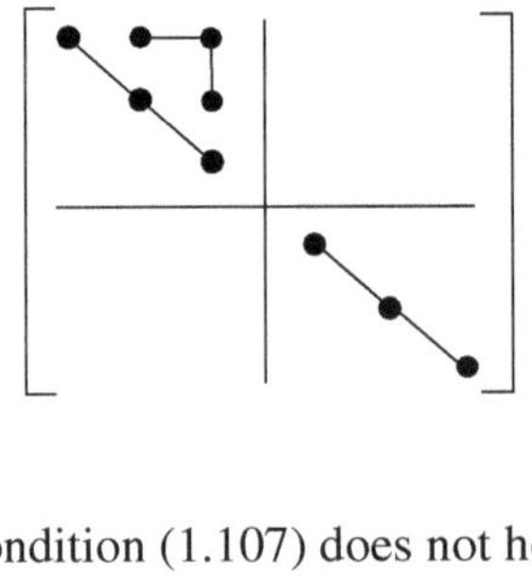

Note that in case of *cubic symmetry* the condition (1.107) does not hold. The corresponding *cubic* or *regular lattice* is shown in item 7 of Table 1.1. A particular example of the cubic symmetry material is *nickel-based single crystal superalloy* widely used in aircraft engines, especially for turbine blades as discussed by Desmorat and Marull [12]. The cubic symmetry is the narrower symmetry case known from crystallography, see Jastrzebski [23], since fully isotropic crystal lattices are unknown.

1.4.8 Isotropic Hooke's Symmetry (2 Constants)

All the aforementioned symmetry groups have equivalences in existing crystal lattice systems. Nevertheless, even narrower than the cubic symmetry called isotropy is frequently used. The isotropy requires the infinite symmetry group which means that all material directions are equivalent in terms of mechanical, thermal, electric, optical, and magnetic properties. In other words it is not possible to distinguish any specific direction. The isotropy is helpful when describing the majority of *polycrystalline materials* either in a virgin state or artificially fabricated as *particulate composites*, *nano-composites*, etc., see item 8 of Table 1.1.

In an isotropic material physical properties are independent of the reference frame. Hence, any optional *reference frame* x, y, z is sufficient for unique definition of material properties. In order to derive mathematical form of the *Hooke law of isotropic material* it is most convenient to apply superposition of strain components $\{\varepsilon\} = \left\{\varepsilon_x, \varepsilon_y, \varepsilon_z, \gamma_{yz}, \gamma_{zx}, \gamma_{xy}\right\}$ caused by subsequent stress components $\{\sigma\} = \left\{\sigma_x, \sigma_y, \sigma_z, \tau_{yz}, \ \tau_{zx}, \tau_{xy}\right\}$ (see Table 1.2). Applying vector-matrix notation the *isotropic Hooke law* takes the form

$$\{\varepsilon\} = \left[\mathbb{E}^{-1}\right]\{\sigma\} \tag{1.110}$$

where the *isotropic compliance matrix* $[\mathbb{E}^{-1}]$ takes the following representation.

$$\left[\mathbb{E}^{-1}\right] = \left[\begin{array}{ccc|ccc} \frac{1}{E} & \left(\frac{1}{E}-\frac{1}{2G}\right) & \left(\frac{1}{E}-\frac{1}{2G}\right) & & & \\ & \frac{1}{E} & \left(\frac{1}{E}-\frac{1}{2G}\right) & & & \\ & & \frac{1}{E} & & & \\ \hline & & & \frac{1}{G} & & \\ & & & & \frac{1}{G} & \\ & & & & & \frac{1}{G} \end{array}\right] \tag{1.111}$$

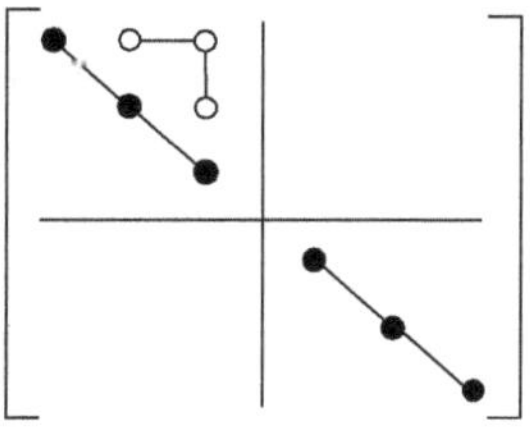

It is clear that the elastic isotropic material is uniquely defined by two independent material constants, the choice of which from among E, G, ν is optional. In the above representation diagonal modules E and G are chosen as independent. Hooke's law can also be transformed to the following inverse relation, (see Ottosen and Ristinmaa [44]):

$$\{\sigma\} = [\mathbb{E}]\{\varepsilon\} \tag{1.112}$$

where the *isotropic stiffness matrix* $[\mathbb{E}]$ is defined as

$$[\mathbb{E}] = \frac{E}{(1+\nu)(1-2\nu)} \left[\begin{array}{ccc|ccc} 1-\nu & \nu & \nu & & & \\ & 1-\nu & \nu & & & \\ & & 1-\nu & & & \\ \hline & & & \frac{1-2\nu}{2} & & \\ & & & & \frac{1-2\nu}{2} & \\ & & & & & \frac{1-2\nu}{2} \end{array} \right] \tag{1.113}$$

Format of elastic stiffness matrix (1.113) involves elements all dependent on both E and ν such that the format equivalent to (1.111) cannot be achieved. Explicit separation of the diagonal matrix elements related to shear deformation and the off-diagonal matrix elements related to extension is possible by use of the format expressed in terms of Lamé's constants λ and μ

$$\left[\begin{array}{ccc|ccc} \lambda+2\mu & \lambda & \lambda & & & \\ & \lambda+2\mu & \lambda & & & \\ & & \lambda+2\mu & & & \\ \hline & & & 2\mu & & \\ & & & & 2\mu & \\ & & & & & 2\mu \end{array} \right]$$

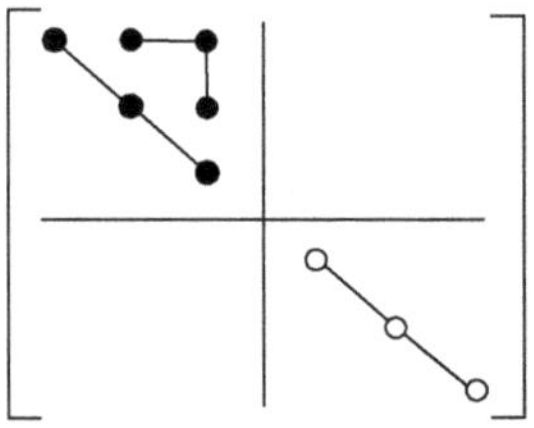

(1.114)

where the classical definitions of Lamé's constants hold

$$\lambda = \frac{E\nu}{(1+\nu)(1-2\nu)} \qquad 2\mu = G \tag{1.115}$$

It is worth to mention that the last format (1.114) can be interpreted by use of Nye graphics (• or ∘) where three off-diagonal first quarter elements and three diagonal third quarter elements are considered as independent.

The considered case of elastic isotropy is the only symmetry case for which it is possible to separate effects of *shape* and *volume changes* when *decomposition* of *strain* and *stress tensors* into *deviators* and *axiators* (1.5), (1.25) is used as

$$\varepsilon_{\mathrm{m}}\mathbf{1} = \frac{1}{3K}\sigma_{\mathrm{m}}\mathbf{1} \qquad \boldsymbol{e} = \frac{1}{2G}\boldsymbol{s} \tag{1.116}$$

Two modules in the above pair of relations called the *bulk modulus* K and the *Kirchhoff modulus* G can be expressed in terms of the *Young modulus* E and the *Poisson ratio* ν

$$K = \frac{E}{3(1-2\nu)} \qquad G = \frac{E}{2(1+\nu)} \tag{1.117}$$

However, in all other cases of material anisotropy (items 1 to 7 in Table 1.1) aforementioned separation of volumetric from shear effects is impossible.

In the particular case of *plane stress state* in the x, y plane strain component ε_z can be expressed in terms of strain components in x, y plane as follows:

$$\sigma_z = 0 \rightarrow \varepsilon_z = -\frac{\nu}{1-\nu} \left(\varepsilon_x + \varepsilon_y\right) \tag{1.118}$$

Finally, *plane stress stiffness matrix* $\mathbb{E}$ can be reduced to the 3×3 matrix

$$\begin{Bmatrix} \sigma_x \\ \sigma_y \\ \tau_{xy} \end{Bmatrix} = \left[\begin{array}{cc|c} \frac{E}{1-\nu^2} & \frac{\nu E}{1-\nu^2} & 0 \\ \frac{\nu E}{1-\nu^2} & \frac{E}{1-\nu^2} & 0 \\ \hline 0 & 0 & \frac{E}{2(1+\nu)} \end{array}\right] \begin{Bmatrix} \varepsilon_x \\ \varepsilon_y \\ \gamma_{xy} \end{Bmatrix} \tag{1.119}$$

1.5 Analogy Between Constitutive Fourth-Order Tensors: The Elastic (Hooke's) and the Yield/Failure (von Mises') of the Same Symmetry

Identification of *material symmetry* in elastic range of deformation (*anisotropy*, *orthotropy*, *transverse isotropy*, *isotropy*, etc.) is a starting point to appropriate description of both the *limit criteria* that control transition from the elastic range into the state connected with energy dissipation (*material damage*, *plastic yield*, *phase change*, etc.) as well as correct constitutive description of deformation processes in nonelastic range. It can be expected that if material in the elastic range exhibits isotropic behavior, then at least in the initial phase of plastic yielding it will approximately save properties of isotropy. The nature of elastic deformation resulting from interatomic distances change in crystal lattice is qualitatively different from the nature of plastic deformation commonly interpreted as plastic microslips considered usually as slips and dislocations between atom layers inside lattice. However, it can be expected that during more advanced plastic deformation certain orientation of plastic slip systems in the particular grains leading to appearance of a *material texture* characterized by an *acquired anisotropy* is observed (metal forming processes like rolling, drawing and press forming, see Mróz and Maciejewski [37]).

On the other hand if material even in *elastic range* is characterized by an *anisotropy* (e.g., *long fiber reinforced composites*, *wood*, *biological tissues*) it can be expected that in *nonelastic range* it will also exhibit *anisotropy*. However, it will be possible decrease of a symmetry class toward more general *plastic anisotropy*, for instance due to gradual evolution of elastic orthotropy. It can be however noticed that in case of dissipative processes different from plasticity (e.g., *material damage* or *failure*) loss of isotropy may be expected just in the elastic range, as observed in *elastic brittle materials* e.g., *ceramics*, *composites*, *concrete*, etc. Additionally, *initiation* and *growth of* other *dissipative processes* connected with *plastic yielding*, *phase change* or other *structural changes* may result in change in the initial symmetry class. For example, in case of *spheroidal graphite cast iron* which generally exhibits *brittle-ductile behavior* a gradual transition from elastic anisotropy caused by directional damage to a state close to isotropy may be observed.

It can be assumed that features of anisotropy present in the elastic range are in general inherited in nonelastic range if some dissipative processes like plastic yield, damage, failure are present. Notice however that even, in the case when in inelastic range material behaves as isotropic, initiation of inelastic range (plasticity, damage, or failure) may provoke a *material symmetry change*. It was previously discussed that in case of damage evolution the fourth-rank *damage effect tensor* $[\mathbb{M}(\boldsymbol{D})]$ may be used to describe *degeneration* of the *elasticity tensor* $[\tilde{\mathbb{E}}] = [\mathbb{M}(\boldsymbol{D})]^{\mathrm{T}}[\mathbb{E}][\mathbb{M}(\boldsymbol{D})]$, in a similar fashion effect of other dissipative phenomena such as *plastic yield*, *structural change* due to *phase transformation* may result in anisotropy nucleation and growth.

Analogy between *crystal unit cells* of *space lattices* and *constitutive matrices* of *elasticity* and *initiation of plasticity* is presented in Table 1.4. In the fundamental book by Love [34] the analogy between crystal symmetry classes and groups from one side and appropriate forms of elastic strain energy function $\mathcal{W} = \frac{1}{2}\{\varepsilon\}^{\mathrm{T}}[\mathbb{E}]\{\varepsilon\}$ from the other, is demonstrated. In this book an extension of the aforementioned analogy also for symmetry of constitutive matrix of plastic yield initiation $[\mathbb{H}]$ appearing in the *von Mises criterion* $\{\boldsymbol{\sigma}\}^{\mathrm{T}}[\mathbb{H}]\{\boldsymbol{\sigma}\} = 1$ is proposed. Unit cells of the four chosen space lattices have been presented following Jastrzebski [23], whereas corresponding constitutive elasticity matrices have schematically been presented applying Nye [42] graphics (symbol • refers to independent element, symbol ∘ refers to dependent element, whereas symbols •—• or ∘—∘ represent pairs of identical matrix elements).

In case of full anisotropy the complete analogy between the Hooke matrix and the von Mises plasticity matrix holds (21 independent matrix elements in both classes). However, when narrower symmetry groups are considered: *orthotropic*, *transversely isotropic* of *tetragonal* or *hexagonal* classes, it is necessary to notice that elastic matrices are usually defined in stress tensor coordinates, whereas plastic constitutive matrices are often defined in the narrower *stress deviator coordinates*.

Reduction of the *tensorial space* to the *deviatoric* one is always equivalent to imposing additional constraints, hence the number of independent elements of *plasticity matrix* is always lower than the corresponding number of independent elements of *elasticity matrix*. Namely, it is clear that the 6-element orthotropic deviatoric Hill's matrix corresponds to the 9-element orthotropic Hooke's matrix. Similarly, the 4-element *transversely isotropic tetragonal* class *Hill's matrix* corresponds to the 6-element Hooke's matrix, when the independence of Hill's matrix of hydrostatic stress is imposed. Finally, the 3-element *transversely isotropic hexagonal* class *Hu–Marin matrix* corresponds to the 5-element transversely isotropic hexagonal class Hooke matrix. Let us note that pairs of identical matrix elements are arranged in the same way in both matrices of elasticity and plasticity.

Nevertheless, some dependent elements in the plasticity matrix (as represented by symbol ∘) correspond to independent elements of elasticity matrix (sketched by symbol •), but general population of both matrices remains unchanged.

Table 1.4 Analogy between chosen symmetry groups: *triclinic*, *orthorhombic*, *tetragonal* and *hexagonal* symmetry of *Hooke's matrix* and plastic yield initiation *von Mises' matrix*

Conventional unit cells of space lattices	Chosen constitutive matrix symmetry	
	Elastic Hooke's matrix $2W=\{\varepsilon\}^{T}[\mathbb{E}]\{\varepsilon\}$	Plastic yield initiation matrix $\{\sigma\}^{T}[\mathbb{\Pi}]\{\sigma\}=1$
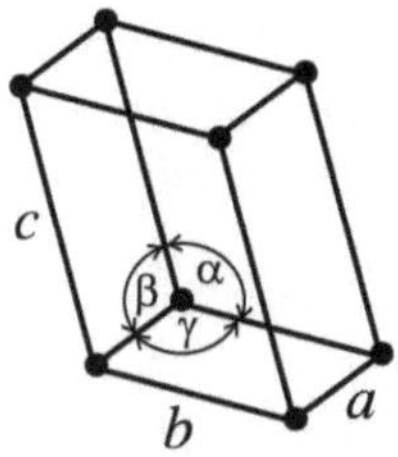 triclinic lattice	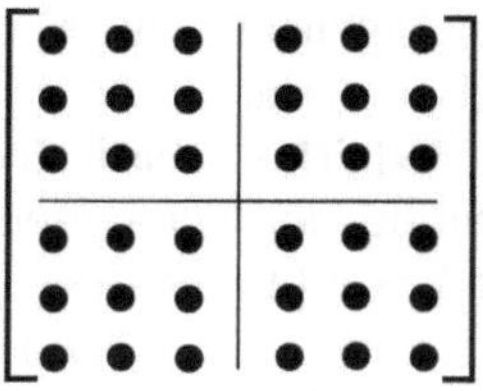 Hooke's (21 constants)	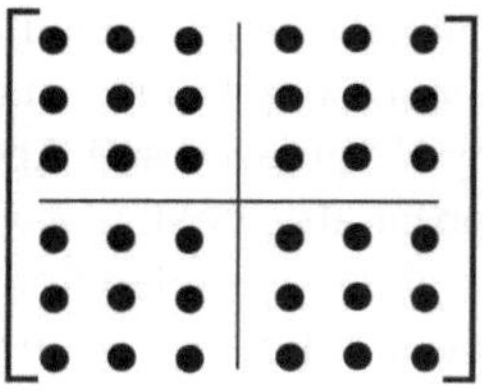 von Mises (21 constants)
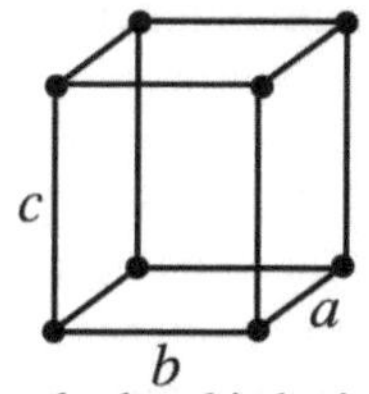 orthorhombic lattice	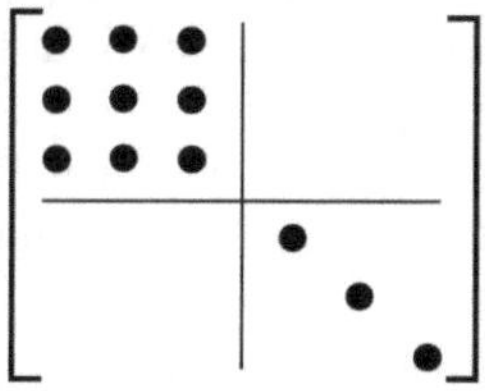 orthotropic Hooke's (9 constants)	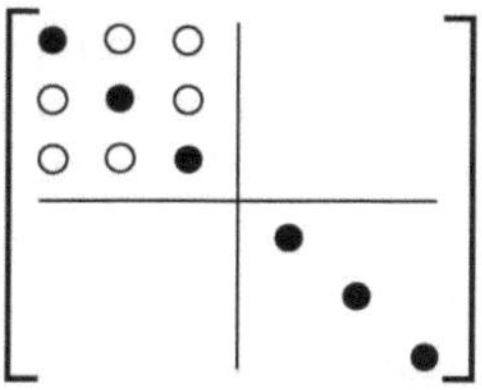 deviatorc Hill's (6 constants)
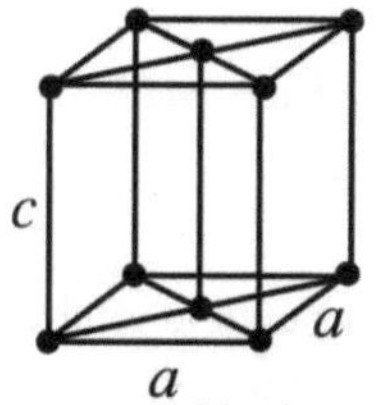 tetragonal lattice	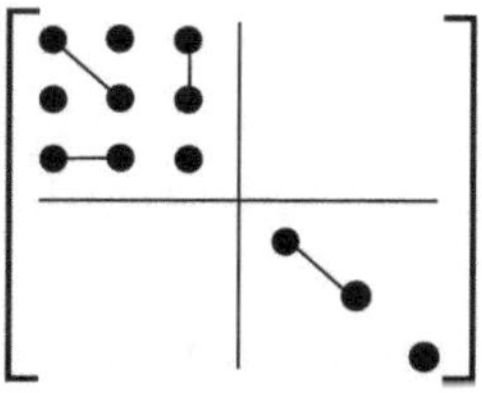 transversely isotropic tetragonal Hooke's (6 constants)	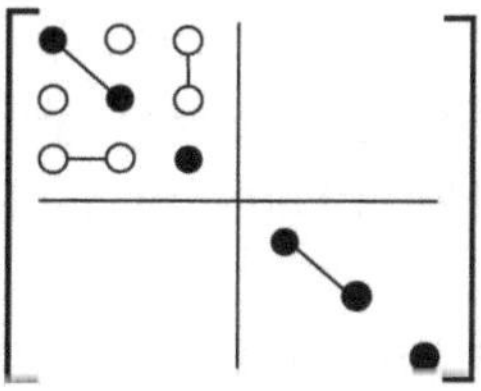 deviatoric transversely isotropic teragonal Hill's (4 constants)
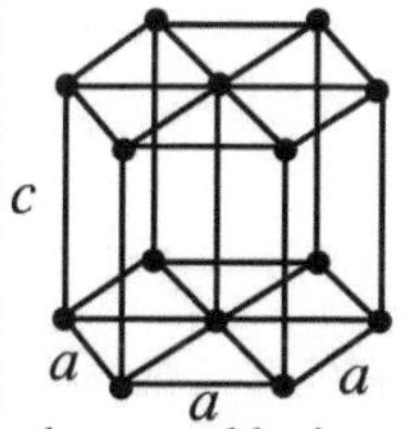 hexagonal lattice	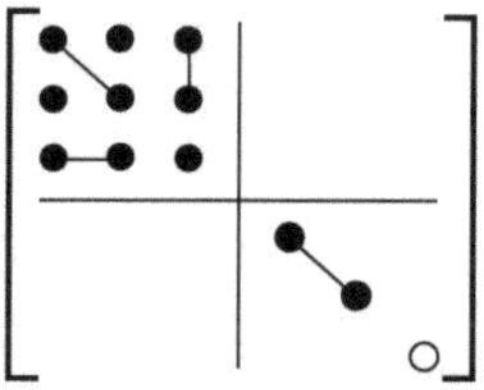 transversely isotropic hexagonal Hooke's (5 constants)	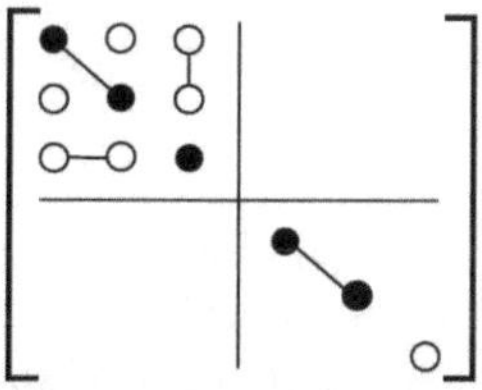 pseudodeviatoric transversely isotropic hexagonal Hu-Marin's (3 constants)

The commonly used term "*transversely isotropic criterion*" may be misleading as long as an additional distinction between the *tetragonal* and the *hexagonal symmetry* is not introduced. The aforementioned distinction is known from the literature dealing with prediction of composite behavior in elastic range and its validation by experiments. For example, Sun and Vaidya [53] examined two types of materials: *Boron/Al composite* and *Graphite/Epoxy composite*, and found that some of them exhibit tetragonal while others hexagonal symmetry classes. However, even this distinction between tetragonal and hexagonal symmetry classes may be insufficient to describe some composite materials, for example, *SiC/Ti unidirectional lamina* examined by Herakovich and Aboudi [19]. This is basically caused by *residual stresses* that appear after cooling-down during *fabrication process*.

The above considerations are limited to the description of initial yield surface only. Generally, it is assumed that during plastic hardening the initial yield surface possessing certain symmetry is rebuilt in an isotropic way, which is generally not true. This question was discussed, e.g., by Malinin and Rżysko [36], who invoked Mursa [41] results for *OTCz Titanium Alloy* that confirms assumption of isotropic nature of plastic hardening. However, Hu and Marin's [22] findings for *Aluminum Alloy* showed anisotropic nature of plastic hardening rather than isotropic.

Nevertheless, the plastic hardening theory is usually taken in an isotropic fashion, e.g., Malinin and Rżysko [36], Ottosen and Ristinmaa [44], Hill [20, 21]. Such approach, although commonly used, may be questionable in light of the aforementioned experimental testing, some of which confirm such assumption, cf. Mursa [41] (*Titanium alloy*) but others contradict it cf. Hu and Marin [22] (*Aluminum alloy*), Kowalewski and Śliwowski [26] (influence of first common invariant).

1.6 Strain Energy and Complementary Energy—The State Potentials for Isotropic or Anisotropic Materials

Material is called *elastic* if its *response* (deformation) is independent of loading history (Fig. 1.6), which means that stress is determined to be strain

$$\sigma_{ij} = \sigma_{ij}(\varepsilon_{kl}) \tag{1.120}$$

or vice versa

$$\varepsilon_{ij} = \varepsilon_{ij}(\sigma_{kl}) \tag{1.121}$$

After the fully closed *loading–unloading cycle* (A-B-A), the initial material state A is recovered, independent of the loading–unloading path, Fig. 1.6b.

When the concept of *strain energy per unit volume* $\mathcal{W}$ [Nm/m^3] is introduced, the following definitions hold:

$$\mathcal{W}(\varepsilon) = \int_0^{\varepsilon} \sigma(\epsilon)\mathrm{d}\epsilon \quad \text{or} \quad \mathcal{W}(\varepsilon_{ij}) = \int_0^{\varepsilon_{ij}} \sigma_{ij}(\epsilon_{kl})\mathrm{d}\epsilon_{ij} \tag{1.122}$$

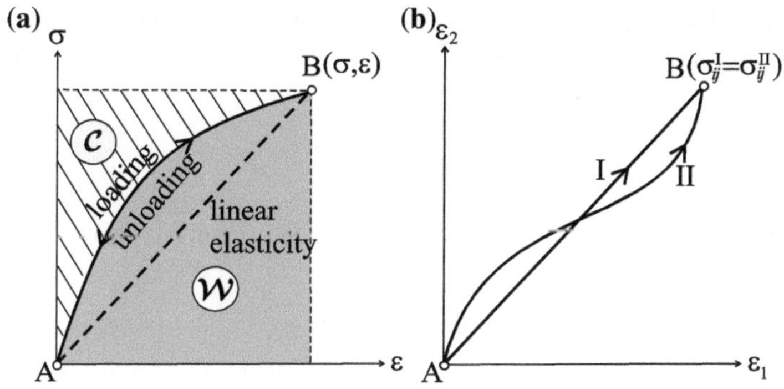

Fig. 1.6 Schematic illustration of elastic material response: **a** strain energy and complementary energy, **b** independence of final state of loading history

in case of the *uniaxial* or the *multiaxial loadings*, respectively. In the following fashion the *complementary energy per unit volume* $\mathcal{C}$ [Nm/m^3] is defined as

$$\mathcal{C}(\sigma) = \int_0^{\sigma} \varepsilon(\varsigma)\mathrm{d}\varsigma \quad \text{or} \quad \mathcal{C}(\sigma_{ij}) = \int_0^{\sigma_{ij}} \varsigma_{ij}(\varepsilon_{kl})\mathrm{d}\varsigma_{ij} \tag{1.123}$$

It is seen from Fig. 1.6 that the following is true:

$$\mathcal{C}(\sigma_{ij}) = \sigma_{ij}\varepsilon_{ij} - \mathcal{W}(\varepsilon_{ij}) \tag{1.124}$$

It should be emphasized that in the considered case of pure elastic material both the *strain energy* $\mathcal{W}$ and *complementary energy* $\mathcal{C}$ are independent of loading path but depend on the current state exclusively.

In a more general case, when the deformation process is accompanied by permanent (irreversible) *changes* in *material microstructure*, for instance, resulting from *plastic yielding*, *damage growth*, or *phase transformation* during *martensitic change* or other *irreversible phenomena*, the strain energy and the complementary energy depend on loading history.

In the elastic material for which strain energy depends on the current state only $\mathcal{W}(\varepsilon_{ij})$ but does not depend on strain path

$$\frac{\partial \sigma_{ij}}{\partial \varepsilon_{kl}} = \frac{\partial \sigma_{kl}}{\partial \varepsilon_{ij}} \tag{1.125}$$

the *strain energy* can be used as an invariant *potential function* for the *stresses*

$$\sigma_{ij} = \frac{\partial \mathcal{W}(\varepsilon_{ij})}{\partial \varepsilon_{ij}} \tag{1.126}$$

In a similar fashion the *complementary energy* that depends on the current state only $\mathcal{C}(\sigma_{ij})$ but does not depend on strain path

$$\frac{\partial \varepsilon_{ij}}{\partial \sigma_{kl}} = \frac{\partial \varepsilon_{kl}}{\partial \sigma_{ij}} \tag{1.127}$$

can be used as an invariant *potential function* for the *strain* as follows:

$$\varepsilon_{ij} = \frac{\partial \mathcal{C}(\sigma_{ij})}{\partial \sigma_{ij}} \tag{1.128}$$

In a general case of *nonlinear elastic material* the strain energy and the complementary energy are not equal to each other, $\mathcal{W} \neq \mathcal{C}$, whereas only in the case of linear elastic material the equality $\mathcal{W} = \mathcal{C}$ holds.

In the above considerations the initial state was treated as stress and strain free, point A ($\sigma = 0, \varepsilon = 0$) in Fig. 1.6. In the more general case a residual stress and/or strain are built-in $\mathrm{A}^{\mathrm{res}}$ ($\sigma = \sigma^{\mathrm{res}}, \varepsilon = \varepsilon^{\mathrm{res}}$). This *residual state* may result from *fabrication process* or prior *loading history* in which some irreversible changes of material structure have occurred (e.g., cyclic plasticity) or certain residual stresses or strains have been built-in (e.g., after cooling-down of *long fiber reinforced composite* characterized by different *thermal properties* of *fiber* and *matrix*). Note that in general this residual state is unknown since the whole history of material, which contains complete information about fabrication, its initial machining, as well as concerning unloading process prior to the appearance of this self-balanced residual stress, is unknown.

Consider the process of elastic deformation of material starting from the *residual state* $\mathrm{A}^{\mathrm{res}}(\sigma^{\mathrm{res}}, \varepsilon^{\mathrm{res}})$ toward the final state $\mathrm{B}(\Delta\sigma, \Delta\varepsilon)$, assuming at the beginning uniaxial tension (see Fig. 1.7).

The *increment* of *elastic strain energy* of material corresponding to applied strain $\Delta\mathcal{W}(\Delta\varepsilon)$ in the presence of *residual stress* $\varepsilon^{\mathrm{res}}$ is equal to

$$\Delta\mathcal{W}(\Delta\varepsilon) = \int_0^{\Delta\varepsilon} \Delta\sigma(\Delta\epsilon)\mathrm{d}(\Delta\epsilon) \tag{1.129}$$

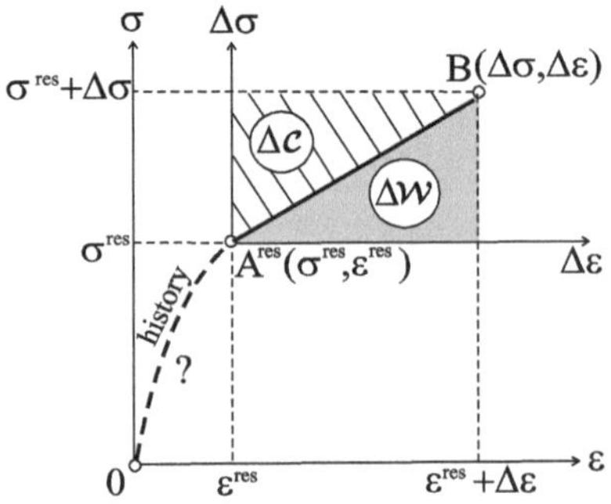

Fig. 1.7 Process of elastic deformation of material with prior residual state included

In the particular case of linear Hooke's law for *isotropic material* it yields

$$\Delta \mathcal{W}(\Delta\varepsilon) = \frac{1}{2}E(\Delta\varepsilon)^2 \tag{1.130}$$

where

$$\Delta\sigma = E\Delta\varepsilon \qquad \Delta\sigma = \sigma - \sigma^{\text{res}} \qquad \Delta\varepsilon = \varepsilon - \varepsilon^{\text{res}} \tag{1.131}$$

obey. In the more general case of *multiaxial deformation state* the strain energy per unit volume of elastic material in the presence of residual stress may be written as

$$\Delta \mathcal{W}(\Delta\varepsilon_{ij}) = \int\limits_0^{\Delta\varepsilon_{ij}} \Delta\sigma_{ij}(\Delta\epsilon_{kl})\mathrm{d}(\Delta\epsilon_{ij}), \tag{1.132}$$

whereas in case, if *linear elastic material* is assumed, the linear relation combining stress and strain increments is furnished as

$$\Delta\sigma_{ij} = E_{ijkl}\Delta\varepsilon_{kl} \tag{1.133}$$

Equation (1.132) represents the increment of elastic energy $\Delta\mathcal{W}$ in the presence of the residual state $\varepsilon_{ij} = \varepsilon_{ij}^{\text{res}} + \Delta\varepsilon_{ij}$, hence

$$\Delta \mathcal{W}(\Delta\varepsilon_{ij}) = \int\limits_0^{\Delta\varepsilon_{ij}} E_{ijkl}\Delta\epsilon_{kl}\mathrm{d}(\Delta\epsilon_{ij}) \tag{1.134}$$

where the *fourth-rank stiffness tensor* E_{ijkl} is used. Note that Eq. (1.134) is true both for *isotropic* and *anisotropic materials* of optional class of symmetry. The stiffness tensor E_{ijkl} comprises complete information defining the elastic material response.

In a similar way, the *complementary energy increment* $\Delta\mathcal{C}$ of elastic material in the presence of residual stress can be written as

$$\Delta\mathcal{C}(\Delta\sigma_{kl}) = \int\limits_0^{\Delta\sigma_{kl}} \Delta\varepsilon_{kl}(\Delta\varsigma_{mn})\mathrm{d}(\Delta\varsigma_{kl}) \tag{1.135}$$

If the *linear elastic material* is assumed we arrive at

$$\Delta\mathcal{C}(\Delta\sigma_{kl}) = \int\limits_0^{\Delta\sigma_{kl}} E_{klmn}^{-1}\Delta\varsigma_{mn}\mathrm{d}(\Delta\varsigma_{kl}) \tag{1.136}$$

where E^{-1}_{klmn} stands for the *compliance tensor* of *elastic* material

$$\Delta\varepsilon_{kl} = E^{-1}_{klmn}\Delta\sigma_{mn} \tag{1.137}$$

The above Eq. (1.137) is an extension of the law of linear elastic material to the case of *existence* of a nonzeroth *residual stress* and *strain* $\Delta\sigma_{ij} = \sigma_{ij} - \sigma^{\rm res}_{ij}$, $\Delta\varepsilon_{kl} = \varepsilon_{kl} - \varepsilon^{\rm res}_{kl}$

$$\sigma_{ij} - \sigma^{\rm res}_{ij} = E_{ijkl}\left(\varepsilon_{kl} - \varepsilon^{\rm res}_{kl}\right) \tag{1.138}$$

or

$$\varepsilon_{kl} - \varepsilon^{\rm res}_{kl} = E^{-1}_{klmn}\left(\sigma_{mn} - \sigma^{\rm res}_{mn}\right) \tag{1.139}$$

When the *vector-matrix notation* is used the fourth-rank elastic tensors E_{ijkl} or E^{-1}_{klmn} can be represented by the symmetric 6×6 matrices: $[\mathbb{E}]$ or $[\mathbb{E}^{-1}]$ called the *stiffness* or the *compliance matrices*, respectively, whereas the tensors $\sigma_{ij} - \sigma^{\rm res}_{ij}$ or $\varepsilon_{kl} - \varepsilon^{\rm res}_{kl}$ take the format of *columnar vectors* of *overstress* or *overstrain*, respectively,

$$\left\{\begin{array}{c} \sigma_{11} - \sigma^{\rm res}_{11} \\ \sigma_{22} - \sigma^{\rm res}_{22} \\ \sigma_{33} - \sigma^{\rm res}_{33} \\ \tau_{23} - \tau^{\rm res}_{23} \\ \tau_{31} - \tau^{\rm res}_{31} \\ \tau_{12} - \tau^{\rm res}_{12} \end{array}\right\} \qquad \left\{\begin{array}{c} \varepsilon_{11} - \varepsilon^{\rm res}_{11} \\ \varepsilon_{22} - \varepsilon^{\rm res}_{22} \\ \varepsilon_{33} - \varepsilon^{\rm res}_{33} \\ \gamma_{23} - \gamma^{\rm res}_{23} \\ \gamma_{31} - \gamma^{\rm res}_{31} \\ \gamma_{12} - \gamma^{\rm res}_{12} \end{array}\right\} \tag{1.140}$$

Hence, when the *Voigt notation* is used Eqs. (1.138) and (1.139) can be written in equivalent fashion

$$\left\{\boldsymbol{\sigma} - \boldsymbol{\sigma}^{\rm res}\right\} = [\mathbb{E}]\left\{\boldsymbol{\varepsilon} - \boldsymbol{\varepsilon}^{\rm res}\right\} \tag{1.141}$$

or

$$\left\{\boldsymbol{\varepsilon} - \boldsymbol{\varepsilon}^{\rm res}\right\} = [\mathbb{E}^{-1}]\left\{\boldsymbol{\sigma} - \boldsymbol{\sigma}^{\rm res}\right\} \tag{1.142}$$

1.7 Elastic Strain Energy as Function of Invariants

The stress and the strain invariants are presented in Sect. 1.1. In the present section the *elastic strain energy per unit volume* $\mathcal{W}$ expressed as the scalar product of both these tensors

$$\mathcal{W} = \frac{1}{2}\sigma_{ij}\varepsilon_{ji} \tag{1.143}$$

will also be presented in terms of invariants. In the case of isotropic material three basic invariants of the strain tensor are sufficient for unique representation of the strain energy, whereas in case of elastic material comprising damage the use of common invariants defining internal material microstructure is necessary (see Sect. 1.3).

1.7.1 Elastic Strain Energy of Isotropic Materials

The simplest example of the scalar function of tensorial argument is the elastic strain energy $\mathcal{W}(\boldsymbol{\varepsilon})$. In the case of isotropic material the strain tensor is uniquely determined in terms of three *basic* or *generic strain invariants* (1.26) as follows:

$$\mathcal{W}(\boldsymbol{\varepsilon}) = \mathcal{W}(J_{1\varepsilon}, J_{2\varepsilon}, J_{3\varepsilon}) \tag{1.144}$$

Constitutive law of *elastic* material (1.126) can be written as follows:

$$\sigma_{ij} = \frac{\partial \mathcal{W}}{\partial \varepsilon_{ij}} = \frac{\partial \mathcal{W}}{\partial J_{1\varepsilon}}\frac{\partial J_{1\varepsilon}}{\partial \varepsilon_{ij}} + \frac{\partial \mathcal{W}}{\partial J_{2\varepsilon}}\frac{\partial J_{2\varepsilon}}{\partial \varepsilon_{ij}} + \frac{\partial \mathcal{W}}{\partial J_{3\varepsilon}}\frac{\partial J_{3\varepsilon}}{\partial \varepsilon_{ij}} \tag{1.145}$$

where

$$\frac{\partial J_{1\varepsilon}}{\partial \varepsilon_{ij}} = \delta_{ij} \qquad \frac{\partial J_{2\varepsilon}}{\partial \varepsilon_{ij}} = \varepsilon_{ij} \qquad \frac{\partial J_{3\varepsilon}}{\partial \varepsilon_{ij}} = \varepsilon_{ik}\varepsilon_{kj} \tag{1.146}$$

hence,

$$\sigma_{ij} = \frac{\partial \mathcal{W}}{\partial J_{1\varepsilon}}\delta_{ij} + \frac{\partial \mathcal{W}}{\partial J_{2\varepsilon}}\varepsilon_{ij} + \frac{\partial \mathcal{W}}{\partial J_{3\varepsilon}}\varepsilon_{ik}\varepsilon_{kj} \tag{1.147}$$

Introducing the *Lamé* elastic *constants* $\lambda = \frac{\nu E}{(1+\nu)(1-2\nu)}$ and $\mu = \frac{E}{2(1+\nu)}$ with

$$\frac{\partial \mathcal{W}}{\partial J_{1\varepsilon}} = \lambda\varepsilon_{kk} \qquad \frac{\partial \mathcal{W}}{\partial J_{2\varepsilon}} = 2\mu \qquad \frac{\partial \mathcal{W}}{\partial J_{3\varepsilon}} = 0 \tag{1.148}$$

we arrive at the classical *Hooke law* of the *isotropic* material

$$\sigma_{ij} = \lambda\varepsilon_{kk}\delta_{ij} + 2\mu\varepsilon_{ij} \tag{1.149}$$

Summing up, the *isotropic elastic Hooke material* is uniquely defined by the *strain energy* which depends on the *first* and the *second basic invariants* of the *strain* tensor

$$\mathcal{W} = \frac{1}{2}\lambda(J_{1\varepsilon})^2 + 2\mu J_{2\varepsilon} \tag{1.150}$$

but does not depend on the *third invariant* $J_{3\varepsilon}$.

1.7.2 Strain or Complementary Energy of Elastic-Damage Material—Common Strain-Damage and Stress-Damage Invariants; the Helmholtz or the Gibbs State Potentials

Theory of invariants allows to determine minimal number the basic invariants from which all other tensorial invariants necessary to obtain a sufficiently general representation of the state equations can be built (cf. e.g., Spencer [52], Rymarz [47]). Usually the strain energy per unit volume $\mathcal{W}(\varepsilon_{ij})$ or the complementary energy per unit volume $\mathcal{C}(\sigma_{ij})$ is taken as the *state potential of elasticity* (see Sect. 1.7.1). As shown in Sect. 1.7.1, in case of elastic isotropy three invariants sufficiently determine both types of energy $\mathcal{W}(J_{i\varepsilon})$ or $\mathcal{C}(J_{i\sigma})$, $i = 1, 2, 3$.

A *scalar function* dependent on a pair of *tensorial arguments*, each of them being the symmetric second-rank tensor, is a more complex case. The representative example of such a case is the *strain energy of damaged material* $\mathcal{W}(\boldsymbol{\varepsilon}, \boldsymbol{D})$. Analogous to the isotropic material (1.144), both tensors $\boldsymbol{\varepsilon}$ or $\boldsymbol{D}$ are determined by their single *basic invariants* $J_{i\varepsilon}$ or J_{iD}, $i = 1, 2, 3$. However, the scalar function dependent on both arguments $\mathcal{W}(\boldsymbol{\varepsilon}, \boldsymbol{D})$ has to be uniquely defined not only by single invariants $J_{i\varepsilon}$ and J_{iD} but also by the *common invariants* $J_{j\varepsilon D}$, $j = 1, 2, 3, 4$. This leads to the format dependent on six single and four common invariants (total 10)

$$\mathcal{W}(\boldsymbol{\varepsilon}, \boldsymbol{D}) = \mathcal{W}(J_{1\varepsilon}, J_{2\varepsilon}, \underline{J_{3\varepsilon}}, J_{1D}, \underline{J_{2D}, J_{3D}}; J_{1\varepsilon D}, J_{2\varepsilon D}, \underline{J_{3\varepsilon D}, J_{4\varepsilon D}}) \tag{1.151}$$

In addition, the strain energy $\mathcal{W}$ has to be a decreasing function with damage growth since energy is released during the *damage nucleation* and *growth*, so it has to be linear with respect to $\boldsymbol{D}$. Hence, the strain energy cannot depend either on the third strain invariant $J_{3\varepsilon}$ and on the two single damage invariants J_{2D}, J_{3D} and also on the two common invariants $J_{3\varepsilon D}$, $J_{4\varepsilon D}$ (underlined arguments in Eq. (1.151)). Based on the above physical reasons the strain energy of elastic damaged material can completely be represented in terms of a combination of five invariants (three single and two common)

$$\mathcal{W}(\boldsymbol{\varepsilon}, \boldsymbol{D}) = \rho\psi(\boldsymbol{\varepsilon}, \boldsymbol{D}) = \rho\psi(J_{1\varepsilon}, J_{2\varepsilon}, J_{1D}, J_{1\varepsilon D}, J_{2\varepsilon D}) \tag{1.152}$$

In this way an invariant representation of the *Helmholtz free energy per unit mass* is furnished and finally applied as the *state potential* that determines the stress state in a unique fashion

$$\boldsymbol{\sigma} = \frac{\partial[\rho\psi(\boldsymbol{\varepsilon}, \boldsymbol{D})]}{\partial\boldsymbol{\varepsilon}} \tag{1.153}$$

Note also that when the representation (1.152) is specified, some *combinations of invariants* are allowed for which the scalar function $\psi(\boldsymbol{\varepsilon}, \boldsymbol{D})$ remains quadratic with respect to $\boldsymbol{\varepsilon}$. Hence, following Murakami and Kamiya [38] the *free energy function* $\rho\psi(\boldsymbol{\varepsilon}, \boldsymbol{D})$ *per unit mass* is furnished as

$$\begin{aligned} \rho\psi(\boldsymbol{\varepsilon}, \boldsymbol{D}) = {} & \tfrac{1}{2}\lambda(J_{1\varepsilon})^2 + 2\mu J_{2\varepsilon} + \eta_1(J_{1\varepsilon})^2 J_{1D} + 2\eta_2 J_{2\varepsilon} J_{1D} \\ & + \eta_3 J_{1\varepsilon} J_{1\varepsilon D} + \eta_4 J_{2\varepsilon D} \end{aligned} \tag{1.154}$$

or

$$\begin{aligned} \rho\psi(\boldsymbol{\varepsilon}, \boldsymbol{D}) = {} & \tfrac{1}{2}\lambda\,(\mathrm{tr}\boldsymbol{\varepsilon})^2 + \mu\mathrm{tr}\,(\boldsymbol{\varepsilon}\cdot\boldsymbol{\varepsilon}) + \eta_1\,(\mathrm{tr}\boldsymbol{\varepsilon})^2\,\mathrm{tr}(\boldsymbol{D}) \\ & + \eta_2\mathrm{tr}\,(\boldsymbol{\varepsilon}\cdot\boldsymbol{\varepsilon})\,\mathrm{tr}(\boldsymbol{D}) + \eta_3\mathrm{tr}(\boldsymbol{\varepsilon})\mathrm{tr}\,(\boldsymbol{\varepsilon}\cdot\boldsymbol{D}) + \eta_4\mathrm{tr}\,(\boldsymbol{\varepsilon}\cdot\boldsymbol{\varepsilon}\cdot\boldsymbol{D}) \end{aligned} \tag{1.155}$$

when the equivalent representation is used, e.g., Skrzypek et al. [50].

Remember that the above formulas (1.154) and (1.155) for the Holmholtz free energy refer to the specific case of elastic *anisotropy* which is *acquired* as the result of *damage nucleation* and *growth*. Hence, in a virgin state where damage does not exist the energy representation of the isotropic elastic material has to be recovered, such that symbol $\boldsymbol{\varepsilon}$ has to be referred to the elastic strain $\boldsymbol{\varepsilon}^{\mathrm{e}}$.

In a general 3D case the following matrix representation of the *constitutive equation* with *total formulation* holds:

$$\begin{Bmatrix} \sigma_{11} \\ \sigma_{22} \\ \sigma_{33} \\ \sigma_{23} \\ \sigma_{13} \\ \sigma_{12} \end{Bmatrix} = \begin{bmatrix} {}^{\mathrm{s}}\widetilde{E}_{11} & {}^{\mathrm{s}}\widetilde{E}_{12} & {}^{\mathrm{s}}\widetilde{E}_{13} & {}^{\mathrm{s}}\widetilde{E}_{14} & {}^{\mathrm{s}}\widetilde{E}_{15} & {}^{\mathrm{s}}\widetilde{E}_{16} \\ & {}^{\mathrm{s}}\widetilde{E}_{22} & {}^{\mathrm{s}}\widetilde{E}_{23} & {}^{\mathrm{s}}\widetilde{E}_{24} & {}^{\mathrm{s}}\widetilde{E}_{25} & {}^{\mathrm{s}}\widetilde{E}_{26} \\ & & {}^{\mathrm{s}}\widetilde{E}_{33} & {}^{\mathrm{s}}\widetilde{E}_{34} & {}^{\mathrm{s}}\widetilde{E}_{35} & {}^{\mathrm{s}}\widetilde{E}_{36} \\ & & & {}^{\mathrm{s}}\widetilde{E}_{44} & {}^{\mathrm{s}}\widetilde{E}_{45} & {}^{\mathrm{s}}\widetilde{E}_{46} \\ & \text{symm.} & & & {}^{\mathrm{s}}\widetilde{E}_{55} & {}^{\mathrm{s}}\widetilde{E}_{56} \\ & & & & & {}^{\mathrm{s}}\widetilde{E}_{66} \end{bmatrix} \begin{Bmatrix} \varepsilon^{\mathrm{e}}_{11} \\ \varepsilon^{\mathrm{e}}_{22} \\ \varepsilon^{\mathrm{e}}_{33} \\ \gamma^{\mathrm{e}}_{23} \\ \gamma^{\mathrm{e}}_{13} \\ \gamma^{\mathrm{e}}_{12} \end{Bmatrix} \tag{1.156}$$

where ${}^{\mathrm{s}}\widetilde{E}_{ij}$ represents *effective elastic-damage secant stiffness matrix*. The damage acquired anisotropy is described by the 6×6 symmetric secant stiffness matrix as follows (cf. Skrzypek et al. [50]):

$$\begin{aligned}
{}^{\mathrm{s}}\widetilde{E}_{11} &= \lambda + 2\mu + 2(\eta_1 + \eta_2)\mathrm{tr}(\boldsymbol{D}) + 2(\eta_3 + \eta_4)D_{11} \\
{}^{\mathrm{s}}\widetilde{E}_{22} &= \lambda + 2\mu + 2(\eta_1 + \eta_2)\mathrm{tr}(\boldsymbol{D}) + 2(\eta_3 + \eta_4)D_{22} \\
{}^{\mathrm{s}}\widetilde{E}_{33} &= \lambda + 2\mu + 2(\eta_1 + \eta_2)\mathrm{tr}(\boldsymbol{D}) + 2(\eta_3 + \eta_4)D_{33} \\
{}^{\mathrm{s}}\widetilde{E}_{12} &= \lambda + 2\eta_1\mathrm{tr}(\boldsymbol{D}) + \eta_3(D_{11} + D_{22}) \\
{}^{\mathrm{s}}\widetilde{E}_{13} &= \lambda + 2\eta_1\mathrm{tr}(\boldsymbol{D}) + \eta_3(D_{11} + D_{33}) \\
{}^{\mathrm{s}}\widetilde{E}_{23} &= \lambda + 2\eta_1\mathrm{tr}(\boldsymbol{D}) + \eta_3(D_{22} + D_{33}) \\
{}^{\mathrm{s}}\widetilde{E}_{44} &= \tfrac{1}{2}\left[2\mu + 2\eta_2\mathrm{tr}(\boldsymbol{D}) + \eta_4(D_{33} + D_{22})\right] \\
{}^{\mathrm{s}}\widetilde{E}_{45} &= \eta_4 D_{12} \\
{}^{\mathrm{s}}\widetilde{E}_{55} &= \tfrac{1}{2}\left[2\mu + 2\eta_2\mathrm{tr}(\boldsymbol{D}) + \eta_4(D_{11} + D_{33})\right] \\
{}^{\mathrm{s}}\widetilde{E}_{46} &= \eta_4 D_{13} \\
{}^{\mathrm{s}}\widetilde{E}_{66} &= \tfrac{1}{2}\left[2\mu + 2\eta_2\mathrm{tr}(\boldsymbol{D}) + \eta_4(D_{11} + D_{22})\right] \\
{}^{\mathrm{s}}\widetilde{E}_{56} &= \eta_4 D_{23} \\
{}^{\mathrm{s}}\widetilde{E}_{14} &= \eta_3 D_{23} \\
{}^{\mathrm{s}}\widetilde{E}_{24} &= {}^{\mathrm{s}}\widetilde{S}_{34} = (\eta_3 + \eta_4)D_{23} \\
{}^{\mathrm{s}}\widetilde{E}_{25} &= \eta_3 D_{13} \\
{}^{\mathrm{s}}\widetilde{E}_{15} &= {}^{\mathrm{s}}\widetilde{S}_{35} = (\eta_3 + \eta_4)D_{13} \\
{}^{\mathrm{s}}\widetilde{E}_{36} &= \eta_3 D_{12} \\
{}^{\mathrm{s}}\widetilde{E}_{16} &= {}^{\mathrm{s}}\widetilde{S}_{26} = (\eta_3 + \eta_4)D_{12}
\end{aligned} \tag{1.157}$$

The alternative formulation based on a concept of the *complementary energy* $\mathcal{C}$ represented by a *scalar function* of the *two tensorial arguments* $\boldsymbol{\sigma}$ and $\boldsymbol{D}$, namely $\mathcal{C}(\boldsymbol{\sigma}, \boldsymbol{D})$, leads to the *Gibbs potential function* per unit mass $\mathcal{G}$ as follows (cf. Hayakawa–Murakami [18], Murakami [40]):

$$\mathcal{C}(\boldsymbol{\sigma}, \boldsymbol{D}) = \rho\mathcal{G}(J_{1\sigma}, J_{2\sigma}, J_{3\sigma}, J_{1D}, J_{2D}, J_{3D}, J_{1\sigma D}, J_{2\sigma D}, J_{3\sigma D}, J_{4\sigma D}) \quad (1.158)$$

where the *crack closure effect* due to compressive stress, originally introduced in Hayakawa–Murakami [18], is omitted.

Repeating the above reasoning for physical nature of the *Gibbs complementary energy* $\mathcal{C}(\boldsymbol{\sigma}, \boldsymbol{D})$, only five of the above aforementioned ten (1.158) common stress and damage invariants can be admitted, namely

$$\mathcal{C}(\boldsymbol{\sigma}, \boldsymbol{D}) = \rho\mathcal{G}(\boldsymbol{\sigma}, \boldsymbol{D}) = \rho\mathcal{G}(J_{1\sigma}, J_{2\sigma}, J_{1D}, J_{1\sigma D}, J_{2\sigma D}) \quad (1.159)$$

Hence, in case of the *elastic isotropic material* in a *virgin state* which changes to *anisotropic* material due to *damage evolution*, the Gibbs state potential takes the following format (cf. Hayakawa and Murakami [18]):

$$\begin{aligned}\rho\mathcal{G}(\boldsymbol{\sigma}, \boldsymbol{D}) = &-\tfrac{\nu}{2E}(\mathrm{tr}\boldsymbol{\sigma})^2 + \tfrac{1+\nu}{2E}\mathrm{tr}\,(\boldsymbol{\sigma}\cdot\boldsymbol{\sigma}) + \vartheta_1\,(\mathrm{tr}\boldsymbol{\sigma})^2\,\mathrm{tr}(\boldsymbol{D})\\ &+\vartheta_2\mathrm{tr}\,(\boldsymbol{\sigma}\cdot\boldsymbol{\sigma})\,\mathrm{tr}(\boldsymbol{D}) + \vartheta_3\mathrm{tr}(\boldsymbol{\sigma})\mathrm{tr}\,(\boldsymbol{\sigma}\cdot\boldsymbol{D}) + \vartheta_4\mathrm{tr}\,(\boldsymbol{\sigma}\cdot\boldsymbol{\sigma}\cdot\boldsymbol{D})\end{aligned} \quad (1.160)$$

which is complementary to (1.155). The matrix representation of secant compliance matrix referring to Hayakawa–Murakami type elastic-plastic-damage material is as follows:

$$\begin{Bmatrix}\varepsilon^{e}_{11}\\ \varepsilon^{e}_{22}\\ \varepsilon^{e}_{33}\\ \varepsilon^{e}_{23}\\ \varepsilon^{e}_{13}\\ \varepsilon^{e}_{12}\end{Bmatrix} = \begin{bmatrix} {}^{s}\widetilde{E}^{-1}_{11} & {}^{s}\widetilde{E}^{-1}_{12} & {}^{s}\widetilde{E}^{-1}_{13} & 0 & 0 & 0\\ & {}^{s}\widetilde{E}^{-1}_{22} & {}^{s}\widetilde{E}^{-1}_{23} & 0 & 0 & 0\\ & & {}^{s}\widetilde{E}^{-1}_{33} & 0 & 0 & 0\\ & & & {}^{s}\widetilde{E}^{44}_{-1} & 0 & 0\\ & \text{symm.} & & & {}^{s}\widetilde{E}^{-1}_{55} & 0\\ & & & & & {}^{s}\widetilde{E}^{-1}_{66}\end{bmatrix} \begin{Bmatrix}\sigma_{11}\\ \sigma_{22}\\ \sigma_{33}\\ \sigma_{23}\\ \sigma_{13}\\ \sigma_{12}\end{Bmatrix} \quad (1.161)$$

where

$$\begin{aligned}
{}^{s}\widetilde{E}^{-1}_{11} &= \tfrac{1}{E} + 2\mathrm{tr}(\boldsymbol{D})(\vartheta_1 + \vartheta_2) + 2D_{11}(\vartheta_3 + \vartheta_4)\\
{}^{s}\widetilde{E}^{-1}_{12} &= -\tfrac{\nu}{E} + 2\vartheta_1\mathrm{tr}(\boldsymbol{D}) + \vartheta_3(D_{11} + D_{22})\\
{}^{s}\widetilde{E}^{-1}_{13} &= -\tfrac{\nu}{E} + 2\vartheta_1\mathrm{tr}(\boldsymbol{D}) + \vartheta_3(D_{11} + D_{33})\\
{}^{s}\widetilde{E}^{-1}_{22} &= \tfrac{1}{E} + 2\mathrm{tr}\boldsymbol{D}(\vartheta_1 + \vartheta_2) + 2D_{22}(\vartheta_3 + \vartheta_4)\\
{}^{s}\widetilde{E}^{-1}_{23} &= -\tfrac{\nu}{E} + 2\vartheta_1\mathrm{tr}(\boldsymbol{D}) + \vartheta_3(D_{22} + D_{33})\\
{}^{s}\widetilde{E}^{-1}_{33} &= \tfrac{1}{E} + 2\vartheta_1\mathrm{tr}(\boldsymbol{D}) + \vartheta_3(D_{22} + D_{33})\\
{}^{s}\widetilde{E}^{-1}_{44} &= \tfrac{1+\nu}{E} + 2\vartheta_2\mathrm{tr}(\boldsymbol{D}) + \vartheta_4(D_{22} + D_{33})\\
{}^{s}\widetilde{E}^{-1}_{55} &= \tfrac{1+\nu}{E} + 2\vartheta_2\mathrm{tr}(\boldsymbol{D}) + \vartheta_4(D_{11} + D_{33})\\
{}^{s}\widetilde{E}^{-1}_{66} &= \tfrac{1+\nu}{E} + 2\vartheta_2\mathrm{tr}(\boldsymbol{D}) + \vartheta_4(D_{11} + D_{22})
\end{aligned} \quad (1.162)$$

Note that the Gibbs complementary energy per unit mass refers to elastic strains ε^{e} and is represented in the stress space by the quadratic function of $\boldsymbol{\sigma}$ linear with respect to $\boldsymbol{D}$, in a similar way as the Helmholtz free energy $\rho\psi(\boldsymbol{\varepsilon}, \boldsymbol{D})$ but defined in the *strain space*.

Four material constants η_i appearing in the *Helmholtz state potential* (1.155) as well as four constants ϑ_i appearing in the *Gibbs state potential* (1.160) $(i = 1, 2, 3, 4)$ act as additional material constants to the elastic constants of the *virgin elastic isotropic material*: λ, μ or E, ν, defining effect of damage on the state equation. Namely, when the *Helmholtz potential function* $\mathcal{W} = \rho\psi(\boldsymbol{\varepsilon}, \boldsymbol{D})$ is used as the *stress potential* we arrive at the *state equation* $\boldsymbol{\sigma} = \mathbb{E}(\boldsymbol{D}) : \boldsymbol{\varepsilon}$

$$\begin{aligned}\boldsymbol{\sigma} = \frac{\partial(\rho\psi)}{\partial\boldsymbol{\varepsilon}} &= [\lambda \mathrm{tr}(\boldsymbol{\varepsilon}) + 2\eta_1 \mathrm{tr}(\boldsymbol{\varepsilon})\mathrm{tr}(\boldsymbol{D}) + \eta_3 \mathrm{tr}\,(\boldsymbol{\varepsilon}\cdot\boldsymbol{D})]\,\mathbf{1} \\ &+ 2\,[\mu + \eta_2 \mathrm{tr}(\boldsymbol{D})]\,\boldsymbol{\varepsilon} + \eta_3 \mathrm{tr}(\boldsymbol{\varepsilon})\boldsymbol{D} + \eta_4(\boldsymbol{\varepsilon}\cdot\boldsymbol{D} + \boldsymbol{D}\cdot\boldsymbol{\varepsilon})\end{aligned} \tag{1.163}$$

On the other hand, when the formulation based on the *Gibbs potential function* is used as the *strain potential* $\mathcal{C} = \rho\mathcal{G}(\boldsymbol{\sigma}, \boldsymbol{D})$ we obtain the state equation in equivalent form $\boldsymbol{\varepsilon} = \mathbb{E}^{-1}(\boldsymbol{D}) : \boldsymbol{\sigma}$

$$\begin{aligned}\boldsymbol{\varepsilon} = \frac{\partial(\rho\mathcal{G})}{\partial\boldsymbol{\sigma}} &= -\tfrac{\nu}{E}\mathrm{tr}\,(\boldsymbol{\sigma})\,\mathbf{1} + \tfrac{1+\nu}{2E}\boldsymbol{\sigma} + 2\vartheta_1 \mathrm{tr}(\boldsymbol{D})\mathrm{tr}(\boldsymbol{\sigma})\mathbf{1} \\ &+ 2\vartheta_2 \mathrm{tr}(\boldsymbol{D})\boldsymbol{\sigma} : \mathbf{1} + \vartheta_3\,[\mathrm{tr}\,(\boldsymbol{\sigma}\cdot\boldsymbol{D})\,\mathbf{1} + \mathrm{tr}(\boldsymbol{\sigma})\boldsymbol{D}] \\ &+ \vartheta_4\,(\boldsymbol{\sigma}\cdot\boldsymbol{D} + \boldsymbol{D}\cdot\boldsymbol{\sigma})\end{aligned} \tag{1.164}$$

Note however that in the case of *elastic damaged material* constitutive matrices stiffness $[\mathbb{E}(\boldsymbol{D})]$ and compliance $[\mathbb{E}^{-1}(\boldsymbol{D})]$ are rebuilt following *damage evolution* such that originally isotropic elastic material *acquires* an *anisotropy*.

The state equation of elastic damaged material (1.155) was calibrated for the *high strength concrete* by Murakami and Kamiya [38], see also Skrzypek [49] as shown in Table 1.5.

Apart from the constants of isotropic elasticity E, ν (λ, μ) additional four constants η_i $(i = 1, 2, 3, 4)$ are shown in Table 1.5.

The state equation of elastic moderate ductility with damage (1.164) was calibrated for *spheroidal graphite cast iron* FCD400 by Hayakawa and Murakami [18], see also Skrzypek [49] as shown in Table 1.6.

Apart from the constants of isotropic elasticity E, ν (λ, μ) additional four constants ϑ_i $(i = 1, 2, 3, 4)$ are shown Table 1.6.

Table 1.5 Calibration of six material constants in the constitutive equation of high strength concrete, after Murakami and Kamiya [38])

E (GPa)	ν (–)	η_1 (MPa)	η_2 (MPa)	η_3 (MPa)	η_4 (MPa)
21.4	0.2	–400	–900	100	–23500

Table 1.6 Callibration of six material constants in the constitutive equation of the spheroidal graphite cast iron FCD400, after Hayakawa and Murakami [18]

E (GPa)	ν (–)	ϑ_1 (MPa^{-1})	ϑ_2 (MPa^{-1})	ϑ_3 (MPa^{-1})	ϑ_4 (MPa^{-1})
169	0.285	-3.95×10^{-1}	4.0×10^{-6}	-4.0×10^{-7}	2.50×10^{-6}

The more extended analysis including: crack closure effect under compressive stress, the initial damage threshold, and the subsequent *damage growth* during the hardening phases can be found in Murakami and Kamiya [38], Hayakawa and Murakami [18], Skrzypek et al. [50], Bielski et al. [4], Kuna-Ciskał and Skrzypek [31].

1.7.3 Strain Energy of the Elastic Orthotropic Materials—The Structural Tensors

So far the case of scalar function of second-order tensors expressed in terms of invariants has been discussed. The more general case of a *scalar function of* a *pair of tensorial arguments* being the second-order and the *structural tensors* is considered in this section. The strain energy of orthotropic material $\mathcal{W} = \mathcal{W}(\boldsymbol{\varepsilon}, \boldsymbol{M}^{(i)})$ is the representative example of such a case.

The constitutive equation of *orthotropic hyperelastic material* is obtained by differentiation of the *strain energy function*, cf. Boehler [5]

$$\begin{aligned}\boldsymbol{\sigma} = \frac{\partial \mathcal{W}}{\partial \boldsymbol{\varepsilon}} = &\frac{\partial \mathcal{W}}{\partial J_1}\boldsymbol{M}^{(1)} + \frac{\partial \mathcal{W}}{\partial J_2}\boldsymbol{M}^{(2)} + \frac{\partial \mathcal{W}}{\partial J_3}\boldsymbol{M}^{(3)} \\ &+ \frac{\partial \mathcal{W}}{\partial J_4}(\boldsymbol{\varepsilon}\cdot\boldsymbol{M}^{(1)} + \boldsymbol{M}^{(1)}\cdot\boldsymbol{\varepsilon}) + \frac{\partial \mathcal{W}}{\partial J_5}(\boldsymbol{\varepsilon}\cdot\boldsymbol{M}^{(2)} + \boldsymbol{M}^{(2)}\cdot\boldsymbol{\varepsilon}) \\ &+ \frac{\partial \mathcal{W}}{\partial J_6}(\boldsymbol{\varepsilon}\cdot\boldsymbol{M}^{(3)} + \boldsymbol{M}^{(3)}\cdot\boldsymbol{\varepsilon})\end{aligned} \tag{1.165}$$

where the following definitions of *common invariants* are used:

$$\begin{aligned}&J_1 = \mathrm{tr}(\boldsymbol{\varepsilon}\cdot\boldsymbol{M}^{(1)}) \quad J_2 = \mathrm{tr}(\boldsymbol{\varepsilon}\cdot\boldsymbol{M}^{(2)}) \quad J_3 = \mathrm{tr}(\boldsymbol{\varepsilon}\cdot\boldsymbol{M}^{(3)}) \\ &J_4 = \mathrm{tr}(\boldsymbol{\varepsilon}\cdot\boldsymbol{\varepsilon}\cdot\boldsymbol{M}^{(1)}) \; J_5 = \mathrm{tr}(\boldsymbol{\varepsilon}\cdot\boldsymbol{\varepsilon}\cdot\boldsymbol{M}^{(2)}) \; J_6 = \mathrm{tr}(\boldsymbol{\varepsilon}\cdot\boldsymbol{\varepsilon}\cdot\boldsymbol{M}^{(3)})\end{aligned} \tag{1.166}$$

and definitions (1.91) hold. Following Boehler [5], in order to determine the *constitutive equation of linear orthotropic material* we choose, (see also Ottosen and Ristinmaa [44])

$$
\begin{aligned}
\frac{\partial \mathcal{W}}{\partial J_1} &= \alpha_1 J_1 + \beta_1 J_2 + \beta_2 J_3 \\
\frac{\partial \mathcal{W}}{\partial J_2} &= \alpha_2 J_1 + \alpha_3 J_2 + \beta_3 J_3 \\
\frac{\partial \mathcal{W}}{\partial J_3} &= \alpha_4 J_1 + \alpha_5 J_2 + \alpha_6 J_3 \\
\frac{\partial \mathcal{W}}{\partial J_4} &= \alpha_7 \qquad \frac{\partial \mathcal{W}}{\partial J_5} = \alpha_8 \qquad \frac{\partial \mathcal{W}}{\partial J_6} = \alpha_9
\end{aligned}
\tag{1.167}
$$

The coefficients $\beta_1, \beta_2, \beta_3$ can be substituted by corresponding coefficients α_2, α_4, α_5 in order to satisfy symmetry of *orthotropic stiffness matrix*

$$
\beta_1 = \alpha_2, \qquad \beta_2 = \alpha_4, \qquad \beta_3 = \alpha_5 \tag{1.168}
$$

The above yields the *constitutive equation* of *linear orthotropic material* by use of *common invariants of strain and structural tensors*

$$
\begin{aligned}
\boldsymbol{\sigma} = & \left[\alpha_1 \mathrm{tr}(\boldsymbol{\varepsilon} \cdot \boldsymbol{M}^{(1)}) + \alpha_2 \mathrm{tr}(\boldsymbol{\varepsilon} \cdot \boldsymbol{M}^{(2)}) + \alpha_4 \mathrm{tr}(\boldsymbol{\varepsilon} \cdot \boldsymbol{M}^{(3)})\right] \boldsymbol{M}^{(1)} \\
& + \left[\alpha_2 \mathrm{tr}(\boldsymbol{\varepsilon} \cdot \boldsymbol{M}^{(1)}) + \alpha_3 \mathrm{tr}(\boldsymbol{\varepsilon} \cdot \boldsymbol{M}^{(2)}) + \alpha_5 \mathrm{tr}(\boldsymbol{\varepsilon} \cdot \boldsymbol{M}^{(3)})\right] \boldsymbol{M}^{(2)} \\
& + \left[\alpha_4 \mathrm{tr}(\boldsymbol{\varepsilon} \cdot \boldsymbol{M}^{(1)}) + \alpha_5 \mathrm{tr}(\boldsymbol{\varepsilon} \cdot \boldsymbol{M}^{(2)}) + \alpha_6 \mathrm{tr}(\boldsymbol{\varepsilon} \cdot \boldsymbol{M}^{(3)})\right] \boldsymbol{M}^{(3)} \\
& + \alpha_7 \left(\boldsymbol{\varepsilon} \cdot \boldsymbol{M}^{(1)} + \boldsymbol{M}^{(1)} \cdot \boldsymbol{\varepsilon}\right) + \alpha_8 \left(\boldsymbol{\varepsilon} \cdot \boldsymbol{M}^{(2)} + \boldsymbol{M}^{(2)} \cdot \boldsymbol{\varepsilon}\right) \\
& + \alpha_9 \left(\boldsymbol{\varepsilon} \cdot \boldsymbol{M}^{(3)} + \boldsymbol{M}^{(3)} \cdot \boldsymbol{\varepsilon}\right)
\end{aligned}
\tag{1.169}
$$

Equation (1.169) can be rewritten in the classical form at $\boldsymbol{\sigma} = \mathbb{E} : \boldsymbol{\varepsilon}$ when the consecutive *tensor products* $\boldsymbol{\varepsilon} \cdot \boldsymbol{M}^{(i)}$ and their *traces* are defined. For instance,

$$
\boldsymbol{\varepsilon} \cdot \boldsymbol{M}^{(1)} = \begin{bmatrix} \varepsilon_{xx} & \varepsilon_{xy} & \varepsilon_{xz} \\ & \varepsilon_{yy} & \varepsilon_{yz} \\ & & \varepsilon_{zz} \end{bmatrix} \cdot \begin{bmatrix} 1 & & \\ & 0 & \\ & & 0 \end{bmatrix} = \begin{bmatrix} \varepsilon_{xx} & 0 & 0 \\ \varepsilon_{xy} & 0 & 0 \\ \varepsilon_{xz} & 0 & 0 \end{bmatrix} \tag{1.170}
$$

from where one finds

$$
\mathrm{tr}(\boldsymbol{\varepsilon} \cdot \boldsymbol{M}^{(1)}) = \varepsilon_{xx} \tag{1.171}
$$

and

$$
\boldsymbol{\varepsilon} \cdot \boldsymbol{M}^{(1)} + \boldsymbol{M}^{(1)} \cdot \boldsymbol{\varepsilon} = \begin{bmatrix} 2\varepsilon_{xx} & \varepsilon_{xy} & \varepsilon_{xz} \\ \varepsilon_{xy} & 0 & 0 \\ \varepsilon_{xz} & 0 & 0 \end{bmatrix} \tag{1.172}
$$

When the remaining products $\boldsymbol{\varepsilon} \cdot \boldsymbol{M}^{(2)}$ and $\boldsymbol{\varepsilon} \cdot \boldsymbol{M}^{(3)}$ are calculated analogously, the coefficients preceding the components of the strain tensor are grouped, and when the engineering notation is consequently used the *state equation* (1.169) can finally be furnished in the following form:

$$\left\{\begin{array}{c}\sigma_{xx}\\ \sigma_{yy}\\ \sigma_{zz}\\ \hline \tau_{yz}\\ \tau_{zx}\\ \tau_{xy}\end{array}\right\}=\left[\begin{array}{ccc|ccc}E_{11}&E_{12}&E_{13}&&&\\ E_{21}&E_{22}&E_{23}&&&\\ E_{31}&E_{32}&E_{33}&&&\\ \hline &&&E_{44}&&\\ &&&&E_{55}&\\ &&&&&E_{66}\end{array}\right]\left\{\begin{array}{c}\varepsilon_{xx}\\ \varepsilon_{yy}\\ \varepsilon_{zz}\\ \hline \gamma_{yz}\\ \gamma_{zx}\\ \gamma_{xy}\end{array}\right\} \tag{1.173}$$

Subsequent elements of *stiffness matrix* of the *orthotropic elastic material* $[\mathbb{E}]$ are expressed in terms of coefficients α_i as follows:

$$\begin{array}{lll}E_{11}=\alpha_1+2\alpha_7 & E_{12}=E_{21}=\alpha_2 & E_{13}=E_{31}=\alpha_4\\ E_{22}=\alpha_3+2\alpha_8 & E_{23}=E_{32}=\alpha_5 & E_{33}=\alpha_6+2\alpha_9\\ E_{44}=\alpha_8+\alpha_9 & E_{55}=\alpha_9+\alpha_7 & E_{66}=\alpha_7+\alpha_8\end{array} \tag{1.174}$$

Note that the above described procedure of *linear orthotropic elasticity* derivation is based on the *theory of invariant representation* which differs from the conventional approach (1.103). More detailed distinction between different ways of formulating the linear elasticity constitutive laws will be presented in Sect. 1.9.

1.8 Remarks on Irreducible Coupling of Volumetric and Shear Response in Anisotropic Materials

In the general case of *full* material *anisotropy* complete mutual coupling between all stress and strain components holds. In fact, the generalized Hooke law (1.39) with the compliance matrix for general anisotropy taken in the form (1.98) leads to (after Rabinovich [45])

$$\begin{aligned}\varepsilon_{11}&=\frac{1}{E_{11}}(\sigma_{11}-\nu_{21}\sigma_{22}-\nu_{31}\sigma_{33}\\ &\quad+\eta_{23(1)}\tau_{23}+\eta_{31(1)}\tau_{31}+\eta_{12(1)}\tau_{12})\\ \varepsilon_{22}&=\frac{1}{E_{22}}(-\nu_{12}\sigma_{11}+\sigma_{22}-\nu_{32}\sigma_{33}\\ &\quad+\eta_{23(2)}\tau_{23}+\eta_{31(2)}\tau_{31}+\eta_{12(2)}\tau_{12})\\ \varepsilon_{33}&=\frac{1}{E_{22}}(-\nu_{13}\sigma_{11}-\nu_{23}\sigma_{22}+\sigma_{33}\\ &\quad+\eta_{23(3)}\tau_{23}+\eta_{31(3)}\tau_{31}+\eta_{12(3)}\tau_{12})\\ \gamma_{23}&=\frac{1}{G_{23}}\big(\eta_{(1)23}\sigma_{11}+\eta_{(2)23}\sigma_{22}+\eta_{(3)23}\sigma_{33}\\ &\quad+\tau_{23}+\mu_{31(23)}\tau_{31}+\mu_{12(23)}\tau_{12}\big)\\ \gamma_{31}&=\frac{1}{G_{31}}\big(\eta_{(1)31}\sigma_{11}+\eta_{(2)31}\sigma_{22}+\eta_{(3)31}\sigma_{33}\end{aligned} \tag{1.175}$$

$$
\begin{aligned}
&+\mu_{23(31)}\tau_{23} + \tau_{31} + \mu_{12(31)}\tau_{12}\big) \\
\gamma_{12} &= \frac{1}{G_{12}}\big(\eta_{(1)12}\sigma_{11} + \eta_{(2)12}\sigma_{22} + \eta_{(3)12}\sigma_{33} \\
&+\mu_{23(12)}\tau_{23} + \mu_{31(12)}\tau_{12} + \tau_{12}\big)
\end{aligned}
$$

Note that in the above equations elastic extensions $\varepsilon_{11}, \varepsilon_{22}, \varepsilon_{33}$ depend not only on all normal stresses $\sigma_{11}, \sigma_{22}, \sigma_{33}$ but also on all shear stresses $\tau_{23}, \tau_{31}, \tau_{12}$ (through the *generalized Young modules* E_{ii} and the *Rabinovich modules* $\eta_{ij(k)}$), resulting in nonzeroth elements of symmetric constitutive matrix of elasticity in its right top block. Moreover, the shear strains $\gamma_{23}, \gamma_{31}, \gamma_{12}$ depend on all shear stresses $\tau_{23}, \tau_{31}, \tau_{12}$ (through the *generalized Kirchhoff modules* G_{ij} and the *Chencov coefficients* $\mu_{ij(kl)}$) as well as on all normal stresses $\sigma_{11}, \sigma_{22}, \sigma_{33}$ such that the left bottom and the right bottom blocks of the elasticity matrix are fully populated. The above remarks lead in consequence to the conclusion that, in all cases different from isotropy, pure volumetric deformation is inseparable from pure shear deformation. In other words, *irreducibility of elasticity equations* (1.175) into uncoupled *law of volume change* and *law of shape change* holds when the *decomposition of strain* and *stress tensors* into axiator and deviator $\boldsymbol{\varepsilon} = \varepsilon_{\mathrm{m}}\mathbf{1} + \boldsymbol{e}$ and $\boldsymbol{\sigma} = \sigma_{\mathrm{m}}\mathbf{1} + \boldsymbol{s}$ is used.

This impossibility is inevitable even in a narrower case of orthotropy (1.103) in spite of the fact that shear stresses are uncoupled to the extensions and, vice versa, normal stresses do not result in shear strains. In order to trace this let us rewrite (1.103) as

$$
\begin{aligned}
\varepsilon_{11} &= \frac{1}{E_{11}}(\sigma_{11} - \nu_{21}\sigma_{22} - \nu_{31}\sigma_{33}) \\
\varepsilon_{22} &= \frac{1}{E_{22}}(-\nu_{12}\sigma_{11} + \sigma_{22} - \nu_{13}\sigma_{33}) \\
\varepsilon_{33} &= \frac{1}{E_{33}}(-\nu_{13}\sigma_{11} - \nu_{23}\sigma_{22} + \sigma_{33}) \\
\gamma_{23} &= \frac{\tau_{23}}{G_{23}} \qquad \gamma_{31} = \frac{\tau_{31}}{G_{31}} \qquad \gamma_{12} = \frac{\tau_{12}}{G_{12}}
\end{aligned}
\tag{1.176}
$$

Calculating the unit *volume change* called *dilatation* $\Theta = \varepsilon_{11} + \varepsilon_{22} + \varepsilon_{33}$ we obtain

$$
\begin{aligned}
\Theta^{\mathrm{orto}} = 3\varepsilon_{\mathrm{m}} &= \frac{1}{E_{11}}(\sigma_{11} - \nu_{21}\sigma_{22} - \nu_{31}\sigma_{33}) \\
&+ \frac{1}{E_{22}}(-\nu_{12}\sigma_{11} + \sigma_{22} - \nu_{13}\sigma_{33}) \\
&+ \frac{1}{E_{33}}(-\nu_{13}\sigma_{11} - \nu_{23}\sigma_{22} + \sigma_{33})
\end{aligned}
\tag{1.177}
$$

or recalling the symmetry of elasticity matrix (1.104) the equivalent form is furnished

$$\Theta^{\mathrm{orto}} = \frac{\sigma_{11}}{E_{11}}(1-\nu_{21}-\nu_{31}) + \frac{\sigma_{22}}{E_{22}}(1-\nu_{12}-\nu_{32}) + \frac{\sigma_{33}}{E_{33}}(1-\nu_{13}-\nu_{23}) \tag{1.178}$$

Note that in case of orthotropy dilatation is expressed not only in terms of the hydrostatic stress $\Theta = \Theta^{\mathrm{izo}}(\sigma_{\mathrm{h}})$ but by the more general function $\Theta = \Theta^{\mathrm{orto}}(\sigma_{11}, \sigma_{22}, \sigma_{33}; E_{ij}, \nu_{ij})$ or $\Theta = \Theta^{\mathrm{orto}}(\sigma_{kk}; \mathbb{E}^{-1}_{ijkl})$.

In the particular case of isotropy when $E_{ij} = E$, $\nu_{ij} = \nu$ the above equations reduce to the classical form

$$\Theta^{\mathrm{izo}} = 3\varepsilon_{\mathrm{m}} = \frac{1-2\nu}{E}(\sigma_{11}+\sigma_{22}+\sigma_{33}) = \frac{3(1-2\nu)}{E}\sigma_{\mathrm{h}} \tag{1.179}$$

or

$$\varepsilon_{\mathrm{m}} = \frac{1}{3K}\sigma_{\mathrm{h}}, \qquad K = \frac{E}{3(1-2\nu)} \tag{1.180}$$

in which dilatation or mean strain ε_{m} depends on hydrostatic stress σ_{h} exclusively.

Contrary to the previous case for *material orthotropy* by use of the following definition of deviatoric strain:

$$\boldsymbol{e} = \boldsymbol{\varepsilon} - \frac{1}{3}\varepsilon_{kk}\mathbf{1} = \mathbb{E}^{-1} : \boldsymbol{\sigma} - \frac{1}{3}\Theta^{\mathrm{orto}}(\sigma_{kk}; \mathbb{E}^{-1}_{ijkl})\mathbf{1} \tag{1.181}$$

we obtain

$$\begin{Bmatrix} e_{11} \\ e_{22} \\ e_{33} \\ e_{23} \\ e_{31} \\ e_{12} \end{Bmatrix} = \left[\begin{array}{ccc|ccc} \frac{1}{E_{11}} & -\frac{\nu_{21}}{E_{11}} & -\frac{\nu_{31}}{E_{11}} & 0 & 0 & 0 \\ -\frac{\nu_{12}}{E_{22}} & \frac{1}{E_{22}} & -\frac{\nu_{32}}{E_{22}} & 0 & 0 & 0 \\ -\frac{\nu_{13}}{E_{33}} & -\frac{\nu_{23}}{E_{33}} & \frac{1}{E_{33}} & 0 & 0 & 0 \\ \hline 0 & 0 & 0 & \frac{1}{G_{23}} & 0 & 0 \\ 0 & 0 & 0 & 0 & \frac{1}{G_{31}} & 0 \\ 0 & 0 & 0 & 0 & 0 & \frac{1}{G_{12}} \end{array}\right] \begin{Bmatrix} \sigma_{11} \\ \sigma_{22} \\ \sigma_{33} \\ \sigma_{23} \\ \sigma_{31} \\ \sigma_{12} \end{Bmatrix} - \tfrac{1}{3}\Theta^{\mathrm{orto}}(\sigma_{kk}; \mathbb{E}^{-1}_{ijkl}) \left\{\begin{array}{c} 1 \\ 1 \\ 1 \\ \hline 0 \\ 0 \\ 0 \end{array}\right\} \tag{1.182}$$

In other words, the *pure shear deformation* obtained by subtracting of the *dilatation* from the full deformation depends also on Θ^{orto}, so separation of these two effects is impossible.

1.9 Cauchy's Elasticity, Hyperelasticity, or Hypoelasticity

In the theory of linear elasticity in case of infinitesimal deformations occurring in isothermal or adiabatic conditions the *constitutive relations* linking tensors of stress and strain can be defined in three equivalent ways:

- According to the *Cauchy formulation* it is assumed that there exists an equilibrium state, called natural state, for which all components of the stress and strain tensors are equal to zero and to which material returns after removing loadings. An environment of natural state obeys unique value relation between stress and strain as

$$\sigma_{ij} = E_{ijkl}\varepsilon_{kl} \tag{1.183}$$

- According to the *Green formulation*, also called *hyperelasticity*, it is postulated an existence of function of elastic strain energy per unit volume $\mathcal{W}$ which is equal to zero in an environment of natural state and such that an increment of work done by stress is equal to an increment of strain energy

$$\sigma_{ij} = \frac{\partial \mathcal{W}}{\partial \varepsilon_{ij}} \qquad \mathcal{W} = \frac{1}{2}\sigma_{ij}\varepsilon_{kl} = \frac{1}{2}E_{ijkl}\varepsilon_{ij}\varepsilon_{kl} \tag{1.184}$$

- According to the third formulation, called *hypoelasticity*, it is postulated an incremental relation of the following form:

$$\mathrm{d}\sigma_{ij} = E_{ijkl}\mathrm{d}\varepsilon_{kl} \qquad \text{or} \qquad \frac{\partial \sigma_{ij}}{\partial t} = E_{ijkl}\frac{\partial \varepsilon_{kl}}{\partial t} \tag{1.185}$$

For all three cases: *Cauchy's*, *hyper-* and *hypoelasticity tensor* E_{ijkl} may depend on temperature but is independent of stress and strain tensors.

Note however that in the general case of nonlinearity constitutive tensors of elasticity or hyperelasticity (1.183) and (1.184) may differ from constitutive tensor of hypoelasticity (1.185). In the first case *tensor representative* matrix $\mathbb{E}$ is the *secant matrix* $[\mathbb{E}] = [_{\text{sec}}\mathbb{E}]$, whereas in the other case it is the tangent matrix $[\mathbb{E}] = [_{\text{tan}}\mathbb{E}]$.

It is worth to mention that although Cauchy, hyper- and hypoformulations of elasticity are alternative in case of theory of infinitesimal deformations, they may lead to essentially different results after entering the *finite deformation range*. Namely, introducing *definitions* of *finite strains*

$$\varepsilon_{ij} = \frac{1}{2}\left(\frac{\partial u_i}{\partial x_j} + \frac{\partial u_j}{\partial x_i}\right) \begin{array}{l} \nearrow \epsilon_{ij} = \dfrac{1}{2}\left(\dfrac{\partial u_i}{\partial X_j} + \dfrac{\partial u_j}{\partial X_i} + \dfrac{\partial u_i}{\partial X_j}\dfrac{\partial u_j}{\partial X_i}\right) \\ \searrow \epsilon_{ij} = \dfrac{1}{2}\left(\dfrac{\partial u_i}{\partial x_j} + \dfrac{\partial u_j}{\partial x_i} + \dfrac{\partial u_i}{\partial x_j}\dfrac{\partial u_j}{\partial x_i}\right) \end{array} \tag{1.186}$$

where ϵ_{ij} and $\in_{ij}$ stand for *Green's* and *Almansi's strain tensors*, respectively, and corresponding stress tensors

$$\sigma_{ij} \longrightarrow \Sigma_{ij} = \frac{\rho_0}{\rho}\frac{\partial X_i}{\partial x_k}\frac{\partial X_j}{\partial x_l}\sigma_{kl} \tag{1.187}$$

where σ_{ij} and Σ_{ij} denote the *Lagrange* and the *second Piola–Kirchhoff stress tensors* instead of formulations (1.183–1.185) we arrive at mutually different formulations

$$\begin{aligned} &\Sigma_{ij} = E_{ijkl}\epsilon_{kl} \\ &\frac{\mathrm{D}\mathcal{W}}{\mathrm{D}t} = \frac{1}{\rho_0}\Sigma_{ij}\frac{\partial \epsilon_{ij}}{\partial t} \\ &\frac{\mathrm{D}\sigma_{ij}}{\mathrm{D}t} - \sigma_{ip}\Omega_{pj} - \sigma_{jp}\Omega_{pi} = E_{ijkl}\dot{\in}_{kl} \end{aligned} \tag{1.188}$$

In case of *hypoelastic material* subjected to finite deformation appropriate constitutive equation $(1.188)_3$ comprises both the symbol of *objective derivative of* the *stress tensor* $\mathrm{D}\sigma_{ij}/\mathrm{D}t$ and an effect of change of stress tensor resulting from rigid rotation which is described by *skew-symmetric spin tensor*

$$\Omega_{ij} = \begin{bmatrix} 0 & \frac{1}{2}\left(\frac{\partial \dot{u}_2}{\partial x_3} - \frac{\partial \dot{u}_3}{\partial x_2}\right) & -\frac{1}{2}\left(\frac{\partial \dot{u}_3}{\partial x_1} - \frac{\partial \dot{u}_1}{\partial x_3}\right) \\ -\frac{1}{2}\left(\frac{\partial \dot{u}_2}{\partial x_3} - \frac{\partial \dot{u}_3}{\partial x_2}\right) & 0 & \frac{1}{2}\left(\frac{\partial \dot{u}_1}{\partial x_2} - \frac{\partial \dot{u}_2}{\partial x_1}\right) \\ \frac{1}{2}\left(\frac{\partial \dot{u}_3}{\partial x_1} - \frac{\partial \dot{u}_1}{\partial x_3}\right) & -\frac{1}{2}\left(\frac{\partial \dot{u}_1}{\partial x_2} - \frac{\partial \dot{u}_2}{\partial x_1}\right) & 0 \end{bmatrix} \tag{1.189}$$

References

1. Abu Al-Rub, R.K., Voyiadjis, G.Z.: On the coupling of anisotropic damage and plasticity models for ductile materials. Int. J. Solids Struct. **40**, 2611–2643 (2003)
2. Berryman, J.G.: Bounds and self-consistent estimates for elastic constants of random polycrystals with hexagonal, trigonal, and tetragonal symmetries. J. Mech. Phys. Solids **53**, 2141–2173 (2005)
3. Betten, J.: Creep Mechanics. Springer, Berlin (2002)
4. Bielski, J., Skrzypek, J., Kuna-Ciskał, H.: Implementation of a model of coupled elastic-plastic unilateral damage material to finite element code. Int. J. Damage Mech. **15**(1), 5–39 (2006)
5. Boehler, J.-P.: A simple derivation of representation for non-polynomial constitutive equations for some cases of anisotropy. Zeitschrift für Angewandte Mathematik und Mechanik **59**, 157–167 (1979)
6. Cauvin, A., Testa, R.B.: Damage mechanics: basic variables in continuum theories. Int. J. Solids Struct. **36**, 747–761 (1999)
7. Chaboche, J.L.: Description Thermodynamique et Phénoménologique de la Viscoplasticité Cyclique avec Endommagement. Thése Univ. Paris VI et Publication ONERA, No. pp. 1978-3 (1978)
8. Chaboche, J.L.: In: Boehler, J.P. (ed.) Mechanical Behavior of Anisotropic Solids. Martinus Nijhoff, Boston (1982)

9. Chen, X.F., Chow, C.L.: On damage strain energy release rate Y. Int. J. Damage Mech. **4, 3**, 251–263 (1995)
10. Cordebois, J.P., Sidoroff, F.: Damage induced elastic anisotropy, Coll. Euromech 115, Villard de Lans, also in Mechanical Behavior of Anisotropic Solids, In: Boehler, J.P. (ed.) Martinus Nijhoff, Boston. pp. 761–774 (1979)
11. Davison, L., Stevens, A.L.: Thermodynamical constitution of spalling elastic bodies. J. Appl. Phys. **44, 2**, 668 (1973)
12. Desmorat, R., Marull, R.: Non quadratic Kelvin modes based plasticity criteria for anisotropic materials. Int. J. Plast. **27**, 327–351 (2011)
13. Egner, H.: On the full coupling between thermo-plasticity and thermo-damage in thermodynamic modeling of dissipative materials. Int. J. Solids Struct. **49**, 279–288 (2012)
14. Egner, H., Skoczeń, B.: Ductile damage development in two-phase metallic materials applied at cryogenic temperatures. Int. J. Plast. **26, 4**, 488–506 (2010)
15. Gambarotta, L., Lagomarsino, S.: A microcrack damage model for brittle materials. Int. J. Solids Struct. **30**, 177–198 (1993)
16. Gan, H., Orozco, C.E., Herkovich, C.T.: A strain-compatible method for micromechanical analysis of multi-phase composites. Int. J. Solids Struct. **37**, 5097–5122 (2000)
17. Ganczarski, A.: Problems of acquired anisotropy and coupled thermo-mechanical fields of CDM, Politechnika Krakowska, nr 25 (2001)
18. Hayakawa, K., Murakami, S.: Thermodynamical modeling of elastic-plastic damage and experimental validation of damage potential. Int. J. Damage Mech. **6**, 333–362 (1997)
19. Herakovich, C.T., Aboudi, J.: Thermal effects in composites. In: Hetnarski, R.B. (ed.) Thermal Stresses V, pp. 1–142. Lastran Corporation, Publishing Division, Rochester (1999)
20. Hill, R.: A theory of the yielding and plastic flow of anisotropic metals. Proc. R. Soc. Lond. **A193**, 281–297 (1948)
21. Hill, R.: The Mathematical Theory of Plasticity. Clarendon Press, Oxford (1950)
22. Hu, Z.W., Marin, J.: Anisotropic loading functions for combined stresses in the plastic range. J. Appl. Mech. **22**, 1 (1956)
23. Jastrzebski, Z.D.: The Nature and Properties of Engineering Materials. Wiley, New York (1987)
24. Kachanov, L.M.: O vremeni razrusheniya v usloviyah polzuchesti, Izvestiya AN SSSR. Otd. Mehn. Nauk. **8**, 26–31 (1958)
25. Kachanov, L.M.: Osnovy mehaniki razrusheniya, Moskva, Nauka, Izdat (1974)
26. Kowalewski, Z.L., Śliwowski, M.: Effect of cyclic loading on the yield surface evolution of 18G2A low-alloy steel. Int. J. Mech. Sci. **39, 1**, 51–68 (1997)
27. Kowalsky, U., Ahrens, H., Dinkler, D.: Distorted yield surfaces-modelling by higher order anisotropic hardening tensors. Comput. Math. Sci. **16**, 81–88 (1999)
28. Krajcinovic, D., Fonseka, G.U.: The continuous damage theory of brittle materials, part I: general theory. Trans. ASME J. Appl. Mech. **48, 4**, 809–815 (1981)
29. Krajcinovic, D.: Damage mechanics. Mech. Mater. **8**, 117–197 (1989)
30. Krajcinovic, D.: Damage Mechanics. Elsevier, Amsterdam (1996)
31. Kuna-Ciskał, H., Skrzypek, J.: CDM based modelling of damage and fracture mechanisms in concrete under tension and compression. Eng. Fract. Mech. **71**, 681–698 (2004)
32. Lacy, T.E., McDowell, D.L., Willice, P.A., Talreja, R.: On the representation of damage evolution in continuum damage mechanics. Int. J. Damage Mech. **6**, 62–95 (1997)
33. Lekhnitskii, S.G.: Theory of Elasticity of an Anisotropic Body. Mir Publishers, in Russian: Nauka 1977, Moscow (1981)
34. Love, A.E.H.: A Treatise on the Mathematical Theory of Elasticity. Dover Publication, New York (1944)
35. Lubarda, V.A., Krajcinovic, D.: Damage tensors and the crack density distribution. Int. J. Solids Struct. **30, 20**, 2859–2877 (1993)
36. Malinin, N.N., Rżysko, J.: Mechanika materiałów. PWN, Warszawa (1981)
37. Mróz, Z., Maciejewski, J.: Failure criteria and compliance variation of anisotropically damaged materials. In: Skrzypek, J.J., Ganczarski, A. (eds.) Anisotropic Behaviour of Damaged Materials, Chap. 3, pp. 75–112. Springer, Berlin (2002)

38. Murakami, S., Kamiya, K.: Constitutive and damage evolution equations of elastic-brittle materials based on irreversible thermodynamics. Int. J. Mech. Sci. **39, 4**, 473–486 (1997)
39. Murakami, S., Ohno, N.: A continuum theory of creep and creep damage. In: Ponter, A.R.S., Hayhurst, D.R. (eds.) Creep in Structures, pp. 422–444. Springer, Berlin (1981)
40. Murakami, S.: Continuum Damage Mechanics. Springer, Berlin (2012)
41. Mursa, K.S.: Examination of orthotropic metal sheets under uniaxial tension (in Russian). Izv. Vys. Ucheb. Zav., Mash. 6 (1972)
42. Nye, J.F.: Physical Properties of Crystals Their Representations by Tensor and Matrices. Clarendon Press, Oxford (1957)
43. Onat, E.T., Leckie, F.A.: Representation of mechanical behavior in the presence of changing internal structure. J. Appl. Mech. Trans. ASME **55**, 1–10 (1988)
44. Ottosen, N.S., Ristinmaa, M.: The Mechanics of Constitutive Modeling. Elsevier, Amsterdam (2005)
45. Rabinovich, A.L.: On the elastic constants and strength of aircraft materials. Trudy Tsentr. Aero-gidrodin. Instr. **582**, 1–56 (1946)
46. Rabotnov, YuN: O razrushenii vsledstvie polzuchesti. Zhurnal prikladnoĭ mehaniki i tehnicheskoĭ fiziki **2**, 113–123 (1963)
47. Rymarz, Cz.: Mechanika ośrodków ciągłych. PWN, Warszawa (1993)
48. Seweryn, A., Mróz, Z.: A non-local failure and damage evolution rule: application to a dilatant crack model. J. de Physique IV France **8**, 257–268 (1998)
49. Skrzypek, J.: Podstawy mechaniki uszkodzeń. Wydawnictwo Politechniki Krakowskiej, Kraków (2006)
50. Skrzypek, J.J., Ganczarski, A.W., Rustichelli, F., Egner, H.: Advanced Materials and Structures for Extreme Operating Conditions. Springer, Berlin (2008)
51. Skrzypek, J., Ganczarski, A.: Modeling of Material Damage and Failure of Structures. Springer, Berlin (1999)
52. Spencer, A.J.M.: Theory of invariants. In: Eringen, C. (ed.) Continuum Physics, pp. 239–353. Academic Press, New York (1971)
53. Sun, C.T., Vaidya, R.S.: Prediction of composite properties from a representative volume element. Compos. Sci. Technol. **56**, 171–179 (1996)
54. Taher, S.F., Baluch, M.H., Al-Gadhib, A.H.: Towards a canonical elastoplastic damage model. Eng. Fract. Mech. **48, 2**, 151–166 (1994)
55. Voyiadjis, G.Z., Kattan, P.I.: Evolution of fabric tensors in damage mechanics of solids with micro-cracks: part I—theory and fundamental concepts. Mech. Res. Commun. **34**, 145–154 (2007)
56. Yun-bing, L., Xing-fu, C.: The order of a damage tensor. Appl. Math. Mech. **10, 3**, 251–258 (1989)

Chapter 2
Constitutive Equations for Isotropic and Anisotropic Linear Viscoelastic Materials

Jacek J. Skrzypek and Artur W. Ganczarski

Abstract In case of isotropic material symmetry, the elastic-viscoelastic correspondence principle is well established to provide the solution of linear viscoelasticity from the coupled fictitious elastic problem by use of the inverse Laplace transformation (Alfrey–Hoff's analogy). Aim of this chapter is to show useful enhancement of the Alfrey–Hoff's analogy to a broader class of material anisotropy for which separation of the volumetric and the shape change effects from total viscoelastic deformation does not occur. Such extension requires use of the vector–matrix notation to description of the general constitutive response of anisotropic linear viscoelastic material (see Pobiedria Izd. Mosk. Univ., (1984) [10]). When implemented to the composite materials which exhibit linear viscoelastic response, the classically used homogenization techniques for averaged elastic matrix, can be implemented to viscoelastic work-regime for associated fictitious elastic Representative Unit Cell of composite material. Next, subsequent application of the inverse Laplace transformation (cf. Haasemann and Ulbricht Technische Mechanik, 30(1–3), 122–135 (2010)) is applied. In a similar fashion, the well-established upper and lower bounds for effective elastic matrices can also be extended to anisotropic linear viscoelastic composite materials. The Laplace transformation is also a convenient tool for creep analysis of anisotropic composites that requires, however, limitation to the narrower class of linear viscoelastic materials. In the space of transformed variable s, instead of time space t, the classical homogenization rules for fictitious elastic composite materials can be applied. For the above reasons in what follows, we shall confine ourselves to the linear viscoelastic materials, isotropic, or anisotropic.

Keywords Linear viscoelasticty · Anisotropic correspondence principle · Anisotropic integral equations of linear viscoelasticity · Fictitious anisotropic elastic problem · Homogenization of linear viscoelastic composites

---

J.J. Skrzypek (✉) · A.W. Ganczarski
Solid Mechanics Division, Institute of Applied Mechanics,
Cracow University of Technology, al. Jana Pawła II 37, 31-864 Kraków, Poland
e-mail: Jacek.Skrzypek@pk.edu.pl

A.W. Ganczarski
e-mail: Artur.Ganczarski@pk.edu.pl

J.J. Skrzypek and A.W. Ganczarski (eds.), *Mechanics of Anisotropic Materials*,
Engineering Materials, DOI 10.1007/978-3-319-17160-9_2

2.1 Selected Uniaxial Models of the Isotropic Linear Viscoelastic Materials

Creep phenomena at elevated temperature are usually treated as *nonlinear creep phenomenon* problems. There exists broad literature in the field of nonlinear creep, for example, *creep anisotropy* Findley et al. [4], survey on constitutive models of nonlinear creep Skrzypek [13], Betten [2], *interaction creep and plasticity* Krempl [6], coupling of creep and damage Skrzypek [14], Skrzypek and Ganczarski [15], *creep fatigue damage* Murakami [7], and *nonconventional creep models* of anisotropic material Altenbach [1] and others.

At the beginning, we confine to the commonly used uniaxial isotropic linear viscoelastic models for which a general differential equation models may be written as

$$p_0\sigma + p_1\dot{\sigma} + p_2\ddot{\sigma} + \cdots p_a\frac{\partial^a\sigma}{\partial t^a} = q_0\varepsilon + q_1\dot{\varepsilon} + \cdots p_b\frac{\partial^b\varepsilon}{\partial t^b} \tag{2.1}$$

where $p_0, p_1, \ldots, q_0, q_1, \ldots$ denote material constants, and constitutive equation is a linear function of the stress σ, strain ε, and their time derivatives $\dot{\sigma}, \ddot{\sigma}$, etc., and $\dot{\varepsilon}, \ddot{\varepsilon}$, etc. In such a case by the use of the Laplace transformation $\mathcal{L}\{f(t)\} = \widehat{f}(s) = \int\limits_0^\infty \mathrm{e}^{-st}\mathrm{d}t$, a linear viscoelastic problem can be reduced to associated fictitious elastic problem in terms of the transformed variable s, $\widehat{\sigma}_{ij}(\boldsymbol{x}, s)$, then the viscoelastic problem $\sigma_{ij}(\boldsymbol{x}, t)$ is obtained by the inverse Laplace transformation. Symbol { } stands here for function argument of the Laplace transformation and should not be confused with the Voigt vector notation.

2.1.1 Maxwell Model

The uniaxial *Maxwell model* (M) consists of a linear elastic spring $\varepsilon^{\mathrm{H}} = \sigma/E$ and a linear viscous dashpot element $\dot{\varepsilon}^{\eta} = \sigma/\eta$ connected in a series, Fig. 2.1. Differentiation of the first formula with time yields $\dot{\varepsilon}^{\mathrm{H}} = \dot{\sigma}/E$. When the additive

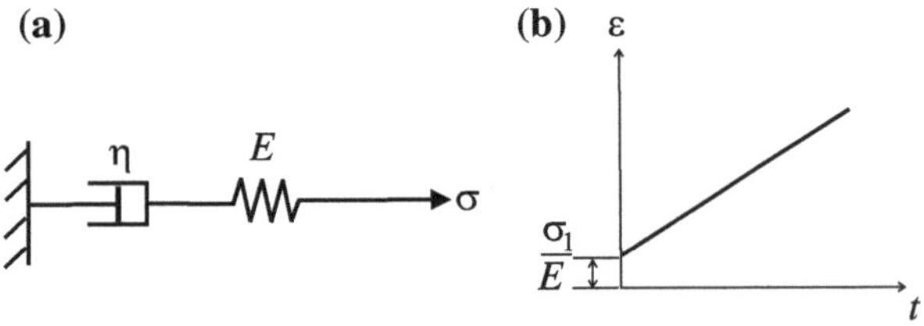

Fig. 2.1 Maxwell's material: **a** mechanical model, **b** creep curve under constant loading

decomposition of the *strain* or the *stain rate* $\dot{\varepsilon} = \dot{\varepsilon}^{\mathrm{H}} + \dot{\varepsilon}^{\eta}$ is used, we arrive at *equation* of the *Maxwell model*, hence

$$\dot{\varepsilon} = \frac{\dot{\sigma}}{E} + \frac{\sigma}{\eta} \quad \text{or} \quad \sigma + \frac{\eta}{E}\dot{\sigma} = \eta\dot{\varepsilon} \tag{2.2}$$

When the integration of above equation at constant stress $\sigma = \sigma_1 = \text{const} \quad (\dot{\sigma} = 0)$ and initial condition $\varepsilon(0) = \sigma_1/E$ is performed, we arrive at the creep function given as, see Fig. 2.1b

$$\varepsilon = \sigma_1 \left(\frac{1}{E} + \frac{1}{\eta} t \right) \tag{2.3}$$

or

$$\varepsilon = \sigma_1 J^{\mathrm{M}}(t), \qquad J^{\mathrm{M}}(t) = \frac{1}{E} + \frac{t}{\eta} \tag{2.4}$$

The time function $J^{\mathrm{M}}(t)$ is the *creep compliance function* of the *Maxwell model.*

2.1.2 Voigt–Kelvin Model

The *Voigt–Kelvin model* (V–K) consists of a linear spring element and a linear dashpot element which are connected in parallel as shown in Fig. 2.2a. Adopting the additive separation of stress into two parts applied to the spring $\sigma^{\mathrm{H}} = E\varepsilon$ and to the dashpot $\sigma^{\eta} = \eta\dot{\varepsilon}$ with $\varepsilon = \varepsilon^{\mathrm{H}} = \varepsilon^{\eta}$, the differential equation of the *V–K model* takes the form

$$\dot{\varepsilon} + \frac{E}{\eta}\varepsilon = \frac{\sigma}{\eta} \tag{2.5}$$

If a constant stress $\sigma = \sigma_1 = \text{const} \quad (\dot{\sigma} = 0)$ is applied to the V–K model, we arrive at *nonhomogeneous differential equation*

$$\dot{\varepsilon} + \frac{E}{\eta}\varepsilon = \frac{\sigma_1}{\eta} \tag{2.6}$$

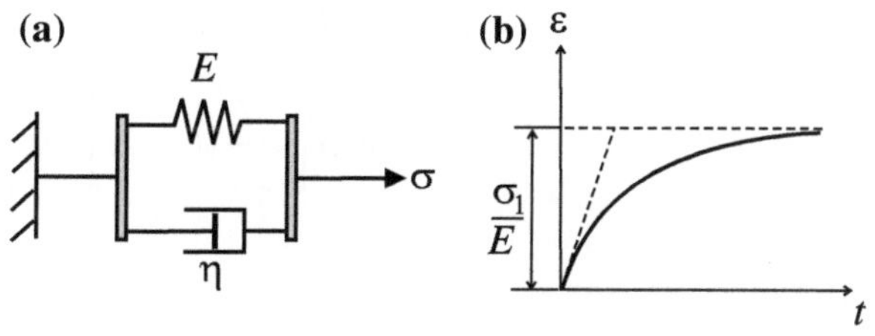

Fig. 2.2 Voigt–Kelvin model: **a** mechanical scheme, **b** creep strain at constant stress input

The homogeneous equation of (2.6) is an equation of separate variables

$$\frac{\dot{\varepsilon}}{\varepsilon} = -\frac{E}{\eta} \tag{2.7}$$

the general integral of which is given by

$$\varepsilon = C \exp\left(-\frac{E}{\eta}t\right) \tag{2.8}$$

When variation of integration constant $C(t)$ with initial condition $\varepsilon(0) = 0$ is done we arrive at the solution of (2.6)

$$\varepsilon = \frac{\sigma_1}{E}\left[1 - \exp\left(-\frac{E}{\eta}t\right)\right] \tag{2.9}$$

or

$$\varepsilon = \sigma_1 J^{\mathrm{VK}}(t), \qquad J^{\mathrm{VK}}(t) = \frac{1}{E}\left[1 - \exp\left(-\frac{E}{\eta}t\right)\right] \tag{2.10}$$

Function $J^{\mathrm{VK}}(t)$ is the *creep compliance function* of the *V–K model*. Note that V–K model does not account for instantaneous elasticity, hence $J^{\mathrm{VK}}(0) = 0$, see Fig. 2.2b.

When the more general case of a time function $\sigma(t)$ is applied and variation of constant $C(t)$ is done in (2.8) we arrive at the differential equation for $C(t)$

$$\dot{C}(t) = \frac{1}{\eta} \exp\left(\frac{E}{\eta}t\right)\sigma(t) \tag{2.11}$$

the general integral of which is expressed in form

$$C(t) = C_1 + \frac{1}{\eta}\int_0^t \exp\left(\frac{E}{\eta}\xi\right)\sigma(\xi)\mathrm{d}\xi \tag{2.12}$$

Substitution of (2.12) to (2.8) with the initial condition $\varepsilon(0) = 0$ yields $C_1 = 0$, such that the following general solution for $\varepsilon(t)$ holds

$$\begin{aligned}\varepsilon(t) &= \frac{1}{\eta}\exp\left(-\frac{E}{\eta}t\right)\int_0^t \exp\left(\frac{E}{\eta}\xi\right)\sigma(\xi)\mathrm{d}\xi \\ &= \frac{1}{\eta}\int_0^t \exp\left[-\frac{E}{\eta}(t-\xi)\right]\sigma(\xi)\mathrm{d}\xi\end{aligned} \tag{2.13}$$

When integration by parts is applied to (2.13), we arrive at so-called *integral representation* of the *V–K model*

$$\varepsilon(t) = \frac{\sigma(t)}{E} - \frac{1}{E}\int_0^t \exp\left[-\frac{E}{\eta}(t-\xi)\right]\dot{\sigma}(\xi)\mathrm{d}\xi \tag{2.14}$$

in which it is clearly seen that the creep function $J^{\mathrm{VK}}(t)$ has two terms: independent of time $J_0 = 1/E$ and dependent on time $\varphi(t) = \frac{1}{E}\exp\left[-\frac{E}{\eta}(t-\xi)\right]$.

Analogous solution may be reached by use of the *Laplace transform method* (2.48). In order to do this the nonhomogeneous V–K equation (2.5) is multiplied both-side by e^{-st} and integrated with respect to variable t in range from 0 to ∞

$$\int_0^\infty \dot{\varepsilon}(t)\mathrm{e}^{-st}\mathrm{d}t + \frac{E}{\eta}\int_0^\infty \varepsilon(t)\mathrm{e}^{-st}\mathrm{d}t = \int_0^\infty \frac{\sigma(t)}{\eta}\mathrm{e}^{-st}\mathrm{d}t \tag{2.15}$$

Consequently, the algebraic equation of the *transformed variable s* is obtained

$$s\widehat{\varepsilon}(s) - \varepsilon(0) + \frac{E}{\eta}\widehat{\varepsilon}(s) = \frac{\widehat{\sigma}(s)}{\eta} \tag{2.16}$$

When the initial condition $\varepsilon(0) = 0$ is used, the solution of (2.16) with respect of transformed variable $\widehat{\varepsilon}(s)$ is given as the following

$$\widehat{\varepsilon}(s) = \frac{1}{s+\frac{E}{\eta}}\widehat{\sigma}(s) \tag{2.17}$$

Applying next the *inverse Laplace transform* and taking advantage of property that multiplication of two transforms in fictitious domain of variable s corresponds to the convolution of two functions in real time space t, we arrive at the solution of linear viscoelastic problem

$$\varepsilon(t) = \frac{1}{\eta}\int_0^t \exp\left[-\frac{E}{\eta}(t-\xi)\right]\sigma(\xi)\mathrm{d}\xi \tag{2.18}$$

identical to (2.13).

2.1.3 Standard Model

The Maxwell and the Voigt–Kelvin two-element uniaxial models described in the Sects. 2.1.1 and 2.1.2 are very simple, although they exhibit strong limitations. The linear creep function at constant stress input corresponding to the Maxwell model does not confirm experiments, whereas the Voigt–Kelvin model is not capable to capture the instantaneous elastic strain effect. Trying to overcome the above objections, the commonly used *three-parameter standard model* is composed of two parts, a spring element (E) and the V–K unit (E_1, η) connected in a series as shown in Fig. 2.3a.

The differential equation of the standard model can be derived in an analogous way as for the Maxwell and the Voigt–Kelvin simple models, such that after necessary rearrangement used, the following is obtained

$$\frac{\eta E}{E_1 + E}\dot{\varepsilon} + \frac{E_1 E}{E_1 + E}\varepsilon = \sigma + \frac{\eta}{E_1 + E}\dot{\sigma} \tag{2.19}$$

The simple creep function, when the standard model is subjected to a step function $\sigma = \sigma_1 = \text{const}$ $(\dot{\sigma} = 0)$ and integrated with the initial condition $\varepsilon(0) = \sigma_1/E$ used, takes one of two equivalent forms

$$\varepsilon = \frac{\sigma_1}{E}\left[\left(1 + \frac{E}{E_1}\right) - \frac{E}{E_1}\exp\left(-\frac{E_1}{\eta}t\right)\right] \tag{2.20}$$

or

$$\varepsilon = \sigma_1 J^{\text{s}}(t), \qquad J^{\text{s}}(t) = \frac{1}{E}\left[\left(1 + \frac{E}{E_1}\right) - \frac{E}{E_1}\exp\left(-\frac{E_1}{\eta}t\right)\right] \tag{2.21}$$

if the time-dependent *creep compliance function* characterizing the *standard model* $J^{\text{s}}(t)$ is used. Note the horizontal asymptote of $\varepsilon(t)$ curve as shown in Fig. 2.3b with the new definition used: $1/H = 1/E + 1/E_1$.

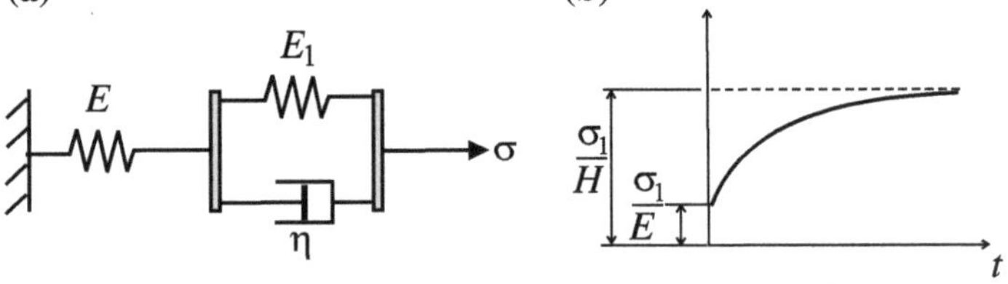

Fig. 2.3 The standard model: **a** mechanical scheme, **b** creep at constant stress input with instantaneous elastic strain built-in

2.1.4 Burgers Model

Although the standard model is free from aforementioned inconvenience of the Maxwell and the Voigt–Kelvin models, it still exhibits a horizontal asymptote (strain stabilization) when $t \to \infty$, which usually is not true; since according to experimental findings, the creep strain shows rather the infinite increase with time. In order to control such behaviors, a more complex *four-parameter Burgers model* which consists of two simple units, the Maxwell unit (E_1, η_1), and the Voigt–Kelvin unit (E_2, η_2) coupled in a series can be used, as presented in Fig. 2.4a. The differential *constitutive equation* of *Burgers' model* may be written in the format

$$\frac{\eta_1\eta_2}{E_2}\ddot{\varepsilon} + \eta_1\dot{\varepsilon} = \frac{\eta_1\eta_2}{E_1E_2}\ddot{\sigma} + \left(\frac{\eta_1}{E_1} + \frac{\eta_1}{E_2} + \frac{\eta_2}{E_2}\right)\dot{\sigma} + \sigma \tag{2.22}$$

Note that the above equation is the second-order linear differential equation with respect to strain and stress but of constant coefficients being functions of four parameters E_1, E_2, η_1 and η_2. It means that all strain and stress functions and their time derivatives are the linear functions, whereas the coefficients in Eq. (2.22) are constants: two *Young's modules* E_1, E_2 and two *viscosity parameters* η_1, η_2.

When the Burgers model is loaded by a step stress input applied at $t = 0$ the integration of Eq. (2.22), with two initial conditions $\varepsilon(0) = \sigma_1/E$, $\dot{\varepsilon}(0) = \sigma_1/\eta_1 + \sigma_1/\eta_2$ used, leads to one of equivalent relationships

$$\varepsilon = \frac{\sigma_1}{E_1}\left\{1 + \frac{E_1}{\eta_1}t + \frac{E_1}{E_2}\left[1 - \exp\left(-\frac{E_2}{\eta_2}t\right)\right]\right\} \tag{2.23}$$

or

$$\varepsilon = \sigma_1 J^{\mathrm{B}}(t), \qquad J^{\mathrm{B}}(t) = \frac{1}{E_1}\left\{1 + \frac{E_1}{\eta_1}t + \frac{E_1}{E_2}\left[1 - \exp\left(-\frac{E_2}{\eta_2}t\right)\right]\right\} \tag{2.24}$$

where the creep *compliance function* characterizing the *Burgers model compliance function* $J^{\mathrm{B}}(t)$ is applied.

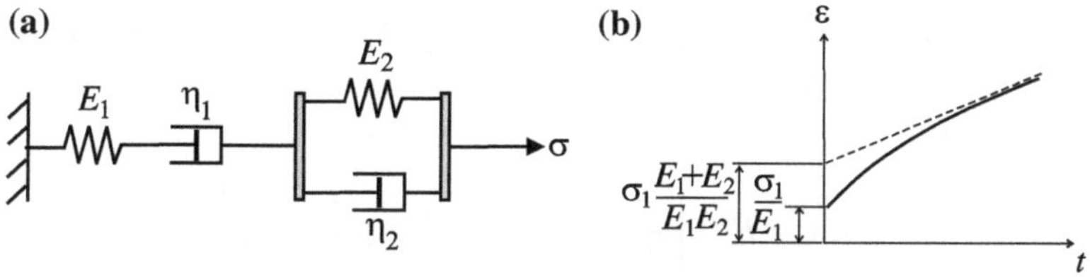

Fig. 2.4 The Burgers model: **a** the mechanical scheme, **b** the simple creep curve at constant stress input applied at $t = 0$

Note that the creep curve described by the Burgers model exhibits a skewed asymptote which corresponds to the unlimited strain increase with decreasing strain rate, which better fits the experimental findings, see Fig. 2.4b.

2.1.5 Creep Compliance and Relaxation Behavior of the Selected Linear Viscoelastic Models

More complex models of linear viscoelastic materials consisting of one Hooke's element and n Voigt–Kelvin's units coupled in a series or one Hooke's element and n Maxwell's units coupled in parallel are analyzed by Betten [2].

Consider now stress relaxation of simple uniaxial linear viscoelastic models discussed in Sects. 2.1.1–2.1.4, subject to a constant strain ε_1 at $t = 0$, from the initial stress level $\sigma_1 = E\varepsilon_1$.

In case of the *Maxwell model*, the *stress relaxation* from the initial level σ_1 at $t = 0$ to $t \to \infty$ is described as follows

$$\sigma(t) = \sigma_1 \exp\left(-\frac{Et}{\eta}\right) \tag{2.25}$$

Note that the rate of stress decrease changes from the initial $\dot{\sigma}(0) = -\sigma_1 E/\eta$ to zero, $\dot{\sigma}(\infty) = 0$. The so-called relaxation time $t_r = \eta/E$ corresponds to the fictitious case if the stress decreases continuously at the initial rate and finally it would reach zero at $t = t_r$.

The Voigt–Kelvin model does not exhibit stress relaxation effect. In this singular case application of the constant strain input, $\varepsilon = \varepsilon_1$ at $t = 0$ can be achieved only by an infinite initial stress response $\sigma(0) \to \infty$, such that the following holds

$$\sigma(t) = \eta\varepsilon_1\delta(t) + E\varepsilon_1 \mathrm{H}(t) \tag{2.26}$$

where the term containing the Heaviside unit function $\mathrm{H}(t)$ describing the constant stress in the spring, followed by the infinite stress input in dashpot described by the δ-Dirac function, appears.

The standard model is free from the above singularity and if it is subject to a constant strain at $t = 0$, the stress continuously decreases from the initial level $E\varepsilon_1(t = 0)$ to the asymptotically approached value $H\varepsilon_1(t \to \infty)$ such that *stress relaxation* function of the *standard model* is written as

$$\sigma(t) = E\varepsilon_1\left[\frac{H}{E} + \left(1 - \frac{H}{E}\right)\exp\left(-\frac{t}{n}\right)\right] \tag{2.27}$$

where the definitions hold: $\dfrac{1}{H} = \dfrac{1}{E} + \dfrac{1}{E_1}$ and $n = \dfrac{\eta}{E + E_1}$.

Table 2.1 Properties of Maxwell, Voigt–Kelvin, standard and Burgers viscoelastic units, after Skrzypek [13]

Model	Creep compliance function $J(t)$
Maxwell	$\frac{1}{E}\left(1+\frac{E}{\eta}t\right)$
V–K	$\frac{1}{E}\left(1-e^{-\frac{E}{\eta}t}\right)$
Standard	$\frac{1}{E}\left(1+\frac{E}{E_1}-\frac{E}{E_1}e^{-\frac{E_1}{\eta}t}\right)$
Burgers	$\frac{1}{E_1}\left[1+\frac{E_1}{\eta_1}t+\frac{E_1}{E_2}\left(1-e^{-\frac{E_2}{\eta_2}t}\right)\right]$
Model	Relaxation modulus $E(t)$
Maxwell	$Ee^{-\frac{E}{\eta}t}$
V–K	$E\left[1+\frac{\eta}{E}\delta(t)\right]$
Standard	$E\left(\frac{E_1}{E_1+E}+\frac{E}{E_1+E}e^{-\frac{E_1+E}{\eta}t}\right)$
Burgers	$\frac{(q_1-q_2r_1)e^{-r_1t}-(q_1-q_2r_2)e^{-r_2t}}{A}$

If the *Burgers model* is subject to a constant strain $\varepsilon = \varepsilon_1$ at $t = 0$, a continuous *stress relaxation* is described by the combination of two exponential functions $\exp(-r_1t)$ and $\exp(-r_2t)$ (cf. Table 2.1), where after Findley et al. [4] the new definitions are used

$$\begin{aligned} p_1 &= \frac{\eta_1}{E_1}+\frac{\eta_1}{E_2}+\frac{\eta_2}{E_2} \quad & p_2 &= \frac{\eta_1\eta_2}{E_1E_2} \quad & q_1 &= \eta_1 \\ q_2 &= \frac{\eta_1\eta_2}{E_2} \quad & r_{1,2} &= \frac{p_1 \mp A}{2p_2} \quad & A &= \sqrt{p_1^2-4p_2} \end{aligned} \tag{2.28}$$

2.2 The Uniaxial Boltzmann Superposition Principle of the Isotropic Linear Viscoelastic Materials

2.2.1 Bending of a Beam Subject to Stationary Load

Summarizing the results of previous subsection, response of the arbitrary *linear viscoelastic material* at step stress input $\sigma = \sigma_1 = \text{const}$ can be written as follows

$$\varepsilon(x,t) = \varepsilon^{e}(x)EJ(t) \tag{2.29}$$

In other words, strain at the arbitrary material point, being a function of the space x and time t variables, can be presented as a product of the *instantaneous elastic strain* $\varepsilon^{e}(x)$ depending on x only and the *creep compliance function* $J(t)$ specifically chosen for given material model dependent on time t only. In the light of comments presented in Sect. 2.1.5, Eq. (2.29) does not apply to the Voigt–Kelvin model in a straightforward manner. The V–K model does not comprise the initial elastic strain.

In particular, a *deflection of beam* made of the *linear viscoelastic material* $w^{ve}(x,t)$ at constant loading can be presented as the product of the elastic deflection $w^{e}(x)$ and the dimensionless creep compliance function $EJ(t)$ as follows

$$w^{ve}(x,t) = w^{e}(x)\,EJ(t) \tag{2.30}$$

For *creep compliance functions* shown in Table 2.1, we arrive at

$$\begin{aligned}
w^{M}(x,t) &= w^{e}(x)\left(1+\frac{E}{\eta}t\right)\\
w^{VK}(x,t) &= w^{e}(x)\left[1-\exp\left(-\frac{E}{\eta}t\right)\right]\\
w^{s}(x,t) &= w^{e}(x)\left[1+\frac{E}{E_1}-\frac{E}{E_1}\exp\left(-\frac{E_1}{\eta}t\right)\right]\\
w^{B}(x,t) &= w^{e}(x)\left\{1+\frac{E_1}{\eta_1}t+\frac{E_1}{E_2}\left[1-\exp\left(-\frac{E_2}{\eta_2}t\right)\right]\right\}
\end{aligned} \tag{2.31}$$

The aforementioned relationships hold for all linear viscoelastic models discussed even though V–K model does not exhibit instantaneous response. This comment holds for all linear viscoelastic models that do not have free elastic spring.

2.2.2 Bending with Tension of a Beam Subject to Nonstationary Load

Consider a prismatic *beam of doubly symmetric cross-section* subject to the axial force and the bending moment being both functions of coordinate x and time t: $N = N(x,t)$, $M = M(x,t)$. Assume also that both external forces N and M can be expressed as products of function dependent of x co-ordinate $N = N(x)$ or $M(x)$ and one common time function $f(t)$: $N(x,t) = N(x)f(t)$ and $M(x,t) = M(x)f(t)$. Supposing for simplicity that viscoelastic deformation fulfils the *Bernoulli hypothesis* of straight and normal segments $\varepsilon(x,z,t) = \lambda(x,t) + z\kappa(x,t)$, it is possible to separate Eq. (2.29) into the *viscoelastic axial elongation* λ^{ve} and the *viscoelastic curvature* κ^{ve}

$$\begin{aligned} \lambda^{\mathrm{ve}}(x,t) &= \frac{N(x)}{A}\int_0^t J(t-\xi)\frac{\partial f(\xi)}{\partial \xi}\mathrm{d}\xi \\ \kappa^{\mathrm{ve}}(x,t) &= \frac{M(x)}{I}\int_0^t J(t-\xi)\frac{\partial f(\xi)}{\partial \xi}\mathrm{d}\xi \end{aligned} \tag{2.32}$$

In a particular case, if both *generalized forces* are applied instantaneously at $t = 0$: $N(x,t) = N(x)\mathrm{H}(t)$ and $M(x,t) = M(x)\mathrm{H}(t)$, remembering that δ-*Dirac function* is defined as $\delta(t) = \dot{\mathrm{H}}(t)$ and applying (see Byron and Fuller [3])

$$\int_0^t J(t-\xi)\delta(\xi)\mathrm{d}\xi = \int_0^t J(\xi)\delta(t-\xi)\mathrm{d}\xi = J(t) \tag{2.33}$$

we finally obtain equations for viscoelastic elongation $\lambda^{\mathrm{ve}}(x,t)$ and curvature $\kappa^{\mathrm{ve}}(x,t)$ in a form

$$\begin{aligned} \lambda^{\mathrm{ve}}(x,t) &= \frac{N(x)}{EA}J(t) \\ \kappa^{\mathrm{ve}}(x,t) &= -w''(x,t) = \frac{M(x)}{EI}J(t) \end{aligned} \tag{2.34}$$

Hence, the axial elongation of the linear viscoelastic beam λ^{ve} is a product of instantaneous (elastic) elongation and the creep compliance function. Analogously, the curvature of the linear viscoelastic beam κ^{ve} is a product of instantaneous curvature and creep compliance function $J(t)$. For simplicity, the conventional beam theory is adopted here.

2.2.3 *Integral Representation of Creep and Relaxation Functions in Case of Arbitrary Loading History*

A general differential equation of uniaxial linear viscoelastic models can be written as follows:

$$p_0\sigma + p_1\dot{\sigma} + p_2\ddot{\sigma} + \cdots = q_0\varepsilon + q_1\dot{\varepsilon} + q_2\ddot{\varepsilon} + \cdots \tag{2.35}$$

where the constant p_i, q_i are coefficients of the linear arbitrary order differential constitutive equation, see Eq. (2.1). Order of Eq. (2.35) is equal to number of viscous elements (dashpots) appearing in the mechanical model. A compact *operator format* can be used instead of Eq. (2.35)

$$\mathrm{P}\sigma(t) = \mathrm{Q}\varepsilon(t) \tag{2.36}$$

where P and Q stand for *linear differential operators* with respect to time acting on stress $\sigma(t)$ and strain $\varepsilon(t)$, respectively

$$\mathrm{P} = p_0 + p_1 \frac{\partial}{\partial t} + p_2 \frac{\partial^2}{\partial t^2} + \cdots$$
$$\mathrm{Q} = q_0 + q_1 \frac{\partial}{\partial t} + q_2 \frac{\partial^2}{\partial t^2} + \cdots \tag{2.37}$$

It is clear that the linear elastic (Hooke's) material is a particular case of linear viscoelastic material in equation of which all time derivatives disappear, whereas $q_0/p_0 = E$.

Note that differential operators P and Q are linear with respect to all derivatives, hence the operator format of Eq. (2.36) can formally be treated as an algebraic equation as follows

$$\frac{\sigma(t)}{\varepsilon(t)} = \mathrm{E}_{\mathrm{ve}}(t) \qquad \mathrm{E}_{\mathrm{ve}}(t) = \frac{\mathrm{Q}(t)}{\mathrm{P}(t)} \tag{2.38}$$

The rational operator $\mathrm{E}_{\mathrm{ve}}(t)$ used in Eq. (2.38) plays a role of the *time-dependent stiffness operator*. As a consequence by contrast to elasticity a fraction $\sigma(t)/\varepsilon(t)$ is not constant but depends on time. Hence, Eq. (2.38) should be read in a symbolic way as follows

$$\frac{\mathrm{Q}(t)}{\mathrm{P}(t)} \bowtie \frac{\sigma(t)}{\varepsilon(t)} \tag{2.39}$$

Class of the *linear viscoelastic materials* is a subclass of the *nonlinear viscoelastic materials*; however, the *Boltzmann superposition principle* holds for the linear viscoelastic materials only. The superposition principle states that resultant response of the system $\varepsilon(t)$ under the "sum" of causes is equal to the "sum" of responses corresponding to causes acting separately. In particular if stress σ_1 is applied at time ξ_1 and, then, stress σ_2 is applied at time ξ_2, the resultant strain $\varepsilon(t)$ at any time $t > \xi_2$ is represented as the sum of the strains resulting from both stresses considered as though each were acting separately

$$\varepsilon\left[\sigma_1\left(t-\xi_1\right)+\sigma_2\left(t-\xi_2\right)\right] = \varepsilon\left[\sigma_1\left(t-\xi_1\right)\right] + \varepsilon\left[\sigma_2\left(t-\xi_2\right)\right] \tag{2.40}$$

In case of arbitrary loading history, stress $\sigma(t)$ can be approximated by a sum of n stress inputs $\Delta\sigma_i$, hence from the Boltzmann principle the strain output holds

$$\varepsilon(t) = \sum_{i=1}^{n} \varepsilon_i(t-\xi_i) = \sum_{i=1}^{n} J(t-\xi_i)\Delta\sigma_i \tag{2.41}$$

If the time step tends to zero, we arrive at the *integral form of uniaxial creep strain* for the linear viscoelastic material

$$\varepsilon(t) = \int_0^t J(t-\xi)\frac{\partial\sigma(\xi)}{\partial\xi}\mathrm{d}\xi = \int_0^t J(t-\xi)\dot{\sigma}(\xi)\mathrm{d}\xi \tag{2.42}$$

If the analogous reasoning is applied to the arbitrary strain history (kinematic control), we arrive at the *integral form of* the *uniaxial stress relaxation* for the linear viscoelastic material

$$\sigma(t) = \int_0^t E(t-\xi)\frac{\partial\varepsilon(\xi)}{\partial\xi}\mathrm{d}\xi = \int_0^t E(t-\xi)\dot{\varepsilon}(\xi)\mathrm{d}\xi \tag{2.43}$$

In above integral equations, $J(t-\xi)$ and $E(t-\xi)$ denote the *creep function* and the *relaxation function* of the material considered, respectively. In practical applications, the alternative forms to (2.42) or (2.43) are more convenient

$$\varepsilon(t) = J_0\sigma(t) + \int_0^t \varphi(t-\xi)\dot{\sigma}(\xi)\mathrm{d}\xi \tag{2.44}$$

or

$$\sigma(t) = E_0\varepsilon(t) - \int_0^t \psi(t-\xi)\dot{\varepsilon}(\xi)\mathrm{d}\xi \tag{2.45}$$

where separation of the instantaneous and time-dependent outputs are distinguished. The general integral forms (2.42) or (2.43) do not comprise explicitly initial conditions, whereas in the forms (2.44) or (2.45) J_0 or E_0 denote *initial* value of *creep* or *relaxation functions* (at $t = 0$) whereas time functions $\varphi(t-\xi)$ or $\psi(t-\xi)$ denote *time-dependent* parts of *creep* or *relaxation functions*. Note that in Eqs. (2.44) and (2.45) symbol ξ denotes time when the stress or the strain inputs are imposed, whereas t denotes the observation time when strain response $\varepsilon(t)$ or stress response $\sigma(t)$ are measured. This approach can be identified as the *linear hereditary model* where kernel function depends on time interval $t-\xi$ by contrast to the *nonlinear hereditary models* where kernel functions depend on t, ξ separately, cf. Rabotnov [11].

2.3 Multiaxial Isotropic Linear Viscoelastic Materials

In what follows, we briefly summarize the fundamentals of the *linear viscoelasticity* in case of material *isotropy*. This will serve as the starting point for further extension of isotropic to anisotropic linear viscoelastic equations. Such extension will further be used as convenient tool for analysis of anisotropic viscoelastic composites.

2.3.1 Differential Representation Under a Multiaxial Stress State

In case of *isotropic linear viscoelastic materials* under the *multiaxial states* of stress and strain, it is convenient to separate *volumetric effect* from the *shape change effect*. Similar to elasticity, such separation is possible only in case of *material isotropy* (see Sect. 1.4.8).

Direct extension of *linear isotropic viscoelastic constitutive equations* (2.36) and (2.37) to the *multiaxial states* takes the form

$$\begin{aligned} \mathrm{P}_1 s_{ij}(t) &= \mathrm{Q}_1 e_{ij}(t) \\ \mathrm{P}_2 \sigma_{kk}(t) &= \mathrm{Q}_2 \varepsilon_{kk}(t) \end{aligned} \tag{2.46}$$

where P_1, Q_1, P_2, and Q_2 are the *linear differential operators* applicable to the separable *shape change* and the *volume change* effects. In the explicit format, equations (2.46) can be rewritten as

$$\begin{aligned} &\left(p_0' + p_1'\frac{\partial}{\partial t} + p_2'\frac{\partial^2}{\partial t^2} + \cdots + p_a'\frac{\partial^a}{\partial t^a}\right) s_{ij}(t) = \\ &\quad \left(q_0' + q_1'\frac{\partial}{\partial t} + q_2'\frac{\partial^2}{\partial t^2} + \cdots + q_b'\frac{\partial^b}{\partial t^b}\right) e_{ij}(t) \\ &\left(p_0'' + p_1''\frac{\partial}{\partial t} + p_2''\frac{\partial^2}{\partial t^2} + \cdots + p_a''\frac{\partial^a}{\partial t^a}\right) \sigma_{kk}(t) = \\ &\quad \left(q_0'' + q_1''\frac{\partial}{\partial t} + q_2''\frac{\partial^2}{\partial t^2} + \cdots + q_b''\frac{\partial^b}{\partial t^b}\right) \varepsilon_{kk}(t) \end{aligned} \tag{2.47}$$

For the purpose of further consideration, it is convenient to transform differential equations (2.47) expressed in terms of physical time t, $f(t)$ to the equivalent algebraic equations expressed in terms of transformed variables s, $\widehat{f}(s)$ according to the *Laplace integral transform*

$$\mathcal{L}\{f(t)\} = \widehat{f}(s) = \int_0^\infty \mathrm{e}^{-st} f(t)\,\mathrm{d}t \tag{2.48}$$

The use of above transformation allows to replace the real initial differential problem of the *linear viscoelastic material* (differential equation and appropriate initial conditions) by the *equivalent elastic* algebraic equation of a *fictitious elastic material.*

In the next step, when the fictitious algebraic problem is solved in elementary way, application of the *inverse Laplace transform* allows to return to the original viscoelastic problem. Described procedure leads to the solution faster than the straightforward integration of a differential equation due to the *Laplace transform pairs* known from literature.

Basic properties of the Laplace transform commonly used in theory of viscoelasticity can be found among others in, e.g., Nowacki [8], Pipkin [9], Findley et al. [4]. Exemplary Laplace transforms for selected elementary functions $f(t)$ are shown in Table 2.2.

By use of the Laplace transformation *equations of transformed isotropic linear viscoelasticity* (2.46) can be expressed in terms of the *transformed variable* s as follows

$$\begin{aligned} \widehat{\mathrm{P}}_1 \widehat{s}_{ij}\,(s) &= \widehat{\mathrm{Q}}_1 \widehat{e}_{ij}\,(s) \\ \widehat{\mathrm{P}}_2 \widehat{\sigma}_{kk}\,(s) &= \widehat{\mathrm{Q}}_2 \widehat{\varepsilon}_{kk}\,(s) \end{aligned} \tag{2.49}$$

Table 2.2 Laplace transforms of frequently used functions

$f(t)$	$\widehat{f}(s)$	$f(t)$	$\widehat{f}(s)$
$\dot{f}(t)$	$s\widehat{f}(s) - f(0)$	$\int_0^t f(\xi)\mathrm{d}\xi$	$\dfrac{\widehat{f}(s)}{s}$
1	$\dfrac{1}{s}$	a	$\dfrac{a}{s}$
$\mathrm{H}(t)$	$\dfrac{1}{s}$	$\mathrm{H}(t-a)$	$\dfrac{\mathrm{e}^{-as}}{s}$
$\delta(t) = \dot{\mathrm{H}}(t)$	1	$\delta(t-a)$	e^{-as}
t	$\dfrac{1}{s^2}$	t^n	$\dfrac{n!}{s^{n+1}}$
e^{-at}	$\dfrac{1}{s+a}$	$t^n\mathrm{e}^{-at}$	$\dfrac{n!}{(s+a)^{n+1}}$
$\mathrm{e}^{-at} - \mathrm{e}^{-bt}$	$\dfrac{b-a}{(s+a)(s+b)}$	$a\mathrm{e}^{-at} - b\mathrm{e}^{-bt}$	$\dfrac{(a-b)s}{(s+a)(s+b)}$
$1 - \mathrm{e}^{-at}$	$\dfrac{a}{s(s+a)}$	$\dfrac{t}{a} - \dfrac{1}{a^2}(1 - \mathrm{e}^{-at})$	$\dfrac{1}{s^2(s+a)}$

or

$$
\begin{aligned}
&\left(p'_0 + p'_1 s + p'_2 s^2 + \cdots + p'_a s^a\right) \widehat{s}_{ij}(s) = \\
&\qquad \left(q'_0 + q'_1 s + q'_2 s^2 + \cdots + q'_b s^b\right) \widehat{e}_{ij}(s) \\
&\left(p''_0 + p''_1 s + p''_2 s^2 + \cdots + p''_a s^a\right) \widehat{\sigma}_{kk}(s) = \\
&\qquad \left(q''_0 + q''_1 s + q''_2 s^2 + \cdots + q''_b s^b\right) \widehat{\varepsilon}_{kk}(s)
\end{aligned}
\tag{2.50}
$$

Based on the analogy between Eq. (2.49) that describe transformed viscoelastic problem and linear isotropic elastic equations

$$
s_{ij} = 2Ge_{ij}, \qquad \sigma_{kk} = 3K\varepsilon_{kk} \tag{2.51}
$$

it is possible to find out the *generalized modules of viscoelasticity* $\mathrm{G_{ve}}(t)$ and $\mathrm{K_{ve}}(t)$ as

$$
\mathrm{G_{ve}}(t) = \frac{\mathrm{Q_1}}{2\mathrm{P_1}} \qquad \mathrm{K_{ve}}(t) = \frac{\mathrm{Q_2}}{3\mathrm{P_2}} \tag{2.52}
$$

which are time-dependent functions of t. Additionally, if the definitions of Young's modulus E and Poisson's ratio ν known from the elasticity are used

$$
E = \frac{9KG}{3K + G}, \qquad \nu = \frac{3K - 2G}{6K + 2G} \tag{2.53}
$$

substitution of (2.52) to (2.53) furnishes the *generalized Young's modulus* $\mathrm{E_{ve}}(t)$ commonly called *relaxation modulus* and *generalized Poisson's ratio* $\nu_{\mathrm{ve}}(t)$ of *isotropic linear viscoelasticity* that can be expressed in terms of time-dependent operators (2.46)

$$
\begin{aligned}
\mathrm{E_{ve}}(t) &= \frac{3\frac{\mathrm{Q_1}}{\mathrm{P_1}}\frac{\mathrm{Q_2}}{\mathrm{P_2}}}{\frac{\mathrm{Q_1}}{\mathrm{P_1}} + 2\frac{\mathrm{Q_2}}{\mathrm{P_2}}} = \frac{3\mathrm{Q_1Q_2}}{\mathrm{P_2Q_1} + 2\mathrm{P_1Q_2}} \\
\nu_{\mathrm{ve}}(t) &= \frac{\frac{\mathrm{Q_2}}{\mathrm{P_2}} - \frac{\mathrm{Q_1}}{\mathrm{P_1}}}{2\frac{\mathrm{Q_2}}{\mathrm{P_2}} + \frac{\mathrm{Q_1}}{\mathrm{P_1}}} = \frac{\mathrm{P_1Q_2} - \mathrm{P_2Q_1}}{\mathrm{P_2Q_1} + 2\mathrm{P_1Q_2}}
\end{aligned}
\tag{2.54}
$$

By contrast to elasticity, the above modules are *time-dependent differential operators* but not material constants.

The *deviatoric* $\mathrm{P_1}, \mathrm{Q_1}$ and the *volumetric* $\mathrm{P_2}, \mathrm{Q_2}$ *differential operators* and the corresponding *transformed operators* $\widehat{\mathrm{P}}_1, \widehat{\mathrm{Q}}_1$ and $\widehat{\mathrm{P}}_2, \widehat{\mathrm{Q}}_2$ for selected isotropic linear viscoelastic models are given in Table 2.3. When the additional assumption of *hydrostatic pressure independence* of the elastic response is used it is necessary to consequently apply $\mathrm{P_2} = 1, \mathrm{Q_2} = 3K$. Note that the above Eqs. (2.52) and (2.54) refer to *isotropic linear viscoelastic material* for which number of independent generalized modules equals 2, namely $\mathrm{G_{ve}}(t)$ and $\mathrm{K_{ve}}(t)$ or equivalently $\mathrm{E_{ve}}(t)$ and $\nu_{\mathrm{ve}}(t)$. In particular case of isotropic elasticity, the above creep modules reduce to two constants $\mathrm{G_{ve}}(t) = G$ and $\mathrm{K_{ve}}(t) = K$ or equivalently $\mathrm{E_{ve}}(t) = E$ and $\nu_{\mathrm{ve}}(t) = \nu$.

Table 2.3 Differential operators for seclected simple linear viscoelastic models (after Findley et al. [4])

Model	Hooke	Maxwell	Voigt–Kelvin	Standard	Burgers
Different equation	$\sigma = E\varepsilon$	$\sigma + \frac{\eta}{E}\dot{\sigma} = \eta\dot{\varepsilon}$	$\sigma = E\varepsilon + \eta\dot{\varepsilon}$	$\sigma + p_1\dot{\sigma} = q_0\varepsilon + q_1\dot{\varepsilon}$	$\sigma + p_1\dot{\sigma} + p_2\ddot{\sigma} = q_1\dot{\varepsilon} + q_2\ddot{\varepsilon}$
Deviatoric operators					
P_1	1	$1 + \frac{\eta'}{E'}\frac{\partial}{\partial t}$	1	$1 + \frac{\eta'}{E_1'+E'}\frac{\partial}{\partial t}$	$1 + p_1'\frac{\partial}{\partial t} + p_2'\frac{\partial^2}{\partial t^2}$
Q_1	$2G$	$\eta'\frac{\partial}{\partial t}$	$E' + \eta'\frac{\partial}{\partial t}$	$\frac{E_1'E'}{E_1'+E'} + \frac{\eta' E'}{E_1'+E'}\frac{\partial}{\partial t}$	$q_1'\frac{\partial}{\partial t} + q_2'\frac{\partial^2}{\partial t^2}$
Volumetric operators					
P_2	1	$1 + \frac{\eta''}{E''}\frac{\partial}{\partial t}$	1	$1 + \frac{\eta''}{E_1''+E''}\frac{\partial}{\partial t}$	$1 + p_1''\frac{\partial}{\partial t} + p_2''\frac{\partial^2}{\partial t^2}$
Q_2	$3K$	$\eta''\frac{\partial}{\partial t}$	$E'' + \eta''\frac{\partial}{\partial t}$	$\frac{E_1''E''}{E_1''+E''} + \frac{\eta'' E''}{E_1''+E''}\frac{\partial}{\partial t}$	$q_1''\frac{\partial}{\partial t} + q_2''\frac{\partial^2}{\partial t^2}$
Transformed deviatoric operators					
$\widehat{\mathrm{P}}_1$	1	$1 + \frac{\eta'}{E'}s$	1	$1 + \frac{\eta'}{E_1'+E'}s$	$1 + p_1's + p_2's^2$
$\widehat{\mathrm{Q}}_1$	$2G$	$\eta' s$	$E' + \eta' s$	$\frac{E_1'E'}{E_1'+E'} + \frac{\eta' E'}{E_1'+E'}s$	$q_1's + q_2's^2$
Transformed volumetric operators					
$\widehat{\mathrm{P}}_2$	1	$1 + \frac{\eta''}{E''}s$	1	$1 + \frac{\eta''}{E_1''+E''}s$	$1 + p_1''s + p_2''s^2$
$\widehat{\mathrm{Q}}_2$	$3K$	$\eta'' s$	$E'' + \eta'' s$	$\frac{E_1''E''}{E_1''+E''} + \frac{\eta'' E''}{E_1''+E''}s$	$q_1''s + q_2''s^2$

2.3.2 Integral Representation Under a Multiaxial Stress State

As mentioned above, in case of isotropic viscoelastic materials number of the material time functions is equal to two. Hence, it is possible to separate effects of *shape change* from the *volume change*. In such a way, we arrive at the *integral form of* the *constitutive equations of isotropic linear viscoelastic material* that generalize the analogous uniaxial equation (2.43) for *multiaxial states*

$$
\begin{aligned}
s_{ij}(t) &= 2\int_0^t \mathrm{G}_{\mathrm{ve}}(t-\xi)\,\frac{\partial e_{ij}(\xi)}{\partial \xi}\mathrm{d}\xi \\
\sigma_{kk}(t) &= 3\int_0^t \mathrm{K}_{\mathrm{ve}}(t-\xi)\,\frac{\partial \varepsilon_{kk}(\xi)}{\partial \xi}\mathrm{d}\xi
\end{aligned}
\tag{2.55}
$$

In a particular case, if the *volume change* is pure *elastic*, Eq. (2.55) take the simplified form

$$
\begin{aligned}
s_{ij}(t) &= 2\int_0^t \mathrm{G}_{\mathrm{ve}}(t-\xi)\,\frac{\partial e_{ij}(\xi)}{\partial \xi}\mathrm{d}\xi \\
\sigma_{kk} &= 3K\varepsilon_{kk}
\end{aligned}
\tag{2.56}
$$

2.4 Elastic-Viscoelastic Correspondence Principle for the Case of Isotropic Materials

Consider at the beginning, a particular case of *isotropic linear viscoelastic behavior* for which *separation of* the *volume change from* the *shape change* holds in a similar fashion as in case of isotropic elastic behavior, see Sect. 1.4.8. Remember however that in a more general case of the anisotropic behavior, linear elastic, and linear viscoelastic, this separation is not possible, see Sect. 1.8.

Analogy between the transformed *equations of linear isotropic viscoelastic materials* (2.49) and conventional *equations of isotropic elasticity* (2.51) leads to the searching of the solutions of viscoelasticity on basis of a priori known *coupled elastic problems*. This analogy is known as the *elastic-viscoelastic correspondence principle*, see Findley et al. [4].

Let us summarize a complete set of *equations of linear isotropic viscoelasticity*, see Skrzypek [13]

- the *equilibrium equations*

$$
\frac{\partial \sigma_{ij}(\boldsymbol{x},t)}{\partial x_i} + b_j(t) = 0 \tag{2.57}
$$

- the *constitutive equations* formed either in the format of *differential operator representation* (2.46)

$$
\begin{aligned}
\mathrm{P}_1 s_{ij}(\boldsymbol{x}, t) &= \mathrm{Q}_1 e_{ij}(\boldsymbol{x}, t) \\
\mathrm{P}_2 \sigma_{kk}(\boldsymbol{x}, t) &= \mathrm{Q}_2 \varepsilon_{kk}(\boldsymbol{x}, t)
\end{aligned}
\tag{2.58}
$$

or in the *integral representation* (2.55)

$$
\begin{aligned}
s_{ij}(\boldsymbol{x}, t) &= 2\int_0^t \mathrm{G}_{\mathrm{ev}}(t-\xi)\,\dot{e}_{ij}(\boldsymbol{x}, \xi)\,\mathrm{d}\xi \\
\sigma_{kk}(\boldsymbol{x}, t) &= 3\int_0^t \mathrm{K}_{\mathrm{ev}}(t-\xi)\,\dot{\varepsilon}_{kk}(\boldsymbol{x}, \xi)\,\mathrm{d}\xi
\end{aligned}
\tag{2.59}
$$

- the *linearized geometric equations*

$$
\varepsilon_{ij}(\boldsymbol{x}, t) = \frac{1}{2}\left[\frac{\partial u_i(\boldsymbol{x}, t)}{\partial x_j} + \frac{\partial u_j(\boldsymbol{x}, t)}{\partial x_i}\right]
\tag{2.60}
$$

- the *boundary conditions* under assumption that boundary between domains of force Γ_P and displacement Γ_U remain unchanged

$$
\begin{aligned}
P_i(\boldsymbol{x}, t) &= \sigma_{ij}(\boldsymbol{x}, t)\, n_j \quad \text{na } \Gamma_P \\
U_i(\boldsymbol{x}, t) &= u_i(\boldsymbol{x}, t) \qquad \text{na } \Gamma_U
\end{aligned}
\tag{2.61}
$$

For simplicity independence of the viscoelastic modules, G_{ve}, K_{ve} from the spatial coordinates holds. In other words, *material homogeneity* is assumed. In a particular case of *composite materials* although that material is inhomogeneous at *microlevel* a *homogenization technique* allows to reduce such problem to *homogeneous at* the *RUC level* of the representative unit cell, see Chap. 3.

When the *Laplace transformation* of the above set of Eqs. (2.57)–(2.61) is done we arrive at the *fictitious coupled elastic problem*

$$
\begin{aligned}
&\frac{\partial \widehat{\sigma}_{ij}(\boldsymbol{x}, s)}{\partial x_i} + \widehat{b}_j(\boldsymbol{x}, s) = 0 \\
&\widehat{s}_{ij}(\boldsymbol{x}, s) = 2s\widehat{\mathrm{G}}\widehat{e}_{ij}(\boldsymbol{x}, s) = \frac{\widehat{\mathrm{Q}}_1}{\widehat{\mathrm{P}}_1}\widehat{e}_{ij}(\boldsymbol{x}, s) \\
&\widehat{\sigma}_{kk}(\boldsymbol{x}, s) = 3s\widehat{\mathrm{K}}\widehat{\varepsilon}_{kk}(s) = \frac{\widehat{\mathrm{Q}}_2}{\widehat{\mathrm{P}}_2}\widehat{\varepsilon}_{kk}(\boldsymbol{x}, s) \\
&\widehat{\varepsilon}_{ij}(\boldsymbol{x}, s) = \frac{1}{2}\left[\frac{\partial \widehat{u}_i(\boldsymbol{x}, s)}{\partial x_j} + \frac{\partial \widehat{u}_j(\boldsymbol{x}, s)}{\partial x_i}\right] \\
&\widehat{P}_i(\boldsymbol{x}, s) = \widehat{\sigma}_{ji}(\boldsymbol{x}, s)\, n_j \quad \text{at } \Gamma_P \\
&\widehat{U}_i(\boldsymbol{x}, s) = \widehat{u}_i(\boldsymbol{x}, s) \qquad \text{at } \Gamma_U
\end{aligned}
\tag{2.62}
$$

in which the *body forces* $\widehat{b}_j(\boldsymbol{x}, s)$, *external forces* $\widehat{P}_i(\boldsymbol{x}, s)$, *displacements* $\widehat{U}_i(\boldsymbol{x}, s)$ as well as *fictitious elastic constants* $\widehat{\mathrm{G}}(s)$ and $\widehat{\mathrm{K}}(s)$ are *functions of* the *transformed variable* s.

Finally, the analogy between viscoelastic and coupled elastic problems can be formulated. Namely if a solution of *coupled fictitious elastic problem* is known (2.62) $\widehat{\sigma}_{ij}\,(\boldsymbol{x}, s)$ and $\widehat{u}_i\,(\boldsymbol{x}, s)$, the solution of corresponding *linear viscoelastic problem* (2.57)–(2.61) can be obtained on the way of the *inverse Laplace transformations* $\sigma_{ij}\,(\boldsymbol{x}, t)$ and $u_i\,(\boldsymbol{x}, t)$. Simultaneously, following relations must hold

$$\widehat{\mathrm{G}}_{\mathrm{ve}} = \frac{\widehat{\mathrm{Q}}_1\,(s)}{2s\widehat{\mathrm{P}}_1\,(s)} \qquad \widehat{\mathrm{K}}_{\mathrm{ve}} = \frac{\widehat{\mathrm{Q}}_2\,(s)}{3s\widehat{\mathrm{P}}_2\,(s)} \tag{2.63}$$

The correspondence principle can be applied only to the boundary problems where the interface between the boundary Γ_P (where the external forces are prescribed) and the boundary Γ_U (where the surface displacements are given) is independent of time, see Findley et al. [4]. The above limitation does not hold in case of some material forming processes, for instance rolling, where the interface between boundaries Γ_P and Γ_U varies with time.

An example of elastic-viscoelastic correspondence principle applied to multiaxial stress and strain states the *thick walled tube* made of the *isotropic standard material* subject to *internal pressure* $p(t) = p\mathrm{H}(t)$ applied instantaneously at $t = 0$ is considered after Findley et al. [4]. Taking advantage of the correspondence principle and recalling *Lamé's solution* for coupled elastic problem

$$u^{\mathrm{e}} = \frac{pa^2}{b^2 - a^2}\frac{1+\nu}{E}\left(\frac{b^2}{r} + \frac{1-\nu}{1+\nu}r\right) \tag{2.64}$$

substitution of (2.54) for E, ν in (2.64) gives

$$\begin{aligned} \frac{1+\nu}{E} &\longrightarrow \frac{\widehat{\mathrm{P}}_1}{\widehat{\mathrm{Q}}_1} \\ \frac{1+\nu}{1-\nu} &\longrightarrow \frac{2\widehat{\mathrm{P}}_2\widehat{\mathrm{Q}}_1 + \widehat{\mathrm{P}}_1\widehat{\mathrm{Q}}_2}{3\widehat{\mathrm{P}}_1\widehat{\mathrm{Q}}_2} \end{aligned} \tag{2.65}$$

In this way, a *fictitious coupled elastic problem* in term of s

$$\widehat{u}(s) = \frac{\widehat{p}(s)a^2}{b^2 - a^2}\frac{\widehat{\mathrm{P}}_1}{\widehat{\mathrm{Q}}_1}\left(\frac{b^2}{r} + \frac{2\widehat{\mathrm{P}}_2\widehat{\mathrm{Q}}_1 + \widehat{\mathrm{P}}_1\widehat{\mathrm{Q}}_2}{3\widehat{\mathrm{P}}_1\widehat{\mathrm{Q}}_2}r\right) \tag{2.66}$$

is find out. Applying *transformed operators* $\widehat{\mathrm{P}}_1, \widehat{\mathrm{P}}_2, \widehat{\mathrm{Q}}_1$, and $\widehat{\mathrm{Q}}_2$ *of standard model* (see Table 2.3) under additional assumption that shape change creep deformation is accompanied by the elastic hydrostatic deformation $\widehat{\mathrm{P}}_2 = 1$, $\widehat{\mathrm{Q}}_2 = 3K$, we find a solution of the fictitious coupled elastic problem

$$\widehat{u}(s) = \frac{pa^2}{b^2 - a^2}\left(\frac{A}{\frac{E_1'}{\eta'}} + \frac{B}{s}\right) \tag{2.67}$$

where A and B are functions of radial coordinate r exclusively

$$A = -\frac{1}{E'}\left(\frac{b^2}{r} + \frac{r}{3}\right), \qquad B = \frac{E_1' + E'}{E_1' E'}\left(\frac{b^2}{r} + \frac{r}{3}\right) + \frac{2r}{9K} \tag{2.68}$$

Finally, the solution of the real linear viscoelastic problem is achieved by use of the *inverse Laplace transform* of (2.67) in the following format

$$\begin{aligned} u(t) = \frac{pa^2}{b^2 - a^2}&\left\{\frac{1}{E'}\left(\frac{b^2}{r} + \frac{r}{3}\right) + \frac{2r}{9K} + \frac{1}{E_1'}\left(\frac{b^2}{r} + \frac{r}{3}\right)\right. \\ &\left.\times\left[1 - \exp\left(-\frac{E_1'}{\eta'}t\right)\right]\right\} \end{aligned} \tag{2.69}$$

2.5 Integral Representation of the Linear Viscoelastic Equations of Anisotropic Materials

The *elastic-viscoelastic correspondence principle* applied *for isotropic material* presented in previous Sect. 2.4, was based on mathematically convenient separation of the volumetric and the shape change effects from total viscoelastic deformation. However, in case of any class of *material anisotropy* such separation does not occur. Hence for sake of generality, we change formulation of the correspondence principle to the uniform fashion that does not employ the above separation. For convenience, the *vector-matrix notation* will be used.

In a general case of *anisotropic linear viscoelastic material*, the *integral form of constitutive equations* is furnished as (see Shu and Onat [12])

$$\varepsilon_{ij}(t) = \int_0^t {}^{\mathrm{ve}}J_{ijkl}(t - \xi)\,\dot{\sigma}_{kl}(\xi)\,\mathrm{d}\xi \tag{2.70}$$

or

$$\sigma_{ij}(t) = \int_0^t {}^{\mathrm{ve}}E_{ijkl}(t - \xi)\dot{\varepsilon}_{kl}(\xi)\mathrm{d}\xi \tag{2.71}$$

where ${}^{\mathrm{ve}}J_{ijkl}(t - \xi)$ defines the *fourth-rank tensor of creep functions*; whereas, ${}^{\mathrm{ve}}E_{ijkl}(t - \xi)$ is the *fourth-rank tensor of relaxation functions* which characterize viscoelastic properties of anisotropic material. Assuming the symmetry conditions: ${}^{\mathrm{ve}}J_{ijkl} = {}^{\mathrm{ve}}J_{klij} = {}^{\mathrm{ve}}J_{jikl} = {}^{\mathrm{ve}}J_{ijlk}$, or ${}^{\mathrm{ve}}E_{ijkl} = {}^{\mathrm{ve}}E_{klij} = {}^{\mathrm{ve}}E_{jikl} = {}^{\mathrm{ve}}E_{ijlk}$, both *tensors of viscoelastic anisotropy* have 21 independent functions. Both constitutive tensor functions ${}^{\mathrm{ve}}J_{ijkl}$ or ${}^{\mathrm{ve}}E_{ijkl}$ depend on current time t (integration limit).

When the *vector-matrix notation* is used, the *general constitutive equation of anisotropic linear viscoelastic material* defined by Eqs. (2.70) and (2.71) takes equivalent *integral form*

$$\{\varepsilon(t)\} = \int_0^t [^{\mathrm{ve}}\mathrm{J}(t-\xi)]\frac{\partial}{\partial\xi}\{\boldsymbol{\sigma}(\xi)\}\,\mathrm{d}\xi \tag{2.72}$$

or

$$\{\boldsymbol{\sigma}(t)\} = \int_0^t [^{\mathrm{ve}}\mathrm{E}(t-\xi)]\frac{\partial}{\partial\xi}\{\varepsilon(\xi)\}\,\mathrm{d}\xi \tag{2.73}$$

When nonabbreviated notation is used introducing matrix of creep compliance functions ${}^{\mathrm{ve}}J_{ij}(t-\xi)$, we arrive at the following *constitutive integral equations of anisotropic linear material*

$$\begin{Bmatrix} \varepsilon_{xx}(t) \\ \varepsilon_{yy}(t) \\ \varepsilon_{zz}(t) \\ \gamma_{yz}(t) \\ \gamma_{zx}(t) \\ \gamma_{xy}(t) \end{Bmatrix} = \int_0^t \left[\begin{array}{ccc|ccc} {}^{\mathrm{ve}}J_{11} & {}^{\mathrm{ve}}J_{12} & {}^{\mathrm{ve}}J_{13} & {}^{\mathrm{ve}}J_{14} & {}^{\mathrm{ve}}J_{15} & {}^{\mathrm{ve}}J_{16} \\ {}^{\mathrm{ve}}J_{21} & {}^{\mathrm{ve}}J_{22} & {}^{\mathrm{ve}}J_{23} & {}^{\mathrm{ve}}J_{24} & {}^{\mathrm{ve}}J_{25} & {}^{\mathrm{ve}}J_{26} \\ {}^{\mathrm{ve}}J_{31} & {}^{\mathrm{ve}}J_{32} & {}^{\mathrm{ve}}J_{33} & {}^{\mathrm{ve}}J_{34} & {}^{\mathrm{ve}}J_{35} & {}^{\mathrm{ve}}J_{36} \\ \hline {}^{\mathrm{ve}}J_{41} & {}^{\mathrm{ve}}J_{42} & {}^{\mathrm{ve}}J_{43} & {}^{\mathrm{ve}}J_{44} & {}^{\mathrm{ve}}J_{45} & {}^{\mathrm{ve}}J_{46} \\ {}^{\mathrm{ve}}J_{51} & {}^{\mathrm{ve}}J_{52} & {}^{\mathrm{ve}}J_{53} & {}^{\mathrm{ve}}J_{54} & {}^{\mathrm{ve}}J_{55} & {}^{\mathrm{ve}}J_{56} \\ {}^{\mathrm{ve}}J_{61} & {}^{\mathrm{ve}}J_{62} & {}^{\mathrm{ve}}J_{63} & {}^{\mathrm{ve}}J_{64} & {}^{\mathrm{ve}}J_{65} & {}^{\mathrm{ve}}J_{66} \end{array}\right] \begin{Bmatrix} \dot\sigma_{xx}(\xi) \\ \dot\sigma_{yy}(\xi) \\ \dot\sigma_{zz}(\xi) \\ \dot\tau_{yz}(\xi) \\ \dot\tau_{zx}(\xi) \\ \dot\tau_{xy}(\xi) \end{Bmatrix} \mathrm{d}\xi \tag{2.74}$$

where $[{}^{\mathrm{ve}}J]_{ij} = [J(t-\xi)]_{ij}$ is the creep compliance matrix. For Eq. (2.73) the inverse relation holds

$$\begin{Bmatrix} \sigma_{xx}(t) \\ \sigma_{yy}(t) \\ \sigma_{zz}(t) \\ \tau_{yz}(t) \\ \tau_{zx}(t) \\ \tau_{xy}(t) \end{Bmatrix} = \int_0^t \left[\begin{array}{ccc|ccc} {}^{\mathrm{ve}}E_{11} & {}^{\mathrm{ve}}E_{12} & {}^{\mathrm{ve}}E_{13} & {}^{\mathrm{ve}}E_{14} & {}^{\mathrm{ve}}E_{15} & {}^{\mathrm{ve}}E_{16} \\ {}^{\mathrm{ve}}E_{21} & {}^{\mathrm{ve}}E_{22} & {}^{\mathrm{ve}}E_{23} & {}^{\mathrm{ve}}E_{24} & {}^{\mathrm{ve}}E_{25} & {}^{\mathrm{ve}}E_{26} \\ {}^{\mathrm{ve}}E_{31} & {}^{\mathrm{ve}}E_{32} & {}^{\mathrm{ve}}E_{33} & {}^{\mathrm{ve}}E_{34} & {}^{\mathrm{ve}}E_{35} & {}^{\mathrm{ve}}E_{36} \\ \hline {}^{\mathrm{ve}}E_{41} & {}^{\mathrm{ve}}E_{42} & {}^{\mathrm{ve}}E_{43} & {}^{\mathrm{ve}}E_{44} & {}^{\mathrm{ve}}E_{45} & {}^{\mathrm{ve}}E_{46} \\ {}^{\mathrm{ve}}E_{51} & {}^{\mathrm{ve}}E_{52} & {}^{\mathrm{ve}}E_{53} & {}^{\mathrm{ve}}E_{54} & {}^{\mathrm{ve}}E_{55} & {}^{\mathrm{ve}}E_{56} \\ {}^{\mathrm{ve}}E_{61} & {}^{\mathrm{ve}}E_{62} & {}^{\mathrm{ve}}E_{63} & {}^{\mathrm{ve}}E_{64} & {}^{\mathrm{ve}}E_{65} & {}^{\mathrm{ve}}E_{66} \end{array}\right] \begin{Bmatrix} \dot\varepsilon_{xx}(\xi) \\ \dot\varepsilon_{yy}(\xi) \\ \dot\varepsilon_{zz}(\xi) \\ \dot\gamma_{yz}(\xi) \\ \dot\gamma_{zx}(\xi) \\ \dot\gamma_{xy}(\xi) \end{Bmatrix} \mathrm{d}\xi \tag{2.75}$$

In particular case of *orthotropic linear viscoelastic material*, Eqs. (2.74) and (2.75) reduce to narrower forms

$$\begin{Bmatrix} \varepsilon_{11}(t) \\ \varepsilon_{22}(t) \\ \varepsilon_{33}(t) \\ \gamma_{23}(t) \\ \gamma_{31}(t) \\ \gamma_{12}(t) \end{Bmatrix} = \int_0^t \left[\begin{array}{ccc|ccc} {}^{\mathrm{ve}}J_{11} & {}^{\mathrm{ve}}J_{12} & {}^{\mathrm{ve}}J_{13} & & & \\ {}^{\mathrm{ve}}J_{21} & {}^{\mathrm{ve}}J_{22} & {}^{\mathrm{ve}}J_{23} & & & \\ {}^{\mathrm{ve}}J_{31} & {}^{\mathrm{ve}}J_{32} & {}^{\mathrm{ve}}J_{33} & & & \\ \hline & & & {}^{\mathrm{ve}}J_{44} & & \\ & & & & {}^{\mathrm{ve}}J_{55} & \\ & & & & & {}^{\mathrm{ve}}J_{66} \end{array}\right] \begin{Bmatrix} \dot\sigma_{11}(\xi) \\ \dot\sigma_{22}(\xi) \\ \dot\sigma_{33}(\xi) \\ \dot\tau_{23}(\xi) \\ \dot\tau_{31}(\xi) \\ \dot\tau_{12}(\xi) \end{Bmatrix} \mathrm{d}\xi \tag{2.76}$$

or

$$\begin{Bmatrix} \sigma_{11}(t) \\ \sigma_{22}(t) \\ \sigma_{33}(t) \\ \tau_{23}(t) \\ \tau_{31}(t) \\ \tau_{12}(t) \end{Bmatrix} = \int_0^t \left[\begin{array}{ccc|ccc} {}^{ve}E_{11} & {}^{ve}E_{12} & {}^{ve}E_{13} & & & \\ {}^{ve}E_{21} & {}^{ve}E_{22} & {}^{ve}E_{23} & & & \\ {}^{ve}E_{31} & {}^{ve}E_{32} & {}^{ve}E_{33} & & & \\ \hline & & & {}^{ve}E_{44} & & \\ & & & & {}^{ve}E_{55} & \\ & & & & & {}^{ve}E_{66} \end{array}\right] \begin{Bmatrix} \dot{\varepsilon}_{11}(\xi) \\ \dot{\varepsilon}_{22}(\xi) \\ \dot{\varepsilon}_{33}(\xi) \\ \dot{\gamma}_{23}(\xi) \\ \dot{\gamma}_{31}(\xi) \\ \dot{\gamma}_{12}(\xi) \end{Bmatrix} \mathrm{d}\xi \qquad (2.77)$$

being extension of equations of orthotropic linear elasticity (1.103). Note that both stresses and strains are functions of time ${}^{ve}\sigma_{ij} ={}^{ve} \sigma_{ij}(t)$, ${}^{ve}\varepsilon_{ij} ={}^{ve} \varepsilon_{ij}(t)$, in similar fashion as elements of creep compliance ${}^{ve}J_{ij} ={}^{ve} J_{ij}(t - \xi)$ and relaxation ${}^{ve}E_{ij} ={}^{ve} E_{ij}(t - \xi)$ matrices.

Applying the *Laplace transform* to Eqs. (2.74) and (2.75), we arrive at the *associated fictitious elastic constitutive equations* in the transformed domain

$$\begin{Bmatrix} \widehat{\varepsilon}_{xx}(s) \\ \widehat{\varepsilon}_{yy}(s) \\ \widehat{\varepsilon}_{zz}(s) \\ \widehat{\gamma}_{yz}(s) \\ \widehat{\gamma}_{zx}(s) \\ \widehat{\gamma}_{xy}(s) \end{Bmatrix} = s \left[\begin{array}{ccc|ccc} \widehat{J}_{11} & \widehat{J}_{12} & \widehat{J}_{13} & \widehat{J}_{14} & \widehat{J}_{15} & \widehat{J}_{16} \\ \widehat{J}_{21} & \widehat{J}_{22} & \widehat{J}_{23} & \widehat{J}_{24} & \widehat{J}_{25} & \widehat{J}_{26} \\ \widehat{J}_{31} & \widehat{J}_{32} & \widehat{J}_{33} & \widehat{J}_{34} & \widehat{J}_{35} & \widehat{J}_{36} \\ \hline \widehat{J}_{41} & \widehat{J}_{42} & \widehat{J}_{43} & \widehat{J}_{44} & \widehat{J}_{45} & \widehat{J}_{46} \\ \widehat{J}_{51} & \widehat{J}_{52} & \widehat{J}_{53} & \widehat{J}_{54} & \widehat{J}_{55} & \widehat{J}_{56} \\ \widehat{J}_{61} & \widehat{J}_{62} & \widehat{J}_{63} & \widehat{J}_{64} & \widehat{J}_{65} & \widehat{J}_{66} \end{array}\right] \begin{Bmatrix} \widehat{\sigma}_{xx}(s) \\ \widehat{\sigma}_{yy}(s) \\ \widehat{\sigma}_{zz}(s) \\ \widehat{\tau}_{yz}(s) \\ \widehat{\tau}_{zx}(s) \\ \widehat{\tau}_{xy}(s) \end{Bmatrix} \qquad (2.78)$$

or

$$\begin{Bmatrix} \widehat{\sigma}_{xx}(s) \\ \widehat{\sigma}_{yy}(s) \\ \widehat{\sigma}_{zz}(s) \\ \widehat{\tau}_{yz}(s) \\ \widehat{\tau}_{zx}(s) \\ \widehat{\tau}_{xy}(s) \end{Bmatrix} = s \left[\begin{array}{ccc|ccc} \widehat{E}_{11} & \widehat{E}_{12} & \widehat{E}_{13} & \widehat{E}_{14} & \widehat{E}_{15} & \widehat{E}_{16} \\ \widehat{E}_{21} & \widehat{E}_{22} & \widehat{E}_{23} & \widehat{E}_{24} & \widehat{E}_{25} & \widehat{E}_{26} \\ \widehat{E}_{31} & \widehat{E}_{32} & \widehat{E}_{33} & \widehat{E}_{34} & \widehat{E}_{35} & \widehat{E}_{36} \\ \hline \widehat{E}_{41} & \widehat{E}_{42} & \widehat{E}_{43} & \widehat{E}_{44} & \widehat{E}_{45} & \widehat{E}_{46} \\ \widehat{E}_{51} & \widehat{E}_{52} & \widehat{E}_{53} & \widehat{E}_{54} & \widehat{E}_{55} & \widehat{E}_{56} \\ \widehat{E}_{61} & \widehat{E}_{62} & \widehat{E}_{63} & \widehat{E}_{64} & \widehat{E}_{65} & \widehat{E}_{66} \end{array}\right] \begin{Bmatrix} \widehat{\varepsilon}_{xx}(s) \\ \widehat{\varepsilon}_{yy}(s) \\ \widehat{\varepsilon}_{zz}(s) \\ \widehat{\gamma}_{yz}(s) \\ \widehat{\gamma}_{zx}(s) \\ \widehat{\gamma}_{xy}(s) \end{Bmatrix} \qquad (2.79)$$

2.6 Application of the Anisotropic Correspondence Principle to the Case of Orthotropic Composite Materials

For a purpose of engineering application, for instance to some *composite materials* in which at least one of phases exhibits *viscoelastic behavior*, it is sufficient to assume the narrower case of *orthotropic linear viscoelastic equations* (2.76) and (2.77). When the *Laplace transformation* is applied to the *integral constitutive equations of the orthotropic linear viscoelastic material* (2.76) and (2.77), we arrive at the associated *fictitious orthotropic elastic equations* in terms of *s*

$$\{\widehat{\sigma}(s)\} = s[\widehat{\mathrm{E}}(s)]\{\widehat{\varepsilon}(s)\} \tag{2.80}$$

or

$$\{\widehat{\varepsilon}(s)\} = s[\widehat{\mathrm{J}}(s)]\{\widehat{\sigma}(s)\} \tag{2.81}$$

When the *vector-matrix notation* is applied the corresponding formulas can be written in an expanded fashion

$$\begin{Bmatrix} \widehat{\sigma}_{11}(s) \\ \widehat{\sigma}_{22}(s) \\ \widehat{\sigma}_{33}(s) \\ \widehat{\tau}_{23}(s) \\ \widehat{\tau}_{31}(s) \\ \widehat{\tau}_{12}(s) \end{Bmatrix} = s \left[\begin{array}{ccc|ccc} \widehat{E}_{11} & \widehat{E}_{12} & \widehat{E}_{13} & 0 & 0 & 0 \\ \widehat{E}_{21} & \widehat{E}_{22} & \widehat{E}_{23} & 0 & 0 & 0 \\ \widehat{E}_{31} & \widehat{E}_{32} & \widehat{E}_{33} & 0 & 0 & 0 \\ \hline 0 & 0 & 0 & \widehat{E}_{44} & 0 & 0 \\ 0 & 0 & 0 & 0 & \widehat{E}_{55} & 0 \\ 0 & 0 & 0 & 0 & 0 & \widehat{E}_{66} \end{array}\right] \begin{Bmatrix} \widehat{\varepsilon}_{11}(s) \\ \widehat{\varepsilon}_{22}(s) \\ \widehat{\varepsilon}_{33}(s) \\ \widehat{\gamma}_{23}(s) \\ \widehat{\gamma}_{31}(s) \\ \widehat{\gamma}_{12}(s) \end{Bmatrix} \tag{2.82}$$

or

$$\begin{Bmatrix} \widehat{\varepsilon}_{11}(s) \\ \widehat{\varepsilon}_{22}(s) \\ \widehat{\varepsilon}_{33}(s) \\ \widehat{\gamma}_{23}(s) \\ \widehat{\gamma}_{31}(s) \\ \widehat{\gamma}_{12}(s) \end{Bmatrix} = s \left[\begin{array}{ccc|ccc} \widehat{J}_{11} & \widehat{J}_{12} & \widehat{J}_{13} & 0 & 0 & 0 \\ \widehat{J}_{21} & \widehat{J}_{22} & \widehat{J}_{23} & 0 & 0 & 0 \\ \widehat{J}_{31} & \widehat{J}_{32} & \widehat{J}_{33} & 0 & 0 & 0 \\ \hline 0 & 0 & 0 & \widehat{J}_{44} & 0 & 0 \\ 0 & 0 & 0 & 0 & \widehat{J}_{55} & 0 \\ 0 & 0 & 0 & 0 & 0 & \widehat{J}_{66} \end{array}\right] \begin{Bmatrix} \widehat{\sigma}_{11}(s) \\ \widehat{\sigma}_{22}(s) \\ \widehat{\sigma}_{33}(s) \\ \widehat{\tau}_{23}(s) \\ \widehat{\tau}_{31}(s) \\ \widehat{\tau}_{12}(s) \end{Bmatrix} \tag{2.83}$$

The above equations generalize *constitutive equations* of isotropic viscoelastic material $(2.62)_{2,3}$ to the case of *material orthotropy*, where $\widehat{E}_{ij} = \widehat{E}_{ij}(s)$ and $\widehat{J}_{ij} = \widehat{J}_{ij}(s)$ stand for the *orthotropic relaxation function matrix* and the *creep compliance matrix*, respectively. Solution of the problem of viscoelastic orthotropy can be obtained on the way of the *Laplace inverse transform* applied to the transformed variables $\widehat{\sigma}_{ij}(s)$, $\widehat{\varepsilon}_{ij}(s)$, which are retransformed to physical variables $\sigma_{ij}(t)$, $\varepsilon_{ij}(t)$.

It should be emphasized that in case of *anisotropic linear viscoelastic materials*, similar to anisotropic elastic materials, it is not possible to separate the constitutive equations into the volumetric change and the shape change uncoupled equations since anisotropy results in full *coupling* between the *volume* and the *shape viscoelastic deformation*.

In case of composite materials, the properties of which exhibit the linear viscoelastic features, a generalization of commonly used homogenization techniques (see Chap. 3) to viscoelastic work-regime can be proposed (cf. Haasemann and Ulbricht [5]). The frequently applied concept of the *representative unit cell* (RUC) originally developed for elastic composites can also be adapted to nonelastic behavior of composites (matrix and/or fiber). To this end, the class of *linear viscoelastic material* occurs to be very convenient when use of the correspondence principle which enables to transform the real *viscoelastic (time-dependent) problem* to a *fictitious elastic (time-independent) problem* in the domain of new variable s (see Table 2.4).

Constitutive equation of viscoelastic material can be formulated twice: at the level of subcell ($\beta\gamma$) where components are described by different viscoelastic equations (matrix, fibers, particles) and at the level of RUC where different properties of material are homogenized in a particular way to yield mean or effective viscoelastic response.

The above described stages of description are briefly presented as follows: at *level of subcell* the *local constitutive equations* of *linear viscoelasticity* hold

$$
{}^{\mathrm{ve}}\varepsilon_{ij}^{(\beta\gamma)}(t) = \int_0^t {}^{\mathrm{ve}}J_{ijkl}^{(\beta\gamma)}(t-\xi)\,{}^{\mathrm{ve}}\dot{\sigma}_{kl}^{(\beta\gamma)}(\xi)\mathrm{d}\xi \tag{2.84}
$$

or

$$
{}^{\mathrm{ve}}\sigma_{ij}^{(\beta\gamma)}(t) = \int_0^t {}^{\mathrm{ve}}E_{ijkl}^{(\beta\gamma)}(t-\xi)\,{}^{\mathrm{ve}}\dot{\varepsilon}_{kl}^{(\beta\gamma)}(\xi)\mathrm{d}\xi \tag{2.85}
$$

where the *local variables*, *microstress*, and *microstrain* are combined by *local constitutive time-dependent fourth-rank tensors* (the *creep compliance tensor* or the *relaxation tensor*) for the *homogeneous constituent material*. In a formal fashion when a homogenization inside the RUC is used, we arrive at

$$
{}^{\mathrm{ve}}\overline{\varepsilon}_{ij}(t) = \int_0^t {}^{\mathrm{ve}}\overline{J}_{ijkl}(t-\xi)\,{}^{\mathrm{ve}}\dot{\overline{\sigma}}_{kl}(\xi)\mathrm{d}\xi \tag{2.86}
$$

or

$$
{}^{\mathrm{ve}}\overline{\sigma}_{ij}(t) = \int_0^t {}^{\mathrm{ve}}\overline{E}_{ijkl}(t-\xi)\,{}^{\mathrm{ve}}\dot{\overline{\varepsilon}}_{kl}(\xi)\mathrm{d}\xi \tag{2.87}
$$

Mean or *effective fourth-rank tensors* of *creep compliance* ${}^{\mathrm{ve}}\overline{J}_{ijkl}$ or *relaxation* ${}^{\mathrm{ve}}\overline{E}_{ijkl}$ are defined at the *level of RUC* in terms of the corresponding *local tensors* ${}^{\mathrm{ve}}\overline{J}_{ijkl}^{(\beta\gamma)}$ or ${}^{\mathrm{ve}}\overline{E}_{ijkl}^{(\beta\gamma)}$ at the *level of subcell* by the use of a *homogenization procedure* in an analogous way as for the elastic composite (3.58). Homogenization of the viscoelastic properties of the composite material is not a trivial problem and is seldom met in literature, cf. e.g., Haasemann and Ulbricht [5].

Application of correspondence principle occurs to be very useful since it makes possible to convert *time-dependent heterogeneous viscoelastic problem* to associated *time-independent elastic problem* for which *homogenization tools* can directly be applied. In particular when elastic-viscoelastic analogy is applied for *anisotropic composites* at the level of RUC, the application of the Laplace transform allows to reduce integral equation of real material (2.86) or (2.87) to coupled set of equations of a *fictitious elastic problem* in space of the transformed variable s

$$^{\mathrm{e}}\widehat{\overline{\varepsilon}}_{ij}(s) = s\widehat{\overline{E}}^{-1}_{ijkl}(s)^{\mathrm{e}}\widehat{\overline{\sigma}}_{kl}(s) \tag{2.88}$$

or

$$^{\mathrm{e}}\widehat{\overline{\sigma}}_{ij}(s) = s\widehat{\overline{E}}_{ijkl}(s)^{\mathrm{e}}\widehat{\overline{\varepsilon}}_{kl}(s) \tag{2.89}$$

Finally, solution of the *anisotropic linear viscoelastic problem* can be obtained by the use of *inverse Laplace transformation* from the transformed domain $^{\mathrm{e}}\widehat{\overline{\varepsilon}}_{ij}(s), {}^{\mathrm{e}}\widehat{\overline{\sigma}}_{ij}(s)$ to the physical domain $^{\mathrm{ve}}\overline{\varepsilon}_{ij}(t), {}^{\mathrm{ve}}\overline{\sigma}_{ij}(t)$. When absolute notation is used we get (see Haasemann and Ulbricht [5])

$$\begin{aligned} {}^{\mathrm{ve}}\overline{\sigma}(t) &= {}^{\mathrm{ve}}\overline{\mathbb{E}}(t) : {}^{\mathrm{ve}}\overline{\varepsilon}(t=0) + \int_0^t {}^{\mathrm{ve}}\overline{\mathbb{E}}(t-\xi) : {}^{\mathrm{ve}}\dot{\overline{\varepsilon}}(\xi)\mathrm{d}\xi \\ &= {}^{\mathrm{ve}}\overline{\mathbb{E}}(t) : {}^{\mathrm{ve}}\overline{\varepsilon}(t=0) + [{}^{\mathrm{ve}}\overline{\mathbb{E}} : {}^{\mathrm{ve}}\dot{\overline{\varepsilon}}](t) \end{aligned} \tag{2.90}$$

Recall definition of the Laplace transform of the function $f(t)$ $(t > 0)$ into the function of transformed variable $\widehat{f}(s)$

$$\mathcal{L}\{f(t)\} = \widehat{f}(s) \overset{\mathrm{def}}{=} \int_0^\infty f(t)e^{-st}\mathrm{d}t \tag{2.91}$$

and definition of the convolution of two functions

$$f(t) \overset{\mathrm{def}}{=} \int_0^t f_1(t-\xi)f_2(\xi)\mathrm{d}\xi \equiv f_1(t) * f_2(t) \tag{2.92}$$

Applying the Laplace transform (2.91) to the convolution integral (2.92), we arrive at the *convolution theorem* (see Findley et al. [4])

$$\mathcal{L}\left\{\int_0^t f_1(t-\xi)f_2(\xi)\mathrm{d}\xi\right\} = \mathcal{L}\{f_1(t) * f_2(t)\} = \widehat{f_1}(s)\widehat{f_2}(s) \tag{2.93}$$

Taking next the Laplace transform of the *integral* form of *constitutive equation of* the *anisotropic linear viscoelasticity* (2.90), we arrive at the equivalent *transformed* algebraic *equation of anisotropic linear elasticity* according to scheme

$${}^{\mathrm{ve}}\overline{\sigma}(t) = {}^{\mathrm{ve}}\overline{\mathbb{E}}(t) : {}^{\mathrm{ve}}\overline{\varepsilon}(t=0) + {}^{\mathrm{ve}}\overline{\mathbb{E}}(t) : {}^{\mathrm{ve}}\dot{\overline{\varepsilon}}(t) \xrightarrow{\mathcal{L}} \widehat{\overline{\sigma}}(s) = s\widehat{\overline{\mathbb{E}}}(s) : \widehat{\overline{\varepsilon}}(s) \tag{2.94}$$

defined by function of the transformed variable s. The *transformed matrix of anisotropic fictitious elasticity* $s\widehat{\overline{\mathbb{E}}}(s)$ is built *at the level of RUC* of considered composite. The above matrix is obtained by the *homogenization of* the *transformed isotropic local matrices* $s\widehat{\widehat{\mathbb{E}}}^{(\beta\gamma)}(s)$ at the *level of composite microstructure* (subcells).

The procedure described above is sketched by the following scheme shown in Table 2.4. The homogenization procedures for the anisotropic elastic composites are well recognized (for details see next chapter). By contrast, there is no unique and direct homogenization procedure to yield the effective creep compliance ${}^{\mathrm{ve}}J_{ijkl}^{(\beta\gamma)}(t)$ and relaxation ${}^{\mathrm{ve}}E_{ijkl}^{(\beta\gamma)}(t)$ tensors (e.g., Haasemann and Ulbricht [5]). Hence to overcome this deficiency, the suggested scheme is as follows: first apply the *Laplace transform* at the *level of subcell* in order to eliminate physical time (left path in Table 2.4), second use a *homogenization* method in order to reach the RUC level for *fictitious elastic RUC of composite* material and finally apply the *inverse Laplace transform* to arrive at the physical *viscoelastic RUC level* (right back path in Table 2.4).

Usually for sake of simplicity of further applications, the *transversely isotropic effective relaxation matrix* $s\widehat{\widehat{\overline{\mathbb{E}}}}(s)$ at the level of RUC is sufficient, whereas at the microlevel (subcell) the isotropic matrices for the constituents (f) fiber and (m) composite matrix $s\widehat{\widehat{\mathbb{E}}}^{(\mathrm{f})}(s)$ and $s\widehat{\widehat{\mathbb{E}}}^{(\mathrm{m})}(s)$ are usually accepted (see also (2.82)). *Elastic-viscoelastic correspondence principle* as applied to *orthotropic viscoelastic*

Table 2.4 Elastic-viscoelastic homogenization method based on the representative unit cell (RUC) applied to the associated elastic material by correspondence principle

$$\left\{\varepsilon^{(\beta\gamma)}\right\} = \int_0^t \left[{}^{\mathrm{ve}}\mathbb{J}^{(\beta\gamma)}\right]\left\{\dot{\sigma}^{(\beta\gamma)}\right\}\mathrm{d}\xi \quad \text{linear viscoelastic} \quad \{\overline{\varepsilon}\} = \int_0^t \left[{}^{\mathrm{ve}}\overline{\mathbb{J}}\right]\{\dot{\overline{\sigma}}\}\,\mathrm{d}\xi$$

$$\Longrightarrow$$

$$\left\{\sigma^{(\beta\gamma)}\right\} = \int_0^t \left[{}^{\mathrm{ve}}\mathbb{E}^{(\beta\gamma)}\right]\left\{\dot{\varepsilon}^{(\beta\gamma)}\right\}\mathrm{d}\xi \quad \text{homogenization ?} \quad \{\overline{\sigma}\} = \int_0^t \left[{}^{\mathrm{ve}}\overline{\mathbb{E}}\right]\{\dot{\overline{\varepsilon}}\}\,\mathrm{d}\xi$$

$$\Downarrow \;\; \mathcal{L} \;\; \Downarrow \qquad\qquad \Uparrow \;\; \mathcal{L}^{-1} \;\; \Uparrow$$

$$\left\{\widehat{\varepsilon}^{(\beta\gamma)}\right\} = s\left[\widehat{\mathbb{J}}^{(\beta\gamma)}\right]\left\{\widehat{\sigma}^{(\beta\gamma)}\right\} \quad \text{associated elastic} \quad \left\{\widehat{\overline{\varepsilon}}\right\} = s\left[\widehat{\overline{\mathbb{J}}}\right]\left\{\widehat{\overline{\sigma}}\right\}$$

$$\Longrightarrow$$

$$\left\{\widehat{\sigma}^{(\beta\gamma)}\right\} = s\left[\widehat{\mathbb{E}}^{(\beta\gamma)}\right]\left\{\widehat{\varepsilon}^{(\beta\gamma)}\right\} \quad \text{homogenization} \quad \left\{\widehat{\overline{\sigma}}\right\} = s\left[\widehat{\overline{\mathbb{E}}}\right]\left\{\widehat{\overline{\varepsilon}}\right\}$$

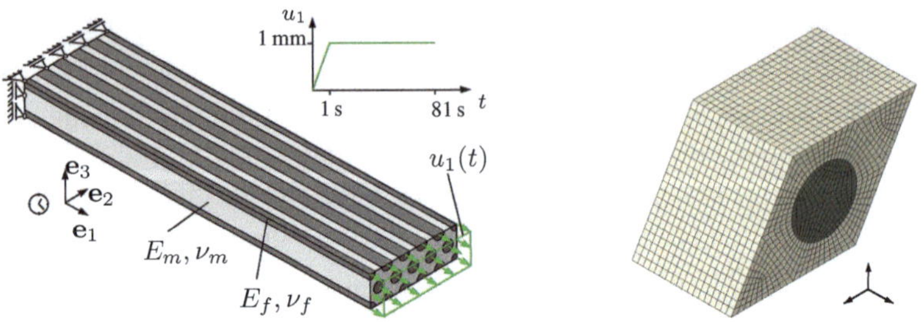

Fig. 2.5 Beam model and RUC of unidirectional composite according to Haasemann and Ulbricht [5]

Table 2.5 Viscoelastic properties of fiber and matrix material following Haasemann and Ulbricht [5]

	Fiber	Matrix
Relaxation function	$70 + 200\,\mathrm{e}^{-0.1t/s}$	$3 + 15\,\mathrm{e}^{-t/s}$
Poisson's ratio	0.2	0.35

and *viscoplastic materials* is applied by Haasemann and Ulbricht [5]. In case of unidirectionally reinforced composite considered by Haasemann and Ulbricht [5] (see Fig. 2.5), both fiber and matrix materials were described as isotropic linear viscoelastic (see Table 2.5); whereas at macroscale, the sane composite material obeyed the transverse isotropy symmetry. The Laplace–Carson transform

$$\mathcal{L}_\mathrm{C}\{f(t)\} = \widehat{f_\mathrm{C}}(s) = s\int_0^\infty \mathrm{e}^{-st} f(t)\,\mathrm{d}t \tag{2.95}$$

was used in order to transform equations of *transversely isotropic linear viscoelastic material* in t domain into corresponding equations of *fictitious linear elastic material* in domain of transformed variable s, in which conventional homogenization techniques of elastic composites were applicable (for details see Chap. 3).

References

1. Altenbach, H.: Classical and non-classical creep models. In: Altenbach, H., Skrzypek, J.J. (eds.) Creep and Damage in Materials and Structures. CISM Courses and Lectures No. 399, pp. 45–96. Springer, Wien (1999)
2. Betten, J.: Creep Mechanics, 3rd edn. Springer, Berlin (2008)
3. Byron, F.W., Fuller, M.D.: Mathematics of elliptic integrals for engineers and physicists. Springer, Berlin (1975)
4. Findley, W.N., Lai, J.S., Onaran, K.: Creep and Relaxation of Nonlinear Viscoplastic Materials. North-Holland, New York (1976)

5. Haasemann, G., Ulbricht, V.: Numerical evaluation of the viscoelastic and viscoplastic behavior of composites. Technische Mechanik **30**(1–3), 122–135 (2010)
6. Krempl, E.: Creep-plastic interaction. In: Altenbach, H., Skrzypek, J.J. (eds.) Creep and Damage in Materials and Structures. CISM Courses and Lectures No. 399, pp. 285–348. Springer, Wien (1999)
7. Murakami, S.: Continuum Damage Mechanics. Springer, Berlin (2012)
8. Nowacki, W.: Teoria pełzania. Arkady, Warszawa (1963)
9. Pipkin, A.C.: Lectures on Viscoelasticity Theory. Springer, Berlin (1972)
10. Pobedrya, B.E.: Mehanika kompozicionnyh materialov. Izd. Mosk. Univ. (1984)
11. Rabotnov, Ju.N.: Creep Problems of the Theory of Creep. North-Holland, Amsterdam (1969)
12. Shu, L.S., Onat, E.T.: On anisotropic linear viscoelastic solids. In: Proceedings of the Fourth Symposium on Naval Structures Mechanics, p. 203. Pergamon Press, London (1967)
13. Skrzypek, J.J.: In Hetnarski, R.B., (ed.) Plasticity and Creep, Theory, Examples, and Problems. Begell House/CRC Press, Boca Raton (1993)
14. Skrzypek, J.: Material models for creep failure analysis and design of structures. In: Altenbach, H., Skrzypek, J.J. (eds.) Creep and Damage in Materials and Structures. CISM Courses and Lectures No. 399, pp. 97–166. Springer, Wien (1999)
15. Skrzypek, J., Ganczarski, A.: Modeling of Material Damage and Failure of Structures. Springer, Berlin (1999)

Chapter 3
Mechanics of Anisotropic Composite Materials

Artur W. Ganczarski, S. Hernik and Jacek J. Skrzypek

Abstract Mechanics of composite materials was in the last decade one of the most rapidly explored engineering area, basically due to huge progress in composite fabrication and use. The main problem referred in this chapter is how to correctly predict averaged effective properties by implementation of numerous homogenization techniques. Useful classification of composites with respect to the format of effective stiffness matrix, based on the analogy between the crystal lattice symmetry and respective configuration of reinforcement in the RUC, is given. Extended section is focused on conventionally used Hill's theorem on upper and lower bounds by Voigt and Reuss' isotropic estimation for approximate determination of stiffness and compliance matrices of anisotropic composite. Consistent application of the Hill theorem to the elements of elastic stiffness or compliance matrices (but not to engineering anisotropy constants) enable to explain some peculiarities of the Poisson ratio diagrams, met in respective bibliography (e.g., Aboudi et al., Micromechanics of Composite Materials, 2013; Sun and Vaidya, Compos. Sci. Technol. 56:171–179, 1996; Gan et al., Int. J. Solids Struct. 37:5097–5122, 2000). The new effective proposal to achieve approximation of the mechanical modules of unidirectionally reinforced composites by the use of hybrid-type rule of weighted average between the Voigt and Reuss upper and lower estimates is proposed. Capability of this averaged interpolation was checked based on selected findings by Gan et al. (Int. J. Solids Struct. 37:5097–5122, 2000) for Boron/Al composite, which show good convergence and enable to treat weighting coefficients as universal ones over the full V_f range.

Keywords Symmetry of composites · Homogenization methods · Consistent use of estimates to averaged matrices · Poisson's ratio peculiarity · Anisotropic hybrid homogenization rule

A.W. Ganczarski (✉) · S. Hernik · J.J. Skrzypek
Solid Mechanics Division, Institute of Applied Mechanics,
Cracow University of Technology, al. Jana Pawła II 37, 31-864 Kraków, Poland
e-mail: Artur.Ganczarski@pk.edu.pl

S. Hernik
e-mail: hernik@mech.pk.edu.pl

J.J. Skrzypek
e-mail: Jacek.Skrzypek@pk.edu.pl

J.J. Skrzypek and A.W. Ganczarski (eds.), *Mechanics of Anisotropic Materials*,
Engineering Materials, DOI 10.1007/978-3-319-17160-9_3

3.1 State of the Art

Essential progress observed in *manufacturing processes* and *application of composite materials* results in necessity to develop methodology of determination of the *effective properties mechanical, thermal*, and others. Among the variety of papers dealing with modeling of effective mechanical properties of composites and their experimental verification, the following group of papers in which a coupling between the topology of *fibrous reinforcement* (or *particle*) *reinforcement* and *material symmetry* of *constitutive model* describing composite can be distinguished, for instance: Sun and Vaidya [30], Gan et al. [9], Liu et al. [19], Würkner et al. [37], Selvadurai and Nikopour [28] and others.

Aforementioned papers deal with the modeling of *unidirectionally reinforced composites* treated as homogeneous orthotropic solids characterized by some effective modules that describe average material properties of the composite. Assuming the *periodic fiber arrangement* inside the matrix usually two types of *Representative Unit Cells* (RUC) that exhibit either the *tetragonal symmetry* (*square array*) or the *hexagonal symmetry* (*hexagonal array*) are considered.

In the significant paper by Sun and Vaidya [30] two *composite* systems: *Boron/Al* and *Graphite/Epoxy* of the respective fixed volume V_f fraction equal to 0.47 and 0.6 are analyzed. Authors find essential scatter in analytical results obtained for two kinds of composites in comparison with earlier data from the literature, namely: Hashin and Rosen [10], Whitney and Riley [35], Chamis [6], Sun and Chen [29], Sun and Zhou [31], Kenaga et al. [15]. In particular, the large scatter is referred to the effective Young modulus, the effective Kirchhoff modulus, and the effective Poisson ratio in the plane of *transverse isotropy*. The obtained material constants, in general, do not confirm the theorem on *upper* and *lower bounds* based on the classical *Voigt* and *Reuss rules*. Especially difficult is to explain the estimated magnitude of the in-plane Poisson ratio exceeding range of two composite components based on either the isotropic characteristic of components in *Boron/Al composite* or the orthotropic characteristic of components in *Graphite/Epoxy composite*.

More systematic analysis of the influence of homogenization methods on estimated effective properties of composites is due to Gan et al. [9]. The authors compare the new *Strain-Compatible Method of Cells* (SCMC) with other homogenization methods such as *Generalized Method of Cells* (GMC) Paley and Aboudi [25] and *micromechanical analysis* using FEM. For numerical simulation, authors used the unidirectionally reinforced Boron/Al composite assuming two types of the representative unit cells based either on a *random topology of* parallel *fibers* or on the hexagonal array for full spectrum of the volume fraction $V_f \in <0, 1>$. The homogenization results are also compared with the classical approximate calculations based on *Voigt/Reuss mixture rules*, Voigt [34], Reuss [27]. The performed analysis confirms applicability of the upper/lower bounds for majority of equivalent material constants except for the in-plane Poisson ratio. However the authors do not precisely distinguish between the tetragonal or the hexagonal symmetry when modeling

Representative Unit Cell (RUC) such that all six modules of orthotropy are treated as independent in spite of clear hexagonal symmetry in fibers topology.

Liu et al. [19] analyze possibility for the Poisson ratio positioned beyond the Voigt/Reuss estimates. Moreover: "It was found that the effective Young modulus in both transverse and longitudinal direction can exceed not only the approximate Voigt estimation, but also the stiffness of the stiffer constituent phase". The authors recommend precautions when applying Voigt/Reuss estimates in cases when one of the components is made of incompressible material.

In the recently published paper by Würkner et al. [37] the effective elastic modules of the composite formed of isotropic *Epoxy matrix* and transversely isotropic *Graphite fibers* are examined for reasonable wide range of volume fraction $V_f \in< 0.1 \div 0.6 >$ see also comments in Sect. 3.5.5 of this chapter. The rhombic array of fibers is used for simulations characterized by different topology angles of RUC. Following cases are considered: $\gamma = 60°$ (*hexagonal array*), $60° < \gamma < 90°$ (*rhombic array*) and $\gamma = 90°$ (*tetragonal array*). The estimated effective modules show satisfactory coincidence with numerical results given by Jiang et al. [14].

The more general approach to modeling of composites reinforced by unidirectional fibers is recently presented by Selvadurai and Nikopour [28]. Authors considered the random parallel identical *Carbon fibers* distribution in the *Epoxy matrix* of a composite. In the light of the numerical analysis performed, it is found that in spite of random fibers distribution it is possible to determine a minimal *Representative Area Element*—RAE (>65 fibers number) that guarantees the property of *transversely isotropic symmetry* of *hexagonal* type (5 independent constants in the elasticity matrix, see Fig. 3.1).

Extensive state-of-the-art review of the micromechanics-based analysis of composite materials, enriched by numerous actual results, both in the field of homogenization techniques and its experimental validation for real long-fiber reinforced composites, are found in recently published excellent monograph by Aboudi et al. [1].

3.2 Analogy Between the Elastic Matrices Symmetry at the Level of Crystal Lattice Unit Cell and the Composite Representative Element

A useful analogy between the *crystal lattice symmetry* at the level of single crystal lattice or crystal grains and the relevant microstructure of composite materials of identical symmetry groups that characterize effective elastic matrices (stiffness or compliance) at the macrolevel is sketched in Fig. 3.1.

Equations of linear elasticity of crystal and *composite materials* are written in (3.1)

$$\{\sigma\}^{(cr)} = [\mathfrak{c}]\,\{\varepsilon\}^{(cr)} \qquad \text{and} \qquad \{\overline{\sigma}\} = [\overline{\mathbb{E}}]\,\{\overline{\varepsilon}\} \tag{3.1}$$

Conventional unit cells of space lattices	Compliance matrix $[\mathbb{E}^{-1}]$	Examplary composits microstructure
(a) 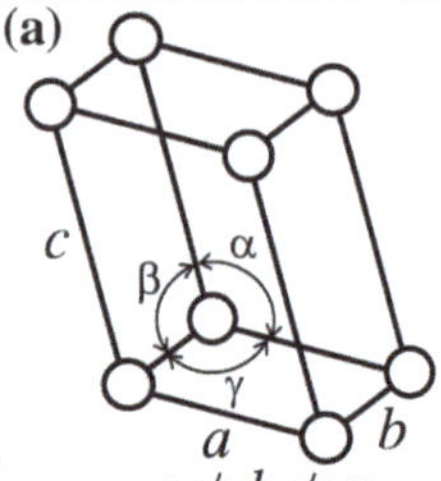 $a \neq b \neq c$ $\alpha \neq \beta \neq \gamma$ triclinic lattice	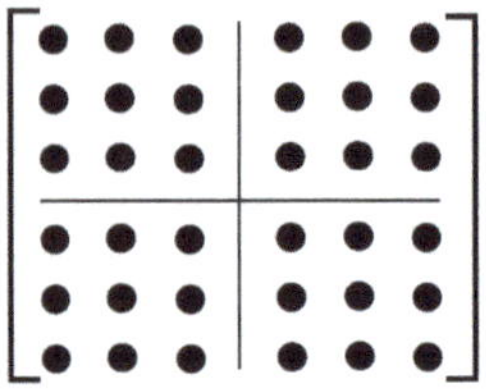 anisotropic Hooke's (21 constants)	 anisotropic fibers arrangement in C/C composite (after Martin-Herrero and Germain [21])
(b) 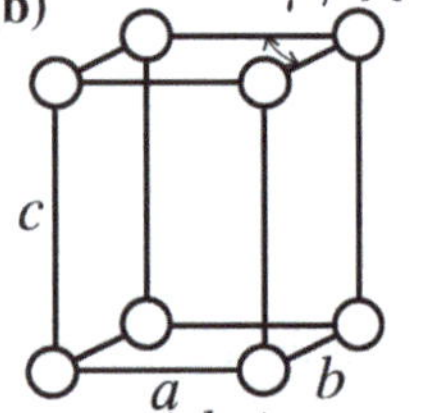 $a \neq b \neq c$ $\alpha = \beta = 90°, \gamma \neq 90°$ monoclinic lattice	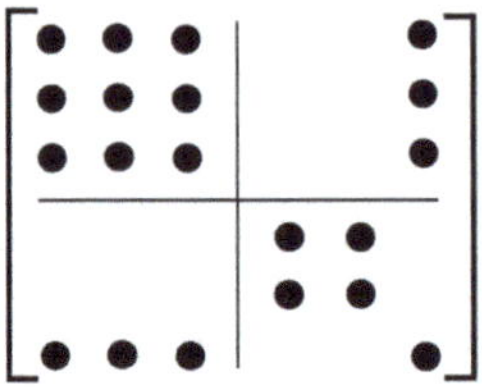 monoclinic or oblique Hooke's anisotropy (13 constants) cf. Nye [23]	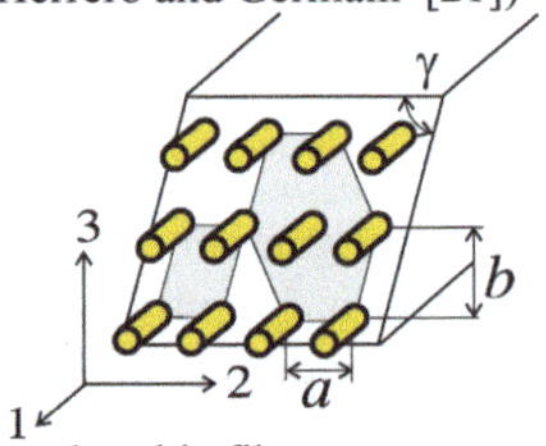 rhombic fiber array (after Würkner et al [37], Li [17])
(c) 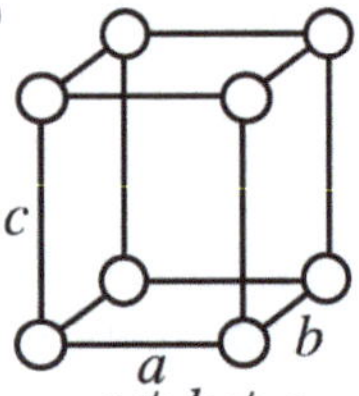 $a \neq b \neq c$ $\alpha = \beta = \gamma = 90°$ orthorhombic lattice	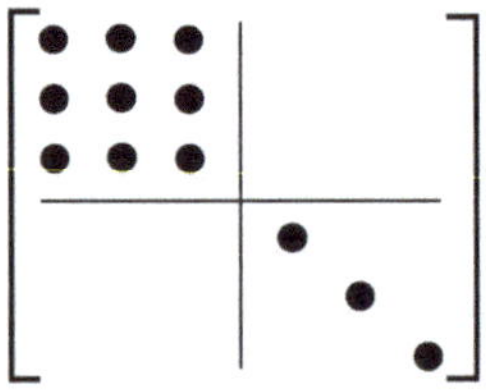 orthotropic Hooke's (9 constants)	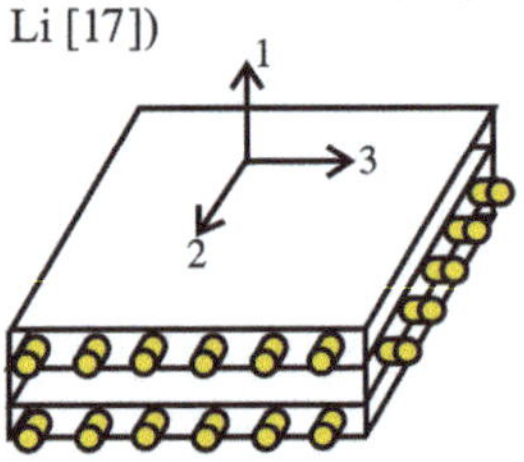 perpendicular fibers arrangement in multi-laminate plate (2D)
(d) 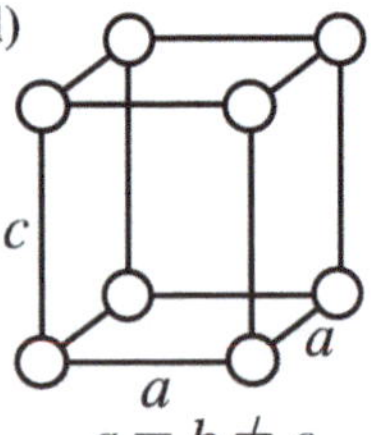 $a = b \neq c$ $\alpha = \beta = \gamma = 90°$ tetragonal lattice	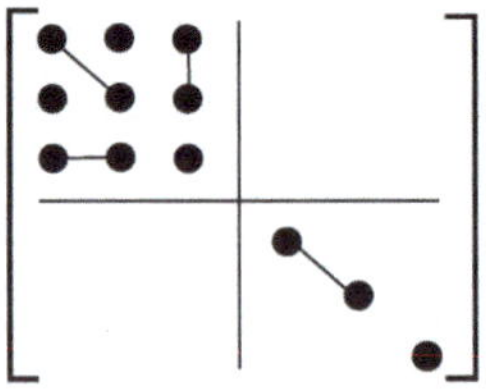 transversely isotropic tetragonal Hooke's (6 constants)	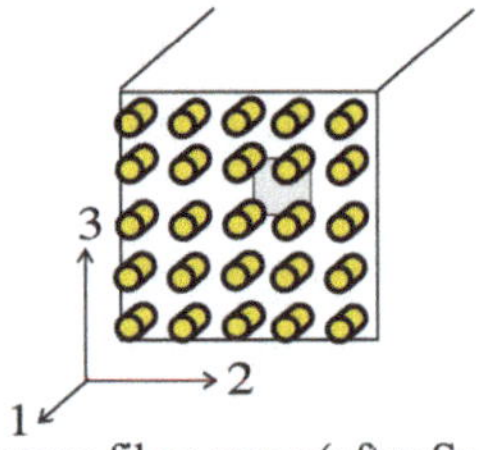 square fiber array (after Sun and Vaidya [30])

Fig. 3.1 Classification of selected composites with respect to the format of compliance matrix $[\mathbb{E}^{-1}]$: **a** anisotropic fiber arrangement, **b** rhombic fiber arrangement, **c** orthotropic fiber arrangement, **d** square fiber arrangement, **e** hexagonal fiber arrangement, **f** regular particle arrangement, **g** random particle arrangement, after Tjong and Ma [33], Martin-Herrero and Germain [21], Nye [23]

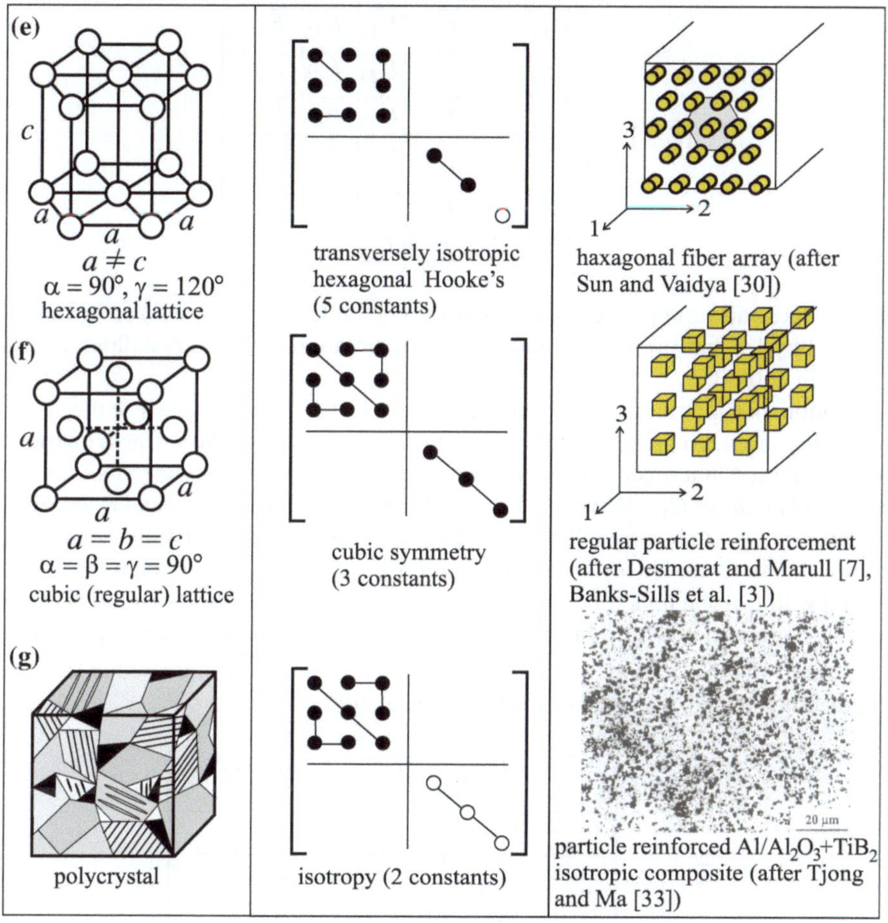

Fig. 3.1 (continued)

where relevant stiffness matrices at the crystal and composite level possessing identical symmetry properties are denoted with $[\mathfrak{s}]$ and $[\mathfrak{c}]$ whereas $\{\boldsymbol{\sigma}\}^{(\mathrm{cr})}$, $\{\varepsilon\}^{(\mathrm{cr})}$, and $\{\overline{\boldsymbol{\sigma}}\}$, $\{\overline{\varepsilon}\}$ stand for stress and strain vectors at the microlevel and the effective stress and strain averaged over the representative element (RVE or RUC) (see Gan et al. [9], Selvadurai and Nikopour [28], etc.). The respective compliance matrices used in Eq. (3.1) can be rewritten in the equivalent fashion

$$\{\varepsilon\}^{(\mathrm{cr})} = [\mathfrak{s}]\,\{\boldsymbol{\sigma}\}^{(\mathrm{cr})} \qquad \text{and} \qquad \{\overline{\varepsilon}\} = [\overline{\mathbb{E}}^{-1}]\,\{\overline{\boldsymbol{\sigma}}\} \tag{3.2}$$

where the *effective compliance matrix* is represented as

$$[\overline{\mathbb{E}}^{-1}] = \left[\begin{array}{ccc|ccc} \overline{E}_{11}^{-1} & \overline{E}_{12}^{-1} & \overline{E}_{13}^{-1} & \overline{E}_{14}^{-1} & \overline{E}_{15}^{-1} & \overline{E}_{16}^{-1} \\ \overline{E}_{21}^{-1} & \overline{E}_{22}^{-1} & \overline{E}_{23}^{-1} & \overline{E}_{24}^{-1} & \overline{E}_{25}^{-1} & \overline{E}_{26}^{-1} \\ \overline{E}_{31}^{-1} & \overline{E}_{32}^{-1} & \overline{E}_{33}^{-1} & \overline{E}_{34}^{-1} & \overline{E}_{35}^{-1} & \overline{E}_{36}^{-1} \\ \hline \overline{E}_{41}^{-1} & \overline{E}_{42}^{-1} & \overline{E}_{43}^{-1} & \overline{E}_{44}^{-1} & \overline{E}_{45}^{-1} & \overline{E}_{46}^{-1} \\ \overline{E}_{51}^{-1} & \overline{E}_{52}^{-1} & \overline{E}_{53}^{-1} & \overline{E}_{54}^{-1} & \overline{E}_{55}^{-1} & \overline{E}_{56}^{-1} \\ \overline{E}_{61}^{-1} & \overline{E}_{62}^{-1} & \overline{E}_{63}^{-1} & \overline{E}_{64}^{-1} & \overline{E}_{65}^{-1} & \overline{E}_{66}^{-1} \end{array}\right] \tag{3.3}$$

The stiffness and compliance matrices at the crystal level in Eqs. (3.1) and (3.2) are denoted by $[\mathfrak{c}]$ and $[\mathfrak{s}]$ in accordance with the notation used in crystallography as shown in Table 3.2.

Compliance matrices are more convenient for further application since they have, generally, simpler representation when compared to the respective stiffness matrices, both expressed in terms of the engineering elasticity constants (*Young modules* E_{ii}, *Kirchhoff modules* G_{ij}, *Poisson ratios* ν_{ij}, *Chencov modules* $\mu_{ij(kl)}$ and *Rabinovich modules* $\eta_{i(jk)}$ as shown in Table 3.1). In a more general case of fully anisotropic composite material, for instance when composite material is at the microlevel reinforced with *Carbon nanotubes* of *irregular arrangement*, the effective continuum of averaged properties is fully anisotropic and characterized by 21 engineering modules where the effective compliance matrix of the composite $[\overline{\mathbb{E}}^{-1}]$ expressed in terms of engineering anisotropy constants is furnished as follows:

$$[\overline{\mathbb{E}}^{-1}] = \left[\begin{array}{ccc|ccc} \frac{1}{E_{11}} & -\frac{\nu_{21}}{E_{11}} & -\frac{\nu_{31}}{E_{11}} & \frac{\eta_{23(1)}}{E_{11}} & \frac{\eta_{31(1)}}{E_{11}} & \frac{\eta_{12(1)}}{E_{11}} \\ -\frac{\nu_{12}}{E_{22}} & \frac{1}{E_{22}} & -\frac{\nu_{32}}{E_{22}} & \frac{\eta_{23(2)}}{E_{22}} & \frac{\eta_{31(2)}}{E_{22}} & \frac{\eta_{12(2)}}{E_{22}} \\ -\frac{\nu_{13}}{E_{33}} & -\frac{\nu_{23}}{E_{33}} & \frac{1}{E_{33}} & \frac{\eta_{23(3)}}{E_{33}} & \frac{\eta_{31(3)}}{E_{33}} & \frac{\eta_{12(3)}}{E_{33}} \\ \hline \frac{\eta_{(1)23}}{G_{23}} & \frac{\eta_{(2)23}}{G_{23}} & \frac{\eta_{(3)23}}{G_{23}} & \frac{1}{G_{23}} & \frac{\mu_{31(23)}}{G_{23}} & \frac{\mu_{12(23)}}{G_{23}} \\ \frac{\eta_{(1)31}}{G_{31}} & \frac{\eta_{(2)31}}{G_{31}} & \frac{\eta_{(3)31}}{G_{31}} & \frac{\mu_{(23)31}}{G_{31}} & \frac{1}{G_{31}} & \frac{\mu_{12(31)}}{G_{31}} \\ \frac{\eta_{(1)12}}{G_{12}} & \frac{\eta_{(2)12}}{G_{12}} & \frac{\eta_{(3)12}}{G_{12}} & \frac{\mu_{(23)12}}{G_{12}} & \frac{\mu_{(31)12}}{G_{12}} & \frac{1}{G_{12}} \end{array}\right] \tag{3.4}$$

In Table 3.1 engineering anisotropy constants are ordered into five groups:

- E_{ii}—*axial* elasticity *modules* (three generalized Young modules)
- G_{ij}—*shear modules* at three anisotropy planes (three generalized Kirchhoff modules)
- ν_{ij}—*transverse strain coefficients* (three generalized Poisson ratios)
- $\mu_{ij(kl)}$—*Chencov modules* (three Chencov modules combining shear in different anisotropy planes)
- $\eta_{i(jk)}$—*Rabinovich modules* (nine Rabinovich modules combining shear and normal strain effects).

It is worth to mention that the symmetry of stress and strain tensors results in appropriate symmetry of the compliance (stiffness) matrix, Lekhnitskii [16].

Table 3.1 Types of engineering modules used in representation of the compliance matrix (3.4)

Engineering modules	Coupling effect		Considered axes or planes (coupling)	Number of components
	Stress component	Strain component		
E_{11}, E_{22}, E_{33}	Axial	Axial	Same axes $1 \rightarrow 1$, etc.	3
G_{12}, G_{32}, G_{31}	Shear	Shear	Same planes $12 \rightarrow 12$, etc.	3
$\nu_{21}, \nu_{31}, \nu_{32}$	Axial	Axial	Transverse directions $1 \rightarrow 2$, etc.	3
$\mu_{31(23)}, \mu_{12(23)}, \mu_{12(31)}$	Shear	Shear	Different planes $13 \rightarrow 23$, etc.	3
$\eta_{23(1)}, \ldots, \eta_{12(3)}$	Shear	Axial	Normal to $23 \rightarrow 1$, etc.	9

$$
\begin{aligned}
\frac{\nu_{ij}}{E_{jj}} &= \frac{\nu_{ji}}{E_{ii}} &&\longrightarrow \nu_{ij} E_{ii} = \nu_{ji} E_{jj} \\
\frac{\eta_{ij(k)}}{E_{kk}} &= \frac{\eta_{(k)ij}}{G_{ij}} &&\longrightarrow \eta_{ij(k)} G_{ij} = \eta_{(k)ij} E_{kk} \\
\frac{\mu_{ij(ki)}}{G_{ki}} &= \frac{\mu_{(ki)ij}}{G_{ji}} &&\longrightarrow \mu_{ij(ki)} G_{ji} = \mu_{(ki)ij} G_{ki}
\end{aligned}
\tag{3.5}
$$

A convenient *analogy* between the crystal lattice symmetry, the effective matrix and respective configuration/orientation of fibers or particles in exemplary unit cells of composites is shown in Fig. 3.1. Before we start to discuss items a–g in Fig. 3.1, a comment should be done that an analogy between the exemplary representative composite microstructure and the conventional unit cell of a crystal lattice is built based on the *identical stiffness matrix format* and symmetry properties at the level of crystal unit representative cells (lattice) and the level of composite representative unit cell (fibers/particles geometry, arrangement, etc.), but not on different physical features.

Such analogy occurs to be helpful in proper description of *symmetry groups* and *classes* of the elastic matrices and proposing their experimental-based identification.

In a general case of anisotropy Eq. (3.4), the respective *triclinic crystal lattice symmetry* ensures fully populated stiffness matrices at both levels considered (crystal lattice vs. microstructure) for instance due to the totally *anisotropic Carbon/Carbon composite* (see Fig. 3.1a Martin-Herrero and Germain [21]).

Composites formed by *stacking layers* (lamina) at different fiber orientation are called *laminates*, the effective properties of which vary with orientation, thickness, and stacking sequence of layers. The effective properties of a unidirectional lamina are classified as orthotropic with different properties in the material directions (cf. Herakovich and Aboudi [12]). In general, the effective properties of such multicomponent systems correspond to *averaged orthotropic continuum* described by nine *orthotropy modules* E_{11}, E_{22}, E_{33}, ν_{21}, ν_{32}, ν_{31}, G_{12}, G_{23}, G_{31}, if elastic range is considered. The corresponding *crystal lattice symmetry* is known as the *orthorhombic* lattice characterized by three different cell edges $a \neq b \neq c$ and identical angles $\alpha = \beta = \gamma = 90°$, Fig. 3.1c.

Unidirectionally reinforced composites with the regular parallel fibers arrangement correspond to the *averaged transversely isotropic continuum* at the macrolevel. However, depending on frequently used periodic fiber arrangements, two of them are specially interesting: *tetragonal (square) fiber array* and *hexagonal fiber array*, as shown in Fig. 3.1d, e, respectively. Corresponding two crystal lattice cells are also presented that exhibit equivalence between the in-plane fiber array over the composite RUC and in-plane atoms in the *Crystal Unit Cell* CUC arrangements. Note that in case of tetragonal transverse isotropy, the number of independent modules is equal to six, whereas in case of hexagonal transverse isotropy this number is reduced to five.

Consider for a moment a more general case called the *monoclinic* or *oblique symmetry*. At the level of *composite RUC* it corresponds to the *rhombic fiber array* as shown in Fig. 3.1b. In this case, periodicity is dependent not only on the distance between layers but also on the angle of slope of the RUC walls $60° < \gamma < 90°$. The corresponding crystal lattice symmetry is known as *monoclinic lattice symmetry*. This case can be recognized as an intermediate between the *triclinic lattice* (Fig. 3.1a) and the *orthorhombic lattice* (Fig. 3.1c). Consequently, the equivalent stiffness matrix describing monoclinic anisotropy is enriched with four nonzeroth independent elements $\overline{E}_{16}^{-1}$, $\overline{E}_{26}^{-1}$, $\overline{E}_{36}^{-1}$, and $\overline{E}_{45}^{-1}$, such that total number of independent modules of the compliance matrix is equal to $13 = 9+4$. Presence of these additional elements is a characteristic feature for *Rabinovich constants* $\eta_{(i)jk}$ and *Chencov constants* $\mu_{ij(kl)}$ responsible for anisotropy (which are not present in orthotropy).

Consider further more detailed two particular fiber arrangements of the monoclinic symmetry (Fig. 3.1b) which easily can be recognized in two fiber arrays of the *tetragonal* or the *hexagonal symmetry* appearing in transversely isotropic *long-fiber-reinforced composites*. In both cases, $a = b$ holds but two particular magnitudes of the slope angle a *rhombic array* of γ are admitted: $\gamma = 90°$ or $\gamma = 60°$ (Fig. 3.2). In the first case when $\gamma = 90°$, rhombic fiber array reduces to the *square fiber array* (at the composite level) and the equivalent representative crystal lattice cell exhibits architecture of *tetragonal symmetry*, as previously shown in Fig. 3.1d. In the second case when $\gamma = 60°$, any arbitrary rhombic array reduces to another *hexagonal fiber array* (at the composite level) with the equivalent crystal lattice cell architecture of hexagonal symmetry, see Fig. 3.1d. In both cases considered in the compliance matrix $\left[\mathbb{E}^{-1}\right]$ of Eq. (3.4) four elements describing the Rabinovich and the Chencov effects $\overline{E}_{16}^{-1} = \overline{E}_{26}^{-1} = \overline{E}_{36}^{-1} = \overline{E}_{45}^{-1} = 0$ disappear such that only nine elements are present in the *orthotropic Hooke law* Fig. 3.1e. However, in case of transverse isotropy, the number of independent modules reduces to either six (square array, $\gamma = 90°$) or five (hexagonal array, $\gamma = 60°$) since in the last case the in-plane modulus equals $\overline{E}_{66}^{-1} = (\overline{E}_{11}^{-1} - \overline{E}_{12}^{-1})/2$ and should be considered as dependent.

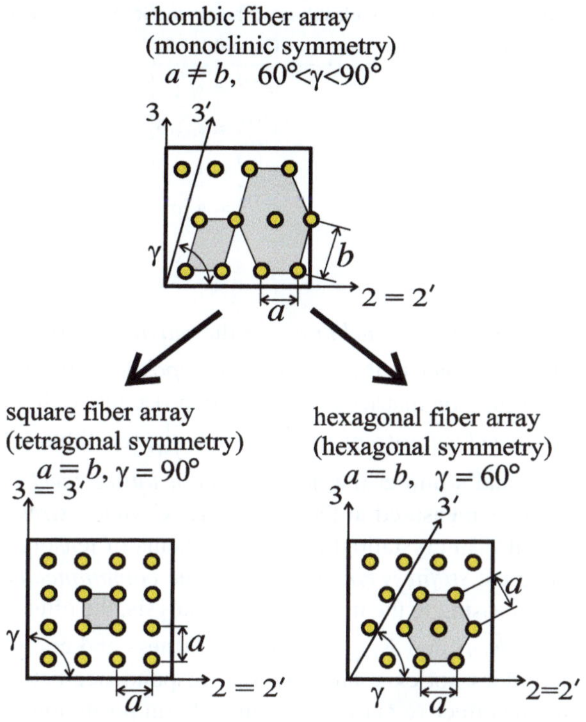

Fig. 3.2 Square or hexagonal fiber arrays as particular cases of rhombic fiber array

Finally, for the narrower case of the tetragonal lattice namely $a = b = c$ and $\alpha = \beta = \gamma = 90°$ the particular *cubic crystal lattice* is recovered (regular lattice). The stiffness or compliance matrices are here characterized by three independent constants: $E_{11}^{-1} = E_{22}^{-1} = E_{33}^{-1}, E_{12}^{-1} = E_{13}^{-1} = E_{23}^{-1}, E_{44}^{-1} = E_{55}^{-1} = E_{66}^{-1}$ see Fig. 3.1f. Such cubic symmetry case is sometimes expected in certain regular particle arrangement, as discussed by Desmorat and Marull [7] and Banks-Sills et al. [3].

To make this classification complete, the particle-reinforced composites of irregular particle shape and their topology should be admitted. In such a case, at the macrolevel, the properties of *isotropy of composite* inside RUC can be admitted, where two independent elastic constants (effective) can satisfactorily be estimated from the Voigt/Reuss rules based on the particle volume fraction V_f only, see Fig. 3.1g.

In schematic representation of the elastic matrices of crystal lattice and composite microstructure, the visualization of matrix elements was adopted after Nye [23] where • depicts independent modules, ∘ dependent modules, whereas •—• or ∘—∘ pairs of identical modules, etc. (see Chap. 2).

As it was aforementioned, a similarity between the symmetry classes of crystals at the crystal lattice level and composite microstructure at the macrolevel has subsidiary meaning only. In fact, the *crystal symmetry* implies format and symmetry of the elastic crystal matrices: stiffness $[\mathfrak{c}_{ij}]$ or compliance $[\mathfrak{s}_{ij}]$ being 2nd rank matrix representation of 4th rank *crystal elasticity tensors* $\mathfrak{c}_{ijkl}$ or $\mathfrak{s}_{ijkl}$. Passing from the

Table 3.2 Equations of elasticity at the crystal level and macrolevel

Notation	Crystal level	Macrolevel
Tensor	$\sigma_{ij}^{(\text{cr})} = \mathfrak{c}_{ijkl}\varepsilon_{kl}^{(\text{cr})}$	$\overline{\sigma}_{ij} = \overline{E}_{ijkl}\overline{\varepsilon}_{kl}$
	$\varepsilon_{ij}^{(\text{cr})} = \mathfrak{s}_{ijkl}\sigma_{kl}^{(\text{cr})}$	$\overline{\varepsilon}_{ij} = \overline{E}_{ijkl}^{-1}\overline{\sigma}_{kl}$
Matrix-vector	$\sigma_{i}^{(\text{cr})} = \mathfrak{c}_{ij}\varepsilon_{j}^{(\text{cr})}$	$\overline{\sigma}_{j} = \overline{E}_{ij}\overline{\varepsilon}_{j}$
	$\varepsilon_{i}^{(\text{cr})} = \mathfrak{s}_{ij}\sigma_{j}^{(\text{cr})}$	$\overline{\varepsilon}_{i} = \overline{E}_{ij}^{-1}\overline{\sigma}_{j}$

atomic level (*crystal lattice*) to the *macrolevel* (*composite RUC*), we arrive at the correspondence to the *equivalent composite matrices* $\overline{E}_{ij}$ or $\overline{E}_{ij}^{-1}$ built as equivalent representation matrices (averaged in procedure of homogenization) of the *composite effective elasticity tensors* $\overline{E}_{ijkl}$ or $\overline{E}_{ijkl}^{-1}$, see Table 3.2. It is necessary to distinguish stress and strain at the *atomic crystal lattice level* $\sigma_{ij}^{(\text{cr})}$ and $\varepsilon_{ij}^{(\text{cr})}$ from analogous variables measured at the level of RUC: *macrostress* and *macrostrain* $\overline{\sigma}_{ij}$ and $\overline{\varepsilon}_{ij}$. Note that in crystallography, components of tensors $\mathfrak{c}_{ijkl}$ and $\mathfrak{s}_{ijkl}$ are traditionally called the *stiffness coefficients* and the *compliance coefficients*. On the other hand, when passing to the macrolevel of analysis, the effective tensor components of composite $\overline{E}_{ijkl}$ and $\overline{E}_{ijkl}^{-1}$ are named stiffness and compliance constants. Mention that there does not exist any direct correspondence between elastic crystal coefficients and the effective elastic constants of composite material at the macrolevel, c.f. Nye [23]. Remember also that during the fabrication process of composite, the residual thermal stresses different in matrix and fibers material have to be built-in into enriched equations of elasticity. Assuming for simplicity that during the fabrication process strains have elastic nature only, the application of conventional equations of thermoelasticity is justified. However, during the final cooling down process of the composite and also in the fabrication phase, some thermoplastic microstructure change in the material can be observed. In such cases, the thermoelastic analysis may occur incorrect (cf. e.g., Herakovich and Aboudi [12]).

3.3 Effective Elastic Matrix Characterization of Composites with Various Symmetries

3.3.1 Triclinic Anisotropic Long-Fiber-Reinforced Composite (Anisotropic Fiber Array, Fig. 3.1a)

Elasticity equation of anisotropic composite material (at the macroscale) written in an arbitrary material frame can be furnished in a following fashion, cf. Eq. (3.4)

$$\begin{Bmatrix} \overline{\varepsilon}_{11} \\ \overline{\varepsilon}_{22} \\ \overline{\varepsilon}_{33} \\ \overline{\gamma}_{23} \\ \overline{\gamma}_{31} \\ \overline{\gamma}_{12} \end{Bmatrix} = \left[\begin{array}{ccc|ccc} \frac{1}{E_{11}} & -\frac{\nu_{21}}{E_{11}} & -\frac{\nu_{31}}{E_{11}} & \frac{\eta_{23(1)}}{E_{11}} & \frac{\eta_{31(1)}}{E_{11}} & \frac{\eta_{12(1)}}{E_{11}} \\ -\frac{\nu_{12}}{E_{22}} & \frac{1}{E_{22}} & -\frac{\nu_{32}}{E_{22}} & \frac{\eta_{23(2)}}{E_{22}} & \frac{\eta_{31(2)}}{E_{22}} & \frac{\eta_{12(2)}}{E_{22}} \\ -\frac{\nu_{13}}{E_{33}} & -\frac{\nu_{23}}{E_{33}} & \frac{1}{E_{33}} & \frac{\eta_{23(3)}}{E_{33}} & \frac{\eta_{31(3)}}{E_{33}} & \frac{\eta_{12(3)}}{E_{33}} \\ \hline \frac{\eta_{(1)23}}{G_{23}} & \frac{\eta_{(2)23}}{G_{23}} & \frac{\eta_{(3)23}}{G_{23}} & \frac{1}{G_{23}} & \frac{\mu_{31(23)}}{G_{23}} & \frac{\mu_{12(23)}}{G_{23}} \\ \frac{\eta_{(1)31}}{G_{31}} & \frac{\eta_{(2)31}}{G_{31}} & \frac{\eta_{(3)31}}{G_{31}} & \frac{\mu_{(23)31}}{G_{31}} & \frac{1}{G_{31}} & \frac{\mu_{12(31)}}{G_{31}} \\ \frac{\eta_{(1)12}}{G_{12}} & \frac{\eta_{(2)12}}{G_{12}} & \frac{\eta_{(3)12}}{G_{12}} & \frac{\mu_{(23)12}}{G_{12}} & \frac{\mu_{(31)12}}{G_{12}} & \frac{1}{G_{12}} \end{array}\right] \begin{Bmatrix} \overline{\sigma}_{11} \\ \overline{\sigma}_{22} \\ \overline{\sigma}_{33} \\ \overline{\tau}_{23} \\ \overline{\tau}_{31} \\ \overline{\tau}_{12} \end{Bmatrix} \quad (3.6)$$

Taking into account the symmetry conditions of the effective compliance matrix $\overline{E}_{ij}^{-1} = \overline{E}_{ji}^{-1}$, see Eq. (3.5), in order to completely determine fully populated 6×6 matrix of elasticity total number of required elements is equal to $n = \frac{(1+6)6}{2} = 21$. However, following the reasoning of Lekhnitskii [16] and others, the maximal number of different from zero but independent matrix elements $\overline{E}_{ij}^{-1}$ equals 18 (see Table 3.1). It follows from requirement that both *effective compliance* $\overline{E}_{ij}^{-1}$ and *stiffness* $\overline{E}_{ij}$ *matrices* have to obey transformation rule by three Euler angles. In such general case of anisotropy, that in crystallography corresponds to *triclinic lattice symmetry*, it is impossible to reduce to zero any matrix elements via some transformation by a rotation of the reference frame with any angles.

3.3.2 Monoclinic or Oblique Anisotropic Long-Fiber Composite (Rhombic Fiber Array, Fig. 3.1b)

Composite systems of the *rhombic*-type *fiber architecture* represent the particular case of generally anisotropic composite geometry in such manner as the *monoclinic crystal lattice symmetry* is the particular case of general *triclinic symmetry* at the crystal lattice level. In such rhombic-type fiber array composites, the axis parallel to the fibers direction can be distinguished (3) being perpendicular to the transverse plane (1, 2). Corresponding equation of elasticity built on the base of *oblique anisotropy compliance matrix* takes the following format

$$\begin{Bmatrix} \overline{\varepsilon}_{11} \\ \overline{\varepsilon}_{22} \\ \overline{\varepsilon}_{33} \\ \overline{\gamma}_{23} \\ \overline{\gamma}_{31} \\ \overline{\gamma}_{12} \end{Bmatrix} = \left[\begin{array}{ccc|ccc} \frac{1}{E_{11}} & -\frac{\nu_{21}}{E_{11}} & -\frac{\nu_{31}}{E_{11}} & & & \frac{\eta_{12(1)}}{E_{11}} \\ -\frac{\nu_{12}}{E_{22}} & \frac{1}{E_{22}} & -\frac{\nu_{32}}{E_{22}} & & & \frac{\eta_{12(2)}}{E_{22}} \\ -\frac{\nu_{13}}{E_{33}} & -\frac{\nu_{23}}{E_{33}} & \frac{1}{E_{33}} & & & \frac{\eta_{12(3)}}{E_{33}} \\ \hline & & & \frac{1}{G_{23}} & \frac{\mu_{31(23)}}{G_{23}} & \\ & & & \frac{\mu_{(23)31}}{G_{31}} & \frac{1}{G_{31}} & \\ \frac{\eta_{(1)12}}{G_{12}} & \frac{\eta_{(2)12}}{G_{12}} & \frac{\eta_{(3)12}}{G_{12}} & & & \frac{1}{G_{12}} \end{array}\right] \begin{Bmatrix} \overline{\sigma}_{11} \\ \overline{\sigma}_{22} \\ \overline{\sigma}_{33} \\ \overline{\tau}_{23} \\ \overline{\tau}_{31} \\ \overline{\tau}_{12} \end{Bmatrix} \quad (3.7)$$

By contrast to generally anisotropic composite matrix Eq. (3.6), in the case of composite of *oblique anisotropy property* number of nonzeroth independent material structural modules equals 13. Among them: three *Young modules* E_{11}, E_{22}, E_{33}; three *Kirchhoff modules* G_{23}, G_{31}, G_{12}; three *Poisson ratios* $\nu_{21}, \nu_{31}, \nu_{32}$; one *Chencov modulus* $\mu_{31(23)}$; and three *Rabinovich modules* $\eta_{12(1)}, \eta_{12(2)}, \eta_{12(3)}$ are present in Eq. (3.7) instead of 21 (18 irreducible) shown in Eq. 3.6. On the other hand, appearance of some Chencov $\mu_{31(23)}$ and Rabinovich $\eta_{12(k)}$ coefficients allows to distinguish formats of the compliance matrices in case of the *rhombic fiber array* in which neither Rabinovoch nor Chencov coefficients are present, when the material orthotropy frame coincides with the effective stress/strain frame.

3.3.3 Orthotropic Composite (Lamina with Perpendicular Fiber Arrangement, Fig. 3.1c)

The narrower case of frequently used composites built of a number of layers which are long-fiber reinforced in an alternate perpendicular layer after layer fashion are called the *orthotropic multi-laminate composites*, commonly also named lamina. In corresponding elasticity matrices, compliance or stiffness, Rabinovich $\eta_{12(k)}$ and Chencov $\mu_{31(23)}$ coefficients (present in previously discussed Eq. 3.7) disappear in Eq. (3.8) such that the number of independent modules of the *effective elastic compliance* $\overline{E}_{ij}^{-1}$ or *stiffness matrix* $\overline{E}_{ij}$ is reduced to $9 = 13 - 4$, namely: 3 *Young modules* E_{11}, E_{22}, E_{33}; 3 *Kirchhoff modules* G_{23}, G_{31}, G_{12}; 3 *Poisson ratios* $\nu_{21}, \nu_{31}, \nu_{32}$. These equivalent anisotropy constants of composite have to be either measured in appropriate 9 tests or estimated by the use of a chosen homogenization method for assumed *perpendicular fiber arrangements* (see for instance Gan et al. [9] for Boron/Al composite)

$$\begin{Bmatrix} \bar{\varepsilon}_{11} \\ \bar{\varepsilon}_{22} \\ \bar{\varepsilon}_{33} \\ \bar{\gamma}_{23} \\ \bar{\gamma}_{31} \\ \bar{\gamma}_{12} \end{Bmatrix} = \left[\begin{array}{ccc|ccc} \frac{1}{E_{11}} & -\frac{\nu_{21}}{E_{11}} & -\frac{\nu_{31}}{E_{11}} & & & \\ -\frac{\nu_{12}}{E_{22}} & \frac{1}{E_{22}} & -\frac{\nu_{32}}{E_{22}} & & & \\ -\frac{\nu_{13}}{E_{33}} & -\frac{\nu_{23}}{E_{33}} & \frac{1}{E_{33}} & & & \\ \hline & & & \frac{1}{G_{23}} & & \\ & & & & \frac{1}{G_{31}} & \\ & & & & & \frac{1}{G_{12}} \end{array} \right] \begin{Bmatrix} \bar{\sigma}_{11} \\ \bar{\sigma}_{22} \\ \bar{\sigma}_{33} \\ \bar{\tau}_{23} \\ \bar{\tau}_{31} \\ \bar{\tau}_{12} \end{Bmatrix} \tag{3.8}$$

Transformation of the relation $\{\varepsilon\} = \left[\mathbb{E}^{-1}\right]\{\sigma\}$ to $\{\sigma\} = [\mathbb{E}]\{\varepsilon\}$ is not a trivial one in case of the elastic orthotropy. It can be done in a numerical fashion by finding the *stiffness matrix* $[\mathbb{E}]$ which is inverse to the *compliance matrix* $\left[\mathbb{E}^{-1}\right]$. Elements of the stiffness matrix $[\mathbb{E}]$ can be explicitly expressed in terms of nine *engineering constants of orthotropic material* determined $E_{11}, E_{22}, E_{33}, G_{23}, G_{13}, G_{12}, \nu_{21}, \nu_{31}$ and ν_{32} as follows (see Ochoa and Reddy [24], Tamma and Avila [32])

$$\begin{Bmatrix} \overline{\sigma}_{11} \\ \overline{\sigma}_{22} \\ \overline{\sigma}_{33} \\ \overline{\sigma}_{23} \\ \overline{\sigma}_{31} \\ \overline{\sigma}_{12} \end{Bmatrix} = \begin{bmatrix} E_{1111} & E_{1122} & E_{1133} & & & \\ E_{2211} & E_{2222} & E_{2233} & & & \\ E_{3311} & E_{3322} & E_{3333} & & & \\ & & & E_{2323} & & \\ & & & & E_{1313} & \\ & & & & & E_{1212} \end{bmatrix} \begin{Bmatrix} \overline{\varepsilon}_{11} \\ \overline{\varepsilon}_{22} \\ \overline{\varepsilon}_{33} \\ \overline{\gamma}_{23} \\ \overline{\gamma}_{31} \\ \overline{\gamma}_{12} \end{Bmatrix} \tag{3.9}$$

where subsequent elements of the stiffness matrix $[\mathbb{E}]$ are given by equations

$$\begin{array}{lll} E_{1111} = \frac{1-\nu_{23}\nu_{32}}{\Delta} E_{11} & E_{1122} = \frac{\nu_{12}+\nu_{13}\nu_{32}}{\Delta} E_{22} & \\ E_{1133} = \frac{\nu_{13}+\nu_{12}\nu_{23}}{\Delta} E_{33} & E_{2222} = \frac{1-\nu_{13}\nu_{31}}{\Delta} E_{22} & \\ E_{2233} = \frac{\nu_{23}+\nu_{21}\nu_{13}}{\Delta} E_{33} & E_{3333} = \frac{1-\nu_{12}\nu_{21}}{\Delta} E_{33} & \\ E_{2323} = G_{23} & E_{1313} = G_{13} & E_{1212} = G_{12} \end{array} \tag{3.10}$$

whereas symbol Δ denotes

$$\Delta = 1 - \nu_{12}\nu_{21} - \nu_{13}\nu_{31} - \nu_{23}\nu_{32} - \nu_{12}\nu_{23}\nu_{31} - \nu_{21}\nu_{13}\nu_{32} \tag{3.11}$$

Note that full orthotropic symmetry and population of both matrices stiffness (3.9) and compliance (3.8) is saved and refers to appropriate combinations of engineering constants but not to engineering constants separately, for instance

$$E_{1122} = \frac{\nu_{21} + \nu_{13}\nu_{32}}{\Delta} E_{22} = \frac{\nu_{12} + \nu_{31}\nu_{23}}{\Delta} E_{11} = E_{2211} \quad \text{etc.} \tag{3.12}$$

Hence only nine *orthotropy modules* are independent.

3.3.4 Unidirectional Long-Fiber Composite—Transversely Isotropic Tetragonal Type (Square Fiber Array, Fig. 3.1d)

Particular case of orthotropic composite is *transversely isotropic symmetry* unidirectional long-fiber-reinforced system in which fibers are built-in with the regular *tetragonal* manner (*square fiber array*, Fig. 3.1d). The effective elasticity matrix of such composite is described with six independent constants: E_{11}, E_{33}, ν_{21}, ν_{32}, G_{23} and G_{12} as shown in Eq. (3.13). At the level of RUC, tetragonal symmetry is observed (4 in-plane axes)

$$\begin{Bmatrix} \bar{\varepsilon}_{11} \\ \bar{\varepsilon}_{22} \\ \bar{\varepsilon}_{33} \\ \bar{\gamma}_{23} \\ \bar{\gamma}_{31} \\ \bar{\gamma}_{12} \end{Bmatrix} = \left[\begin{array}{ccc|ccc} \frac{1}{E_{11}} & -\frac{\nu_{21}}{E_{11}} & -\frac{\nu_{21}}{E_{11}} & & & \\ -\frac{\nu_{12}}{E_{22}} & \frac{1}{E_{22}} & -\frac{\nu_{32}}{E_{22}} & & & \\ -\frac{\nu_{12}}{E_{22}} & -\frac{\nu_{23}}{E_{22}} & \frac{1}{E_{22}} & & & \\ \hline & & & \frac{1}{G_{23}} & & \\ & & & & \frac{1}{G_{12}} & \\ & & & & & \frac{1}{G_{12}} \end{array} \right] \begin{Bmatrix} \bar{\sigma}_{11} \\ \bar{\sigma}_{22} \\ \bar{\sigma}_{33} \\ \bar{\tau}_{23} \\ \bar{\tau}_{31} \\ \bar{\tau}_{12} \end{Bmatrix} \tag{3.13}$$

3.3.5 *Unidirectional Long-Fiber Composite—Transversely Isotropic Hexagonal Type (Hexagonal Fiber Array Fig. 3.1e)*

In the another case of *unidirectionally reinforced composites*, when in the system fibers are row after row shifted by the half-distance, at the level of RUC the *hexagonal symmetry* property holds (six symmetry axes). Hence, only five from among mechanical constants are independent, since $G_{23} = \frac{E_{22}}{2(1+\nu_{23})}$

$$\begin{Bmatrix} \bar{\varepsilon}_{11} \\ \bar{\varepsilon}_{22} \\ \bar{\varepsilon}_{33} \\ \bar{\gamma}_{23} \\ \bar{\gamma}_{31} \\ \bar{\gamma}_{12} \end{Bmatrix} = \left[\begin{array}{ccc|ccc} \frac{1}{E_{11}} & -\frac{\nu_{21}}{E_{11}} & -\frac{\nu_{21}}{E_{11}} & & & \\ -\frac{\nu_{12}}{E_{22}} & \frac{1}{E_{22}} & -\frac{\nu_{32}}{E_{22}} & & & \\ -\frac{\nu_{12}}{E_{22}} & -\frac{\nu_{23}}{E_{22}} & \frac{1}{E_{22}} & & & \\ \hline & & & \frac{2(1+\nu_{23})}{E_{22}} & & \\ & & & & \frac{1}{G_{12}} & \\ & & & & & \frac{1}{G_{12}} \end{array} \right] \begin{Bmatrix} \bar{\sigma}_{11} \\ \bar{\sigma}_{22} \\ \bar{\sigma}_{33} \\ \bar{\tau}_{23} \\ \bar{\tau}_{31} \\ \bar{\tau}_{12} \end{Bmatrix} \tag{3.14}$$

The two types of transversely isotropic composites dependent on the fiber arrangement of either tetragonal or hexagonal symmetry are not always consistently examined which may lead to some erroneous conclusions (cf. Sun and Vaidya [30]).

3.3.6 *Regular Particle-Reinforced Composite—Cubic Symmetry (Regular Particles Arrangement, Fig. 3.1f)*

It is commonly assumed that the composites reinforced with a randomly distributed particles of irregular size and shape can be treated at the level of RVE as the isotropic continuum. However, in case of some *regular particle reinforcement* by repeating identical shape and size particles, the equivalent composite continuum exhibits the *cubic symmetry* (Fig. 3.1f). Among the crystal materials of cubic (regular) symmetry long list can be mentioned: Pyrites (cubic), Fluor Spar, Rock-salt, Potassium

Chloride (cf. Love [20]) or Tantalum, Aluminum, Gold, Copper, Germanium, α–iron, Magnesium Oxide (Magnesia), and Spinel ($MgAl_2O_4$) (cf. Berryman [5]). All cubic symmetry materials are characterized by three independent compliance modules: $\overline{E}_{11}^{-1}, \overline{E}_{12}^{-1}$ and $\overline{E}_{44}^{-1}$ where $\overline{E}_{44}^{-1} \neq (\overline{E}_{11}^{-1} - \overline{E}_{12}^{-1})/2$ or equivalently $G \neq \frac{E}{2(1+\nu)}$. In a similar way, the composite reinforced with three-directional mutually perpendicular short-fiber of the cubic symmetry is described by three independent engineering constants E, ν, and G

$$\begin{Bmatrix} \overline{\varepsilon}_{11} \\ \overline{\varepsilon}_{22} \\ \overline{\varepsilon}_{33} \\ \overline{\gamma}_{23} \\ \overline{\gamma}_{31} \\ \overline{\gamma}_{12} \end{Bmatrix} = \left[\begin{array}{ccc|ccc} \frac{1}{E} & -\frac{\nu}{E} & -\frac{\nu}{E} & & & \\ -\frac{\nu}{E} & \frac{1}{E} & -\frac{\nu}{E} & & & \\ -\frac{\nu}{E} & -\frac{\nu}{E} & \frac{1}{E} & & & \\ \hline & & & \frac{1}{G} & & \\ & & & & \frac{1}{G} & \\ & & & & & \frac{1}{G} \end{array}\right] \begin{Bmatrix} \overline{\sigma}_{11} \\ \overline{\sigma}_{22} \\ \overline{\sigma}_{33} \\ \overline{\tau}_{23} \\ \overline{\tau}_{31} \\ \overline{\tau}_{12} \end{Bmatrix} \tag{3.15}$$

3.3.7 Isotropic Composite (Random Particle Arrangement, Fig. 3.1g)

Irregular particle-reinforced composite in which the distribution shape and orientation of particles are fully disordered (chaotic) can be described at the level of the repeating RVE by the effective elasticity matrix (stiffness or compliance) characterized by two independent modules: $\overline{E}_{11}^{-1}, \overline{E}_{12}^{-1}$ ($\overline{E}_{44}^{-1} = (\overline{E}_{11}^{-1} - \overline{E}_{12}^{-1})/2$ or equivalently $G = \frac{E}{2(1+\nu)}$). In the *isotropic composite* with irregular particle reinforcement, no characteristic material frame can be distinguished inside RVE (infinite number of symmetry axes)

$$\begin{Bmatrix} \overline{\varepsilon}_{11} \\ \overline{\varepsilon}_{22} \\ \overline{\varepsilon}_{33} \\ \overline{\gamma}_{23} \\ \overline{\gamma}_{31} \\ \overline{\gamma}_{12} \end{Bmatrix} = \left[\begin{array}{ccc|ccc} \frac{1}{E} & -\frac{\nu}{E} & -\frac{\nu}{E} & & & \\ -\frac{\nu}{E} & \frac{1}{E} & -\frac{\nu}{E} & & & \\ -\frac{\nu}{E} & -\frac{\nu}{E} & \frac{1}{E} & & & \\ \hline & & & \frac{2(1+\nu)}{E} & & \\ & & & & \frac{2(1+\nu)}{E} & \\ & & & & & \frac{2(1+\nu)}{E} \end{array}\right] \begin{Bmatrix} \overline{\sigma}_{11} \\ \overline{\sigma}_{22} \\ \overline{\sigma}_{33} \\ \overline{\tau}_{23} \\ \overline{\tau}_{31} \\ \overline{\tau}_{12} \end{Bmatrix} \tag{3.16}$$

More general approach to describe particle-reinforced composites in which size/shape and topology of particles are ordered with the specific symmetries may lead to various symmetry classes of elastic matrices (cf. Banks-Sills et al. [3]).

3.4 Bounds for Effective Elastic Properties of Unidirectionally (Long Fiber) Reinforced Composites of Tetragonal or Hexagonal Symmetry

3.4.1 Nature of Homogenization Problem in Modeling of Heterogeneous Composites—Voigt and Reuss' Concept

Composite materials described in Sect. 3.3 have to be considered as two- or multi-component systems at the microlevel (microcomposites) or the nanolevel (nanocomposites). Composite materials are in essence nonhomogeneous or in fact *heterogeneous materials* due to different properties of the system constituents (components) commonly recognized as the *matrix* (most frequently *metallic*, *ceramic* or *polymer*) and the *reinforcing fibers* or *particles* (for instance long fibers made of *ceramic* or *metallic materials* and others) although the constituent materials are essentially homogeneous. At microscale, on boundaries between the components of different materials a jump of mechanical, thermal, and other properties arise. Averaging methods inside the *representative element* (RVE) or the representative cell (RUC) used for analysis of *multicomponent composite materials* known as *homogenization methods* are based on the assumption that it is possible to determine approximate values of the effective properties of the equivalent homogeneous composite (heterogeneous in fact) as well as uniform macrostress and macrostrain (nonuniform in fact at the microlevel), cf. Fig. 3.3. It is necessary to accept existence of the repeating *Representative Volume Element*—RVE (cf. e.g., Sun and Vaidya [30], Gan et al. [9], Würkner et al. [37], Bayat and Aghdam [4]), or the Representative Unit Cell—RUC

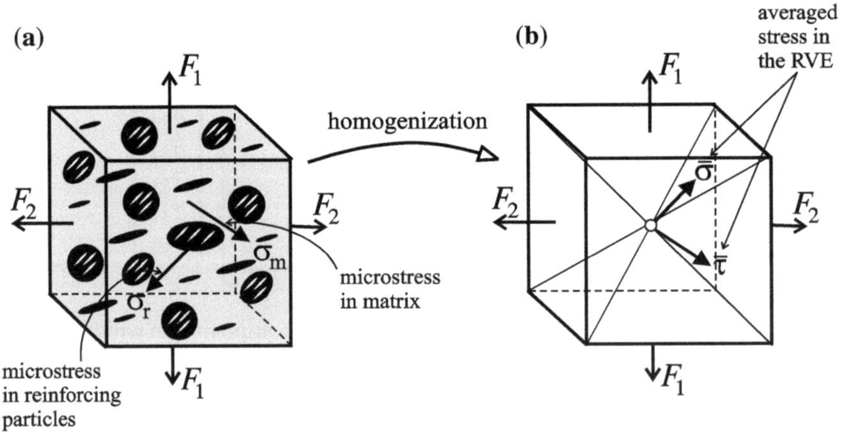

Fig. 3.3 Representative volume element RVE: **a** heterogeneous material at microscale, **b** homogeneous material at macroscale

(cf. e.g., Li and Wongsto [18], Li [17], Wongsto and Li [36], Pidaparti and May [26], Banks-Sills et al. [3], Herakovich and Aboudi [12]) which are subsequently divided into the subcells, Fig. 3.5. The RVE size or the RUC size and geometry have to be sufficiently large in order to properly catch an essence of composite system properties and behavior at the macroscale. Simultaneously, they have to be sufficiently small but repeatedly noticeable to assure that the representation of a uniform deformation field described by the displacement $\boldsymbol{u}$ and the gradient $\nabla \boldsymbol{u}$ such that the averaged (effective) strain $\overline{\varepsilon} = \frac{1}{2}(\nabla^{\mathrm{T}} \boldsymbol{u} + \nabla \boldsymbol{u})$ is justified (cf. Gan et al. [9]). Note that component material at the microlevel (or nanolevel) is usually isotropic; however, a multiphase composite can be either isotropic (for majority of particular composites) or anisotropic (for instance in case of fibrous composites reinforced with directionally oriented fiber beam).

The differences between the RVE (Representative Volume Element) and RUC (Repeating Unit Cell) concept are discussed in details by Drago and Pindera [8]. The authors claim that the concept of RVE is addressed to the statistically homogeneous material at an appropriate scale. Moreover it is assumed that the strain and stress are uniform throughout the RVE. On the other hand Drago and Pindera assume the periodicity in the material, both in strain and stress fields. However most researchers assume that the RUC is the periodic RVE and use its interchangeably [1, 30].

Traditionally it is assumed that the *particle-reinforced composites* in a disordered manner (e.g., with dispersed micro or nanoparticles as well as short micro or nanowires) show isotropic symmetry after homogenization (at the level of RVE). However, the above reasoning has to be accepted with necessary care. If repeatable shape and regular orientation of reinforcing particles are ensured throughout the matrix volume, in spite of the isotropic properties of both phases—*matrix* and *reinforcement* it may happen that after homogenization the averaged material modules at the macroscale (composite level) exhibit other than isotropic symmetry properties. Such problem was analyzed by Banks-Sills et al. [3] with respect to the *Glass-Epoxy composite*, by the use for simulation particles of various but regular geometries: spherical, cylindrical, cubic and rectangular parallelepiped. To be more precise the following unusual remark can be cited: "An interesting surprise for rotated particles was the existence of unusual material constants which cause normal deformations to produce orthogonal shear stresses and vice versa effect of Rabinovich's coefficients and shear deformations to produce orthogonal shear stresses and vice versa effect of Chencov's coefficients", cf. Banks-Sills et al. [3].

Only in the specific case if reinforcing particles are repeatedly spherical and do not exhibit same characteristic spatial distribution the assumption about isotropic symmetry at the macroscale (RVE-level) is reasonable to accept. In such specific case the classical mixture rules can be applied in order to achieve averaging methods: the *Voigt* [34] or the *Reuss estimates* [27]. In the simplest case of two isotropic constituent phase materials, Voigt and Reuss' *rules of mixture* are simply based on the volume fraction of matrix V_1 and reinforcement V_2

$$
\begin{aligned}
{}^{\mathrm{V}}\overline{p} &= p_1 c_1 + p_2 c_2 && \text{Voigt's rule} \\
\frac{1}{{}^{\mathrm{R}}\overline{p}} &= \frac{c_1}{p_1} + \frac{c_2}{p_2} && \text{Reuss' rule}
\end{aligned}
\tag{3.17}
$$

Symbols p_1 and p_2 stand for elastic constants of constituent materials, matrix and reinforcement (particles), for instance Young modules E_1 and E_2 and Kirchhoff modules G_1 and G_2 whereas ${}^{\mathrm{V}}\overline{p}$ and ${}^{\mathrm{R}}\overline{p}$ denote the corresponding effective modules $\overline{E}$ and $\overline{G}$ averaged at the RVE level. Symbols c_1 and c_2 stand for *volume fraction* of both phases (V_{f} and $1 - V_{\mathrm{f}}$) with *irregular particles distribution* throughout the RVE ignoring effect of local concentration density, size and shape of particles and their orientation and mutual interaction, see Fig. 3.3a. After homogenization, the *averaged* (effective) *stress* $\overline{\sigma}$ and $\overline{\tau}$ are met in RVE instead of different *microstresses* in constituent materials: matrix σ_{m} and reinforcement σ_{r} (see Fig. 3.3b).

The mixture rules Voigt and Reuss' (3.17) lead to different estimates of averaged material constants of homogenized isotropic continuum $\overline{E}$ and $\overline{G}$. In case of *Voigt estimate* compatibility of strains in both phase materials is assumed, whereas in case of *Reuss' estimate* compatibility of stresses is postulated. The first approach leads to discontinuity of stress at the boundary between constituents whereas the second approach causes strain discontinuity. In other words, the Voigt approximation can be treated as equivalent to kinematically admissible approach in contrast to the Reuss approximation which is statically admissible. In fact at the microlevel of *heterogeneous composite* both stress and strain continuity hold such that the Voigt and the Reuss approximations can serve as *upper* and *lower estimates* for the *effective stiffness matrix* elements of anisotropic composite systems (cf. Herakovich [11], Gan et al. [9]). In the impressive monograph "Micromechanics of composite materials," Aboudi et al. [1] analyze the effective engineering constants of the Glass/Epoxy fibrous composite $E_{11}, E_{22} = E_{33}, \nu_{12} = \nu_{13}, \nu_{23}, G_{12} = G_{13}, G_{23}$ as functions of fiber volume fraction V_{f}. This findings generally confirm the upper and lower bounds by Voigt and Reuss' isotropic estimates except for the transverse Poisson ratio ν_{23} for which an excess of the bounds is observed.

In order to simply explain the essence of Voigt and Reuss' estimates, consider elementary one-dimensional two-component mechanical systems sketched in Fig. 3.4 representing: (a) Voigt, (b) Reuss' and (c) the effective homogeneous elements.

In case of *Voigt scheme*, Fig. 3.4a, two bars of A_1 and A_2 cross-sectional areas that represent matrix and reinforcement (particle) of the same length l are jointed in parallel ($\bar{l} = l_1 = l_2$ and $\overline{A} = A_1 + A_2$). Loading force F is separated between matrix and reinforcement $F = F_1 + F_2$ whereas identical elongation of both constituents is equal to the averaged elongation of substituting homogeneous system, Fig. 3.4c: $\Delta\bar{l} = \Delta l_1 = \Delta l_2$. Hence, when the Hooke law is applied to schemes a) and c) we arrive at distribution of force between matrix and reinforcement

$$
F_1 = \frac{E_1 A_1}{\overline{E}(A_1 + A_2)} F \qquad F_2 = \frac{E_2 A_2}{\overline{E}(A_1 + A_2)} F \tag{3.18}
$$

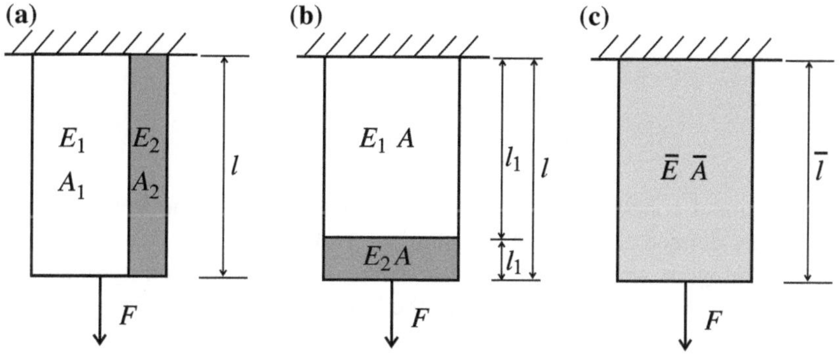

Fig. 3.4 Uniaxial mechanical models for mixture rules application in composites: **a** Voigt rule, **b** Reuss' rule, **c** effective homogeneous material

Finally, introducing definitions of volume fractions $V_1 = A_1/\overline{A}$ and $V_2 = A_2/\overline{A}$ the Voigt-based effective Young modulus ${}^{\mathrm{V}}\overline{E}$ is furnished

$$\overline{E} = E_1 V_1 + E_2 V_2 = {}^{\mathrm{V}}\overline{E} \tag{3.19}$$

In case of *Reuss' scheme*, Fig. 3.4b, two bars of different lengths l_1 and l_2 and $A_1 = A_2 = \overline{A}$ representing matrix and reinforcement materials are joined in series and loaded by identical force $F = F_1 = F_2$ whereas averaged elongation of substitutive system Fig. 3.4c is the sum of component elongations $\Delta\bar{l} = \Delta l_1 + \Delta l_2$. Again, when Hooke law is applied to schemes (b) and (c) the following must hold

$$\frac{F\bar{l}}{\overline{E}\,\overline{A}} = \frac{Fl_1}{E_1\overline{A}} + \frac{Fl_2}{E_2\overline{A}} \tag{3.20}$$

Finally applying definitions of volume fractions $V_1 = l_1/\bar{l}$ and $V_2 = l_2/\bar{l}$ we arrive at the Reuss-based effective Young modulus ${}^{\mathrm{R}}E$ in the format

$$\frac{1}{\overline{E}} = \frac{V_1}{E_1} + \frac{V_2}{E_2} = \frac{1}{{}^{\mathrm{R}}\overline{E}} \tag{3.21}$$

In order to make further considerations easier we introduce original notation used by Hill in [13]. In this way equations describing the *uniaxial Voigt* and *Reuss' models* can be rewritten in the new following formats. In case of Voigt model the identity of strains in both phases ${}^{\mathrm{V}}\overline{\varepsilon} = \varepsilon_1 = \varepsilon_2$ holds. Hence the following set of equations describe the uniaxial Voigt model

$$
\begin{aligned}
{}^{\mathrm{V}}\overline{\sigma} &= {}^{\mathrm{V}}c_1\sigma_1 + {}^{\mathrm{V}}c_2\sigma_2 \\
{}^{\mathrm{V}}\overline{\sigma} &= {}^{\mathrm{V}}\overline{E}\,{}^{\mathrm{V}}\overline{\varepsilon} \\
{}^{\mathrm{V}}\overline{E}\,{}^{\mathrm{V}}\overline{\varepsilon} &= {}^{\mathrm{V}}c_1E_1\varepsilon_1 + {}^{\mathrm{V}}c_2E_2\varepsilon_2 \\
{}^{\mathrm{V}}\overline{E} &= {}^{\mathrm{V}}c_1E_1 + {}^{\mathrm{V}}c_2E_2
\end{aligned}
\tag{3.22}
$$

where fractional concentrations by volume of the phases in the Voigt model (see Fig. 3.4a) are defined as ${}^{\mathrm{V}}c_1 = A_1/\overline{A}$ and ${}^{\mathrm{V}}c_2 = A_2/\overline{A}$; (${}^{\mathrm{V}}c_1 + {}^{\mathrm{V}}c_2 = 1$).

In the analogous way in case of *uniaxial Reuss' model* the identity of stresses in both phases ${}^{\mathrm{R}}\overline{\sigma} = \sigma_1 = \sigma_2$ holds, hence the basic set of equations is

$$
\begin{aligned}
{}^{\mathrm{R}}\overline{\varepsilon} &= {}^{\mathrm{R}}c_1\varepsilon_1 + {}^{\mathrm{R}}c_2\varepsilon_2 \\
{}^{\mathrm{R}}\overline{\varepsilon} &= \frac{{}^{\mathrm{R}}\overline{\varepsilon}}{{}^{\mathrm{R}}\overline{E}} \\
\frac{{}^{\mathrm{R}}\overline{\sigma}}{{}^{\mathrm{R}}\overline{E}} &= {}^{\mathrm{R}}c_1\frac{\sigma_1}{E_1} + {}^{\mathrm{V}}c_2\frac{\sigma_2}{E_2} \\
\frac{1}{{}^{\mathrm{R}}\overline{E}} &= \frac{{}^{\mathrm{R}}c_1}{E_1} + \frac{{}^{\mathrm{R}}c_2}{E_2}
\end{aligned}
\tag{3.23}
$$

where fractional concentrations by volume of the phases in the Reuss model (see Fig. 3.4b) are defined as ${}^{\mathrm{R}}c_1 = l_1/\overline{l}$ and ${}^{\mathrm{R}}c_2 = l_2/\overline{l}$; (${}^{\mathrm{R}}c_1 + {}^{\mathrm{R}}c_2 = 1$).

In fact both pairs ${}^{\mathrm{V}}c_1$, ${}^{\mathrm{V}}c_2$ and ${}^{\mathrm{R}}c_1$, ${}^{\mathrm{R}}c_2$ can be interpreted as common volume fraction of both phases V_{f} and $1 - V_{\mathrm{f}}$ in the uniaxial models of the same material, hence it must hold

$$
c_1 = {}^{\mathrm{V}}c_1 = {}^{\mathrm{R}}c_1 = V_{\mathrm{f}} \qquad c_2 = {}^{\mathrm{V}}c_2 = {}^{\mathrm{R}}c_2 = 1 - V_{\mathrm{f}}
\tag{3.24}
$$

Note that the Poisson effect is ignored in aforementioned considerations.

3.4.2 General 3D Formulation of Voigt and Reuss' Homogenization Estimates

On the RVE level, that represents heterogeneous material, the definitions of either the *averaged stress* or the *averaged strain* tensors can be written down

$$
\overline{\boldsymbol{\sigma}} = \frac{1}{V_{\mathrm{RVE}}}\int_{V_{\mathrm{RVE}}} \boldsymbol{\sigma}\,\mathrm{d}V
\tag{3.25}
$$

or

$$
\overline{\boldsymbol{\varepsilon}} = \frac{1}{V_{\mathrm{RVE}}}\int_{V_{\mathrm{RVE}}} \boldsymbol{\varepsilon}\,\mathrm{d}V
\tag{3.26}
$$

where V_{RVE} denotes a *volume of* the chosen *RVE*, see Aboudi et al. [1].

Average values of stress and strain $\overline{\boldsymbol{\sigma}}$ and $\overline{\boldsymbol{\varepsilon}}$ in RVE are given in terms of $\overline{\boldsymbol{\sigma}}_1, \overline{\boldsymbol{\sigma}}_2$ and $\overline{\boldsymbol{\varepsilon}}_1, \overline{\boldsymbol{\varepsilon}}_2$ in the phases by the following relations

$$\overline{\boldsymbol{\sigma}} = c_1 \overline{\boldsymbol{\sigma}}_1 + c_2 \overline{\boldsymbol{\sigma}}_2 \qquad \overline{\boldsymbol{\varepsilon}} = c_1 \overline{\boldsymbol{\varepsilon}}_1 + c_2 \overline{\boldsymbol{\varepsilon}}_2 \tag{3.27}$$

Since the elastic material is assumed for both phases the obvious relations must hold at any point in the phases

$$\boldsymbol{\sigma}_1 = \mathbb{E}_1 : \boldsymbol{\varepsilon}_1 \quad \text{and} \quad \boldsymbol{\sigma}_2 = \mathbb{E}_2 : \boldsymbol{\varepsilon}_2 \tag{3.28}$$

and

$$\boldsymbol{\varepsilon}_1 = \mathbb{E}_1^{-1} : \boldsymbol{\sigma}_1 \quad \text{and} \quad \boldsymbol{\varepsilon}_2 = \mathbb{E}_2^{-1} : \boldsymbol{\sigma}_2 \tag{3.29}$$

if the inverse format is used.

Substitution of (3.28) and (3.29) into (3.27) with the assumption that phases are uniform and isotropic ($\boldsymbol{\sigma}_{1,2} = \overline{\boldsymbol{\sigma}}_{1,2}, \boldsymbol{\varepsilon}_{1,2} = \overline{\boldsymbol{\varepsilon}}_{1,2}$) the analogous relations hold between the average quantities

$$\overline{\boldsymbol{\sigma}} = c_1 \mathbb{E}_1 : \overline{\boldsymbol{\varepsilon}}_1 + c_2 \mathbb{E}_2 : \overline{\boldsymbol{\varepsilon}}_2 \qquad \overline{\boldsymbol{\varepsilon}} = c_1 \mathbb{E}_1^{-1} : \overline{\boldsymbol{\sigma}}_1 + c_2 \mathbb{E}_2^{-1} : \overline{\boldsymbol{\sigma}}_2 \tag{3.30}$$

where consistently $\overline{\boldsymbol{\varepsilon}}_1$ and $\overline{\boldsymbol{\varepsilon}}_2$, as well as $\overline{\boldsymbol{\sigma}}_1$ and $\overline{\boldsymbol{\sigma}}_2$, stand for uniform strain and uniform stress fields in each of the phases in RVE, respectively.

A distribution of the two-phase materials in the RVE is obviously not necessarily random, but must be *structurally representative distribution* for composite material at the *macrolevel*. In the light of above remark a unique relationship between the *average strains in* the *phases* $\overline{\boldsymbol{\varepsilon}}_1, \overline{\boldsymbol{\varepsilon}}_2$ upon the *average overall strain in RVE* $\overline{\boldsymbol{\varepsilon}}$ can be furnished by the use of *strain concentration tensors* $\mathbb{A}_1$ and $\mathbb{A}_2$

$$\overline{\boldsymbol{\varepsilon}}_1 = \mathbb{A}_1 : \overline{\boldsymbol{\varepsilon}} \qquad \overline{\boldsymbol{\varepsilon}}_2 = \mathbb{A}_2 : \overline{\boldsymbol{\varepsilon}} \tag{3.31}$$

where the obvious condition holds $c_1 \mathbb{A}_1 + c_2 \mathbb{A}_2 = \mathbb{I}$ with $\mathbb{I}$ being the *unit tensor*. By combining Eq. (3.31) with Eq. (3.30) we arrive at

$$\overline{\boldsymbol{\sigma}} = (c_1 \mathbb{E}_1 : \mathbb{A}_1 + c_2 \mathbb{E}_2 : \mathbb{A}_2) : \overline{\boldsymbol{\varepsilon}} = \overline{\mathbb{E}} : \overline{\boldsymbol{\varepsilon}} \tag{3.32}$$

where $\overline{\mathbb{E}}$ stands for the effective stiffness tensor of the overall composite.

Equivalently reverse unique relationships between the *average stresses in* the *phases* $\overline{\boldsymbol{\sigma}}_1, \overline{\boldsymbol{\sigma}}_2$ upon the *average stress in RVE* $\overline{\boldsymbol{\sigma}}$

$$\overline{\boldsymbol{\sigma}}_1 = \mathbb{B}_1 : \overline{\boldsymbol{\sigma}} \qquad \overline{\boldsymbol{\sigma}}_2 = \mathbb{B}_2 : \overline{\boldsymbol{\sigma}} \tag{3.33}$$

must hold if the *stress concentration tensors* $\mathbb{B}_1$ and $\mathbb{B}_2$ which satisfy the relation $c_1 \mathbb{B}_1 + c_2 \mathbb{B}_2 = \mathbb{I}$, are introduced. Again combining Eq. (3.33) with the second of Eq. (3.30) we arrive at

$$\overline{\varepsilon} = (c_1 \mathbb{E}_1^{-1} : \mathbb{B}_1 + c_2 \mathbb{E}_2^{-1} : \mathbb{B}_2) : \overline{\boldsymbol{\sigma}} = \overline{\mathbb{E}}^{-1} : \overline{\boldsymbol{\sigma}} \tag{3.34}$$

where $\overline{\mathbb{E}}^{-1}$ is the *effective compliance tensor of* the *composite*.

The first homogenization rule was introduced by Voigt (1889) [34] as average constants of polycrystals. Assuming the strain concentration is constant $\mathbb{A}_1 = \mathbb{A}_2 = \mathbb{I}$ and *strain* is *uniform* $\overline{\varepsilon}_1 = \overline{\varepsilon}_2 = \overline{\varepsilon}$, it follows:

$$\overline{\mathbb{E}} = c_1 \mathbb{E}_1 + c_2 \mathbb{E}_2 \tag{3.35}$$

Equation (3.35) provides the *effective stiffness matrix* elements *of* the *composite* in terms of the *volume-averaged stiffness* of individual phases.

By contrast, Reuss (1929) [27] assumed that constituents of the composite are subjected to a *uniform stress* equal to the *average stress in RVE* $\mathbb{B}_1 = \mathbb{B}_2 = \mathbb{I}$ in Eq. (3.33) and effective compliance is given by a rule of mixture as follows:

$$\overline{\mathbb{E}}^{-1} = c_1 \mathbb{E}_1^{-1} + c_2 \mathbb{E}_2^{-1} \tag{3.36}$$

Note that in fact neither the Voigt nor the Reuss assumption is correct. The implied stress due to Voigt causes tractions at phase boundaries not satisfying equilibrium $\overline{\sigma}_1 \neq \overline{\sigma}_2$. On the other hand the implied strain due to Reuss' causes discontinuity of strain at the interface between matrix and particle $\overline{\varepsilon}_1 \neq \overline{\varepsilon}_2$.

3.4.3 Theorem of Lower and Upper Bounds by Voigt and Reuss' Estimation

Hill theorem, which is called the *theorem of lower and upper bounds*, allows to connect a constitutive description at two scales: *micro level* at the point level and the *meso level*, where the representative volume element RVE is defined. After Auriault et al. [2], it is assumed that:

- the *global variables* are the volume means of the local stress and strains, and that the conservation and constitutive equations have the same structure at microscopic and mesoscales,
- the assumption of *energetic consistency*, known as the Hill principle, which imposes equality on the energy contained within the medium, whether it is expressed in local variables or using variables defined at mesoscale.

According to the second assumption, the *equivalence of energy* at micro and RVE level leads to the following formula:

$$\int_V \boldsymbol{\sigma} : \boldsymbol{\varepsilon} \mathrm{d}V = \int_V \overline{\boldsymbol{\sigma}} : \overline{\varepsilon} \mathrm{d}V = V \overline{\boldsymbol{\sigma}} : \overline{\varepsilon} \tag{3.37}$$

where $V = V_{\mathrm{RVE}}$ is used for brevity. Hence, when the Hooke law is applied, both at micro level $\boldsymbol{\sigma} = \mathbb{E} : \boldsymbol{\varepsilon}$ and mesoscale $\overline{\boldsymbol{\sigma}} = \overline{\mathbb{E}} : \overline{\boldsymbol{\varepsilon}}$, the previous equation can be rewritten as

$$\int_V \boldsymbol{\varepsilon} : \mathbb{E} : \boldsymbol{\varepsilon} \mathrm{d}V = V \overline{\boldsymbol{\varepsilon}} : \overline{\mathbb{E}} : \overline{\boldsymbol{\varepsilon}} \tag{3.38}$$

According to the *Hill–Mandel relation* and Eq. (3.37) the following equality holds:

$$\overline{\boldsymbol{\sigma}} : \overline{\boldsymbol{\varepsilon}} = \left(\frac{1}{V} \int_V \boldsymbol{\sigma} \mathrm{d}V \right) : \left(\frac{1}{V} \int_V \boldsymbol{\varepsilon} \mathrm{d}V \right) = \frac{1}{V} \int_V \boldsymbol{\sigma} : \boldsymbol{\varepsilon} \mathrm{d}V = (\overline{\boldsymbol{\sigma} : \boldsymbol{\varepsilon}}) \tag{3.39}$$

Let us consider the Representative Volume Element bounded by surface S in which uniform strain field $\overline{\boldsymbol{\varepsilon}} = \text{const}$ accompanies linear displacement field $\boldsymbol{u} = \overline{\boldsymbol{\varepsilon}} \cdot \boldsymbol{x}$, hence the external work can be rewritten down as follows:

$$L_z = \frac{1}{2} \int_S \boldsymbol{t} \cdot \boldsymbol{u} \mathrm{d}S = \frac{1}{2} \int_S \boldsymbol{t} \cdot \overline{\boldsymbol{\varepsilon}} \cdot \boldsymbol{x} \mathrm{d}S = \frac{1}{2} \overline{\boldsymbol{\varepsilon}} \cdot \int_S \boldsymbol{t} \cdot \boldsymbol{x} \mathrm{d}S \tag{3.40}$$

Applying the traction boundary condition in following form $\boldsymbol{t} = \boldsymbol{\sigma} \cdot \boldsymbol{n}$, where $\boldsymbol{n}$ stands for a normal vector to the surface, and the Gauss theorem of divergence, the Eq. (3.40) can be rewritten as follows:

$$L_z = \frac{1}{2} \overline{\boldsymbol{\varepsilon}} \cdot \int_V \operatorname{div}(\boldsymbol{\sigma} \cdot \boldsymbol{x}) \, \mathrm{d}V = \frac{1}{2} \overline{\boldsymbol{\varepsilon}} \cdot \int_V [\operatorname{div}(\boldsymbol{\sigma}) \cdot \boldsymbol{x} + \boldsymbol{\sigma} \cdot \operatorname{div}(\boldsymbol{x})] \, \mathrm{d}V \tag{3.41}$$

The uniform stress accompanying the uniform strain leads to $\operatorname{div}(\boldsymbol{\sigma}) = 0$ hence the external work (3.41) reduces to

$$L_z = \frac{1}{2} \overline{\boldsymbol{\varepsilon}} \cdot \int_V \boldsymbol{\sigma} \mathrm{dV} = \frac{1}{2} \overline{\boldsymbol{\sigma}} : \overline{\boldsymbol{\varepsilon}} = \frac{1}{2} (\overline{\boldsymbol{\sigma} : \boldsymbol{\varepsilon}}) \tag{3.42}$$

when the Hill–Mandel relation is applied. Applying assumption, that the constitutive relations at both scales are the same:

$$\overline{\boldsymbol{\sigma}} = \overline{\mathbb{E}} : \overline{\boldsymbol{\varepsilon}}, \qquad \boldsymbol{\sigma} = \mathbb{E} : \boldsymbol{\varepsilon} \tag{3.43}$$

the Eq. (3.42) can be rewritten as follows:

$$\overline{\boldsymbol{\sigma}} : \overline{\boldsymbol{\varepsilon}} = (\overline{\boldsymbol{\sigma} : \boldsymbol{\varepsilon}}) = \frac{1}{V} \int_V \boldsymbol{\varepsilon} : \mathbb{E} : \boldsymbol{\varepsilon} \mathrm{dV} = \overline{\boldsymbol{\varepsilon}} : \overline{\mathbb{E}} : \overline{\boldsymbol{\varepsilon}} \tag{3.44}$$

Let us defined a new *fictitious stress* field $\widehat{\boldsymbol{\sigma}}$, where the Hooke law can be defined as:

$$\widehat{\boldsymbol{\sigma}} = \mathbb{E} : \overline{\boldsymbol{\varepsilon}} \tag{3.45}$$

The real, effective fields (e.g., stress $\boldsymbol{\sigma}$ and strain $\boldsymbol{\varepsilon}$) must fulfil the *theorem of minimal potential energy*, hence the energy based on a fictitious stress field $\widehat{\boldsymbol{\sigma}}$ must be greater than effective one, so the following inequality is true:

$$\left(\overline{\boldsymbol{\sigma} : \boldsymbol{\varepsilon}}\right) = \frac{1}{V} \int\limits_V \boldsymbol{\sigma} : \boldsymbol{\varepsilon} \mathrm{d}V \le \frac{1}{V} \int\limits_V \widehat{\boldsymbol{\sigma}} : \overline{\boldsymbol{\varepsilon}} \mathrm{d}V \tag{3.46}$$

Input of the Eq. (3.44) to the left-hand side of above inequality and the definition of fictitious stress (3.45) on the right-hand side, yields the inequality:

$$\frac{1}{V} \int\limits_V \boldsymbol{\varepsilon} : \mathbb{E} : \boldsymbol{\varepsilon} \mathrm{d}V = \overline{\boldsymbol{\varepsilon}} : \overline{\mathbb{E}} : \overline{\boldsymbol{\varepsilon}} \le \int\limits_V \overline{\boldsymbol{\varepsilon}} : \mathbb{E} : \overline{\boldsymbol{\varepsilon}} \mathrm{d}V = \overline{\boldsymbol{\varepsilon}} : \overline{\boldsymbol{\varepsilon}} : \left(\frac{1}{V} \int\limits_V \mathbb{E} \mathrm{d}V \right) \tag{3.47}$$

After some rearrangements the inequality (3.47) can be rewritten as follows:

$$\overline{\mathbb{E}} \le \frac{1}{V} \int\limits_V \mathbb{E} \mathrm{d}V \tag{3.48}$$

Inequality (3.48) means that the *effective stiffness tensor on RVE* level is the *lower bound of mean constitutive tensor* on micro level, where the mean operation is calculated over the volume of Representative Volume Element.

Consider the two-phase continuum, where the total volume of RVE is a sum of two volumes $V = V_1 \cup V_2$. Next, it is assumed that for the both phases constitutive law is Hooke equation, where the material behavior is defined by the tensors $\mathbb{E}_1$ and $\mathbb{E}_2$. Hence it is possible to change the continuous formulation described by Eq. (3.48) to the discrete form as follows, compare (3.32):

$$\overline{\mathbb{E}} \le c_1 \mathbb{E}_1 + c_2 \mathbb{E}_2 = {}^{\mathrm{V}}\overline{\mathbb{E}} \tag{3.49}$$

where $c_1 = V_1/V$, $c_2 = V_2/V$ and $c_1 + c_2 = 1$. The right-hand side of above equation is well-known relation called *Voigt estimation*, which means that Voigt formula is a *lower bound of* the *effective stiffness matrix* components.

On the other hand it is assumed that across entire boundary S the uniform boundary conditions $\boldsymbol{t} = \overline{\boldsymbol{\sigma}} \cdot \boldsymbol{n}$ hold, where $\overline{\boldsymbol{\sigma}}$ is a uniform stress in the representative volume RVE. In this case the *work of external forces* is as follows

$$L_z = \frac{1}{2}\int_S \boldsymbol{t}\cdot\boldsymbol{u}\mathrm{d}S = \frac{1}{2}\int_S \overline{\boldsymbol{\sigma}}\cdot\boldsymbol{n}\cdot\boldsymbol{u}\mathrm{d}S \tag{3.50}$$

Consider the theorem of divergence:

$$L_z = \frac{1}{2}\overline{\boldsymbol{\sigma}} : \int_V \frac{1}{2}\left(\nabla\boldsymbol{u} + \nabla^T\boldsymbol{u}\right)\mathrm{d}V = \frac{1}{2}\overline{\boldsymbol{\sigma}} : \int_V \boldsymbol{\varepsilon}\mathrm{d}V = \frac{1}{2}\overline{\boldsymbol{\sigma}} : \overline{\boldsymbol{\varepsilon}} \tag{3.51}$$

According to Eq. (3.44) and substituting Hooke law $\overline{\boldsymbol{\varepsilon}} = \mathbb{E}^{-1} : \overline{\boldsymbol{\sigma}}$ the work of internal forces can be evaluated as follows:

$$\int_V \boldsymbol{\sigma} : \boldsymbol{\varepsilon}\mathrm{d}V = V\overline{\boldsymbol{\sigma}} : \overline{\boldsymbol{\varepsilon}} = V\overline{\boldsymbol{\sigma}} : \overline{\mathbb{E}}^{-1} : \overline{\boldsymbol{\sigma}} \tag{3.52}$$

Consider now a new *fictitious strain* field $\widehat{\boldsymbol{\varepsilon}} = \overline{\mathbb{E}}^{-1} : \overline{\boldsymbol{\sigma}}$ defined in an analogous fashion as fictitious stress (3.45). On the base of theorem of minimum of potential energy, the inequality as follows must be true:

$$\int_V \boldsymbol{\sigma} : \boldsymbol{\varepsilon}\mathrm{d}V = V\overline{\boldsymbol{\sigma}} : \boldsymbol{\varepsilon} \leq \int_V \overline{\boldsymbol{\sigma}} : \widehat{\boldsymbol{\varepsilon}}\mathrm{d}V \tag{3.53}$$

According to Hooke law applied to the term of right-hand side in above equation and taking into account uniform stress $\overline{\boldsymbol{\sigma}}$, it can be evaluated, compare (3.47):

$$\overline{\mathbb{E}}^{-1} \leq \frac{1}{V}\int_V \mathbb{E}^{-1}\mathrm{d}V \tag{3.54}$$

Consider a similar continuum like previous one, where the whole volume of RVE is a sum of two volumes $V = V_1 \cup V_2$. Next, it is assumed that for the both phases constitutive law is Hooke equation, where the material behavior is defined by the tensors $\mathbb{E}_1$ and $\mathbb{E}_2$. Therefore it is possible to change the continuous formulation described by Eq. (3.54) to discrete form as follows, compare (3.49):

$$\overline{\mathbb{E}}^{-1} \leq c_1\mathbb{E}_1^{-1} + c_2\mathbb{E}_2^{-1} = {}^{\mathrm{R}}\overline{\mathbb{E}}^{-1} \tag{3.55}$$

where $c_1 = V_1/V$, $c_2 = V_2/V$ and $c_1 + c_2 = 1$. The right-hand side of above equation is well-known relation called *Reuss' estimation*, which means that Reuss' formula is a *lower bound of* the *effective compliance matrix* components, or equivalently the *upper bound of* the elements of the *effective stiffness matrix*, because the product $\overline{\mathbb{E}} : \overline{\mathbb{E}}^{-1}$ is equal identity tensor $\mathbb{I}$.

3.5 Micromechanics-Based Homogenization Methods

3.5.1 *Effective Elastic Stiffness Matrices of Unidirectional Composites*

It is incorrect to directly apply the Voigt and the Reuss rules to anisotropic composites since these simple isotropic mixture rules are based on volume fraction of matrix and reinforcement materials V_{m} and V_{r}, but not on true constituents geometry and topology. Hence, Voigt and Reuss' approximations are insufficient for correct estimation of the effective modules of stiffness or compliance matrices of true composite system, for instance with *long-fiber-reinforced composite architecture* of various symmetry. Temporary *micromechanics-based homogenization models* take into account not only the volume fraction of constituents, but also their configuration, geometry and other factors such as *built-in residual stresses* due to fabrication methods. Among them the following homogenization methods are frequently used: the method of *Concentric Cylinder Assembly* (CCA), Hashin and Rosen [10], the *Mori–Tanaka Method* (MT), Mori and Tanaka [22], the *Generalized Method of Cells* (GMC), Paley and Aboudi [25] or *Strain Compatible Method of Cells* (SCMC), Gan et al. [9]. Not going deeply in details, all micromechanics based homogenization methods assume existence of periodically repeating *Representative Volume Element* (RVE) or *Representative Unit Cell* (RUC) the size and geometry of which must capture the essence of the true composite behavior on the macroscale, and which can be mapped into a point of a homogeneous continuum characterized by the displacement field $\boldsymbol{u}$ and the gradient $\nabla\boldsymbol{u}$. Two common GMC and SCMC assumptions are: displacements continuity inside the cell and across the subcell boundaries, and constant strain within the subcells $\varepsilon^{(\beta\gamma)}$, Fig. 3.5. However, by contrast to SCMC method the GMC method does not account for coupling between the transverse shear stresses and the transverse normal stresses, cf. Gan et al. [9]. However, both *microstresses* and *microstrains* averaged at the RUC level $\overline{\boldsymbol{\sigma}}$ and $\overline{\boldsymbol{\varepsilon}}$ are periodical and repeatable at the macroscale (cf. Sun and Vaidya [30]).

Exemplary 2D cross-section of the square representative unit cell hl of *unidirectionally long-fiber-reinforced composite* with RUC domain lh divided into subcells built of different material $h_\beta^{(k)} l_\gamma^{(k)}$, where $k = \mathrm{m}$ and $k = \mathrm{r}$ stand for matrix and reinforcing fiber, is presented in Fig. 3.5. At the subcell level $h_\beta l_\gamma$ the *local elasticity equation* holds combining local variables in the subcell, *microstrain* $\varepsilon^{(\beta\gamma)}$ and *microstress* $\boldsymbol{\sigma}^{(\beta\gamma)}$

$$\sigma_{ij}^{(\beta\gamma)} = E_{ijkl}^{(\beta\gamma)} \varepsilon_{kl}^{(\beta\gamma)} \tag{3.56}$$

where $E_{ijkl}^{(\beta\gamma)}$ denotes *local stiffness tensor in subcell* $(\beta\gamma)$, different for the matrix material $E_{ijkl}^{(\mathrm{m})}$ and the reinforcing fiber material $E_{ijkl}^{(\mathrm{r})}$. Effective strain $\overline{\varepsilon}_{ij}$ and effective stress $\overline{\sigma}_{ij}$ averaged inside RUC are defined by approximate equations

Fig. 3.5 2D (x_2, x_3 plane) dicretization of the RUC cross-section in unidirectional composite of fiber direction coincident with x_1 direction with size $h \times l$ divided into subcells $h_\beta^{(m)} l_\gamma^{(m)}$ (matrix) and $h_\beta^{(r)} l_\gamma^{(r)}$ (single fiber) of *circular shape* approximated by sufficiently dense *square* subcells

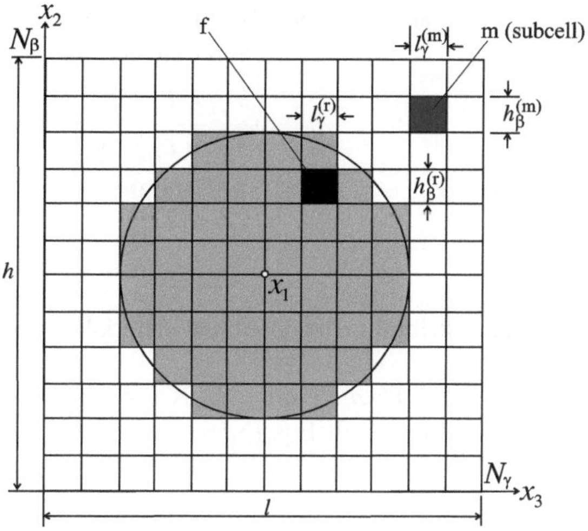

$$
\begin{aligned}
\overline{\varepsilon}_{ij} &= \frac{1}{hl} \sum_{\beta=1}^{N_\beta} \sum_{\gamma=1}^{N_\gamma} h_\beta l_\gamma \varepsilon_{ij}^{(\beta\gamma)} \\
\overline{\sigma}_{ij} &= \frac{1}{hl} \sum_{\beta=1}^{N_\beta} \sum_{\gamma=1}^{N_\gamma} h_\beta l_\gamma \sigma_{ij}^{(\beta\gamma)}
\end{aligned}
\tag{3.57}
$$

When the inverse formulas for local variables are taken from (3.57), namely $\varepsilon_{ij}^{(\beta\gamma)}(\overline{\varepsilon}_{ij})$ and $\sigma_{ij}^{(\beta\gamma)}(\overline{\sigma}_{ij})$, and substituted next to local equation of elasticity (3.56) at subcell we arrive at the equation of elasticity at RUC level that combines average stress and average strain

$$
\overline{\sigma}_{ij} = \underbrace{\frac{1}{hl} \sum_{\beta=1}^{N_\beta} \sum_{\gamma=1}^{N_\gamma} h_\beta l_\gamma E_{ijkl}^{(\beta\gamma)} A_{klmn}^{(\beta\gamma)}}_{\overline{E}_{ijmn}} \overline{\varepsilon}_{mn}
\tag{3.58}
$$

where A_{klmn} is so called tensorial concentration operator the components of which $A_{klmn}^{(\beta\gamma)}$ allow to separate properties of constitutive material matrix and reinforcement (fiber). If the new definition $\overline{E}_{ijmn}$ over RUC for averaged stiffness tensor is introduced (cf. Eq. 3.58) the *averaged elasticity equation in RUC* is furnished

$$
\overline{\sigma}_{ij} = \overline{E}_{ijmn} \overline{\varepsilon}_{mn}
\tag{3.59}
$$

Note that in the averaged elasticity equation (3.59) $\overline{E}_{klmn}$ stands for effective stiffness tensor of composite expressed in terms of the local elasticity tensors in subcells $\mathbb{E}^{(\beta\gamma)}$ and the concentration tensor $A_{klmn}^{(\beta\gamma)}$ represented by the matrix of concentration factors

Table 3.3 Local and averaged elasticity equations

Notation	Subcell level $(\beta\gamma)$	RUC level
Index	$\sigma_{ij}^{\beta\gamma} = E_{ijkl}^{\beta\gamma} \varepsilon_{kl}^{\beta\gamma}$	$\overline{\sigma}_{ij} = \overline{E}_{ijkl} \overline{\varepsilon}_{kl}$
	$\varepsilon_{ij}^{\beta\gamma} = E_{ijkl}^{-1\beta\gamma} \sigma_{kl}^{\beta\gamma}$	$\overline{\varepsilon}_{ij} = \overline{E}_{ijkl}^{-1} \overline{\sigma}_{kl}$
Vector/matrix	$\{\sigma^{(\beta\gamma)}\} = [\mathbb{E}^{(\beta\gamma)}]\{\varepsilon^{(\beta\gamma)}\}$	$\{\overline{\sigma}\} = [\overline{\mathbb{E}}]\{\overline{\varepsilon}\}$
	$\{\varepsilon^{(\beta\gamma)}\} = [\mathbb{E}^{(\beta\gamma)}]^{-1}\{\sigma^{(\beta\gamma)}\}$	$\{\overline{\varepsilon}\} = [\overline{\mathbb{E}}^{-1}]\{\overline{\sigma}\}$

defining distribution of constituent materials (subcell level) over the RUC (composite level).

If the *vector/matrix notation* is used and the homogenized stiffness matrix is defined in RUC both Eqs. (3.56) and (3.59) can be rewritten in format shown in Table 3.3.

3.5.2 *Effective Stiffness Matrices of Unidirectional Composites Characterized by Regular Fiber Configuration—Square Array Versus Hexagonal Array*

Final format of the *effective elastic stiffness matrix of composite* $\overline{\mathbb{E}}$ depends not only on the selected homogenization method (for instance Reuss', Voigt, GMC, SCMC etc.) but also on a choice of the *Representative Unit Cell* RUC. In fact a proper choice of RUC geometry should follow true fiber topology in the considered composite. Two basic regular fiber arrays repeating (periodic) at the macroscale of the unidirectional composite are of particular interest: the *square array* and the *hexagonal array* (Fig. 3.6). The rhombic array (Fig. 3.2) is not commonly used in practice, and it will not be considered here.

In case of *tetragonal symmetry* (square array) fibers are arranged in parallel rows and series being equally spaced by distance a in the matrix material of composite (Fig. 3.6a). Such fiber configuration in unidirectional composite is used by Tamma and Avila [32], Würkner et al. [37] and others. By contrast, in case of *hexagonal symmetry* (hexagonal array) fibers are distributed in position of parallel rows equally spaced with distance a but neighboring rows are shifted each to the other with distance $a/2$ (Fig. 3.6b). Hexagonal symmetry fiber topology is used for example by Herakovich and Aboudi [12], Sun and Vaidya [30] and other authors. Configuration of fibers in composite at the macroscale is a starting point for appropriate selection of the Representative Unit Cell (RUC) geometry for numerical simulation employed by the use of homogenization methods in order to find the effective properties of a composite.

When either the tetragonal or the hexagonal symmetry fiber configurations are employed various Representative Unit Cells can be defined (cf. Sun and Vaidya [30] and others).

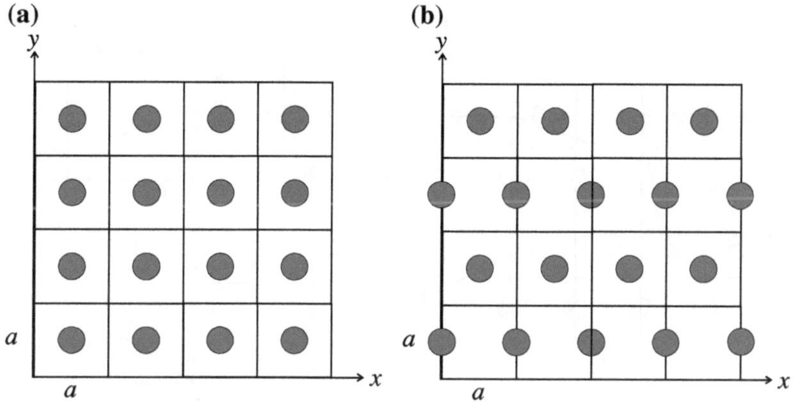

Fig. 3.6 Two configurations of fibers in unidirectionally reinforced transversely isotropic composite (macroscale): **a** tetragonal symmetry, **b** hexagonal symmetry

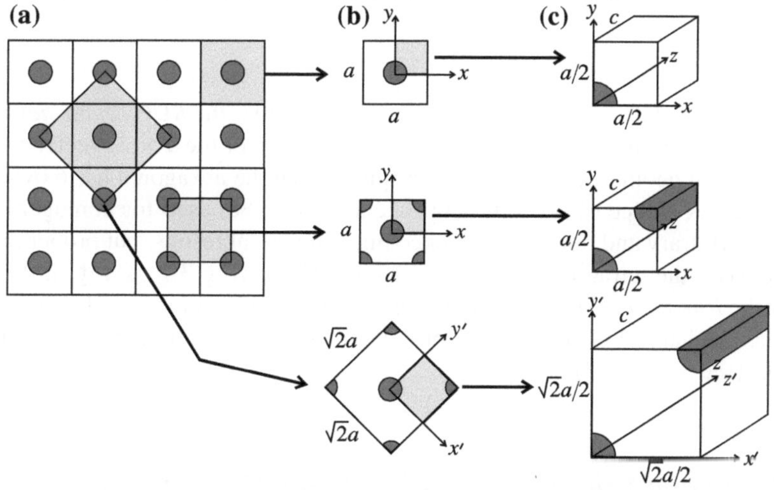

Fig. 3.7 Representative Unit Cells (RUCs) of tetragonal symmetry (square fiber array): **a** various choices of repeating RUCs at the macrolevel, **b** three shape geometires and fiber dispositions in RUCs, **c** three sub-RUCs and fiber geometry with additional symmetry used

In case of the fibers topology that exhibits *tetragonal symmetry* (*square fiber arrays*, Fig. 3.7a) three different representative unit cells are used (see Fig. 3.7b, c). Due to the transverse isotropy property, in fact the 2D analysis is sufficient whereas the choice of one of the three cells presented in Fig. 3.7c for numerical simulation is insignificant.

On the other hand if the fibers topology is governed by the *hexagonal symmetry* (*hexagonal arrays*), Fig. 3.8a, the other two Representative Unit Cells can be established as shown in Fig. 3.8b. Note that the *RUCs of the tetragonal symmetry*

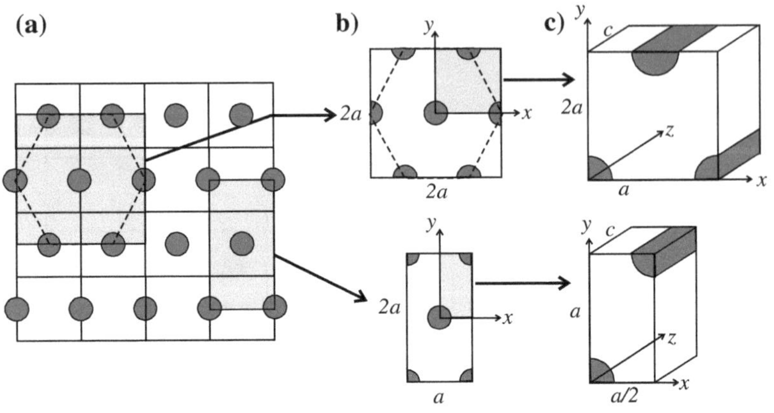

Fig. 3.8 Representative Unit Cells RUCs of hexagonal symmetry: **a** various choices of repeating RUCs at the macrolevel, **b** two different geometries of repeating RUCs—square and rectangular and corresponding fibers configuration, **c** shape and fibers disposition in two sub-RUCs with additional symmetry used

(Fig. 3.7b) have 4 axes of geometrical symmetry whereas the *RUCs of the hexagonal symmetry* (Fig. 3.8b) inscribed into the $2a \times 2a$ square or the $a \times 2a$ rectangle have only 2 axes of geometrical symmetry, in spite of that the hexagonal has 6 own symmetry axes. A choice of subcells (Fig. 3.7c and Fig. 3.8c) used for homogenization is in fact arbitrary and does not influence final numerical results, but proper distinction between the tetragonal and the hexagonal RUCs (Fig. 3.7 vs. Fig. 3.8) should follow the *true fibers arrangement* during composite fabrication in order to properly estimate mechanical characteristics of the composite which meet the experimental findings (see Sect. 3.5.3).

3.5.3 Sun and Vaidya Findings for Boron/Al Composite

In what follows let us inspect some results presented in Sun and Vaidya [30] for *transversely isotropic Boron/Aluminum composite* by the use of *FEM micromechanics-based homogenization* models when compared to other methods, c.f. Hashin and Rosen [10], Chamis [6] for various fibers topology (square array vs. hexagonal array) and some experimental evidence. The isotropic material properties of both constituents: *Boron fiber* and *Aluminum matrix* used by the authors are recalled in Table 3.4.

Key to distinct elastic response of the transversely isotropic (unidirectional) composite of either the *tetragonal* or the *hexagonal symmetry* is a number of independent material constants. In general case of the composite that exhibits plane isotropy property of tetragonal type, the averaged composite material is characterized by six

Table 3.4 Material properties of isotropic constituents of the unidirectional Boron fibers reinforcement in the Aluminum matrix, after Sun and Vaidya [30]

Constituent material	E (GPa)	ν
Boron fiber	379.3	0.1
Aluminum matrix	68.3	0.3

independent constants: E_{11}, E_{33}, ν_{21}, ν_{32}, G_{23}, and G_{12}. By contrast, in case of plane isotropy of hexagonal type the number of independent material constants is reduced to five since the shear modulus in the isotropy plane G_{23} is coupled with the transverse Young modulus E_{22} and corresponding Poisson ratio ν_{23} by the relationship which holds for isotropic media

$$G_{23} = \frac{E_{22}}{2(1 + \nu_{23})} \tag{3.60}$$

Let us examine the data given in Table 3.5 based on Sun and Vaidya [30] in the light of above constraint. It is visible that in case of micromechanics-based FEM model with the hexagonal array used by Sun and Vaidya [30] as well as its simulations by Hashin and Rosen [10] the *transversely isotropic hexagonal symmetry* roughly holds Eq. (3.60). However, when the RUC of *tetragonal symmetry* (square array) is used by Sun and Vaidya [30] or Chamis [6] composite exhibits the tetragonal symmetry property.

Note also that in literature a big scatter of both the material properties of the constituents of the same type (Boron/Al) and results based on different homogenization methods are met.

Table 3.5 Comparison of selected elastic material modules for the Boron/Al composite, obtained in various numerical experiments by different authors for the same Boron/Al composite material ($V_f = 0.47$), after Sun and Vaidya [30]

Material constants of composite Boron/Al ($V_f = 0.47$)	FEM Sun and Vaidya [30]		Numerical simulations	
	Square array	Hexagonal array	Chamis [6]	Hashin and Rosen [10]
E_{11} (GPa)	215	215	214	215
$E_{22} = E_{33}$ (GPa)	144	136.5	156	139.1
G_{23} (GPa)	45.9	52.5	43.6	54.6
$G_{12} = G_{13}$ (GPa)	57.2	54.0	62.6	53.9
ν_{32}	0.29	0.34	0.31	0.31
$\nu_{21} = \nu_{31}$	0.19	0.19	0.20	0.195

3.5.4 Interpretation of the Theorem of Lower and Upper Bounds in the Light of Gan et al. [9] and Aboudi et al. [1] Findings for Boron/Al Composite

Sun and Vaidya findings presented in Sect. 3.5.3 are limited to the selected volume fraction ($V_f = 0.47$). For further analysis it is convenient to discuss Gan et al. [9] findings concerning the similar Boron/Al long-fiber composite Table 3.6, but obtained for a complete volume fraction spectrum $V_f \in [0, 1]$.

Note large discrepancy between the input data used for numerical experiments by Sun and Vaidya [30] (Table 3.4) and that used by Gan et al. [9] (Table 3.6). In numerical experiments based on homogenization methods FEM, GMC and SCMS (see Sect. 3.5.1) Gan et al. [9] compared various round Boron fiber arrangements in the RUC: *unidirectional random* (disordered) *disposition*, the *single fiber* centered *in the square cell* and the *hexagonal symmetry array*, but applying the *general orthotropy symmetry group* (9 material constants explored), see Fig. 3.1. Results obtained from numerical experiments FEM, GMC and SCMC (Table 3.7) closely resemble data governed by the transverse symmetry group, but in case of GMC method a higher divergence is met. Further distinction between the tetragonal or the hexagonal symmetry group can be done by checking the condition (3.60). An analysis performed in Table 3.8 leads to the conclusion that the considered composite exhibits the *tetragonal symmetry class* when *GMC* and *SCMC homogenization methods* are involved since the condition (3.60) does not hold. By contrast, when the *micromechanics-based FEM* was implemented the results obtained satisfy the requirement of the *hexagonal symmetry class* (condition (3.60) is satisfied) where only 5 material constants are essentially independent (see Fig. 3.1 and Eq. 3.13 vs. 3.14).

Examine closer the selected Gan et al. [9] findings from numerical experiments based on the regular hexagonal fibers packing in the Boron/Al composite by the use of SCMC homogenization method, compared with the Voigt and Reuss models relied upon the volume fractions of the phases only. Inspection of these results obtained in numerical experiment for long-fiber-reinforced composite characterized by *transversely isotropic tetragonal* or *hexagonal symmetry* performed in light of the *Hill theorem on lower and upper bounds by Voigt* and *Reuss isotropic estimates* will be subject of further considerations.

Chronologically, Voigt (1889) and Reuss (1929) had proposed estimates for engineering constants E and G in polycrystals a long time before Hill proved famous theorem on lower and upper bounds for the *averaged stiffness matrix* $[\overline{\mathbb{E}}]$ or the *com-*

Table 3.6 Material properties of constituents of the unidirectional Boron/Al composite, after Gan et al. [9]

Constituent material	E (GPa)	ν
Boron fiber	413.7	0.2
Aluminum matrix	55.16	0.3

Table 3.7 Approximation of material constants by orthotropic numerical experiment for unidirectional long-fiber Boron/Al composite with random fibers arrangement in-plane transverse to fibers beam direction and 30 × 30 number od subcells in the RVE, after Gan et al. [9]

Material constants of Boron composite/Al ($V_f = 0.5$)	Homogenization methods (random arrangement of fibers)		
	FEM	GMC	SCMC
E_{11} (GPa)	234.7	234.7	234.7
E_{22} (GPa)	138.5	117.9	131.0
E_{33} (GPa)	137.3	113.1	128.6
G_{23} (GPa)	54.78	37.78	57.70
G_{21} (GPa)	60.48	42.94	58.51
G_{31} (GPa)	60.99	40.53	58.77
ν_{21}	0.2361	0.2446	0.2387
ν_{31}	0.2369	0.2492	0.2405
ν_{32}	0.3078	0.3289	0.3182

Table 3.8 Comparison of the shear modules in the transverse plane obtained from the experiment by Gan et al. [9] for random unidirectional fibers dispersion in RVE (orthotropy) of Boron/Al composite with the expected magnitude under the hexagonal-type transverse isotropy constraint (3.60)

	FEM	GMC	SCMC
G_{23} (GPa)	54.78	37.78	57.70
$G_{23} = \frac{E_{22}}{2(1+\nu_{32})}$	52.7	43.45	49.23
% of divergence	−3.9	13.0	−17.2

pliance matrix $[\overline{\mathbb{E}}^{-1}]$ addressed to heterogeneous media (see Hill [13]). Recently, scientists involved in the composite mechanics field and development of reliable homogenization methods, commonly employ the Hill theorem originated for multiphase media, to estimate numerically the effective stiffness or compliance matrices for composite materials. Very often they need to find the engineering constants which are conventional input data for existing FEM-based codes addressed to anisotropic composites. As a consequence, magnitudes of the Young modules E_{11}, E_{22}, E_{33} and the Kirchhoff modules G_{12}, G_{23}, G_{31} counted from $[\overline{\mathbb{E}}]$ or $[\overline{\mathbb{E}}^{-1}]$ lay inside the Voigt and the Reuss estimates. Contrary, the magnitudes of Poisson ratios ν_{12}, ν_{23}, ν_{31}, that may exceed both estimates, even though the Hill theorem on lower and upper bounds holds for all elements of averaged stiffness $[\overline{\mathbb{E}}]$ or compliance $[\overline{\mathbb{E}}^{-1}]$ matrices. Such peculiarity occurs although the theorem on Voigt and Reuss estimates is fulfilled, if consistently applied to all elements of elastic matrices $[\overline{\mathbb{E}}]$ or $[\overline{\mathbb{E}}^{-1}]$, but not to the engineering constants evaluated from the appropriate formulas. Note that the engineering constants are measured from experiments. Such peculiarity can be observed for instance in results for the *Boron/Al composites* presented by Gan et al. [9] Fig. 3.9, as well as the another *Glass/Epoxy composites* by Aboudi et al. [1], see Fig. 3.10.

Presented in the Fig. 3.9 *Voigt* and *Reuss' bounds* are obtained in two different ways. First, the *"loose" bounds* are obtained by extracting the averaged Poisson ratios from the appropriate stiffness matrix element, which will be discussed further. Second, the *"tight" bounds* are obtained in the way of straightforward use of Voigt and Reuss' mixture rules to Poisson ratios of both phases. It is evident that the exemplary results by Gan et al. [9] obtained by application of SCMC method exceed both "loose" as well as "tight" systems of bounds. Similar behavior is typical also for another *Glass/Epoxy composite* system discussed by Aboudi et al. [1], see Fig. 3.10. The Voigt and the Reuss bounds used here are enriched by other "loose" bounds of *Concentric Cylindrical Assemblage model* (CCA^+, CCA^-) which turn out to be much broader. Although such a broad bound systems are admitted, the Poisson ratios

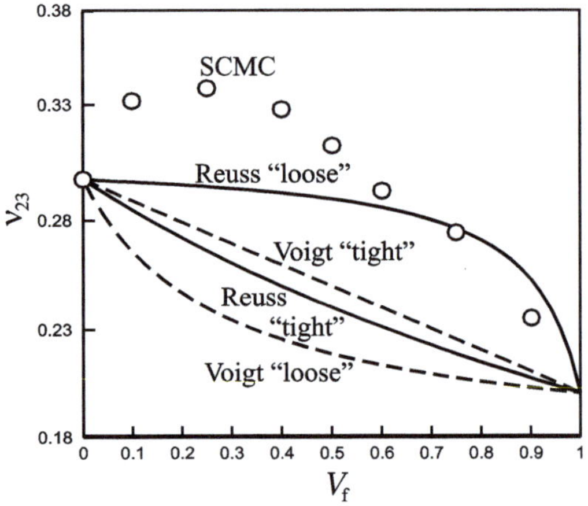

Fig. 3.9 Peculiarity of the Poisson ratio ν_{23} diagrams for the long-fiber *Boron/Al system*, after Gan et al. [9]

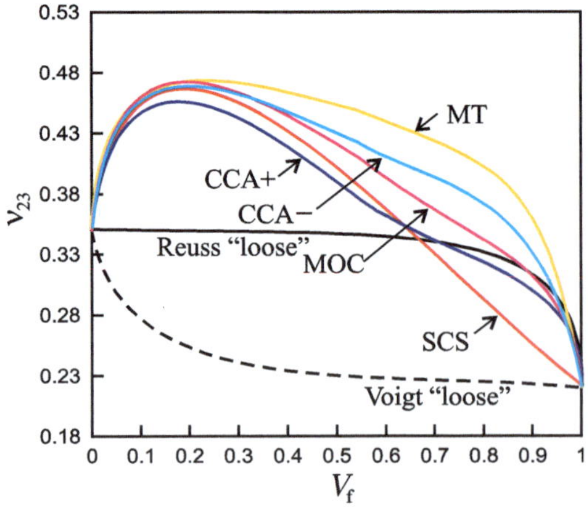

Fig. 3.10 Peculiarity of the Poisson ratio ν_{23} diagrams for the long-fiber *Glass/Epoxy composite*, system after Aboudi et al. [1]

ν_{23} obtained by use of the *Mori–Tanaka method* (MT), the *Micromechanics-based Method of Cells* (MMC) and the *Self Consistent Scheme* (SCS) exceed these bounds.

Note that in both cases the "loose" Voigt and Reuss bounds are shown by two curved diagrams versus V_f in Figs. 3.9 and 3.10, although the Voigt estimate is in fact linear (from definition). It can be understood when the mixture rules, Voigt and Reuss', are consistently applied to the *stiffness modulus of* $E_{11}^{r/m}$ both *phases*, matrix and fiber reinforcement

$$E_{11}^{m/r} \stackrel{\text{def}}{=} \frac{E^{m/r}(1-\nu^{m/r})}{(1+\nu^{m/r})(1-2\nu^{m/r})} \tag{3.61}$$

namely

$$\begin{aligned} {}^{V}\overline{E}_{11} &= E_{11}^{m}(1-V_r) + E_{11}^{r} V_r \\ \frac{1}{{}^{R}\overline{E}_{11}} &= \frac{(1-V_r)}{E_{11}^{m}} + \frac{V_r}{E_{11}^{r}} \end{aligned} \tag{3.62}$$

Bars in Eq. (3.62) over the symbol refer to the composite as a whole, superscripts ^{V} and ^{R} refer to the *Voigt* and *Reuss' estimates* whereas symbols $^{m/r}$ refer to the constituents (matrix and fiber reinforcement). Symbols ${}^{V/R}\overline{E}_{11}$ are given by the following formulas

$$\begin{aligned} {}^{V}\overline{E}_{11} &= \frac{{}^{V}\overline{E}(1-{}^{V}\overline{\nu})}{(1+{}^{V}\overline{\nu})(1-2\,{}^{V}\overline{\nu})} \\ {}^{R}\overline{E}_{11} &= \frac{{}^{R}\overline{E}(1-{}^{R}\overline{\nu})}{(1+{}^{R}\overline{\nu})(1-2\,{}^{R}\overline{\nu})} \end{aligned} \tag{3.63}$$

Solution of the above equation system (3.63) for the magnitudes of averaged *Poisson ratio* $\overline{\nu}$ with the *Young modules* ${}^{V}\overline{E}$, ${}^{R}\overline{E}$ *averaged* straightforwardly *by* the use of appropriate *mixture rules* for Voigt and Reuss' estimates

$$\begin{aligned} {}^{V}\overline{E}(V_f) &= E^{m}(1-V_f) + E^{r} V_f \\ \frac{1}{{}^{R}\overline{E}(V_f)} &= \frac{1-V_f}{E^{m}} + \frac{V_f}{E^{r}} \end{aligned} \tag{3.64}$$

yields the following formula for the "loose" Poisson ratio bounds

$${}^{V/R}\overline{\nu} = \frac{\sqrt{\left(1-\frac{{}^{V/R}\overline{E}_{11}}{{}^{V/R}\overline{E}}\right)^2 - 8\left(1-\frac{{}^{V/R}\overline{E}_{11}}{{}^{V/R}\overline{E}}\right)\frac{{}^{V/R}\overline{E}_{11}}{{}^{V/R}\overline{E}}} + 1 - \frac{{}^{V/R}\overline{E}_{11}}{{}^{V/R}\overline{E}}}{4\frac{{}^{V/R}\overline{E}_{11}}{{}^{V/R}\overline{E}}} \tag{3.65}$$

Alternatively, applying the Voigt or the Reuss mixture rules directly to the Poisson ratios of both phases, matrix ν^{m} and fiber ν^{r} other two "tight" Poisson ratio bounds are found

$$
\begin{aligned}
{}^{\mathrm{V}}\overline{\nu} &= \nu^{\mathrm{m}}(1 - V_{\mathrm{f}}) + \nu^{\mathrm{r}} V_{\mathrm{f}} \\
\frac{1}{{}^{\mathrm{R}}\overline{\nu}} &= \frac{1 - V_{\mathrm{f}}}{\nu^{\mathrm{m}}} + \frac{V_{\mathrm{f}}}{\nu^{\mathrm{r}}}
\end{aligned}
\tag{3.66}
$$

where the "tight" Voigt bound preserves linearity.

Concluding, both "loose" bounding diagrams in Figs. 3.9 and 3.10 exhibit non-linear property since the magnitudes of Poisson ratios were obtained in an artificial paths: either by extracting them from Eq. (3.63) or by straightforward application of the mixture rules to engineering Poisson ratios for which linear "tight" Voigt estimate is saved (3.66).

Finally, when the *Hill theorem of lower and upper bounds* is consistently applied to the elements of *elastic stiffness* or *compliance matrices* then and only then all *effective matrix elements of* a *composite* considered lay inside the lower and upper bounds or at most at one of the bounds. In fact, if the results by Gan et al. [9], originally presented in terms of the *engineering anisotropic constants* E_{11}, E_{22}, G_{23}, G_{12}, ν_{12} and ν_{23} are consistently transformed to the space of elements of compliance matrix $\overline{E}_{11}^{-1}$, $\overline{E}_{33}^{-1}$, $\overline{E}_{12}^{-1}$, $\overline{E}_{23}^{-1}$, $\overline{E}_{44}^{-1}$, $\overline{E}_{55}^{-1}$, the results obtained by use of the SCMC method follow the Hill theorem upper and lower bounds as shown in Fig. 3.11.

3.5.5 Approximation of Mechanical Modules of Long-Fiber Unidirectionally Reinforced Composites by the Use of a Hybrid Rule Between Voigt and Reuss Estimates

Mention at the beginning that classical mixture rules by Voigt (3.17_1) and Reuss (3.17_2) apply a random dispersion of composite constituents over RVE. It is obvious that the Voigt and the Reuss estimates converge at appropriate magnitudes of modules of matrix and reinforcement for *volume fraction* $V_{\mathrm{f}} = 0$ or $V_{\mathrm{f}} = 1$, respectively. This question should be carefully considered in light of fabrication procedure. Namely, assuming identical fibers of circular cross-section regularly packed over the RUC either according to square or hexagonal arrays we arrive at two different *maximal fiber packing limits* V_{fmax}, see Fig. 3.12. It is seen that maximal fiber packing for the *square array* $V_{\mathrm{fmax}}^{\mathrm{sq}} \cong 78.5\,\%$ is much lower than analogous maximal fiber packing for the *hexagonal array* $V_{\mathrm{fmax}}^{\mathrm{hex}} \cong 90.7\,\%$. Even higher maximal fiber packing can be achieved by using fibers of either various diameters or noncircular cross-section (*square* cross-section *fibers* or *honey-comb* cross-section *fibers* joined by thin matrix layers). Hence, the homogenization results according to Voigt or Reuss for surroundings $V_{\mathrm{f}} \cong 1$ have only theoretical sense. Analogous objections can be formulated to homogenization results for surroundings $V_{\mathrm{f}} \cong 0$ where there is difficult to talk about a composite.

Consider now in detail results by Gan et al. [9]. In what follows in order to formulate a *weighted homogenization rule* based on a tensorial interpolation between lower and upper bounds it will be more convenient to consistently formulate the Voigt and Reuss estimates in application to stiffness or compliance matrix components but

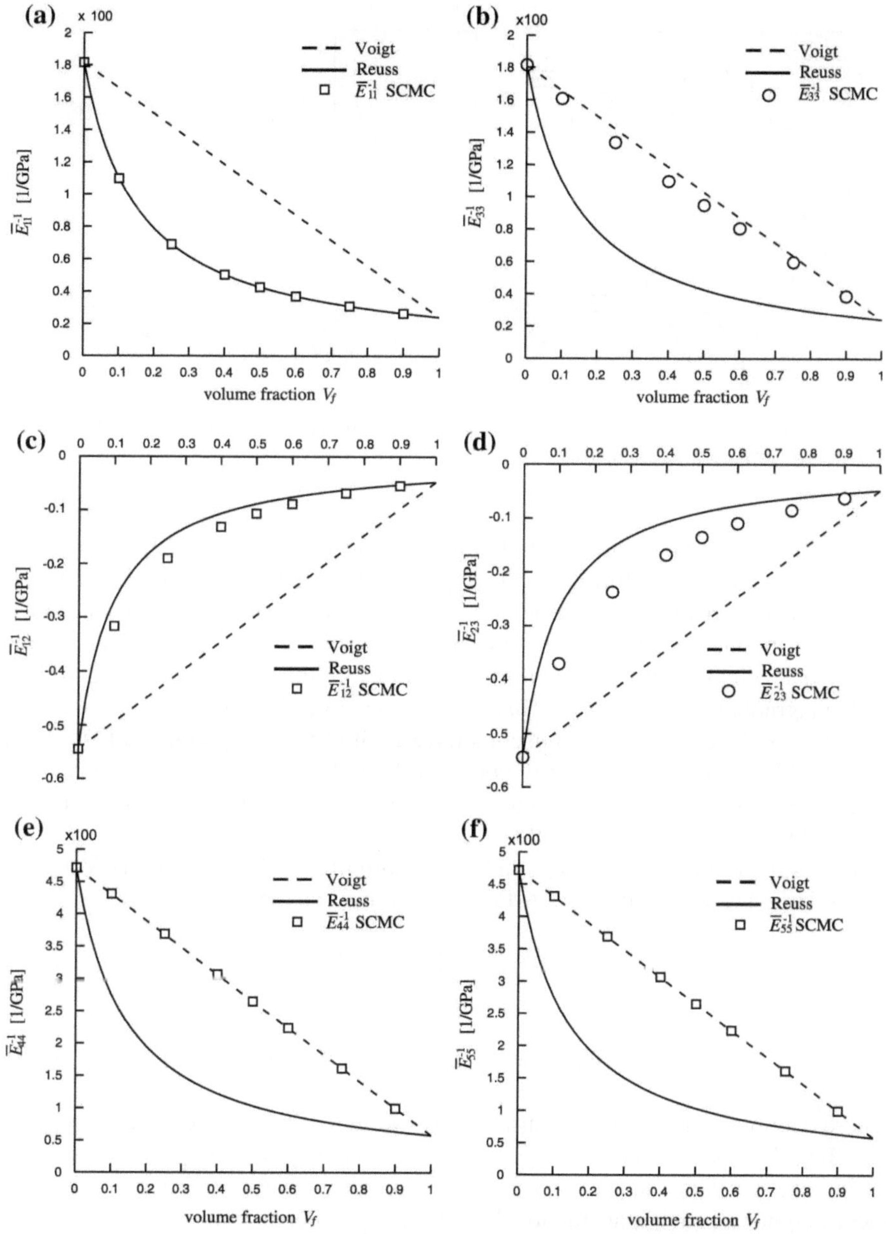

Fig. 3.11 Interpretation of the Gan et al. [9] results in the space of elements of effective compliance matrix $[\overline{\mathbb{E}}^{-1}]$ obtained on the base of diagrams of engineering constants of the Boron/Al composite in light of theorem of upper and lower bounds by the Voigt and Reuss estimates

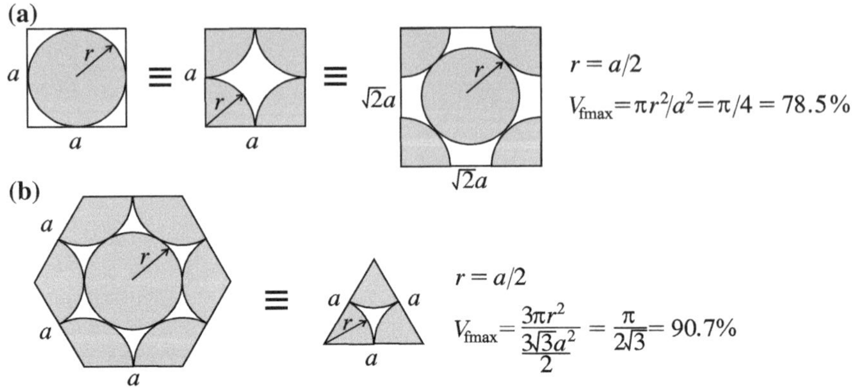

Fig. 3.12 Illustration of maximal fiber packing for identical fibers of circular cross-section in case of: **a** square array, **b** hexagonal array

not to the engineering constants (what is commonly done). The weighted homogenization rule allows to formulate the approximate method to estimate elements of effective elasticity matrices (stiffness or compliance) by the use of the values of lower and upper bounds and performing interpolation between them with the use of new *tensor-like rule of mixture* (a *hybrid formulation*). In this way it will be possible to build diagrams for all *orthotropic matrix components* $[\mathbb{E}^{-1}]$ in the full range of volume fraction $V_{\mathrm{f}} \in [0, 1]$ assuming coincidence with known experimentally obtained matrix $[{}^{\exp}\mathbb{E}^{-1}]$ for one arbitrarily chosen volume fraction V_{f}^{0}. Additionally, coincidence with known matrices of pure constituents: $V_{\mathrm{f}} = 0$ for matrix material and $V_{\mathrm{f}} = 1$ for fiber material must hold.

Let us rewrite the scalar Voigt and Reuss formulas (3.17) into matrix Voigt and Reuss formulas, respectively to stiffness or compliance matrices

$$^{\mathrm{V}}[\overline{\mathbb{E}}] = c_1[{}^{\mathrm{r}}\mathbb{E}] + c_2[{}^{\mathrm{m}}\mathbb{E}] \tag{3.67}$$

or

$$^{\mathrm{R}}[\overline{\mathbb{E}}^{-1}] = c_1[{}^{\mathrm{r}}\mathbb{E}^{-1}] + c_2[{}^{\mathrm{m}}\mathbb{E}^{-1}] \tag{3.68}$$

where common fractional concentrations by volume of the phases according Voigt and Reuss' rules $c_1 = V_{\mathrm{f}}$ and $c_2 = 1 - V_{\mathrm{f}}$ as previously shown for uniaxial models (3.24), see Aboudi et al. [1]. This simplification means that orientation of reinforcement is ignored, such that fractional concentrations depend on volume fraction V_{f} only. Symbols $[{}^{\mathrm{r/m}}\mathbb{E}]$ and $[{}^{\mathrm{r/m}}\mathbb{E}^{-1}]$ stand for elements of stiffness or compliance matrices for reinforcing fiber or matrix, respectively. As a matter of fact c_1 and c_2 must account for both volume fraction and reinforcement orientation, therefore for determination of them advanced homogenization schemes are required (e.g., FEM, GMC, SCMC, CCM and others).

In what follows a simple rule of *weighted average between*, the Voigt and the Reuss *upper and lower estimates* is proposed. Such approach is based on *tensorial interpolation between upper and lower estimates* which enables to avoid application of numerous cumbersome homogenization methods, for instance micromechanics-based FEM, GMC, SCMC, CCM etc.

To this end, we define weighting vector α_k built of weighting coefficients for subsequent elements of stiffness or compliance matrices. For brevity we confine ourselves to the compliance matrix only. Hence, the proposed *hybrid* or *weighting homogenization rule* takes the following format

$$\begin{aligned}
\overline{\overline{E}}_{11}^{-1}(V_f) &= \alpha_1^{\mathrm{V}}\overline{E}_{11}^{-1}(V_f) + (1-\alpha_1)^{\mathrm{R}}\overline{E}_{11}^{-1}(V_f)\\
\overline{\overline{E}}_{22}^{-1}(V_f) &= \alpha_2^{\mathrm{V}}\overline{E}_{22}^{-1}(V_f) + (1-\alpha_2)^{\mathrm{R}}\overline{E}_{22}^{-1}(V_f)\\
\overline{\overline{E}}_{33}^{-1}(V_f) &= \alpha_3^{\mathrm{V}}\overline{E}_{33}^{-1}(V_f) + (1-\alpha_3)^{\mathrm{R}}\overline{E}_{33}^{-1}(V_f)\\
\overline{\overline{E}}_{23}^{-1}(V_f) &= \alpha_4^{\mathrm{V}}\overline{E}_{23}^{-1}(V_f) + (1-\alpha_4)^{\mathrm{R}}\overline{E}_{23}^{-1}(V_f)\\
\overline{\overline{E}}_{13}^{-1}(V_f) &= \alpha_5^{\mathrm{V}}\overline{E}_{13}^{-1}(V_f) + (1-\alpha_5)^{\mathrm{R}}\overline{E}_{13}^{-1}(V_f)\\
\overline{\overline{E}}_{12}^{-1}(V_f) &= \alpha_6^{\mathrm{V}}\overline{E}_{12}^{-1}(V_f) + (1-\alpha_6)^{\mathrm{R}}\overline{E}_{12}^{-1}(V_f)\\
\overline{\overline{E}}_{44}^{-1}(V_f) &= \alpha_7^{\mathrm{V}}\overline{E}_{44}^{-1}(V_f) + (1-\alpha_7)^{\mathrm{R}}\overline{E}_{44}^{-1}(V_f)\\
\overline{\overline{E}}_{55}^{-1}(V_f) &= \alpha_8^{\mathrm{V}}\overline{E}_{55}^{-1}(V_f) + (1-\alpha_8)^{\mathrm{R}}\overline{E}_{55}^{-1}(V_f)\\
\overline{\overline{E}}_{66}^{-1}(V_f) &= \alpha_9^{\mathrm{V}}\overline{E}_{66}^{-1}(V_f) + (1-\alpha_9)^{\mathrm{R}}\overline{E}_{66}^{-1}(V_f)
\end{aligned} \tag{3.69}$$

Additionally, independence of the weighting coefficients α_k of the volume fraction V_f over the whole range of $V_f \in [0, 1]$ is assumed. This assumption refers to definition of convex set of two vectors. If the magnitudes of stiffness or compliance elements are known at certain point $V_f = V_f^0$

$$[\overline{\overline{\mathbb{E}}}^{-1}(V_f^0)] = [{}^{\mathrm{exp}}\mathbb{E}^{-1}(V_f^0)] \tag{3.70}$$

then it is possible to determine unknown *vector of weighting coefficients* α_k for the compliance. Applying these coefficients over the whole range of volume fraction $V_f \in [0, 1]$ the sought elements of compliance matrix can be determined.

3.5.6 Capability of the Proposed Hybrid-Type Rule Versus Experimental Evidence in Light of Fiber Array Symmetry: Tetragonal or Hexagonal

The *weighting average homogenization rules* defined in the previous Sect. 3.5.5 by Eqs. (3.69) are rather simple and effective ones that allow to easily predict unknown

Table 3.9 Values of weight coefficients according to compliance matrix components for Voigt and Reuss' homogenization and SCMC [9]

$[\mathbb{E}^{-1}]$	Interpolation points			α_k
	$[^{V}\mathbb{E}^{-1}] \times 10^{-2}$ (GPa^{-1})	$[^{R}\mathbb{E}^{-1}] \times 10^{-2}$ (GPa^{-1})	$[^{SCMC}\mathbb{E}^{-1}] \times 10^{-2}$ (GPa^{-1})	
E_{11}^{-1}	1.01	0.42	0.909	0.833
E_{22}^{-1}	1.01	0.42	0.909	0.833
E_{33}^{-1}	1.01	0.42	0.463	0.075
E_{23}^{-1}	−0.29	−0.09	−0.116	0.142
E_{13}^{-1}	−0.29	−0.09	−0.116	0.142
E_{12}^{-1}	−0.29	−0.09	−0.300	1.050
E_{44}^{-1}	2.59	1.01	2.130	0.707
E_{55}^{-1}	2.59	1.01	2.268	0.794
E_{66}^{-1}	2.59	1.01	2.130	0.707

constitutive modules of the composite system over the whole range of the volume fraction $V_f \in [0, 1]$ providing that they are known for one V_f^0. Efficiency of this method is tested by the use of the results of numerical simulation by *SCMS homogenization method* [9]. To this end nine weighting coefficients α_k for the *orthotropic Boron/Al composite* are calculated by interpolation between Voigt and Reuss' estimates shown in Fig. 3.13. Magnitudes of the weighting coefficients α_k are established at the point $V_f^0 = 0.513$ by comparison with the homogenization results SCMC by Gan et al. [9]. Obviously the weighting homogenization rule must give correct results at the end points $V_f = 0$ and $V_f = 1$. Calculated weighting coefficients and set of predictions $^{V}\mathbb{E}^{-1}$, $^{R}\mathbb{E}^{-1}$, $^{SCMC}\mathbb{E}^{-1}$ for nine elements of compliance matrix are presented in Table 3.9. The results of the weighting homogenization rule are verified with the results given by Gan et al. [9] based on SCMC method, that fully confirm the assumption that weighting coefficients α_k can be treated as universal ones for the composite tested over the full range of volume fraction as shown by curves of weight rule • versus SCMC homogenization □.

3.5.7 Interpretation of Results Obtained by Weighting Homogenization in Terms of Engineering Constants

Nevertheless the formulated in previous subsection *"hybrid" mixture rules* based on *weighting interpolation* between Voigt and Reuss' estimates have to be formulated in the space of elements of elasticity matrix (compliance or stiffness), where Hill theorem of lower and upper estimates by Voigt and Reuss holds, it is usually necessary to express the results in terms of *engineering orthotropy constants*. The reason for such representation results from usually applied homogenization techniques to engineering constants, but not to elasticity elements. This system of engineering

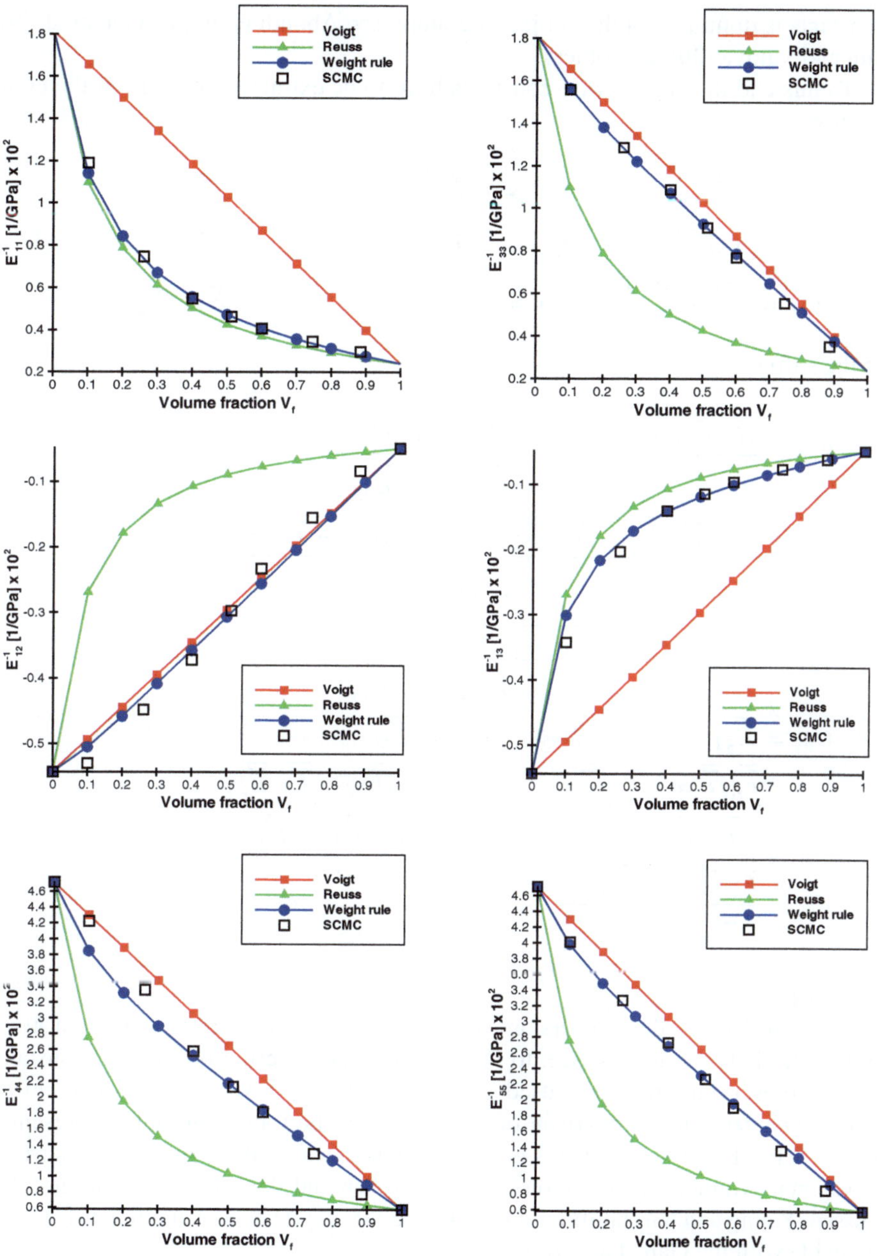

Fig. 3.13 Comparison of SCMC [9], Voigt, Reuss and proposed hybrid rule according to compliance matrix coefficients

constants is dominant in the subject literature, see Aboudi et al. [1], Gan et al. [9], Sun and Vaidya [30], and others.

To this end the engineering constants have to be extracted either from the compliance

$$
\begin{array}{lll}
E_{11} = 1/\overline{\overline{E}}_{11}^{-1} & E_{22} = 1/\overline{\overline{E}}_{22}^{-1} & E_{33} = 1/\overline{\overline{E}}_{11}^{-1} \\
G_{44} = 1/\overline{\overline{E}}_{44}^{-1} & G_{55} = 1/\overline{\overline{E}}_{55}^{-1} & G_{66} = 1/\overline{\overline{E}}_{66}^{-1} \\
\nu_{12} = -\overline{\overline{E}}_{12}^{-1}/\overline{\overline{E}}_{11}^{-1} & \nu_{13} = -\overline{\overline{E}}_{13}^{-1}/\overline{\overline{E}}_{11}^{-1} & \nu_{23} = -\overline{\overline{E}}_{23}^{-1}/\overline{\overline{E}}_{22}^{-1}
\end{array}
\tag{3.71}
$$

or the stiffness

$$
\begin{aligned}
E_{11} &= \frac{2\overline{\overline{E}}_{12}\overline{\overline{E}}_{13}\overline{\overline{E}}_{23} + \overline{\overline{E}}_{11}\overline{\overline{E}}_{22}\overline{\overline{E}}_{33} - \overline{\overline{E}}_{33}\overline{\overline{E}}_{12}^{2} - \overline{\overline{E}}_{11}\overline{\overline{E}}_{23}^{2} - \overline{\overline{E}}_{22}\overline{\overline{E}}_{13}^{2}}{\overline{\overline{E}}_{22}\overline{\overline{E}}_{33} - \overline{\overline{E}}_{23}^{2}} \\
E_{22} &= \frac{2\overline{\overline{E}}_{12}\overline{\overline{E}}_{13}\overline{\overline{E}}_{23} + \overline{\overline{E}}_{11}\overline{\overline{E}}_{22}\overline{\overline{E}}_{33} - \overline{\overline{E}}_{33}\overline{\overline{E}}_{12}^{2} - \overline{\overline{E}}_{11}\overline{\overline{E}}_{23}^{2} - \overline{\overline{E}}_{22}\overline{\overline{E}}_{13}^{2}}{\overline{\overline{E}}_{11}\overline{\overline{E}}_{33} - \overline{\overline{E}}_{13}^{2}} \\
E_{33} &= \frac{2\overline{\overline{E}}_{12}\overline{\overline{E}}_{13}\overline{\overline{E}}_{23} + \overline{\overline{E}}_{11}\overline{\overline{E}}_{22}\overline{\overline{E}}_{33} - \overline{\overline{E}}_{33}\overline{\overline{E}}_{12}^{2} - \overline{\overline{E}}_{11}\overline{\overline{E}}_{23}^{2} - \overline{\overline{E}}_{22}\overline{\overline{E}}_{13}^{2}}{\overline{\overline{E}}_{11}\overline{\overline{E}}_{22} - \overline{\overline{E}}_{12}^{2}} \\
G_{44} &= \overline{\overline{E}}_{44} \qquad G_{55} = \overline{\overline{E}}_{55} \qquad G_{66} = \overline{\overline{E}}_{66} \\
\nu_{12} &= \frac{\overline{\overline{E}}_{12}\overline{\overline{E}}_{33} - \overline{\overline{E}}_{13}\overline{\overline{E}}_{23}}{\overline{\overline{E}}_{22}\overline{\overline{E}}_{33} - \overline{\overline{E}}_{23}^{2}} \qquad \nu_{13} = \frac{\overline{\overline{E}}_{13}\overline{\overline{E}}_{22} - \overline{\overline{E}}_{12}\overline{\overline{E}}_{23}}{\overline{\overline{E}}_{22}\overline{\overline{E}}_{33} - \overline{\overline{E}}_{23}^{2}} \\
\nu_{23} &= \frac{\overline{\overline{E}}_{23}\overline{\overline{E}}_{11} - \overline{\overline{E}}_{13}\overline{\overline{E}}_{12}}{\overline{\overline{E}}_{11}\overline{\overline{E}}_{33} - \overline{\overline{E}}_{13}^{2}}
\end{aligned}
\tag{3.72}
$$

matrices.

In what follows the conversion of results shown in the previous section given in the elasticity modules space, to the system of engineering orthotropic constants is done preserving previously used assumption of the *transversely isotropic hexagonal symmetry*. The comparison of engineering orthotropic constants is presented in Fig. 3.14. The figure contains only four plots, because the transversely isotropic hexagonal symmetry assumption has been proven and Poisson's ratios are not discussed. Young and Kirchhoff modules obtained from proposed weighting rule (3.69) coincide with the Gan et al. [9] results.

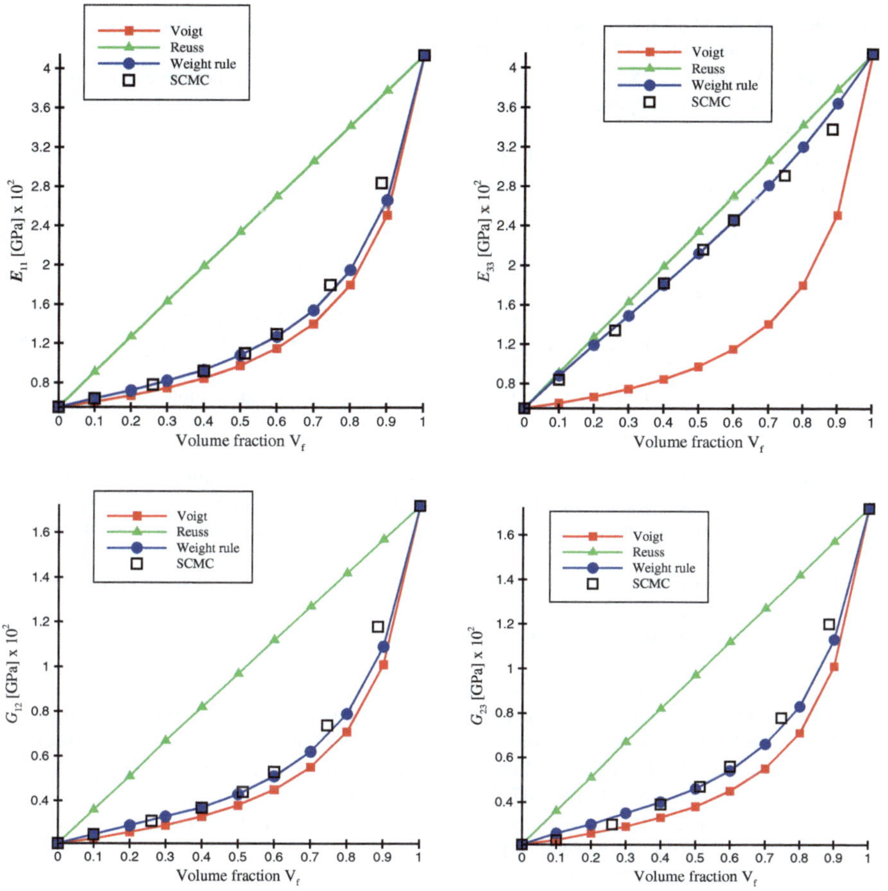

Fig. 3.14 Comparison of SCMC [9], Voigt, Reuss' and proposed hybrid estimates in the engineering orthotropic constants domain

References

1. Aboudi, J., Arnold, S.M., Bednarcyk, B.A.: Micromechanics of Composite Materials. Elsevier, Amsterdam (2013)
2. Auriault, J.-L., Boutin, C., Geindreau, C.: Homogenization of Coupled Phenomena in Heterogenous Media, pp. 39–44. Wiley-ISTE, New York (2009)
3. Banks-Sills, L., Leiderman, V., Fang, D.: On the effect of particle shape and orientation on elastic properties of metal matrix composites. Compos. Part B **28B**, 465–481 (1997)
4. Bayat, M., Aghdam, M.M.: A micromechanics-based analysis of effects of square and hexagonal fiber arrays in fibrous composites using DQEM. Eur. J. Mech.—A/Solids **32**, 32–40 (2012)
5. Berryman, J.G.: Bounds and self-consistent estimates for elastic constants of random polycrystals with hexagonal, trigonal, and tetragonal symmetries. J. Mech. Phys. Solids **53**, 2141–2173 (2005)
6. Chamis, C.C.: Simplified composite micromechanics equations for hygral, thermal and mechanical properties. SAMPE Q. **15**, 14–23 (1984)

7. Desmorat, R., Marull, R.: Non-quadratic Kelvin models based plasticity criteria for anisotropic materials. Int. J. Plast. **27**, 328–351 (2011)
8. Drago, A., Pindera, M.-J.: Micro-macromechanical analysis of heterogeneous materials: macroscopically homogeneous vs periodic microstructures. Compos. Sci. Technol. **67**, 1246–1263 (2007)
9. Gan, H., Orozco, C.E., Herkovich, C.T.: A strain-compatible method for micromechanical analysis of multi-phase composites. Int. J. Solids Struct. **37**, 5097–5122 (2000)
10. Hashin, Z., Rosen, B.W.: The elastic moduli of fiber-reinforced materials. J. Appl. Mech. **31**, 223–232 (1964)
11. Herakovich, C.T.: Mechanics of Fibrous Composites. Wiley, New York (1998)
12. Herakovich, C.T., Aboudi, J.: Thermal effects in composites. In: Hetnarski, R.B. (ed.) Stresses, Thermal, pp. 1–142. Lastran Corporation-Publishing Division, Rochester (1999)
13. Hill, R.: Elastic properties of reinforced solids: some theoretical principles. J. Mech. Phys. Solids **11**, 357–372 (1963)
14. Jiang, C.P., Xu, Y.L., Chueng, Y.K., Lo, S.H.: A rigorous analytical method for doubly periodic cylindrical inclusions under longitudinal shear and its application. Mech. Mater. **36**, 225–237 (2004)
15. Kenaga, D., Doyle, J.F., Sun, C.T.: The characterization of boron/aluminium in the nonlinear ranges as an orthotropic elastic plastic material. J. Compos. Mater. **27**, 516–531 (1987)
16. Lekhnitskii, S.G.: Theory of Elasticity of an Anisotropic Body (trans: English). Mir Publishers, Moscow (1981)
17. Li, S.: Boundary conditions for unit cells from periodic microstructures and their implications. Compos. Sci. Technol. **68**, 1962–1974 (2008)
18. Li, S., Wongsto, A.: Unit cells for micromechanical analyses of particle-reinforced composites. Mech. Mater. **36**, 543–572 (2004)
19. Liu, B., Feng, X., Zhang, S.-M.: The effective Young's modulus of composites beyond the Voigt estimation due to the Poisson effect. Compos. Sci. Technol. **69**, 2198–2204 (2009)
20. Love, A.E.H.: Mathematical Theory of Elasticity. Dover Publications, New York (1944)
21. Martin-Herrero, J., Germain, C.: Microstructure reconstruction of fibrous C/C composites from X-ray microtomography. Carbon **45**, 1242–1253 (2007)
22. Mori, T., Tanaka, K.: Average stress in matrix and average elastic energy of materials with misfitting inclusions. Acta Metall. **21**, 571–574 (1973)
23. Nye, J.F.: Własności fizyczne kryształów. PWN, Warszawa (1962)
24. Ochoa, O.O., Reddy, J.N.: Finite Element Analysis of Composite Laminates. Kluwer Academic Publishers, Dordrecht (1992)
25. Paley, M., Aboudi, J.: Micromechanical analysis of composites by generalized method of cells. Mech. Mater. **14**, 127–139 (1992)
26. Pidaparti, R.M.V., May, A.W.: A micromechanical analysis to predict the cord-rubber composite properties. Compos. Struct. **34**, 361–369 (1996)
27. Reuss, A.: Berechnung der Flissgrenze von Mischkristallen auf Grund der Plastizitätsbedingung für Einkristalle. Z. angew. Math. Mech. **9**, 49–58 (1929)
28. Selvadurai, A.P.S., Nikopour, H.: Transverse elasticity of a unidirectionally reinforced composite with an irregular fiber arrangement: experiments, theory and computations. Compos. Struct. **94**, 1973–1981 (2012)
29. Sun, C.T., Chen, J.L.: A micromechanical model for plastic behavior of fibrous composites. Compos. Sci. Technol. **40**, 115–129 (1990)
30. Sun, C.T., Vaidya, R.S.: Prediction of composite properties from a representative volume element. Compos. Sci. Technol. **56**, 171–179 (1996)
31. Sun, C.T., Zhou, S.G.: Failure of quasi-isotropic composite laminates with free edges. J. Reinf. Plast. Compos. **7**, 515–557 (1988)
32. Tamma, K.K., Avila, A.F.: An integrated micro/macro modelling and computational methodology for high temperature composites. In: Hetnarski, R.B. (ed.) Thermal Stresses, pp. 143–256. Lastran Corporation-Publishing Division, Rochester (1999)

33. Tjong, S.C., Ma, Z.Y.: Microstructural and mechanical characteristics of in situ metal matrix composites. Mater. Sci. Eng. **R29**(3–4), 49–113 (2000)
34. Voigt, W.: Über die Beziehungen zwischen beiden Elastizitätskonstanten isotroper Körpär. Wied. Ann. **38**, 573–587 (1889)
35. Whitney, J.M., Riley, M.B.: Elastic properties of fiber reinforced composite materials. AIAA J. **4**, 1537–1542 (1966)
36. Wongsto, A., Li, S.: Micromechanical FE analysis of UD fibre-reinforced composites with fibers distributed at random over the transverse cross-section. Compos. Part A **36**, 1246–1266 (2005)
37. Würkner, M., Berger, H., Gabbert, U.: On numerical evaluation of effective material properties for composite structures with rhombic arrangements. Int. J. Eng. Sci. **49**, 322–332 (2011)

Chapter 4
General Concept of Limit Surfaces—Convexity and Normality Rules, Material Stability

Artur W. Ganczarski and Jacek J. Skrzypek

Abstract General nature of yield or failure criteria terminating elastic range of isotropic or anisotropic materials is summarized. As shown, the hydrostatic pressure sensitivity of anisotropic materials can be captured either by first stress and second common deviatoric invariant direct use (Tsai–Wu), or by the second common stress invariant in an indirect fashion (von Mises). Tension/compression asymmetry in anisotropic materials is accounted for either by presence of first common invariant (only translation, Tsai–Wu) or third common invariant (distortion, Kowalsky). Comparison of two ways to capture anisotropic response, more rigorous explicit common invariants formulations or implicit approaches based on extension of traditional isotropic criteria in terms of transformed invariants (Barlat, Khan) capable of capturing a complete distortion, is shown. Convexity requirement of limit surfaces is discussed and compared for two material behaviors by the use of Drucker's material stability postulate extended to multi-dissipative response or Sylvester's stability condition based on positive definiteness of the tangent stiffness or compliance matrices of hyperelastic material. Generalized Drucker's postulate based on elastic–plastic stiffness matrix is also shown.

Keywords Isotropic/anisotropic yield/failure criteria · Hydrostatic pressure dependence · Tension–compression asymmetry · Elastic/plastic material stability · Convexity of limit surface · Drucker's versus Sylvester's stability

A.W. Ganczarski (✉) · J.J. Skrzypek
Solid Mechanics Division, Institute of Applied Mechanics,
Cracow University of Technology, al. Jana Pawła II 37, 31-864 Kraków, Poland
e-mail: Artur.Ganczarski@pk.edu.pl

J.J. Skrzypek
e-mail: Jacek.Skrzypek@pk.edu.pl

J.J. Skrzypek and A.W. Ganczarski (eds.), *Mechanics of Anisotropic Materials*,
Engineering Materials, DOI 10.1007/978-3-319-17160-9_4

4.1 Termination of the Elastic Range in Cases of Isotropic or Anisotropic Materials

General classification of *anisotropic* initial *yield/failure criteria* requires clear distinction between two approaches met in subject literature. First approach, developed by Sayir, Goldenblat and Kopnov, Spencer, Boehler, Tsai and Wu, Życzkowski, Voyiadjis, and others, is directly based on a concept of *common invariants of stress and structural* anisotropy *tensors*. Second mixed-type approach, developed by Barlat, Khan, Cazacu, Kyriakides, Yoon to mention only some, is based on the extension of classical isotropic yield criteria to anisotropy, by application of *linear transformation of Cauchy's stress*. Approach based on common invariants concept, although more rigorous from invariant theory point of view, is for practical reasons usually limited to *first* and *second common invariants*, such that *distortion effect* is hard to handle. Mixed-type approach, although less formalized in viewpoint of the theory of common invariants, occurs to be very useful for practical description of the materials that exhibit *strong limit surface distortion*. In other words, in the common invariants-based approach the existence of *first common invariant* is necessary to describe *strength differential effect*. On the other hand, in the mixed-type approach, strength differential effect is captured by a *modified third invariant*. It is clear that second invariant (either common or modified stress invariant) has to be present in both formulations.

General *tensorially polynomial anisotropic* plastic flow or failure *criterion* is based on a consistent concept of common invariants of the stress tensor $\boldsymbol{\sigma}$ and of the *structural tensors of plastic or failure anisotropy* $\mathbb{III}$, e.g., $\Pi_{ij}\sigma_{ij}$, $\Pi_{ijkl}\sigma_{ij}\sigma_{kl}$, $\Pi_{ijklmn}\sigma_{ij}\sigma_{kl}\sigma_{mn}$, etc. Structural tensors of plastic/failure anisotropy $\Pi^{\mathrm{p/f}}_{ij}$ second rank, $\Pi^{\mathrm{p/f}}_{ijkl}$ fourth-rank and $\Pi^{\mathrm{p/f}}_{ijklmn}$ sixth rank, different for plasticity (p) or failure (f) initiations, are satisfactory to describe basic transformation modes of limit surfaces due to *plastic* or *damage hardening processes*, namely: *isotropic change of size of limit surface*, its *translation* and *rotation*, as well as *distortion* due to a curvature change (cf. Sayir [28], Kowalsky et al. [20]). The basic *postulates of material stability* either in a *Drucker's* sense for ductile materials (cf. Drucker [8]), or the positive definiteness of the Hessian matrix $[_{\tan}\mathbb{E}]_{mn}\varepsilon_m\varepsilon_n > 0$ in a *Sylvester's* sense for brittle materials (cf. Kuna-Ciskał and Skrzypek [21]), imply restriction, which allow the plastic yield or failure initiation surfaces to be always closed and convex surfaces in the stress space.

Traditionally in case of *ductile materials* (e.g., majority of metals, alloys, intermetallics), second rank tensors Π_{ij} of plastic anisotropy are usually neglected, since the hydrostatic stress does not influence yield initiation criterion (Cazacu and Barlat [4]). Additionally, tension or compression asymmetry is negligible ($k_{\mathrm{t}} \approx k_{\mathrm{c}}$). On the other hand, in case of *brittle materials* (e.g., concrete, ceramic materials, rocks, composite materials, etc.), where initiation of failure or damage manifests mostly or prior to other dissipative phenomena such that first stress invariant plays important role, *tension or compression asymmetry* is essential ($k_{\mathrm{t}} \neq k_{\mathrm{c}}$). Hence, the first (linear) common invariant $\Pi_{ij}\sigma_{ij}$ cannot be omitted (e.g., Tsai–Wu criterion [32]). Moreover, the third (cubic) invariant $\Pi_{ijklmn}\sigma_{ij}\sigma_{kl}\sigma_{mn}$, which describes limit surface distortion, can play an essential role if consecutive hardening phenomena due to advanced plasticity and damage or other microstructure changes occur (e.g., Kowalsky et al. [20]). However, when only initiation of plastic or failure mechanisms is considered, also this term is consequently omitted.

There exists a wide class of advanced engineering materials (e.g., Mg-Th, Mg-Li, Ti-Ni *superalloys*, etc.) that exhibit *strength differential effect* but do not exhibit *hydrostatic pressure sensitivity*. The initial yield criteria describing such materials have to comprise second and third invariants, either of stress tensor J_2, J_3 (Raniecki and Mróz [27]), or common invariants J_2^0, J_3^0 (Barlat et al. [2], Cazacu and Barlat [4], Plunkett et al. [26]). Symbols J_2^0, J_3^0 denote *generalizations* to orthotropy *of classical stress invariants* J_2 and J_3, in the orthotropic Drucker-based yield condition, by the use of a *linear transformation* $\mathbb{L}$ *of the Cauchy stress tensor* $\boldsymbol{S} = \mathbb{L} : \boldsymbol{\sigma}$.

Recently, a generalized form of anisotropic yield/fracture criteria was proposed by Khan and Liu [17] and Khan et al. [18]. These criteria are capable of capturing different types of *limit surfaces* (*quadratic Hill-type*, *nonquadratic Tresca* or *maximum shear stress type* and *intermediate type loci*). They are based on *equi-biaxial tension* loading *condition*, and include tension or compression asymmetry ratio of fracture, and are successfully verified for wide class of materials (e.g., Ti–6Al–4V alloy).

Classical *orthotropic Hill criterion* [13], despite obvious advantages and wide technical applications, is limited however by some constraints of applicability. These constraints are described in detail in Chaps. 5 and 6. However for the purpose of the present preliminary introduction, they will be pointed out briefly.

First limitation of applicability range of the classical *Hill criterion* is established through the inequality bounding magnitudes of the engineering orthotropy constants k_x, k_y, and k_z, in order to avoid *ellipticity loss of the limit surface* in the stress space (e.g., Ottosen and Ristinmaa [25], Ganczarski and Skrzypek [11]). Such limit bounds put upon the orthotropy limits usually hold in case if the degree of material orthotropy is moderate. For example, if the material ensures the *transverse isotropy symmetry*, it is shown that the orthotropy degree bounded by the inequality $k_{\max}/k_{\min} < 2$ guarantees ellipticity of the limit surface to be saved. However, if the orthotropy bound is violated, the Hill criterion becomes useless, when the degeneration of the cylindrical (elliptic) surface into two *concave hyperbolic cylinders* occurs, what is inadmissible in the light of *Drucker's* or *Sylvester's stability postulates*. In a case of high orthotropy degree (observed for majority of the *long fiber reinforced composites*, for instance, Boron/Al, SiC/Ti, Glass/ Epoxy, Graphite/Epoxy, etc., e.g., Herakovich and Aboudi [12], Sun and Vaidya [31], and others), the concept other than Hill is proposed. This new approach suggests formulation of limit criterion based on the nine-parameter von Mises condition, but enhanced by the Hu–Marin-type biaxial orthotropic loading conditions (Hu and Marin [14], Skrzypek and Ganczarski [30]). It will be demonstrated that, even in a case of arbitrarily strong orthotropy (for instance, $k_{\max}/k_{\min} \approx 9$, in case if brass Ł62 is tested) the property of ellipticity is saved.

Second limitation of applicability range of the Hill criterion arises when transverse isotropy property is considered. It will be shown that, if reduction of Hill criterion to the transverse isotropy symmetry is performed, the four-parameter form that satisfies the *tetragonal symmetry class* is furnished (e.g., Voyiadjis and Thiagarajan [33], Sun and Vaidya [31]). This type of symmetry is of particular importance in case of unidirectional fiber reinforced composites. In such a case moduli, k_x, k_y, k_z, and k_{xy} are considered as independent (z orthotropy axis), which makes impossible to reduce the classical Hill criterion to the isotropic von Mises condition in the plane of transverse isotropy. To avoid this irreducibility, new *Hu–Marin-based transversely isotropic criterion* exhibiting *hexagonal symmetry class* is proposed instead of the deviatoric transversely isotropic Hill criterion exhibiting tetragonal symmetry. It enables to achieve coincidence with the *isotropic von Mises condition* in the transverse isotropy plane, preserving cylindricity regardless of the magnitude of orthotropy degree.

Finally, it will be demonstrated that, for some composite materials it is necessary to further modify the three-parameter Hu–Marin-type criterion to the new four-parameter *intermediate* type criterion *between* the classical *Hill and* hexagonal *Hu–Marin's concepts*, taking advantage of the bulge test, that differs essentially from both the Hu–Marin hexagonal type criterion and the isotropic von Mises criterion in the isotropy plane. *Bulge tests* have been performed and described by Jackson et al. [16] with equipment used by Lankford et al. [22]. This new criterion is capable of properly describing the SiC/Ti long fiber reinforced composite examined by Herakovich and Aboudi [12].

4.2 Survey of Pressure Sensitive or Insensitive Yield Criteria

Invariant description of any limit surface (*initial yield* or *failure*) has to be performed by the use of *irreducible set of invariants* being arguments of a scalar function defining limit surface. For *isotropic materials*, equation of limit surface is a *scalar function of three stress invariants* (cf. Iyer [15], etc.)

$$f\left[\mathrm{tr}(\boldsymbol{\sigma}),\ \frac{1}{2}\mathrm{tr}(\boldsymbol{s}\cdot\boldsymbol{s}),\ \frac{1}{3}\mathrm{tr}(\boldsymbol{s}\cdot\boldsymbol{s}\cdot\boldsymbol{s})\right] \tag{4.1}$$

For *anisotropic materials* equation of limit surface is a *scalar function of common* stress and structural anisotropy tensor *invariants* (cf. Sayir [28], etc.)

$$f[\boldsymbol{\sigma}:\overset{\langle 2\rangle}{\mathbb{III}},\ \boldsymbol{\sigma}:\overset{\langle 4\rangle}{\mathbb{III}}:\boldsymbol{\sigma},\ \boldsymbol{\sigma}:\overset{\langle 6\rangle}{\mathbb{III}}:\boldsymbol{\sigma}:\boldsymbol{\sigma},\ldots] \tag{4.2}$$

where only even ranks of anisotropy tensors are taken into account.

In some cases of anisotropic alloys exhibiting *tension/compression asymmetry*, it is convenient to consider a *scalar function of* selected (mixed) *stress invariants* and *common invariants* (cf. Khan and Liu [17], etc.)

$$f\left[\operatorname{tr}(\boldsymbol{\sigma}),\ \frac{1}{2}\operatorname{tr}(\boldsymbol{s}\cdot\boldsymbol{s}),\ \frac{1}{3}\operatorname{tr}(\boldsymbol{s}\cdot\boldsymbol{s}\cdot\boldsymbol{s});\ \ldots,\ \boldsymbol{\sigma} : \overset{\langle 4\rangle}{\mathbb{I}} : \boldsymbol{\sigma},\ \ldots\right] \tag{4.3}$$

The following effects are of particular importance when describing features of the limit surfaces:

- hydrostatic pressure dependence
- tension/compression asymmetry
- anisotropic behavior

In order to properly capture all features considered, the limit criteria have to include the second stress invariant (either $J_{2\sigma}$ or J_{2s}) in case of isotropy or the second common invariant (either $\boldsymbol{\sigma} : \overset{\langle 4\rangle}{\mathbb{I}} : \boldsymbol{\sigma}$ or $\boldsymbol{s} : \overset{\langle 4\rangle}{\mathbb{I}} : \boldsymbol{s}$) in case of anisotropy. This is a direct consequence of the necessity to include total or pure shear elastic energy. The presence of the first and the third invariants (the stress invariants $J_{1\sigma}$, J_{3s} or the common invariants $\boldsymbol{\sigma} : \overset{\langle 2\rangle}{\mathbb{I}}$, $\boldsymbol{\sigma} : \overset{\langle 6\rangle}{\mathbb{I}} : \boldsymbol{\sigma} : \boldsymbol{\sigma}$) is necessary to capture *dependence on hydrostatic pressure* and *tension/compression asymmetry*.

In general, materials can be classified into two groups: hydrostatic pressure dependent and hydrostatic pressure independent materials, alternatively called *pressure sensitive* and *pressure insensitive materials*. Traditionally, *ductile materials* (majority of metals) can be considered as hydrostatic pressure independent. On the other hand, *brittle materials* (rocks, ceramics, etc.) should be treated as hydrostatic pressure dependent ones. Hydrostatic pressure dependence of isotropic or anisotropic limit criteria can be captured in the two different manners:

- *direct dependence on* the *hydrostatic pressure* through both first and second invariants:
  - the first stress invariant plus the second deviatoric invariant

$$f\left[\operatorname{tr}(\boldsymbol{\sigma}),\ \frac{1}{2}\operatorname{tr}(\boldsymbol{s})\right] \qquad \text{isotropy} \tag{4.4}$$

  - the first common invariant plus the second common deviatoric invariant

$$f\left[\boldsymbol{\sigma} : \overset{\langle 2\rangle}{\mathbb{I}},\ \boldsymbol{s} : \overset{\langle 4\rangle}{\mathbb{I}} : \boldsymbol{s}\right] \qquad \text{anisotropy} \tag{4.5}$$

- *indirect dependence on* the *hydrostatic pressure* through the second invariants with the first invariants ignored
  - the second stress invariant

$$f\left[\frac{1}{2}\mathrm{tr}(\boldsymbol{\sigma}\cdot\boldsymbol{\sigma})\right] \qquad (\boldsymbol{\sigma}=\boldsymbol{s}+\sigma_{\mathrm{h}}\mathbf{1}) \qquad \text{isotropy} \tag{4.6}$$

  - the second common stress invariant

$$f\left(\boldsymbol{\sigma}:\overset{\langle 4\rangle}{\mathbb{\Pi}}:\boldsymbol{\sigma}\right) \qquad (\boldsymbol{\sigma}=\boldsymbol{s}+\sigma_{\mathrm{h}}\mathbf{1}) \qquad \text{anisotropy} \tag{4.7}$$

There exists a broad class of engineering materials which do not exhibit any dependence on the hydrostatic pressure, neither direct nor indirect. This means that in case of the isotropic hydrostatic pressure independent materials, the corresponding limit surfaces have to include the *second deviatoric invariant* exclusively. In case of anisotropy, limit surfaces can include the *second common deviatoric invariant* and additionally the first common deviatoric invariant. However in all cases considered, equation of limit surface has to include the *second stress* or the *second common invariants* which results from the quadratic form of energy representation. Exemplary equations of limit surfaces that found confirmation in engineering materials are presented in Table 4.1 according to aforementioned classification.

Table 4.1 shows comparison between the pairs of appropriate isotropic and anisotropic criteria that correspond to the direct dependence on hydrostatic pressure (*Drucker–Prager criterion* vs. *Tsai–Wu criterion*), the indirect dependence on hydrostatic pressure (*Beltrami criterion* vs. *von Mises criterion*) and independence of the hydrostatic pressure (*Huber–von Mises criterion* vs. *Hill criterion*). The oldest criterion based on total elastic energy formulated by Beltrami in 1885 is invoked in this table although it has no experimental evidence.

Tension/compression asymmetry, also called *strength differential effect* is included in a natural way in limit criteria for anisotropic materials. In case of limit criteria for isotropic materials, this effect manifests through the presence of first stress invariant ($J_{1\sigma}$) and/or the third stress invariant (J_{3s}), as shown in Fig. 4.1. Note that in case of the first stress invariant dependent surface compressive and tensile meridians are in identical distance from the center of limit curve, but axis of the limit surface is

Table 4.1 Hydrostatic pressure dependence of initial yield/failure criteria

	Dependence on σ_{h}		
	Direct	Indirect	Lack of dependence
Isotropy	Drucker–Prager $\alpha J_{1\sigma}+\sqrt{J_{2s}}=k$	Beltrami $\sqrt{3J_{2\sigma}}=k$	Huber–von Mises $\sqrt{3J_{2s}}=k$
Anisotropy	Tsai–Wu $\boldsymbol{\sigma}:\overset{\langle 2\rangle}{\mathbb{\Pi}}+\boldsymbol{s}:\overset{\langle 4\rangle}{\mathbb{\Pi}}{}^{\mathrm{H}}:\boldsymbol{s}=1$	von Mises $\boldsymbol{\sigma}:\overset{\langle 4\rangle}{\mathbb{\Pi}}:\boldsymbol{\sigma}=1$	Hill $\boldsymbol{s}:\overset{\langle 4\rangle}{\mathbb{\Pi}}{}^{\mathrm{H}}:\boldsymbol{s}=1$

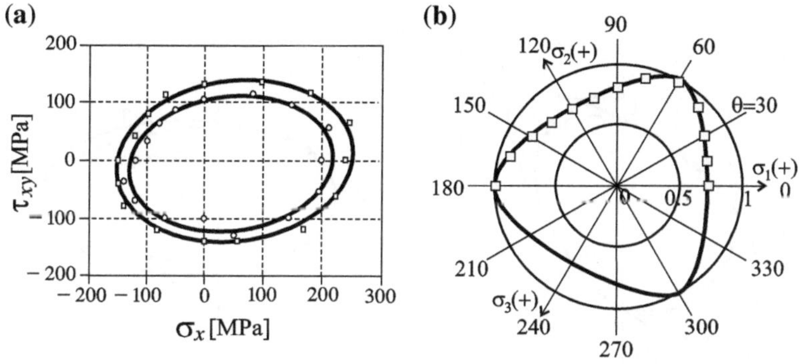

Fig. 4.1 Tension/compression asymmetry caused by **a** First stress invariant—low carbon steel 18G2A subjected to monotonic plastic offsets ○ − $\varepsilon_{\text{off}} = 1 \times 10^{-5}$, □ − $\varepsilon_{\text{off}} = 5 \times 10^{-5}$ after Kowalewski and Śliwowski [19], **b** third stress invariant—TiNi alloy after Raniecki and Mróz [27]

shifted (Fig. 4.1a). In other case, when *limit surface* is *third stress invariant dependent* function, compressive and tensile meridians are not in identical distance from the center of limit curve but the axis of the limit surface remains at the position of hydrostatic axis (Fig. 4.1b).

It was aforementioned that the limit criteria have to include appropriate second invariants. However, limit surfaces based on the second invariants exclusively (stress invariants or common invariants) are capable to capturing neither *tension/compression asymmetry* nor shape change due to *distortion*. By contrast, the limit criteria based on the second and the third invariants (stress invariants or common invariants) are capable of capture both tension/compression asymmetry and distortion. Table 4.2 shows comparison between the pairs of selected isotropic and *anisotropic criteria* that correspond to the lack of tension/compression asymmetry and distortion (*Huber–von Mises' criterion* vs. *Hill criterion*), *tension/compression asymmetry* with distortion ruled out (*Drucker–Prager criterion* vs. *Tsai–Wu criterion*) and *tension/compression asymmetry with distortion* accounted for (Drucker criterion vs. Kowalsky et al. criterion). To illustrate classification described in Table 4.2 a comparison between *asymmetry without distortion* and *asymmetry with distortion* accounted for is presented in Fig. 4.2. In the case of isotropy Fig. 4.2a the Drucker criterion is compared with the Drucker–Prager criterion. In case of *Drucker–Prager*

Table 4.2 Effect of first and third invariants on tension/compression asymmetry and distortion of limit surfaces

	Lack of asymmetry and distortion	Asymmetry without distortion	Asymmetry and distortion
Isotropy	Huber–von Mises $\sqrt{3J_{2s}} = k$	Drucker–Prager $\alpha J_{1\sigma} + \sqrt{J_{2s}} = k$	Drucker $J_{2s}^3 - cJ_{3s}^2 = k^6$
Anisotropy	Hill $\boldsymbol{s} : \overset{\langle 4\rangle}{\mathbb{\Pi}}{}^{\mathrm{H}} : \boldsymbol{s} = 1$	Tsai–Wu $\boldsymbol{\sigma} : \overset{\langle 2\rangle}{\mathbb{\Pi}} + \boldsymbol{s} : \overset{\langle 4\rangle}{\mathbb{\Pi}}{}^{\mathrm{H}} : \boldsymbol{s} = 1$	Kowalsky et al. $\boldsymbol{s} : \overset{\langle 4\rangle}{\mathbb{\Pi}} : \boldsymbol{s} + \boldsymbol{s} : \overset{\langle 6\rangle}{\mathbb{\Pi}} : \boldsymbol{s} : \boldsymbol{s} = 1$

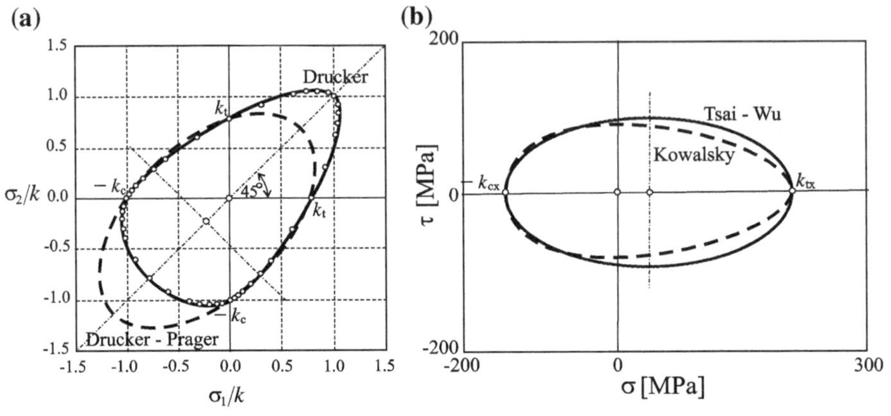

Fig. 4.2 Comparision of tension/compression asymmetry and distortion of limit curves in case of: **a** isotropy (Drucker–Prager criterion vs. Drucker criterion), **b** anisotropy (Tsai–Wu citerion vs. Kowalsky et al. criterion)

criterion based on the first and the second stress invariants tension/compression asymmetry appears independently from shape distortion. By contrast when the *Drucker criterion* is used, both effects are coupled through the third invariant so they appear simultaneously. In the case of anisotropy Fig. 4.2b the Tsai–Wu criterion is compared with the Kowalsky et al. criterion. The *Tsai–Wu criterion* accounts for tension/compression asymmetry without distortion (only translation through the first common invariant accounted for). By contrast when the *Kowalsky et al.* the six order *criterion* is used, the *tension/compression asymmetry* and *shape distortion* are coupled in an anisotropic fashion through the third common invariant.

In general, a material anisotropy can be captured by use of the two approaches. In the first, mathematically consistent approach called the *explicit anisotropy approach* the system of stress invariants $J_{1\sigma}$, J_{2s}, J_{3s} is substituted by the corresponding system of common invariants $\boldsymbol{\sigma} : \overset{\langle 2 \rangle}{\mathbb{H}}$, $\boldsymbol{s} : \overset{\langle 4 \rangle}{\mathbb{H}} : \boldsymbol{s}$, $\boldsymbol{s} : \overset{\langle 6 \rangle}{\mathbb{H}} : \boldsymbol{s} : \boldsymbol{s}$ according to the Goldenblat, Kopnov, and Sayir concept when formulating anisotropic limit criteria. In the other, currently dynamically developed approach called the *implicit anisotropy approach* by Barlat and Khan either the second J_{2s} and the third J_{3s} stress invariants are substituted by the corresponding *transformed deviatoric invariants* J^0_{2s}, J^0_{3s} or the stress deviator is transformed by use of the two independent *4-rank transformation tensors* $\boldsymbol{\Sigma} = \overset{\langle 4 \rangle}{\mathbb{C}} : \boldsymbol{s}$ and $\boldsymbol{\Sigma}' = \overset{\langle 4 \rangle}{\mathbb{C}'} : \boldsymbol{s}$ and next they are inserted to one of well know isotropic criteria, either Drucker or Hosford, respectively. These linear transformations correspond to mapping of the deviatoric Cauchy stress tensor $\boldsymbol{\sigma}$ to other two deviatoric stresses $\boldsymbol{\Sigma}$, $\boldsymbol{\Sigma}'$ referring to the material anisotropy (orthotropy) frame.

The implicit approach is able to capture the full material orthotropy with distortion effect included by use of two 4th rank orthotropic transformation tensors $\overset{\langle 4 \rangle}{\mathbb{C}}$, $\overset{\langle 4 \rangle}{\mathbb{C}'}$ containing $2 \times 9 = 18$ independent material constants by contrast to the explicit common invariant based approach which requires in case of material orthotropy 4-rank tensor $\overset{\langle 4\rangle}{\mathbb{\Pi}}$ and 6-rank tensor $\overset{\langle 6\rangle}{\mathbb{\Pi}}$ containing $9 + 56 = 65$ material constants. Although the explicit approach is more mathematically rigorous than the implicit one, simultaneously it is much more cumbersome and open to misunderstandings. Both approaches, the explicit and the implicit, are alternative ones and obviously they lead to different approximations.

Comparison of the explicit and the implicit approaches to capture anisotropy is schematically presented for selected criteria in Table 4.3. A major difficulty for the limit yield/failure description is caused by the *coupling* between *anisotropy and* strong *tension/compression asymmetry* as discussed by Khan et al. [18]. Such significant coupling can lead to a complete distortion of the limit surface (possible lack of any axis of symmetry) as it is presented in Fig. 4.3 based on Luo et al. [23] experimental findings for AZ31B Mg alloy fitted by Plunkett et al. [26].

Table 4.3 Explicit or implicit anisotropy of limit surfaces

	Explicit	Implicit	
Isotropy	Huber–von Mises $\mathbf{s} : \overset{\langle 4\rangle}{\mathbb{I}} : \mathbf{s} = k^2$	Raniecki–Mróz $J_{2s}^{3/2} - cJ_{3s} = k^3$	Cazacu et al. [5] $\sum_{i=1}^{3} (\lvert s_i\rvert - \widehat{k} s_i)^a = 2k^a$
Anisotropy	Hill $\mathbf{s} : \overset{\langle 4\rangle}{\mathbb{\Pi}}{}^{\mathrm{H}} : \mathbf{s} = 1$	Cazacu and Barlat $(J_2^0)^{3/2} - cJ_3^0 = k^3$ where J_2^0, J_3^0 transformed common invariants	Plunkett et al. $\sum_{i=1}^{3} (\lvert\Sigma_i\rvert - \widehat{k}\Sigma_i)^a +$ $\sum_{i=1}^{3} (\lvert\Sigma_i'\rvert - \widehat{k}'\Sigma_i')^a = 2k^a$ where $\boldsymbol{\Sigma} = \overset{\langle 4\rangle}{\mathbb{C}} : \mathbf{s}$, $\boldsymbol{\Sigma}' = \overset{\langle 4\rangle}{\mathbb{C}'} : \mathbf{s}$

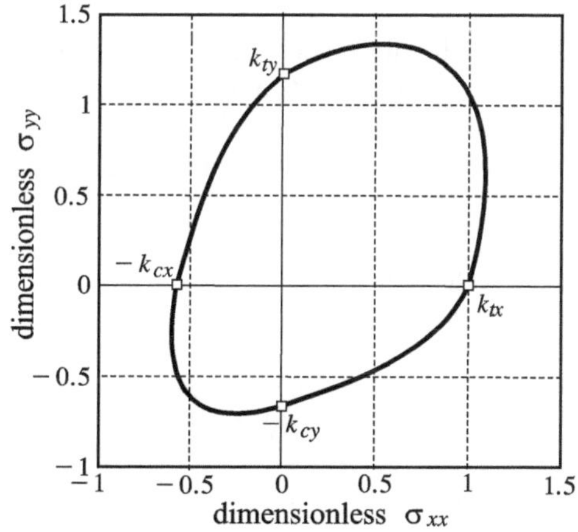

Fig. 4.3 Fitting of Luo et al. [23] experimental data for AZ31B Mg alloy by the use of the implicit anisotropic yield criterion by Cazacu and Barlat [4]

4.3 Drucker's Postulate for Stability of Ductile Materials—Requirements for Convexity and Normality of the Yield Surface

Let us confime first to only one situation when termination of the elastic range is caused by the *initiation of plastic flow mechanism*. When formulating equations of the *initial yield surface* and its further rebuilding due to consecutive process of strain growth exceeding the elastic range, it is convenient to restrict further considerations to processes that guarantee the *convexity of* initial and subsequent *yield surfaces* as well as the *associated flow rule*. This means that the *plastic potential function* g is considered to be equal to the *yield function* f. Such simplified approach is useful for description of majority of metals by contrast to the *nonassociated flow rules* ($g \neq f$) applicable to majority of brittle materials.

Consider first arbitrary *stress cycle* ABCD that consists of elastic loading AB from the initial state inside current yield surface (σ^*_{ij}) to point belonging to this surface (σ_{ij}), subsequent elementary loading BC corresponding to stress increment $\mathrm{d}\sigma_{ij}$ during which the yield surface is rebuilt $f_i \rightarrow f_{i+1}$, and final unloading CD to the initial stress level (σ^*_{ij}) as shown in Fig. 4.4. Note however that this process describes the closed cycle only in stress space $\sigma^{\mathrm{D}}_{ij} = \sigma^*_{ij}$ but the final state D corresponds to changed strain state $\varepsilon^{\mathrm{D}}_{ij} = \varepsilon^{\mathrm{e}}_{ij} + \mathrm{d}\varepsilon^{\mathrm{p}}_{ij}$. The strain increment $\mathrm{d}\varepsilon^{\mathrm{p}}_{ij}$ stands for permanent and irreversible plastic strain change connected with rebuilding of the subsequent yield surface.

According to the *Drucker postulate*, work per unit volume done by stress quasi-cycle on total deformation ABCD is nonnegative

$$\mathcal{W} = \oint\limits_{\mathrm{ABCD}} \left(\sigma_{ij} - \sigma^*_{ij}\right) \mathrm{d}\varepsilon_{ij} \geq 0 \tag{4.8}$$

The additional load carried by the material over a complete stress quasi-cycle is called the external agency. In other words, when the work done by an external agency over the *stress quasi-cycle* would be negative a subsequent equilibrium state would have been reached in a spontaneous way associated with an energy dissipation. According

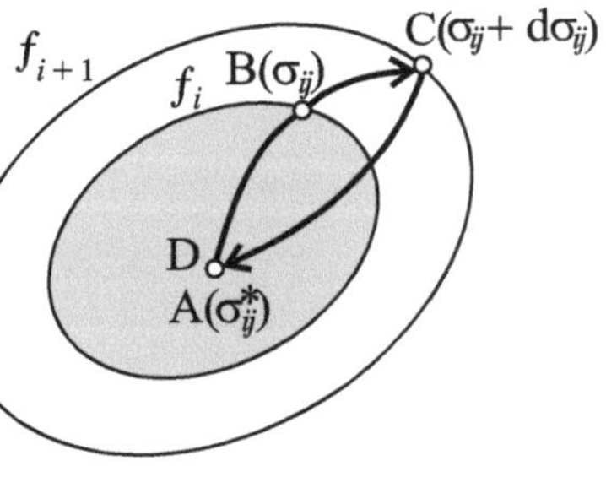

Fig. 4.4 Ilustration of the closed stress quasi-cycle

to small strain theory the *additive decomposition of* the total *strain increment* into the *reversible* and *irreversible terms* is done $\mathrm{d}\varepsilon_{ij} = \mathrm{d}\varepsilon^{\mathrm{e}}_{ij} + \mathrm{d}\varepsilon^{\mathrm{p}}_{ij}$ we arrive at

$$\mathcal{W} = \oint_{\mathrm{ABCD}} \left(\sigma_{ij} - \sigma^*_{ij}\right) \mathrm{d}\varepsilon^{\mathrm{e}}_{ij} + \oint_{\mathrm{ABCD}} \left(\sigma_{ij} - \sigma^*_{ij}\right) \mathrm{d}\varepsilon^{\mathrm{p}}_{ij} \tag{4.9}$$

The first of above integrals is equal to zero since, according to the Hooke law $\varepsilon^{\mathrm{e}}_{ij} = E^{-1}_{ijkl}\sigma_{kl}$ holds. Providing that the compliance tensor is constant $E^{-1}_{ijkl} = \mathrm{const}$ we arrive at

$$\begin{aligned} \oint_{\mathrm{ABCD}} \left(\sigma_{ij} - \sigma^*_{ij}\right) E^{-1}_{ijkl}\mathrm{d}\sigma_{kl} &= E^{-1}_{ijkl} \oint_{\mathrm{ABCD}} \left(\sigma_{ij} - \sigma^*_{ij}\right) \mathrm{d}\sigma_{kl} \\ &= E^{-1}_{ijkl} \left(\oint_{\mathrm{ABCD}} \sigma_{ij}\mathrm{d}\sigma_{kl} - \sigma^*_{ij} \oint_{\mathrm{ABCD}} \mathrm{d}\sigma_{kl} \right) \\ &= E^{-1}_{ijkl} \left(\frac{\sigma_{ij}\sigma_{mn}}{2} - \sigma^*_{ij}\sigma_{mn} \right) \Bigg|^{\sigma^{\mathrm{A}}_{kl}}_{\sigma^{\mathrm{D}}_{kl} = \sigma^{\mathrm{A}}_{kl}} = 0 \end{aligned} \tag{4.10}$$

Hence, keeping in mind that plastic strain is different from zero $\varepsilon^{\mathrm{p}}_{ij} \neq 0$ only along path BC the Drucker postulate (4.8) is finally expressed by inequality for the following simple integral (not circular integral)

$$\mathcal{W}^{\mathrm{p}} = \int_{\mathrm{BC}} \left(\sigma_{ij} - \sigma^*_{ij}\right) \mathrm{d}\varepsilon^{\mathrm{p}}_{ij} \geq 0 \tag{4.11}$$

This means that the work done by the external agency on plastic strain is nonnegative and corresponds to *rebuilding of yield surface* $f_i \longrightarrow f_{i+1}$. Applying expansion of $\mathcal{W}^{\mathrm{p}}$ in the Taylor series around the initial point $\sigma_{ij} = \sigma^*_{ij}$ we arrive at

$$\begin{aligned} \mathcal{W}^{\mathrm{p}} &= \mathcal{W}^{\mathrm{p}} \left(\sigma_{ij} = \sigma^*_{ij}\right) + \frac{\mathrm{d}\mathcal{W}^{\mathrm{p}} \left(\sigma_{ij} = \sigma^*_{ij}\right)}{1!} + \frac{\mathrm{d}^2\mathcal{W}^{\mathrm{p}} \left(\sigma_{ij} = \sigma^*_{ij}\right)}{2!} + \cdots \\ &= 0 + \left(\sigma_{ij} - \sigma^*_{ij}\right) \mathrm{d}\varepsilon^{\mathrm{p}}_{ij} + \frac{1}{2}\mathrm{d}\sigma_{ij}\mathrm{d}\varepsilon^{\mathrm{p}}_{ij} + \cdots \geq 0 \end{aligned} \tag{4.12}$$

When the two first nonzero terms of expansion series (4.12) are saved, we find an inequality

$$\left(\sigma_{ij} - \sigma^*_{ij}\right) \mathrm{d}\varepsilon^{\mathrm{p}}_{ij} + \frac{1}{2}\mathrm{d}\sigma_{ij}\mathrm{d}\varepsilon^{\mathrm{p}}_{ij} \geq 0 \tag{4.13}$$

which must hold for arbitrary initial stress state σ^*_{ij}, inside or on the current yield surface. Therefore, the inequality (4.13) that expresses the *condition of stability* of elastic–plastic material in *Drucker's* sense can be finally furnished in the form of two following inequalities

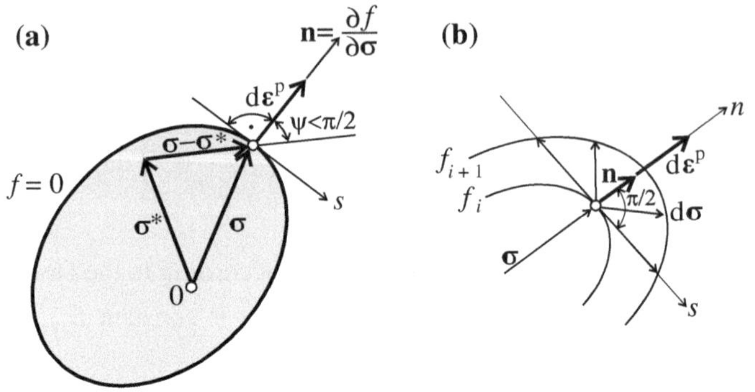

Fig. 4.5 Interpretation of Drucker's postulate consequences: **a** convexity, **b** normality

$$(\sigma_{ij} - \sigma^*_{ij})\mathrm{d}\varepsilon^{\mathrm{p}}_{ij} \geq 0 \qquad \text{and} \qquad \mathrm{d}\sigma_{ij}\mathrm{d}\varepsilon^{\mathrm{p}}_{ij} \geq 0 \tag{4.14}$$

The above inequalities must simultaneously hold which in mathematical sense corresponds to nonnegative value of the first and the second energy differential $\mathcal{W}^{\mathrm{p}}$ in the neighborhood of the initial point $\sigma_{ij} = \sigma^*_{ij}$.

Conditions (4.14) can be interpreted in a geometric way regarding *convexity of* a *yield surface* and *normality of* vector of *plastic strain increment.*

The first of inequalities (4.14) can interpreted as nonnegative value of the scalar product of two vectors $(\boldsymbol{\sigma} - \boldsymbol{\sigma}^*)$ and $\mathrm{d}\boldsymbol{\varepsilon}^{\mathrm{p}}$. Hence, the angle ψ between these two vectors in the stress space σ_{ij} has to be either acute or right angle $\psi \leq \pi/2$ (Fig. 4.5a). This condition holds for each vector $\boldsymbol{\sigma}^*$ which is located on or within the yield surface. This implies that the yield surface must be convex surface in the stress space f. It is called in literature the *convexity postulate of yield surface* f.

The second of inequalities (4.14) can be interpreted as the scalar product of two vectors $\mathrm{d}\boldsymbol{\sigma}$ and $\mathrm{d}\boldsymbol{\varepsilon}^{\mathrm{p}}$ which must also be nonnegative for arbitrarily chosen stress increment $\mathrm{d}\boldsymbol{\sigma}$ (Fig. 4.5b). This requirement must hold for arbitrary vectors $\mathrm{d}\boldsymbol{\sigma}$ connected with transition of f_i surface into f_{i+1} surface (Fig. 4.5b) which belong to the half-space tangent to ith surface, hence the only one possible vector $\mathrm{d}\boldsymbol{\varepsilon}^{\mathrm{p}}$ which always ensures condition (4.14) must be normal to this surface f, $\boldsymbol{n} = \boldsymbol{n}_f$

$$\mathrm{d}\boldsymbol{\varepsilon}^{\mathrm{p}} = \lambda\boldsymbol{n} = \lambda\frac{\partial f}{\partial \boldsymbol{\sigma}} = \lambda\,\mathbf{grad}\,f \tag{4.15}$$

Symbol λ is the scalar multiplier the magnitude of which ensures that the new point at the stress trajectory belongs to the new yield surface f. The condition (4.15) determines direction of the plastic strain increment $\mathrm{d}\boldsymbol{\varepsilon}^{\mathrm{p}}$ consistent with the gradient $\boldsymbol{n} = \partial f/\partial\boldsymbol{\sigma} = \lambda\,\mathbf{grad}\,f$ which is normal to the yield surface. This requirement is equivalent to the so-called flow rule associated with the yield surface.

Drucker's postulate of stability (4.14) assumes that both *convexity* and *normality rules* must hold. In case when the normality does not hold, it is possible to choose

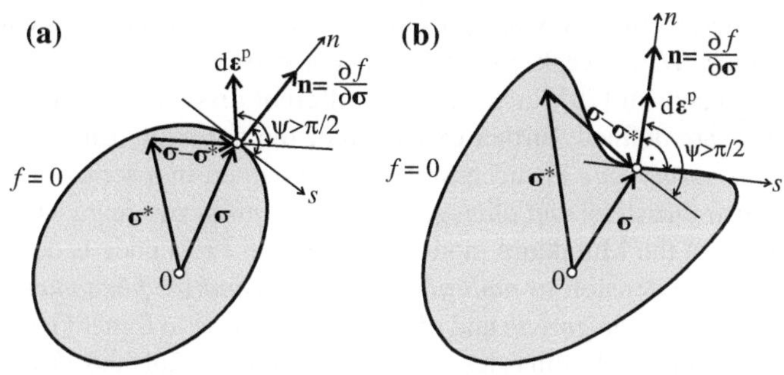

Fig. 4.6 Illustration of a case when Drucker's stability postulate is not satisfied: **a** convexity of the surface f saved but normality of $d\varepsilon^p$ does not hold, **b** lack of convexity of the surface f but normality of $d\varepsilon^p$ saved

Fig. 4.7 Graphical interpretation of nonassociated flow rule

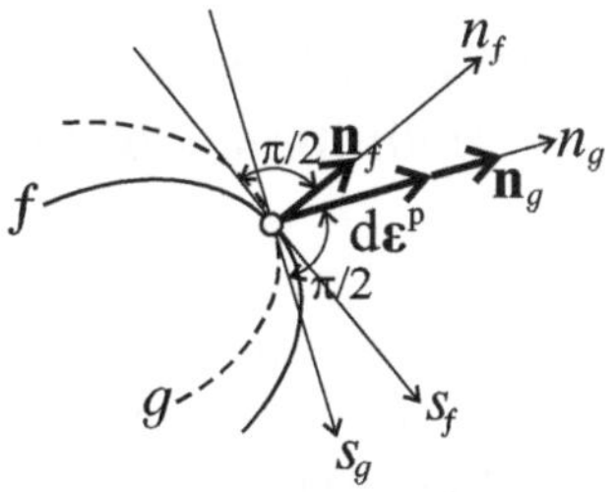

$\sigma^* \neq \sigma$ that the angle between vectors $(\sigma - \sigma^*)$ and $d\varepsilon^p$ is greater than $\pi/2$ such that the scalar product of these two vectors is negative. This means that in this case Drucker's postulate is not satisfied (Fig. 4.6a). By contrast when the convexity of a yield surface is violated it is possible to choose σ^* such that the scalar product is negative $(\sigma - \sigma^*)d\varepsilon^p < 0$ (Fig. 4.6b). Both above negative examples (Fig. 4.6a, b) are an indirect proof that violation even one of normality or convexity conditions means violation of Drucker's postulate as a whole.

Approach based on the associated plastic flow rule is applicable for majority of metals but acceptance of this rule is not necessary in case of nonmetallic materials (soils, rocks, some composites). In such case, the so-called *nonassociated flow rule* is applicable. For nonassociated flow rule the direction of plastic strain increment is determined from the gradient to other surface g which does not coincide with the yield surface f, $g(\sigma) \neq f(\sigma)$: $\boldsymbol{n} = \boldsymbol{n}_g \neq \boldsymbol{n}_f$ (Fig. 4.7). As consequence, in case of nonassociated flow rule we arrive at

$$d\varepsilon^p = \lambda \frac{\partial g}{\partial \sigma} = \lambda \mathbf{grad} g \tag{4.16}$$

instead of (4.15). For such materials Drucker's stability postulate does not obey.

In classical formulation of Drucker's material stability postulate [8, 9], the single dissipation process connected with plastic flow is considered. However, when *multi-*

dissipative material *processes* are present an extension of the classical normality rule can be done. When weak-coupling concept between the dissipative processes is applied Simo and Yu [29], in which two different dissipation surfaces plastic $f_{\rm p}$ and damage $f_{\rm d}$ are defined, further extension of the normality rule to the so-called *generalized normality rule* is successfully implemented in a series of papers dealing with *coupled plasticity and damage dissipation processes*, e.g., Murakami [24]. Implementation of the Murakami model to ABAQUS FEM code is done by Bielski et al. [3]. Further extension to *multiple-coupled dissipative phenomena* connected with *plastic flow*, *damage growth* and *phase change* is due to Egner [10]. A comparison between existing evolution rules proposed by Abu Al-Rub and Voyiadjis [1] and Chaboche [6] with recently developed formulations has been done. More advanced discussion on multi-dissipative response description can be found in further chapters.

4.4 Stability Postulate for Elastic Materials—Positive Definiteness of the Tangent Stiffness Matrix

Following Chen and Han [7], we consider the *stability criterion of hyperelastic material* for which all deformations are reversible such that stability requires the work done by the external agency in a cycle to be zero. For an elastic material, both the stress state and the strain state in (4.8) return back to σ^*_{ij} and ε^*_{ij} as shown in Fig. 4.8. Over such a cycle the Drucker stability postulate becomes an equality

$$\oint \left(\sigma_{ij} - \sigma^*_{ij}\right) \mathrm{d}\varepsilon_{ij} = 0 \tag{4.17}$$

since no permanent strain over such cycle occurs. Note that the above equality holds for the elastic material by contrast to previously formulated Drucker's inequality formulated for elastic–plastic material (4.8). Choosing next the initial state to be

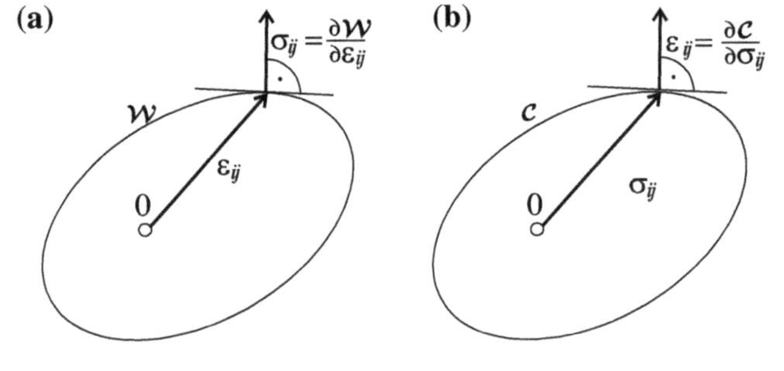

Fig. 4.8 Illustration of normality of: **a** stress vector σ_{ij} to normal hypersurface of constant strain energy per unit volume $\mathcal{W} = \text{const}$, **b** strain vector ε_{ij} to normal hypersurface of constant complementary energy $\mathcal{C} = \text{const}$

stress and strain free, we reduce (4.8) to

$$\oint \sigma_{ij} \mathrm{d}\varepsilon_{ij} = 0 \tag{4.18}$$

which must hold irrespectively to the path. In other words, the integrand in (4.18) is the exact differential such that the *elastic strain energy* per unit volume $\mathcal{W}$ serves as the *potential function for stress*

$$\mathcal{W}\left(\varepsilon_{ij}\right) = \int \sigma_{ij} \mathrm{d}\varepsilon_{ij} \quad \text{and} \quad \sigma_{ij} = \frac{\partial \mathcal{W}}{\partial \varepsilon_{ij}} \tag{4.19}$$

In an analogous way, we may prove that the *complementary energy* per unit volume $\mathcal{C}$ serves as the *potential function for strain*

$$\mathcal{C}\left(\sigma_{ij}\right) = \int \varepsilon_{ij} \mathrm{d}\sigma_{ij} \quad \text{and} \quad \varepsilon_{ij} = \frac{\partial \mathcal{C}}{\partial \sigma_{ij}} \tag{4.20}$$

Note however that in general case of *inelastic material*, the strain energy $\mathcal{W}$ may depend not only on the current strain state $\mathcal{W}(\varepsilon_{ij})$ but also on the strain history $\mathcal{W}(\varepsilon_{ij}, f(\varepsilon_{ij}))$, e.g., as a result of irreversible microstructure change due to plastic flow or microdamage growth. Similarly in such general case, the complementary energy per unit volume $\mathcal{C}$ may depend not only on current stress state $\mathcal{C}(\sigma_{ij})$ but also on stress history $\mathcal{C}(\sigma_{ij}, f(\sigma_{ij}))$. This means that neither $\mathcal{W}$ nor $\mathcal{C}$ can be directly chosen as potential functions for stress and strain, respectively.

The *strain energy per unit volume* $\mathcal{W}(\varepsilon_{ij})$ and the *complementary energy per unit volume* $\mathcal{C}(\sigma_{ij})$ are being interpreted as hypersurfaces of constant value energy in six-dimensional spaces of strain ε_{ij} and stress σ_{ij} respectively. Assuming independence of both considered energies of loading histories in respective spaces, the derivatives of both scalar functions $\partial\mathcal{W}/\partial\varepsilon_{ij}$ and $\partial\mathcal{C}/\partial\sigma_{ij}$ with respect to their arguments are being interpreted as the gradients of corresponding hypersurfaces, which are the vectors normal oriented outward to these hypersurfaces (Fig. 4.8).

Thus *stress increment* $\dot{\sigma}_{ij}$ can be furnished in terms of strain increment $\dot{\varepsilon}_{ij}$ as follows

$$\dot{\sigma}_{ij} = \frac{\partial \sigma_{ij}}{\partial \varepsilon_{kl}} \dot{\varepsilon}_{kl} = \frac{\partial^2 \mathcal{W}}{\partial \varepsilon_{ij} \partial \varepsilon_{kl}} \dot{\varepsilon}_{kl} \tag{4.21}$$

where definition of the potential function for stress (4.19_2) is involved.

The following definitions of material stability are further explored (see Chen and Han [7]):

- The work done by the added stress increment on the strain increment is positive and the following inequality is called *stability in small*

$$\dot{\sigma}_{ij} \dot{\varepsilon}_{ij} > 0 \tag{4.22}$$

- The work done over the cycle by application and removal of stress increment on strain increment is nonnegative and the following inequality is called *stability on cycle*

$$\oint \dot{\sigma}_{ij}\dot{\varepsilon}_{ij} \geq 0 \tag{4.23}$$

For numerical applications in order to find *stability criterion for hyperelastic materials*, the first inequality (4.22) occurs the effective tool to this end (Fig. 4.9).

Substitution of (4.21) for $\dot{\sigma}_{ij}$ into (4.22) leads to the following stability condition

$$\frac{\partial^2 \mathcal{W}}{\partial \varepsilon_{ij}\partial \varepsilon_{kl}}\dot{\varepsilon}_{ij}\dot{\varepsilon}_{kl} > 0 \tag{4.24}$$

The above inequality written in nonabbreviated form reads as

$$\begin{Bmatrix} \dot{\varepsilon}_{xx} \\ \dot{\varepsilon}_{yy} \\ \dot{\varepsilon}_{zz} \\ \dot{\gamma}_{xy} \\ \dot{\gamma}_{yz} \\ \dot{\gamma}_{zx} \end{Bmatrix}^{\mathrm{T}} \begin{bmatrix} \frac{\partial^2 \mathcal{W}}{\partial \varepsilon_{xx}^2} & \frac{\partial^2 \mathcal{W}}{\partial \varepsilon_{xx}\partial \varepsilon_{yy}} & \frac{\partial^2 \mathcal{W}}{\partial \varepsilon_{xx}\partial \varepsilon_{zz}} & \frac{\partial^2 \mathcal{W}}{\partial \varepsilon_{xx}\partial \gamma_{xy}} & \frac{\partial^2 \mathcal{W}}{\partial \varepsilon_{xx}\partial \gamma_{yz}} & \frac{\partial^2 \mathcal{W}}{\partial \varepsilon_{xx}\partial \gamma_{zx}} \\ & \frac{\partial^2 \mathcal{W}}{\partial \varepsilon_{yy}^2} & \frac{\partial^2 \mathcal{W}}{\partial \varepsilon_{yy}\partial \varepsilon_{zz}} & \frac{\partial^2 \mathcal{W}}{\partial \varepsilon_{yy}\partial \gamma_{xy}} & \frac{\partial^2 \mathcal{W}}{\partial \varepsilon_{yy}\partial \gamma_{yz}} & \frac{\partial^2 \mathcal{W}}{\partial \varepsilon_{yy}\partial \gamma_{zx}} \\ & & \frac{\partial^2 \mathcal{W}}{\partial \varepsilon_{zz}^2} & \frac{\partial^2 \mathcal{W}}{\partial \varepsilon_{zz}\partial \gamma_{xy}} & \frac{\partial^2 \mathcal{W}}{\partial \varepsilon_{zz}\partial \gamma_{yz}} & \frac{\partial^2 \mathcal{W}}{\partial \varepsilon_{zz}\partial \gamma_{zx}} \\ & & & \frac{\partial^2 \mathcal{W}}{\partial \gamma_{xy}^2} & \frac{\partial^2 \mathcal{W}}{\partial \gamma_{xy}\partial \gamma_{yz}} & \frac{\partial^2 \mathcal{W}}{\partial \gamma_{xy}\partial \gamma_{zx}} \\ & & & & \frac{\partial^2 \mathcal{W}}{\partial \gamma_{yz}^2} & \frac{\partial^2 \mathcal{W}}{\partial \gamma_{yz}\partial \gamma_{zx}} \\ & & & & & \frac{\partial^2 \mathcal{W}}{\partial \gamma_{zx}^2} \end{bmatrix} \begin{Bmatrix} \dot{\varepsilon}_{xx} \\ \dot{\varepsilon}_{yy} \\ \dot{\varepsilon}_{zz} \\ \dot{\gamma}_{xy} \\ \dot{\gamma}_{yz} \\ \dot{\gamma}_{zx} \end{Bmatrix} > 0 \tag{4.25}$$

The Voigt notation used above allows to represent the fourth-rank elasticity tensor by its representation matrix identified as the *tangent stiffness matrix*

$$[{}_{\tan}E]_{mn} = \left[\frac{\partial^2 \mathcal{W}}{\partial \varepsilon_m \partial \varepsilon_n}\right] \tag{4.26}$$

The above reasoning dealing with the incremental formulation of constitutive equation based on quadratic form for strain energy $\mathcal{W}$ can easily be converted into

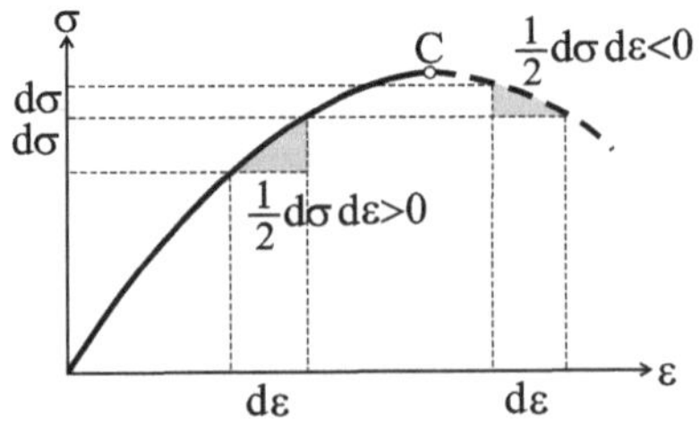

Fig. 4.9 Illustration of material stability loss in case of hiper elastic material

dual formulation when *incremental form of constitutive equation* based on the complementary energy is used alternatively. The strain increment $\dot{\varepsilon}_{ij}$ expressed in terms of the stress increment $\dot{\sigma}_{ij}$ takes the following form

$$\dot{\varepsilon}_{ij} = \frac{\partial^2 \mathcal{C}}{\partial \sigma_{ij} \partial \sigma_{kl}} \dot{\sigma}_{kl} \tag{4.27}$$

which substituted to the condition of stability in small (4.22), leads to inequality

$$\frac{\partial^2 \mathcal{C}}{\partial \sigma_{ij} \partial \sigma_{kl}} \dot{\sigma}_{ij} \dot{\sigma}_{kl} > 0 \tag{4.28}$$

The above inequality written in nonabbreviated form reads as

$$\begin{Bmatrix} \dot{\sigma}_{xx} \\ \dot{\sigma}_{yy} \\ \dot{\sigma}_{zz} \\ \dot{\tau}_{xy} \\ \dot{\tau}_{yz} \\ \dot{\tau}_{zx} \end{Bmatrix}^{\mathrm{T}} \begin{bmatrix} \frac{\partial^2 \mathcal{C}}{\partial \sigma_{xx}^2} & \frac{\partial^2 \mathcal{C}}{\partial \sigma_{xx} \partial \sigma_{yy}} & \frac{\partial^2 \mathcal{C}}{\partial \sigma_{xx} \partial \sigma_{zz}} & \frac{\partial^2 \mathcal{C}}{\partial \sigma_{xx} \partial \tau_{xy}} & \frac{\partial^2 \mathcal{C}}{\partial \sigma_{xx} \partial \tau_{yz}} & \frac{\partial^2 \mathcal{C}}{\partial \sigma_{xx} \partial \tau_{zx}} \\ & \frac{\partial^2 \mathcal{C}}{\partial \sigma_{yy}^2} & \frac{\partial^2 \mathcal{C}}{\partial \sigma_{yy} \partial \sigma_{zz}} & \frac{\partial^2 \mathcal{C}}{\partial \sigma_{yy} \partial \tau_{xy}} & \frac{\partial^2 \mathcal{C}}{\partial \sigma_{yy} \partial \tau_{yz}} & \frac{\partial^2 \mathcal{C}}{\partial \sigma_{yy} \partial \tau_{zx}} \\ & & \frac{\partial^2 \mathcal{C}}{\partial \sigma_{zz}^2} & \frac{\partial^2 \mathcal{C}}{\partial \sigma_{zz} \partial \tau_{xy}} & \frac{\partial^2 \mathcal{C}}{\partial \sigma_{zz} \partial \tau_{yz}} & \frac{\partial^2 \mathcal{C}}{\partial \sigma_{zz} \partial \tau_{zx}} \\ & & & \frac{\partial^2 \mathcal{C}}{\partial \tau_{xy}^2} & \frac{\partial^2 \mathcal{C}}{\partial \tau_{xy} \partial \tau_{yz}} & \frac{\partial^2 \mathcal{C}}{\partial \tau_{xy} \partial \tau_{zx}} \\ & & & & \frac{\partial^2 \mathcal{C}}{\partial \tau_{yz}^2} & \frac{\partial^2 \mathcal{C}}{\partial \tau_{yz} \partial \tau_{zx}} \\ & & & & & \frac{\partial^2 \mathcal{C}}{\partial \tau_{zx}^2} \end{bmatrix} \begin{Bmatrix} \dot{\sigma}_{xx} \\ \dot{\sigma}_{yy} \\ \dot{\sigma}_{zz} \\ \dot{\tau}_{xy} \\ \dot{\tau}_{yz} \\ \dot{\tau}_{zx} \end{Bmatrix} > 0 \tag{4.29}$$

The above representation matrix can be identified as the tangent compliance matrix

$$[{}_{\tan}E^{-1}]_{mn} = \left[\frac{\partial^2 \mathcal{C}}{\partial \sigma_m \partial \sigma_n} \right] \tag{4.30}$$

Summing up conditions (4.24–4.25) and (4.28–4.29) state that both surfaces of constant strain energy $\mathcal{W} = \text{const}$ and complementary energy $\mathcal{C} = \text{const}$ determined in the strain space $\mathcal{W}(\varepsilon_{ij})$ or the stress space $\mathcal{C}(\sigma_{ij})$, respectively, are convex. These are so-called *convexity postulates* for *surfaces of constant strain energy* $\mathcal{W}(\varepsilon_{ij})$ or *complementary energy* $\mathcal{C}(\sigma_{ij})$ which can be proved in a following way.

4.5 Convexity of Surfaces of Constant Strain Energy or Complementary Energy

In what follows the convexity of surfaces of constant strain energy $\mathcal{W} = \text{const}$ and constant complementary energy $\mathcal{C} = \text{const}$ will be proved (convexity requirement).

Consider two strain states of *linear* or *nonlinear elasticity* $\varepsilon_{ij}^{(1)}$ and $\varepsilon_{ij}^{(2)}$. The strain energy values corresponding to these strains can be denoted as $\mathcal{W}(\varepsilon_{ij}^{(1)})$ and $\mathcal{W}(\varepsilon_{ij}^{(2)})$. The corresponding increment of elastic strain energy $\mathcal{W}(\varepsilon_{ij}^{(1)}) - \mathcal{W}(\varepsilon_{ij}^{(2)})$ can be expanded in Taylor's series around the state $\varepsilon_{ij}^{(1)}$

$$\begin{aligned}\mathcal{W}(\varepsilon_{ij}^{(2)}) - \mathcal{W}(\varepsilon_{ij}^{(1)}) &= \left.\frac{\partial \mathcal{W}}{\partial \varepsilon_{mn}}\right|_{\varepsilon_{ij}^{(1)}} \left(\varepsilon_{ij}^{(2)} - \varepsilon_{ij}^{(1)}\right) \\ &+ \frac{1}{2}\left.\frac{\partial^2 \mathcal{W}}{\partial \varepsilon_{mn}\partial \varepsilon_{kl}}\right|_{\varepsilon_{ij}^{(1)}} \left(\varepsilon_{ij}^{(2)} - \varepsilon_{ij}^{(1)}\right)\left(\varepsilon_{kl}^{(2)} - \varepsilon_{kl}^{(1)}\right) + \cdots\end{aligned} \tag{4.31}$$

Limiting ourselves to the first two terms only we arrive at

$$\begin{aligned}\mathcal{W}(\varepsilon_{ij}^{(2)}) - \mathcal{W}(\varepsilon_{ij}^{(1)}) &= \left.\frac{\partial \mathcal{W}}{\partial \varepsilon_{mn}}\right|_{\varepsilon_{ij}^{(1)}} \left(\varepsilon_{ij}^{(2)} - \varepsilon_{ij}^{(1)}\right) \\ &+ \frac{1}{2}\left.\frac{\partial^2 \mathcal{W}}{\partial \varepsilon_{mn}\partial \varepsilon_{kl}}\right|_{\varepsilon_{ij}^{(1)}} \left(\varepsilon_{ij}^{(2)} - \varepsilon_{ij}^{(1)}\right)\left(\varepsilon_{kl}^{(2)} - \varepsilon_{kl}^{(1)}\right)\end{aligned} \tag{4.32}$$

The second term on the right hand side is positive taking into account (4.24) hence neglecting it we arrive at the following inequality

$$\mathcal{W}(\varepsilon_{ij}^{(2)}) - \mathcal{W}(\varepsilon_{ij}^{(1)}) > \left.\frac{\partial \mathcal{W}}{\partial \varepsilon_{ij}}\right|_{\varepsilon_{ij}^{(1)}} \left(\varepsilon_{ij}^{(2)} - \varepsilon_{ij}^{(1)}\right) \tag{4.33}$$

or in an equivalent format

$$\frac{\mathcal{W}(\varepsilon_{ij}^{(2)}) - \mathcal{W}(\varepsilon_{ij}^{(1)})}{\left(\varepsilon_{ij}^{(2)} - \varepsilon_{ij}^{(1)}\right)} > \left.\frac{\partial \mathcal{W}}{\partial \varepsilon_{mn}}\right|_{\varepsilon_{ij}^{(1)}} \tag{4.34}$$

Geometric interpretation of the inequality (4.34) is the following: the left hand side represents the *hyperplane secant* passing through points $\varepsilon_{ij}^{(1)}$ and $\varepsilon_{ij}^{(2)}$, whereas the right hand side presents the *hyperplane tangent* at point $\varepsilon_{ij}^{(1)}$ to the surface $\mathcal{W}(\varepsilon_{ij}) =$ const. For the sake of simplicity, consider first the unidimensional case when the strain energy is a function of one independent variable $\mathcal{W}(\varepsilon)$ inequality (4.34) reads as

$$\frac{\mathcal{W}(\varepsilon^{(2)}) - \mathcal{W}(\varepsilon^{(1)})}{\left(\varepsilon^{(2)} - \varepsilon^{(1)}\right)} > \left.\frac{\partial \mathcal{W}}{\partial \varepsilon}\right|_{\varepsilon^{(1)}} \tag{4.35}$$

It is shown in Fig. 4.10 that the strain energy $\mathcal{W}(\varepsilon)$ being a quadratic function of strain ε exhibits property that the slope of the secant is always greater than slope of the tangent both passing through the initial point. In other words, the tangent always

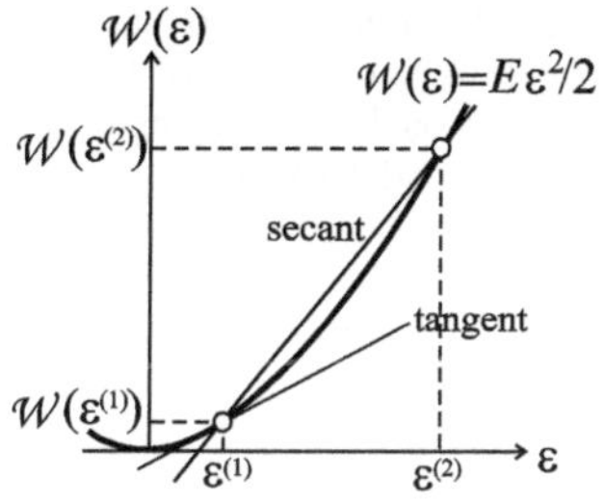

Fig. 4.10 Illustration of convexity of strain energy function $\mathcal{W}(\varepsilon)$ when it depends on only one independent variable ε

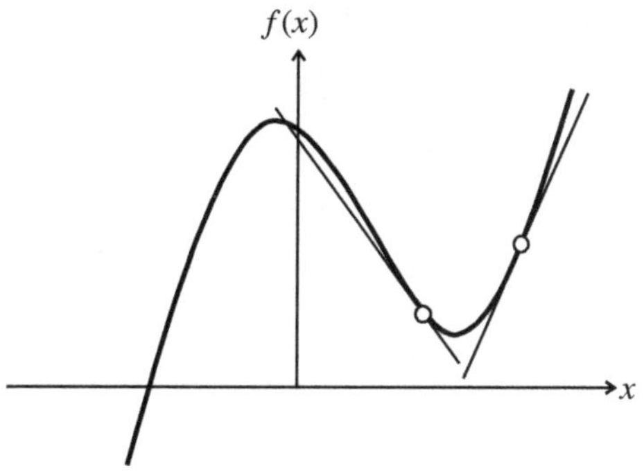

Fig. 4.11 Illustration of lack of convexity for cubic function

"slides" at outside of the energy function curve $\mathcal{W}(\varepsilon)$ never crossing it by contrast to the other case of a *cubic function* for which above condition is not satisfied Fig. 4.11.

A case of energy function of two strain arguments is illustrated in Fig. 4.12. Note that it is possible to define convexity of a function $f(x)$ on the basis of positive definiteness of its second derivative, see Fig. 4.13.

Generalization of above conclusion for function of many arguments, for instance the strain energy $\mathcal{W}(\varepsilon_{ij})$, *condition of* convexity is equivalent to *positive definiteness*

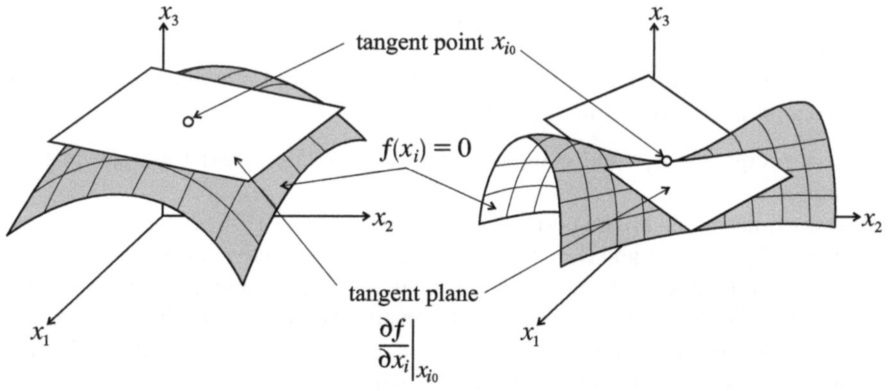

Fig. 4.12 Illustration of convexity or concavity of surface

Fig. 4.13 Illustration of convexity or concavity of a function on the basis of sign of its second derivative

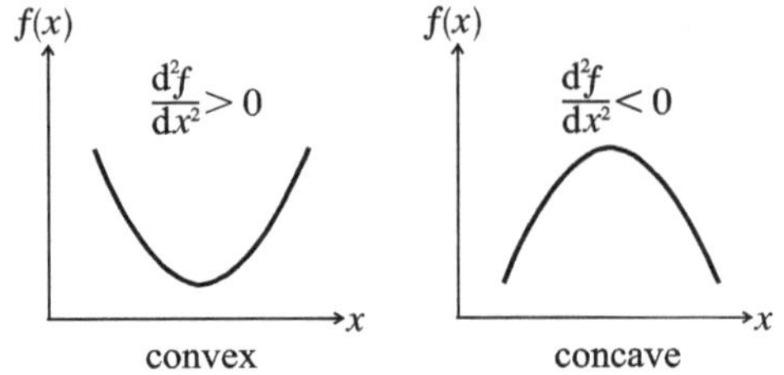

of the Hessian $\mathcal{H}_{ij}$ being (6 × 6) matrix of fourth-rank tensor components $\mathcal{H}_{ijkl} = \partial^2\mathcal{W}/\partial\varepsilon_{ij}\partial\varepsilon_{kl}$ (4.24)

$$\mathcal{H}_{ij} = \left[\frac{\partial^2\mathcal{W}}{\partial\varepsilon_i\partial\varepsilon_j}\right] = \begin{bmatrix} \frac{\partial^2\mathcal{W}}{\partial\varepsilon_1^2} & \frac{\partial^2\mathcal{W}}{\partial\varepsilon_1\partial\varepsilon_2} & \frac{\partial^2\mathcal{W}}{\partial\varepsilon_1\partial\varepsilon_3} & \frac{\partial^2\mathcal{W}}{\partial\varepsilon_1\partial\gamma_4} & \frac{\partial^2\mathcal{W}}{\partial\varepsilon_1\partial\gamma_5} & \frac{\partial^2\mathcal{W}}{\partial\varepsilon_1\partial\gamma_6} \\ & \frac{\partial^2\mathcal{W}}{\partial\varepsilon_2^2} & \frac{\partial^2\mathcal{W}}{\partial\varepsilon_2\partial\varepsilon_3} & \frac{\partial^2\mathcal{W}}{\partial\varepsilon_2\partial\gamma_4} & \frac{\partial^2\mathcal{W}}{\partial\varepsilon_2\partial\gamma_5} & \frac{\partial^2\mathcal{W}}{\partial\varepsilon_2\partial\gamma_6} \\ & & \frac{\partial^2\mathcal{W}}{\partial\varepsilon_3^2} & \frac{\partial^2\mathcal{W}}{\partial\varepsilon_3\partial\gamma_4} & \frac{\partial^2\mathcal{W}}{\partial\varepsilon_3\partial\gamma_5} & \frac{\partial^2\mathcal{W}}{\partial\varepsilon_3\partial\gamma_6} \\ & & & \frac{\partial^2\mathcal{W}}{\partial\gamma_4^2} & \frac{\partial^2\mathcal{W}}{\partial\gamma_4\partial\gamma_5} & \frac{\partial^2\mathcal{W}}{\partial\gamma_4\partial\gamma_6} \\ & & & & \frac{\partial^2\mathcal{W}}{\partial\gamma_5^2} & \frac{\partial^2\mathcal{W}}{\partial\gamma_5\partial\gamma_6} \\ & & & & & \frac{\partial^2\mathcal{W}}{\partial\gamma_6^2} \end{bmatrix} \tag{4.36}$$

what is a proof for convexity of the surface $\mathcal{W}(\varepsilon_{ij})$. In analogous way, it is possible to prove convexity of surface $\mathcal{C}(\sigma_{ij})$.

Summarizing fulfillment of the stability postulate for elastic material guarantees following *properties of elastic deformation*:

- Strain energy $\mathcal{W}(\varepsilon_{ij})$ and complementary energy $\mathcal{C}(\sigma_{ij})$ exist and are positive definite.
- Stress σ_{ij} and strain ε_{ij} are normal to respective surfaces $\mathcal{W} = \text{const}$ or $\mathcal{C} = \text{const}$.
- Surfaces $\mathcal{W} = \text{const}$ and $\mathcal{C} = \text{const}$ are convex in respective spaces of ε_{ij} or σ_{ij}.
- Positive definiteness of tangent stiffness matrix $[{}_{\tan}\mathbb{E}]$ and compliance matrix $[{}_{\tan}\mathbb{E}^{-1}]$ guarantees unique inverse of constitutive equations which means that for any constitutive relation of type $\sigma_{ij} = \mathcal{F}(\varepsilon_{ij})$ based on function $\mathcal{W}$ there always exists its unique inverse $\varepsilon_{ij} = \mathcal{F}^{-1}(\sigma_{ij})$.

4.6 Discussion: Criterion of Positive Definiteness of the Tangent Stiffness Matrix in Sylvester's Sense Versus Stability in Drucker's Sense

In Sects. 4.3 and 4.4 different versions of *stability postulate* for *elastic–plastic* or *hyperelastic materials* are presented which are similar to each other by common analysis of elementary work done on closed stress cycle. In both cases, detailed analysis of the work term leads to formulation of normality rule and convexity of respective surface (see Table 4.4). This is however the end of similarities between both stability postulates. Namely, one of the essential differences between both stability postulates is that *Drucker's postulate* saves sense in case of *elastic–plastic deformation* only. From the mathematical point of view, Drucker's postulate is weaker constraint than the *Sylvester stability postulate of hyperelastic material.* As a matter of fact, requirement of nonnegative elementary work of plastic strain according to Drucker's postulate for elastic–plastic material

$$\mathrm{d}\sigma_{ij}\mathrm{d}\varepsilon^{\mathrm{p}}_{ij} \geq 0 \tag{4.37}$$

can be satisfied even for a case when only one scalar product, for instance $\mathrm{d}\sigma_x \mathrm{d}\varepsilon^{\mathrm{p}}_x$ is positive and dominant over the others

Table 4.4 Schematic formulation of the Drucker stability criterion for hyperelastic material versus the Sylvester conditions for minors of the Hessian matrix: total versus incremental formulations of constitutive equations

Definitions	Constitutive relations
Stress potential definition	$\sigma_{ij} = \dfrac{\partial \mathcal{W}}{\partial \varepsilon_{ij}}$
Total form of constitutive equation	$\sigma_{ij} = _{\sec} E_{ijkl}\varepsilon_{kl}$
Incremental form of constitutive equation	$\dot{\sigma}_{ij} = \dfrac{\partial^2 \mathcal{W}}{\partial \varepsilon_{ij}\partial \varepsilon_{kl}}\dot{\varepsilon}_{kl}$
Tangent stiffness tensor	$_{\tan}E_{ijkl} \overset{\text{def}}{=} \dfrac{\partial^2 \mathcal{W}}{\partial \varepsilon_{ij}\partial \varepsilon_{kl}}$
Drucker's stability postulate in small	$\dot{\sigma}_{ij}\dot{\varepsilon}_{ij} > 0$
positive definiteness of quadratic form (stability criterion)	$\dfrac{\partial^2 \mathcal{W}}{\partial \varepsilon_{ij}\partial \varepsilon_{kl}}\dot{\varepsilon}_{kl} > 0$
Hessian matrix	$[\mathcal{H}_{ij}] \overset{\text{def}}{=} \left[\dfrac{\partial^2 \mathcal{W}}{\partial \varepsilon_i \partial \varepsilon_j}\right]$
Sylvester's stability condition (all minors $k \times k$ of matrix $[\mathcal{H}]$)	$\det[\mathcal{H}_{ij}]_k > 0$ $(k = 1, 2, \ldots, n)$

$$\begin{aligned}&\mathrm{d}\sigma_x\mathrm{d}\varepsilon_x^{\mathrm{p}} - |\mathrm{d}\sigma_y\mathrm{d}\varepsilon_y^{\mathrm{p}}| - |\mathrm{d}\sigma_z\mathrm{d}\varepsilon_z^{\mathrm{p}}| - |\mathrm{d}\tau_{zx}\mathrm{d}\gamma_{zx}^{\mathrm{p}}| - |\mathrm{d}\tau_{xy}\mathrm{d}\gamma_{xy}^{\mathrm{p}}|\\ &\quad -|\mathrm{d}\tau_{yz}\mathrm{d}\gamma_{yz}^{\mathrm{p}}| \geq 0\end{aligned} \tag{4.38}$$

By contrast stability postulate for hyperelastic material is stronger constraint because it requires *positive definiteness of the quadratic form* (4.24) which in Voigt's notation is as follows

$$[_{\tan}\mathbb{E}]_{mn}\dot{\varepsilon}_m\dot{\varepsilon}_n > 0 \tag{4.39}$$

According to the *Sylvester criterion*, necessary and sufficient condition for positive definiteness of the quadratic form (4.39) is as follows:

$$\det[_{\tan}\mathbb{E}]_k > 0 \qquad k = 1, 2, \ldots, 6 \tag{4.40}$$

for arbitrary arguments $\dot{\varepsilon}_n$ and $\dot{\varepsilon}_m$ where $[_{\tan}\mathbb{E}]_k$ denote minors (sub-matrices of dimensions $k \times k$) of the tangent stiffness matrix $[^{\tan}\mathbb{E}]$ and symbol det[ ] stands for determinant. Hence, the Sylvester criterion (4.40) performs in nonabbreviated form system of six inequalities (for sub-determinants) which together guarantee stability of *hyperelastic material*

$$\begin{aligned}
&{}_{\tan}E_{11} > 0\\
&\det\begin{bmatrix} {}_{\tan}E_{11} & {}_{\tan}E_{12}\\ & {}_{\tan}E_{22}\end{bmatrix} > 0\\
&\det\begin{bmatrix} {}_{\tan}E_{11} & {}_{\tan}E_{12} & {}_{\tan}E_{13}\\ & {}_{\tan}E_{22} & {}_{\tan}E_{23}\\ & & {}_{\tan}E_{33}\end{bmatrix} > 0\\
&\vdots\\
&\det\begin{bmatrix} {}_{\tan}E_{11} & {}_{\tan}E_{12} & {}_{\tan}E_{13} & {}_{\tan}E_{14} & {}_{\tan}E_{15} & {}_{\tan}E_{16}\\ & {}_{\tan}E_{22} & {}_{\tan}E_{23} & {}_{\tan}E_{24} & {}_{\tan}E_{25} & {}_{\tan}E_{26}\\ & & {}_{\tan}E_{33} & {}_{\tan}E_{34} & {}_{\tan}E_{35} & {}_{\tan}E_{36}\\ & & & {}_{\tan}E_{44} & {}_{\tan}E_{45} & {}_{\tan}E_{46}\\ & & & & {}_{\tan}E_{55} & {}_{\tan}E_{56}\\ & & & & & {}_{\tan}E_{66}\end{bmatrix} > 0
\end{aligned} \tag{4.41}$$

Finally it is worth to notice that the essential difference between *Drucker's stability postulate* (4.37) formulated *for elastic–plastic material* and the *Sylvester's stability postulate for hyperelastic material* based on positive definiteness of tangent stiffness matrix (4.40) vanishes when the constitutive law in an incremental form based on tangent stiffness matrix (4.15) is used as follows

$$\mathrm{d}\varepsilon_{ij}^{\mathrm{p}} = \lambda\frac{\partial f}{\partial \sigma_{ij}} = {}^{\mathrm{ep}}_{\tan}E_{ijkl}^{-1}\mathrm{d}\sigma_{kl} \tag{4.42}$$

After introducing (4.42) into criterion of nonnegative plastic work (4.37), we arrive at *generalized stability Drucker's postulate* based on criterion of positive semi-definiteness[1] of *elastic–plastic stiffness matrix*

$$ {}^{\mathrm{ep}}_{\tan}E^{-1}_{ijkl}\mathrm{d}\sigma_{ij}\mathrm{d}\sigma_{kl} \geq 0 \tag{4.43}$$

Generalized Drucker's postulate formulated in such a way (4.42) is however stronger condition than conventional Drucker's postulate (4.37), and it takes form analogous to Sylvester's criterion (4.41) in which sub-determinants of *tangent elastic stiffness matrix* $[{}_{\tan}\mathbb{E}]$ have been formally substituted by *sub-determinants of tangent elastic–plastic stiffness matrix* $[{}^{\mathrm{ep}}_{\tan}\mathbb{E}]$ (see Kuna-Ciskał and Skrzypek [21]).

$$\begin{aligned}
& {}^{\mathrm{ep}}_{\tan}E_{11} > 0 \\
& \det\begin{bmatrix} {}^{\mathrm{ep}}_{\tan}E_{11} & {}^{\mathrm{ep}}_{\tan}E_{12} \\ & {}^{\mathrm{ep}}_{\tan}E_{22} \end{bmatrix} > 0 \\
& \det\begin{bmatrix} {}^{\mathrm{ep}}_{\tan}E_{11} & {}^{\mathrm{ep}}_{\tan}E_{12} & {}^{\mathrm{ep}}_{\tan}E_{13} \\ & {}^{\mathrm{ep}}_{\tan}E_{22} & {}^{\mathrm{ep}}_{\tan}E_{23} \\ & & {}^{\mathrm{ep}}_{\tan}E_{33} \end{bmatrix} > 0 \\
& \vdots \\
& \det\begin{bmatrix} {}^{\mathrm{ep}}_{\tan}E_{11} & {}^{\mathrm{ep}}_{\tan}E_{12} & {}^{\mathrm{ep}}_{\tan}E_{13} & {}^{\mathrm{ep}}_{\tan}E_{14} & {}^{\mathrm{ep}}_{\tan}E_{15} & {}^{\mathrm{ep}}_{\tan}E_{16} \\ & {}^{\mathrm{ep}}_{\tan}E_{22} & {}^{\mathrm{ep}}_{\tan}E_{23} & {}^{\mathrm{ep}}_{\tan}E_{24} & {}^{\mathrm{ep}}_{\tan}E_{25} & {}^{\mathrm{ep}}_{\tan}E_{26} \\ & & {}^{\mathrm{ep}}_{\tan}E_{33} & {}^{\mathrm{ep}}_{\tan}E_{34} & {}^{\mathrm{ep}}_{\tan}E_{35} & {}^{\mathrm{ep}}_{\tan}E_{36} \\ & & & {}^{\mathrm{ep}}_{\tan}E_{44} & {}^{\mathrm{ep}}_{\tan}E_{45} & {}^{\mathrm{ep}}_{\tan}E_{46} \\ & & & & {}^{\mathrm{ep}}_{\tan}E_{55} & {}^{\mathrm{ep}}_{\tan}E_{56} \\ & & & & & {}^{\mathrm{ep}}_{\tan}E_{66} \end{bmatrix} > 0
\end{aligned} \tag{4.44}$$

Sylvester's stability criterion in the format analogous to (4.41) or (4.44) was implemented as failure criterion in other elastic-damage material to predict secondary link-type or wing-type secondary crack initiation and subsequent growth stages in the plane-stress concrete specimen with a pre-load crack, subject to tension or compression (cf. Kuna-Ciskał and Skrzypek [21]). Note however that in numerical simulation of crack-growth response in elastic-damage material, the localized strain-damage field is met at the crack tip. Consequently, local formulation of the constitutive equations is no longer sufficient to assure convergence and avoid mesh-dependence, such that more advanced nonlocal material model has to be used (cf. Skrzypek et al. [30]) in such case.

---

[1] Inequality in a weak form with respect to accounting for possible perfectly plastic deformation.

References

1. Abu Al-Rub: R.K., Voyiadjis, G.Z.: On the coupling of anisotropic damage and plasticity models for ductile materials. Int. J. Solids Struct. **40**, 2611–2643 (2003)
2. Barlat, F., Lege, D.J., Brem, J.C.: A six-component yield function for anisotropic materials. Int. J. Plast. **7**, 693–712 (1991)
3. Bielski, J., Skrzypek, J., Kuna-Ciskał, H.: Implementation of a model of coupled elastic-plastic unilateral damage material to finite element code. Int. J. Damage Mech. **15**, 5–39 (2006)
4. Cazacu, O., Barlat, F.: A criterion for description of anisotropy and yield differential effects in pressure-insensitive materials. Int. J. Plast. **20**, 2027–2045 (2004)
5. Cazacu, O., Planckett, B., Barlat, F.: Orthotropic yield criterion for hexagonal close packed metals. Int. J. Plast. **22**, 1171–1194 (2006)
6. Chaboche, J.-L.: Thermodynamic formulation of constitutive equations and application to the viscoplasticity and viscoelasticity of metals and polymers. Int. J. Solids Struct. **34**(18), 2239–2254 (1997)
7. Chen, W.F., Han, D.J.: Plasticity for Structural Engineers. Springer, Berlin (1995)
8. Drucker, D.C.: A more fundamental approach to plastic stress-strain relations, Proceedings of the 1st US National Congress of Applied Mechanics, Chicago, 487–491 (1951)
9. Drucker, D.C.: On the postulate of stability of material in the mechanics of continua. J Méchanique **3**, 235–249 (1964)
10. Egner, H.: On the full coupling between thermo-plasticity and thermo-damage in thermodynamic modeling of dissipative materials. Int. J. Solids Struct. **49**, 279–288 (2012)
11. Ganczarski, A., Skrzypek, J.: Mechanics of Novel Materials (in Polish). Wydawnictwo, Politechniki Krakowskiej, Poland (2013)
12. Herakovich, C.T., Aboudi, J.: Thermal effects in composites. In: Hetnarski, R.B. (ed.) Thermal Stresses V, pp. 1–142. Publications Division, Lastran Corp (1999)
13. Hill, R.: A theory of the yielding and plastic flow of anisotropic metals. Proc. R. Soc. Lond. **A193**, 281–297 (1948)
14. Hu, Z.W., Marin, J.: Anisotropic loading functions for combined stresses in the plastic range. J. Appl. Mech. **22**, 1 (1956)
15. Iyer, S.K.: Viscoplastic model development to account for strength differential: application to aged Inconel 718 at elevated temperature. Ph.D thesis, The Pennsylvania State University (2000)
16. Jackson, L.R., Smith, K.F., Lankford, W.T.: Plastic flow in anisotropic steel sheet. Am. Inst. Min. Metall. Eng. **2440**, 1–15 (1948)
17. Khan, A.S., Liu, H.: Strain rate and temperature dependent fracture criteria for isotropic and anisotropic metals. Int. J. Plast. **37**, 1–15 (2012)
18. Khan, A.S., Yu, S., Liu, H.: Deformation enhanced anisotropic responses of Ti-6Al-4V alloy, Part II: A stress rate and temperature dependent anisotropic yield criterion. Int. J. Plast. **38**, 14–26 (2012)
19. Kowalewski, Z.L., Śliwowski, M.: Effect of cyclic loading on the yield surface evolution of 18G2A low-alloy steel. Int. J. Mech. Sci. **39**(1), 51–68 (1997)
20. Kowalsky, U.K., Ahrens, H., Dinkler, D.: Distorted yield surfaces—modeling by higher order anisotropic hardening tensors. Comput. Math. Sci. **16**, 81–88 (1999)
21. Kuna-Ciskał, H., Skrzypek, J.: CDM based modelling of damage and fracture mechanisms in concrete under tension and compression. Eng. Fract. Mech. **71**, 681–698 (2004)
22. Lankford, W.T., Low, J.R., Gensamer, M.: The plastic flow of aluminium alloy sheet under combined loads. Transactions on AIME 171, 574; TP 2238, Meteorological Technology, August 1947
23. Luo, X.Y., Li, M., Boger, R.K., Agnew, S.R., Wagoner, R.H.: Hardening evolution of AZ31B Mg sheet. Int. J. Plast. **23**, 44–86 (2007)
24. Murakami, S.: Continuum Damage Mechanics. Springer, Berlin (2012)
25. Ottosen, N.S., Ristinmaa, M.: The Mechanics of Constitutive Modeling. Elsevier, Amsterdam (2005)

26. Plunkett, B., Cazacu, O., Barlat, F.: Orthotropic yield criteria for description of the anisotropy in tension and compression of sheet metal. Int. J. Plast. **24**, 847–866 (2008)
27. Raniecki, B., Mróz, Z.: Yield or martensitic phase transformation conditions and dissipative functions for isotropic, pressure-insensitive alloys exhibiting SD effect. Acta Mech. **195**, 81–102 (2008)
28. Sayir, M.: Zur Fließbedingung der Plastiztätstheorie. Ingenierarchiv **39**, 414–432 (1970)
29. Simo, J.C., Yu, J.W.: Strain- and stress-based continuum damage models: I-Formulation II-Computational aspects. Int. J. Solid Struct. **23**, 821–869 (1987)
30. Skrzypek, J., Ganczarski, A.: Anisotropic initial yield and failure criteria including temperature effect. In: Hetnarski, R. 2013, Encyclopedia of Thermal Stresses. (2014). doi:10.1007/978-94-007-2739-7, © Springer Science+Business Media Dordrecht
31. Sun, C.T., Vaidya, R.S.: Prediction of composite properties from a representative volume element. Compos. Sci. Technol. **56**, 171–179 (1996)
32. Tsai, S.T., Wu, E.M.: A general theory of strength for anisotropic materials. Int. J. Numer. Methods Eng. **38**, 2083–2088 (1971)
33. Voyiadjis, G.Z., Thiagarajan, G.: An anisotropic yield surface model for directionally reinforced metal-matrix composites. Int. J. Plast. **11**, 867–894 (1995)

Chapter 5
Termination of Elastic Range of Pressure Insensitive Materials—Isotropic and Anisotropic Initial Yield Criteria

Artur W. Ganczarski and Jacek J. Skrzypek

Abstract In this chapter basic features of isotropic versus anisotropic initial yield criteria are discussed. Two ways to account for anisotropy are presented: the explicit and implicit formulations. The explicit description of anisotropy is rigorously based on well-established theory of common invariants (Sayir, Goldenblat–Kopnov, von Mises, Hill). The implicit approach involves linear transformation tensor of the Cauchy stress that accounts for anisotropy to enhance the known isotropic criteria to be able to capture anisotropy, hydrostatic pressure insensitivity, and asymmetry of the yield surface (Barlat, Plunckett, Cazacu, Khan). The advantages and differences of both formulations are critically presented. Possible convexity loss of the classical Hill'48 yield surface in the case of strong orthotropy is examined and highlighted in contrast to unconditionally stable von Mises–Hu–Marin's criterion. Various transitions from the orthotropic yield criteria to the transversely isotropic ones are carefully distinguished in the light of irreducibility or reducibility to the isotropic Huber–von Mises criterion in the transverse isotropy plane and appropriate symmetry class of tetragonal symmetry (classical Hill's formulation) or hexagonal symmetry (hexagonal Hill's or von Mises–Hu–Marin's). The new hybrid formulation applicable for some engineering materials based on additional bulge test is also proposed.

Keywords Pressure insensitive criteria · von Mises anisotropic criterion · Convexity loss for strong orthotropy · Degeneration of Hill's surface to isotropic von Mises · Unconditionally stable Hu–Marin's criterion · Implicit versus explicit formulation

---

A.W. Ganczarski (✉) · J.J. Skrzypek
Solid Mechanics Division, Institute of Applied Mechanics,
Cracow University of Technology, al. Jana Pawła II 37, 31-864 Kraków, Poland
e-mail: Artur.Ganczarski@pk.edu.pl

J.J. Skrzypek
e-mail: Jacek.Skrzypek@pk.edu.pl

J.J. Skrzypek and A.W. Ganczarski (eds.), *Mechanics of Anisotropic Materials*,
Engineering Materials, DOI 10.1007/978-3-319-17160-9_5

5.1 Isotropic Initial Yield Criteria of Pressure Insensitive Materials

In case of isotropic materials limit criteria for elastic range are independent of reference frame. For this reason the isotropic initial yield criteria can be written down in the reduced frame of simple stress tensor invariants $f(J_{i\sigma}, \Pi_i) = 0$ instead of the stress tensor components frame $f(\sigma_{ij}, \Pi_i) = 0$. By contrast, in case of anisotropic materials the stress components frame has to be applied and the common both stress and structural tensor invariants should be used (see Table 5.2). Such a simplification means reduction of the six-dimensional stress space to the three-dimensional space spanned by arbitrary set of three stress invariants ($J_{i\sigma}$; $i = 1, 2, 3$). Symbol Π_i denotes scalar material constants defining termination of the elastic behavior through the yield initiation in a form of micro-slips in ductile material $\Pi_i = k_i^{\mathrm{p}}$ (yield stresses) or through the local microcracks in brittle material $\Pi_i = k_i^{\mathrm{d}}$ (failure limits).

The number of independent material constants Π_i depends on the number of parameters in the equation of limit surface (*yield or failure initiation*) which have to be identified from independent *strength tests*: e.g., the *uniaxial tension* (k_{t}), the *uniaxial compression* (k_{c}), and the *pure shear* (k_{s}). In the simplest case, when conditions of initial yielding or failure are identical for tension and compression and simultaneously the shear is not independent constant the number of material constants reduces to one parameter $k_{\mathrm{t}} = k_{\mathrm{c}} = k$ which corresponds to yield or failure initiation, whereas $k_{\mathrm{s}} = \frac{k}{\sqrt{3}}$. Such a limitation is true for majority of *ductile materials* (metals and metallic alloys). However, in case of *brittle materials* that exhibit different limit stress points for tension and compression (both yield and failure), the limit surface is to be characterized by at least two independent constants $k_{\mathrm{t}} \neq k_{\mathrm{c}}$ and such property is called *strength differential effect.*

Assuming narrower case of the experimentally confirmed for majority of metals independence of yield initiation from hydrostatic pressure $J_{1\sigma}$, we arrive at the limit surface equation being function of the second and the third stress deviator invariants

$$f\,(J_{2s}, J_{3s}; k_i) = 0 \tag{5.1}$$

Such a narrower class of materials is called *hydrostatic pressure insensitive isotropic materials.*

The above condition depends on both the second and the third stress deviator invariants J_{2s}, J_{3s} but it is independent of the first stress invariant $J_{1\sigma}$. It simply means that the cylindrical limit surface possesses the axis equally inclined to the principal stress axes ($\sigma_1, \sigma_2, \sigma_3$) called the *hydrostatic axis* (Fig. 5.1).

For purpose of further geometric illustration of considered surfaces it is convenient to apply the *Haigh–Westergaard coordinates* [21, 63] ξ, ρ, and θ which represent, respectively: distance along the hydrostatic axis measured from the origin to the current stress point (effect of $J_{1\sigma}$), distance in the *deviatoric plane* measured from the hydrostatic axis and the stress point considered (effect of J_{2s}), and the polar coordinate of the stress point in the deviatoric plane (effect of J_{3s}) (Fig. 5.1). Hence the following definitions of the Haigh–Westergaard coordinates hold

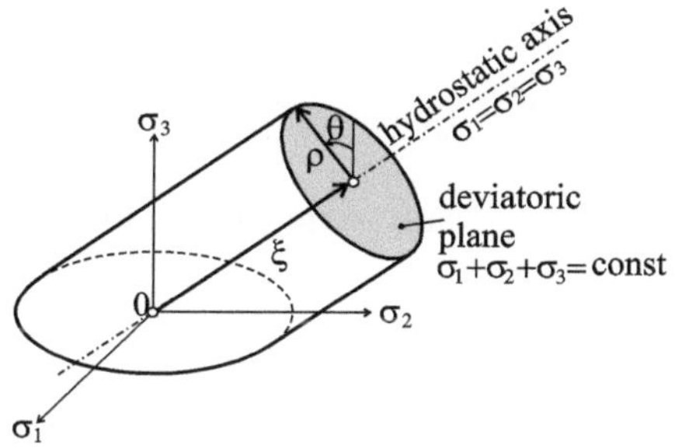

Fig. 5.1 Cylindrical yield surface in the Haigh–Westergaard coordinates

$$\xi = \frac{J_{1\sigma}}{\sqrt{3}}, \quad \rho = \sqrt{2J_{2s}}, \quad \cos(3\theta) = \frac{3\sqrt{3}}{2}\frac{J_{3s}}{(J_{2s})^{3/2}} \quad \text{for} \quad 0 \leq \theta \leq \frac{\pi}{3} \tag{5.2}$$

Roughly speaking, dependence on the first coordinate ξ stands for noncylindricity, the second one ρ comprises size and the third one θ describes asymmetry of the yield surface.

For further consideration it is also convenient to use a concept of the generating curve of limit surface conventionally called the meridian. *Meridians of* the *limit surface* either yield or failure are curves being intersections of the surface by planes of θ = const containing the hydrostatic axis. In case of *rotationally symmetric limit surfaces* all meridians are identical. In a particular case of *cylindrical surface* all cross sections by planes ξ = const (deviatoric planes) are identical and hence meridians are straight lines.

In more general case of cylindrical but *nonrotationally symmetric surface*, which depends on either the third invariant J_{3s} or alternatively the third coordinate θ, three of all meridians are of the particular importance (Fig. 5.2)

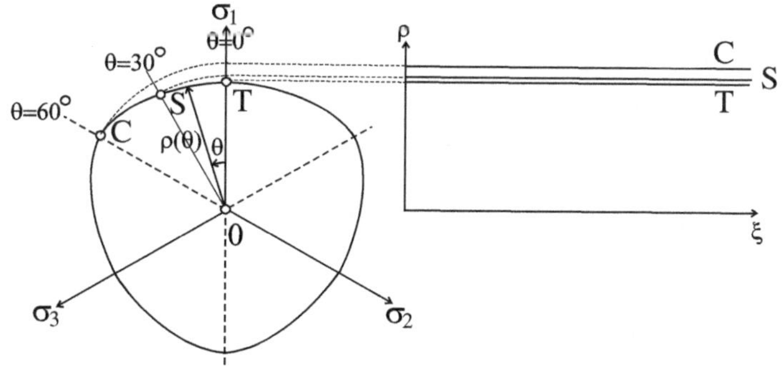

Fig. 5.2 Cross section of the cylindrical limit surface in deviatoric plane ξ = const; points T, S, and C correspond to the tensile k_t, the shear k_s, and the compressive k_c yield points, $k_t \neq k_c$

1. *Tensile meridian* T ($\theta = 0°$)
2. *Shear meridian* S ($\theta = 30°$)
3. *Compressive meridian* C ($\theta = 60°$)

Hence the equation of nonrotationally symmetric cylindrical surface (5.1) can be written as

$$f(\rho, \theta; k_i) = 0 \tag{5.3}$$

where the independence of the position at the hydrostatic axis ξ is obvious.

Note that in a general case Eq. (5.3) represents cylindrical surface, the cross section of which is not necessarily circular $\rho(\theta)$. This property is called the *strength differential effect* or the *tension and compression asymmetry* $k_t \neq k_c$. Summarizing, for isotropic materials considered the 60° *symmetry property* must be fulfilled which means that the curve in the *deviatoric plane* is completely described by the form for the sector $0 \leq \theta \leq \frac{\pi}{3}$ and this form is repeated in the remaining sectors (Fig. 5.2), for details see Chen and Han [9], Ottosen and Ristinmaa [47]. In case of majority of metals yield point stresses for compression and tension do not differ $k_t = k_c = k$ which means that no strength differential effect exists. In other words in the Haigh–Westergaard space arbitrary cross section of a cylindrical yield surface done by any deviatoric plane has to pass through six skeletal points: T_i ($\theta = 0°, 120°, 240°$) and C_i ($\theta = 60°, 180°, 300°$) at constant distance from the origin equal to $\sqrt{\frac{2}{3}}k$. Simultaneously, each of sectorial curve has to pass through three points corresponding to pure shear S_i ($\theta = 30°, 150°, 270°$), see Ottosen and Ristinmaa [47], (Fig. 5.3).

In the simple case of majority of metals and steels the additional assumption of independence of the cross section from the angle θ or alternatively from the

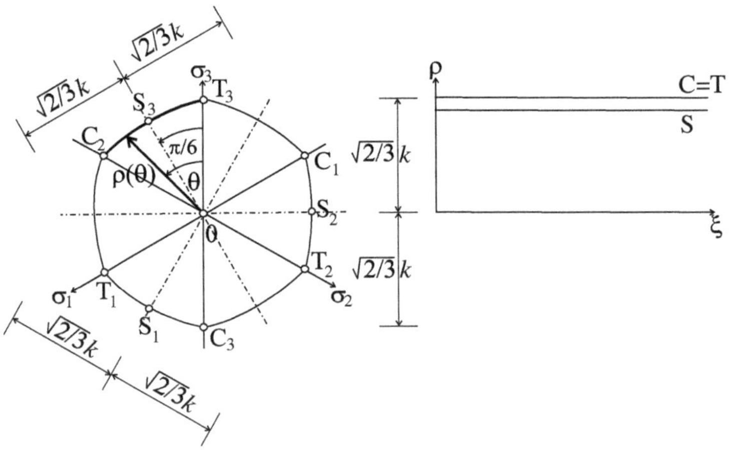

Fig. 5.3 The 60° symmetry property of the yield surface (5.3) in the deviatoric plane

third invariant J_{3s} can be done. In such a case *limit surface* is the *cylindrical* and *rotationally symmetric* simultaneously as follows:

$$f(\rho;k) = \sqrt{3J_{2s}} - k = \sqrt{\frac{3}{2}s_{ij}s_{ij}} - k = \sqrt{\frac{3}{2}}\rho - k = 0 \tag{5.4}$$

In such a way we arrive at the limit case of the unit shear strain energy-based classical *isotropic von Mises criterion* occasionally called the *Huber–von Mises criterion* anticipated by Huber [31], extended by von Mises [43] and interpreted physically by Hencky [22], cf. Ottosen and Ristinmaa [47]. When the engineering notation is used the isotropic von Mises criterion takes the explicitly deviatoric form

$$\left(\sigma_y - \sigma_z\right)^2 + \left(\sigma_z - \sigma_x\right)^2 + \left(\sigma_x - \sigma_y\right)^2 + 6\left(\tau_{yz}^2 + \tau_{zx}^2 + \tau_{xy}^2\right) = 2k^2 \tag{5.5}$$

or

$$\sigma_1^2 - \sigma_1\sigma_2 + \sigma_2^2 - \sigma_2\sigma_3 + \sigma_3^2 - \sigma_1\sigma_3 = k^2 \tag{5.6}$$

if principal stresses are used.

In a more general case when the yield criterion depends on both the second and the third Haigh–Westergaard coordinates $f(\rho, \theta)$ or alternatively on both the second and the third deviatoric stress invariants $f(J_{2s}, J_{3s})$ we met the historically earlier cylindrical criterion proposed by Tresca [60]

$$f(\rho, \theta; k) = \sqrt{2}\rho \sin\left(\theta - \frac{\pi}{3}\right) - k = 0 \qquad 0 \le \theta \le \frac{\pi}{3} \tag{5.7}$$

When the principal stresses are used the classical form of the *Tresca criterion*

$$f(\sigma_1, \sigma_2, \sigma_3; k) = \max\left(|\sigma_1 - \sigma_2|, |\sigma_2 - \sigma_3|, |\sigma_3 - \sigma_1|\right) - k = 0 \tag{5.8}$$

clearly corresponds to the hypothesis of maximum shear stress. The Tresca criterion can also by presented in terms of the second and the third stress deviator invariants (1.15), cf. Reuss [49]

$$f(J_{2s}, J_{3s}) = 4J_{2s}^3 - 27J_{3s}^2 - 9k^2J_{2s}^2 + 6k^4J_{2s} - k^6 = 0 \tag{5.9}$$

The Tresca *initial yield surface* is *cylindrical* but *not rotationally symmetric* built on the regular hexagon and the hydrostatic axis Fig. 5.4. It is clear that the Tresca yield surface represents a regular prism inscribed into the Huber–von Mises circular cylinder and possessing six joint meridians seen here as six skeletal points T_1, T_2, T_3 and C_1, C_2, C_3 (Fig. 5.4). The Tresca initial yield surface exhibits the 60° *symmetry property*.

The Tresca limit surface suffers from the existence of edges (tension T_1, T_2, T_3 and compression C_1, C_2, C_3 meridians) in which the normality rule does not hold Fig. 5.5. In order to avoid this deficiency the Hosford and Backhofen [27] and Hosford [28, 29] limit surface can be introduced

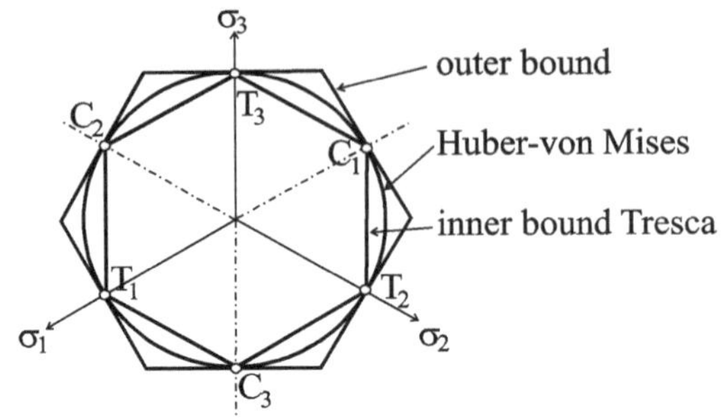

Fig. 5.4 Cylindrical initial yield criteria in the deviatoric plane

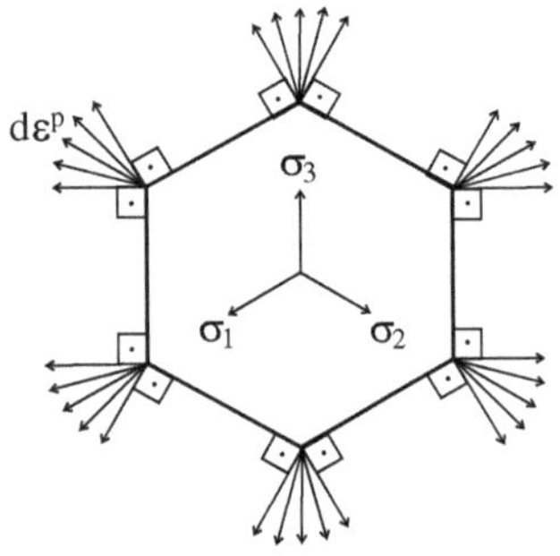

Fig. 5.5 Nonuniqueness of plastic strain increment direction in case of skeletal points for Tresca yield surface

$$|\sigma_1 - \sigma_2|^m + |\sigma_2 - \sigma_3|^m + |\sigma_3 - \sigma_1|^m = 2k^m \tag{5.10}$$

The discussed criterion is commonly called the *Hosford criterion* (1964) although it was earlier suggested by *Hershey* [24] and *Davies* [11]. The exponent m used in the Hosford criterion is an additional material constant that should be chosen according to experimental evidence. The range of this constant exhibits certain limitations and particular cases. It can theoretically change in range $1 \leq m < \infty$, cf. Cazacu and Barlat [7]. In the cases $1 < m < 2$ or $4 < m < \infty$ the initial yield curves are located between the Tresca and the Huber–von Mises loci, whereas for $m = 1$ and $m \to \infty$ or for $m = 2$ and $m = 4$ the Tresca or the Huber–von Mises yield loci are recovered, respectively. If $2 < m < 4$ the yield curve slightly exceeds the Huber–von Mises loci as shown in Fig. 5.6. If $0 < m < 1$ is chosen a *concave yield curve* is met, which is inadmissible from the *Drucker stability postulate* point of view. According to Hershey, magnitudes $m = 6$ and $m = 8$ well fit experimental findings.

Concluding, the *Tresca* initial yield criterion is the *inner bound* for all *limit curves* of the isotropic materials without the strength differential effect. Note however that there exists wide class of materials which exhibits the strength differential effect hence the Tresca does not have to be treated as the inner bound nevertheless the convexity condition resulting from the Drucker postulate is not violated, see Cazacu and Barlat [7].

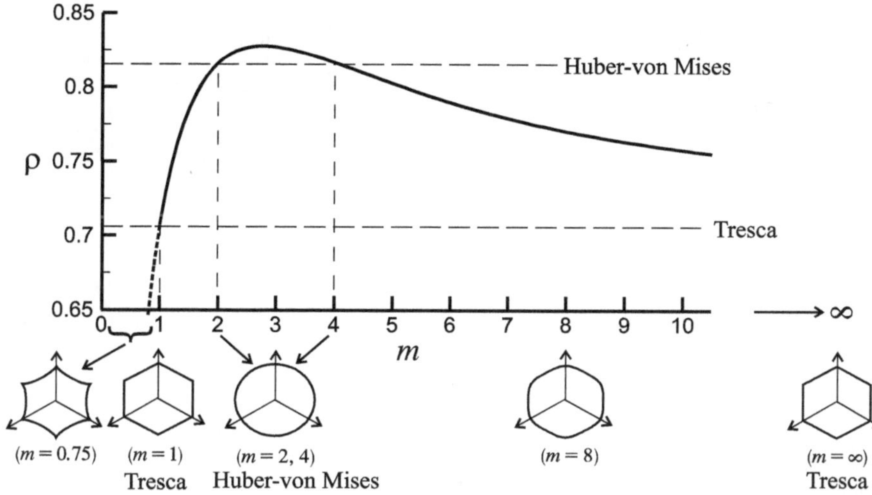

Fig. 5.6 The second Haigh–Westergaard coordinate ρ versus exponent m in Hosford's criterion (5.10)

It is also possible to derive the *outer bound* for all *limit curves* of the isotropic materials without the strength differential effect which does not violate convexity according to Drucker's postulate. To this end the *criterion of maximal deviatoric stress* proposed by Schmidt [53], Ishlinsky [32], and Hill [26] can be used

$$f(\sigma_1, \sigma_2, \sigma_3; k) = \max\left[|\sigma_1 - \sigma_{\text{h}}|, |\sigma_2 - \sigma_{\text{h}}|, |\sigma_3 - \sigma_{\text{h}}|\right] - \frac{2}{3}k = 0 \qquad (5.11)$$

The above equation when rigorously expressed in the Haigh–Westergaard space takes the alternative form

$$\begin{aligned} f(\rho, \theta; k) = \max &\left[\left|\sqrt{\tfrac{2}{3}}\rho\cos\theta\right|, \left|\sqrt{\tfrac{2}{3}}\rho\cos\left(\theta + \tfrac{2\pi}{3}\right)\right|, \right. \\ &\left.\left|\sqrt{\tfrac{2}{3}}\rho\cos\left(\theta - \tfrac{2\pi}{3}\right)\right|\right] - \tfrac{2}{3}k = 0 \end{aligned} \qquad (5.12)$$

In this space the outer bound represents a regular prism circumscribed onto the Huber–von Mises circular cylinder and possessing six joint meridians seen here as six skeletal points T_1, T_2, T_3 and C_1, C_2, C_3 (Fig. 5.4). By contrast to Tresca's inner bound now six meridians do not coincide with the outer bound prism edges but lie in the middle of walls (Fig. 5.4).

Summarizing the above considerations, the *postulate of inner and outer bounds* of limit surfaces of initial yield in isotropic and tension/compression materials (no strength differential effect included) by Tresca (inner bound) and the criterion of maximal deviatoric stress (outer bound) define the admissible range for all cylindrical

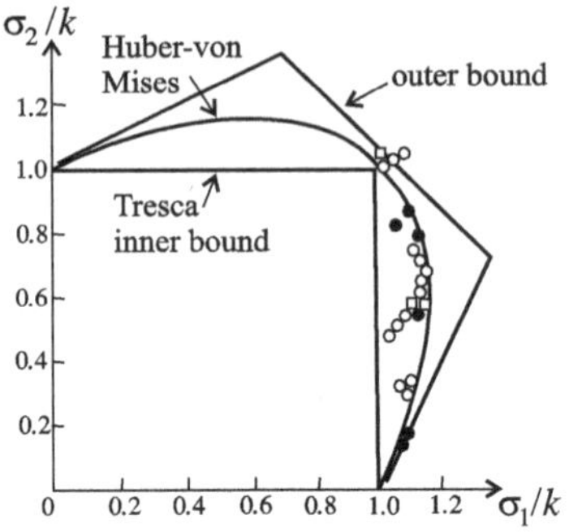

Fig. 5.7 Experimental findings for thin-walled tubes made of steel (○), copper (•), and nickel (□), after Lode [40]

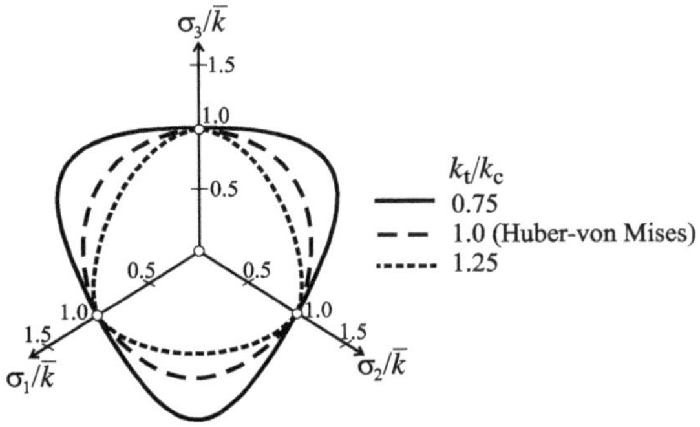

Fig. 5.8 Yield surfaces exhibiting strength differential effect $k_t/k_c = 0.75$, 1.0 (Huber–von Mises), 1.25; described by (5.13), after Cazacu and Barlat [7]

limit surfaces for the class of metals and steels. It directly results from both the 60° symmetry property in the Haigh–Westergaard space as well as the *Drucker convexity assumption*. Hence, all initial yield surfaces of real tension/compression asymmetry insensitive materials have to include tensile T_1, T_2, T_3 and compressive C_1, C_2, C_3 meridians being straight lines equidistant form the hydrostatic axis $\sqrt{\frac{2}{3}}k$. For instance, the Lode [40] experimental findings for thin-walled tubes made of steel, copper, and nickel confirm suitability of the Huber–von Mises and the Tresca criteria for prediction of yield initiation in case of ductile materials under the plane stress state ($\sigma_3 = 0$), see Fig. 5.7. Limit surface dependent on the second and the third stress invariants with the *strength differential effect* accounted for, was used by Raniecki and Mróz [48] when applied to initial yield or phase change surfaces in *NiTi shape memory alloys*

$$f(J_{2s}, J_{3s}) = (J_{2s})^{3n/2} - c\,(J_{3s})^n - k^{3n} = 0 \tag{5.13}$$

Raniecki and Mróz's criterion (5.13) includes three material constants c, k, n and it is an extension of the *Cazacu and Barlat* [7] *criterion* for $n = 1$ for describing *asymmetry* in *yielding initiation* in *pressure insensitive isotropic materials*, Fig. 5.8

$$f(J_{2s}, J_{3s}) = (J_{2s})^{3/2} - cJ_{3s} - k^3 = 0 \tag{5.14}$$

On the other hand, substituting $n = 2$ to Eq. (5.13) we arrive at *Drucker's criterion* [12]

$$f(J_{2s}, J_{3s}) = (J_{2s})^3 - c\,(J_{3s})^2 - k^6 = 0 \tag{5.15}$$

5.2 Von Mises Anisotropic Criterion

In a general case of material anisotropy, extension of the isotropic yield initiation criteria (Table 5.3) to the *anisotropic yield/failure behavior* (Table 5.4), by the use of *common invariants* of the stress tensor and of the structural tensors of plastic anisotropy (cf. Hill [25], Sayir [52], Betten [5], Życzkowski [65]), can be shown in a general fashion

$$f\left(\Pi, \Pi_{ij}\sigma_{ij}, \Pi_{ijkl}\sigma_{ij}\sigma_{kl}, \Pi_{ijklmn}\sigma_{ij}\sigma_{kl}\sigma_{mn}, \ldots\right) = 0 \tag{5.16}$$

where Einstein's summation convention holds.

In such a case, initiation of plastic flow or failure is governed by the *structural tensors of material anisotropy* of even-ranks: $\overset{<0>}{\mathbb{\Pi}} = \Pi$, $\overset{<2>}{\mathbb{\Pi}} = \Pi_{ij}$, $\overset{<4>}{\mathbb{\Pi}} = \Pi_{ijkl}$, $\overset{<6>}{\mathbb{\Pi}} = \Pi_{ijklmn}, \ldots$, etc., instead of the scalar constants k_i as it is known for isotropic materials. Equation (5.16) owns a general representation, but its practical identification is limited by a large number of required material tests and, additionally, because the components of the structural tensors are temperature dependent, which makes identification much more complicated (cf., e.g., Herakovich and Aboudi [23], Tamma and Avila [59]). Hence, a general form (5.16) is usually more specified and limited for engineering needs.

In a particular case when a general tensorially polynomial form of Eq. (5.16) is assumed (cf. Sayir [52], Kowalsky et al. [37], Życzkowski [65], Ganczarski and Skrzypek [18]) the *polynomial anisotropic yield criterion* is furnished

$$(\Pi_{ij}\sigma_{ij})^{\alpha} + (\Pi_{ijkl}\sigma_{ij}\sigma_{kl})^{\beta} + (\Pi_{ijklmn}\sigma_{ij}\sigma_{kl}\sigma_{mn})^{\gamma} + \cdots - 1 = 0 \tag{5.17}$$

where, if the Voigt notation is used the *structural anisotropy tensors* take corresponding *matrix* forms

$$[\,\overset{<2>}{\mathbb{\Pi}}\,] = \begin{bmatrix} \pi_{11} & \pi_{12} & \pi_{13} \\ & \pi_{22} & \pi_{23} \\ & & \pi_{33} \end{bmatrix} \tag{5.18}$$

and

$$\overset{<4>}{[\ \mathbb{\Pi}\ ]} = \left[\begin{array}{ccc|ccc} \Pi_{11} & \Pi_{12} & \Pi_{13} & \Pi_{14} & \Pi_{15} & \Pi_{16} \\ & \Pi_{22} & \Pi_{23} & \Pi_{24} & \Pi_{25} & \Pi_{26} \\ & & \Pi_{33} & \Pi_{34} & \Pi_{35} & \Pi_{36} \\ \hline & & & \Pi_{44} & \Pi_{45} & \Pi_{46} \\ & & & & \Pi_{55} & \Pi_{56} \\ & & & & & \Pi_{66} \end{array}\right] \tag{5.19}$$

The even-rank structural anisotropy tensors Π_{ij}, Π_{ijkl}, Π_{ijklmn}, ..., in Eq. (5.17) are normalized by the common constant Π and α, β, γ..., etc., are arbitrary exponents of a polynomial representation. In a narrower case if $\alpha = 1, \beta = 1/2, \gamma = 1/3$, and limiting an infinite form (5.17) to the equation that contains only three common invariants, we arrive at the narrower form known as the *Goldenblat and Kopnov criterion* [19]

$$\Pi_{ij}\sigma_{ij} + (\Pi_{ijkl}\sigma_{ij}\sigma_{kl})^{1/2} + (\Pi_{ijklmn}\sigma_{ij}\sigma_{kl}\sigma_{mn})^{1/3} - 1 = 0 \tag{5.20}$$

which satisfies the dimensional homogeneity of three polynomial components.

Equation (5.20), when limited only to three common invariants of the stress tensor $\boldsymbol{\sigma}$ and structural anisotropy tensors of even orders: 2nd Π_{ij}, 4th Π_{ijkl}, and 6th Π_{ijklmn} is not the most general one, in the meaning of the representation theorems, which determine the most general irreducible representation of the scalar and tensor functions that satisfy the invariance with respect to change of coordinates and material symmetry properties (cf., e.g., Spencer [56], Rymarz [51], Rogers [50]). However, 2nd, 4th, and 6th order structural anisotropy tensors, which are used in (5.20) or in case if $\alpha = 1, \beta = 1, \gamma = 1$ and the deviatoric stress representation used by Kowalsky et al. [37]

$$h^{(1)}_{ij} s_{ij} + h^{(2)}_{ijkl} s_{ij} s_{kl} + h^{(3)}_{ijklmn} s_{ij} s_{kl} s_{mn} - h^{(0)} = 0 \tag{5.21}$$

are found satisfactory for describing fundamental *transformation modes* of limit surfaces caused by plastic or failure processes, namely: isotropic *change of size*, kinematic *translation* and *rotation*, as well as surface *distortion* (cf. Betten [5], Kowalsky et al. [37]).

In what follows, we shall reduce class of the limit surface from the general tensorially polynomial representation to the forms independent of both the first $\Pi_{ij}\sigma_{ij}$ and the third $\Pi_{ijklmn}\sigma_{ij}\sigma_{kl}\sigma_{mn}$ common invariants, but preserving the most general representation for the second common invariant, according to von Mises [43, 44]. In such a case the 4th *rank tensor of material anisotropy* Π_{ijkl} is, in general, defined by 21 anisotropy modules (but 18 of them independent), since the anisotropy 6×6 matrix $[\mathbb{\Pi}]_{ij}$ (5.19) can completely be populated. Further reduction of the number of modules to 15 will be achieved, when the insensitivity of general von Mises quadratic form with respect to the change of hydrostatic stress will be assumed. In such a way the *general* tensorial *von Mises criterion* will be reduced to the deviatoric von Mises form defined by 15 anisotropy modules. A choice of 15 anisotropy modules considered as independent is, in general, not unique (cf. Szczepiński [58], Ganczarski and Skrzypek [17]). However, the 15-parameter *deviatoric von Mises criterion* is sensitive to the change of sign of shear stresses, which may be considered as questionable (cf., e.g., Malinin and Rżysko [42]). Simplest way to avoid a doubtful physical explanation for existence of terms linear for shear stresses τ_{ij}, a reduction of the 15-parameter von Mises equation to the 9-parameter orthotropic von Mises criterion can be done. This form does not satisfy the deviatoric property, but when the constraints of independence of the hydrostatic stress is consistently applied, it is easily reduced to the deviatoric form, known as orthotropic Hill's criterion, with only 6 independent moduli of orthotropy (cf. Hill [25]).

Limiting ourselves to plastic yield initiation in ductile materials, a consecutive reduction of the general tensorially polynomial anisotropic criterion (5.20) to the form dependent only on the 4th rank common invariant $\sigma_{ij}\Pi_{ijkl}\sigma_{kl}$ holds, as it was proposed in the *von Mises criterion for anisotropic yield initiation* (item D8 in Table 6.3) (cf. von Mises [43, 44]).

$$\sigma_{ij}\Pi_{ijkl}\sigma_{kl} - 1 = 0 \tag{5.22}$$

When the more convenient Voigt's vector–matrix notation is used, the form equivalent to (5.22) is obtained

$$\{\boldsymbol{\sigma}\}^{\mathrm{T}}\,[\overset{<4>}{\mathbb{\Pi}}]\,\{\boldsymbol{\sigma}\} - 1 = 0 \tag{5.23}$$

where only one fourth-rank tensor of plastic anisotropy $\mathbb{\Pi}$ is saved.

Anisotropic von Mises criterion (5.22) or (5.23), being an initial yield criterion of anisotropic material is an extension of the *isotropic Huber–von Mises criterion* (5.4). This is more clear when the Huber–von Mises condition is rewritten in a following fashion

$$\sigma_{ij}\Pi_{ijkl}^{\mathrm{HMH}}\sigma_{kl} - 1 = 0 \tag{5.24}$$

where $\Pi_{ijkl}^{\mathrm{HMH}}$ stands for the isotropic fourth-rank structural tensor whose representation matrix is

$$[\mathbb{\Pi}^{\mathrm{HMH}}] = \frac{1}{k^2}\left[\begin{array}{ccc|ccc} 1 & -\frac{1}{2} & -\frac{1}{2} & 0 & 0 & 0 \\ & 1 & -\frac{1}{2} & 0 & 0 & 0 \\ & & 1 & 0 & 0 & 0 \\ \hline & & & 3 & 0 & 0 \\ & & & & 3 & 0 \\ & & & & & 3 \end{array}\right] \tag{5.25}$$

Note however that condition (5.24) comprises stress tensor components σ_{ij} but not stress deviator components s_{ij} as commonly used. However, Eq. (5.24) takes analogous form when stress deviator components s_{ij} are used, namely

$$s_{ij}\Pi_{ijkl}^{\mathrm{HMH}}s_{kl} - 1 = 0 \tag{5.26}$$

since when decomposition of the stress tensor into the deviatoric and the hydrostatic part is done $\sigma_{ij} = s_{ij} + \frac{1}{3}\sigma_{kk}\delta_{ij}$ we arrive at

$$s_{ij}\Pi^{\mathrm{HMH}}_{ijkl}s_{kl} + \left(2s_{ij} + \frac{1}{3}\sigma_{mm}\delta_{ij}\right)\underline{\left(\Pi^{\mathrm{HMH}}_{ijkl}\delta_{kl}\right)\frac{1}{3}}\sigma_{nn} - 1 = 0 \tag{5.27}$$

However, the underlined term in (5.27) is identically equal to zero since the following holds

$$\begin{aligned}\Pi^{\mathrm{HMH}}_{11} + \Pi^{\mathrm{HMH}}_{12} + \Pi^{\mathrm{HMH}}_{13} &= 1 - \tfrac{1}{2} - \tfrac{1}{2} = 0\\ \Pi^{\mathrm{HMH}}_{21} + \Pi^{\mathrm{HMH}}_{22} + \Pi^{\mathrm{HMH}}_{23} &= -\tfrac{1}{2} + 1 - \tfrac{1}{2} = 0\\ \Pi^{\mathrm{HMH}}_{31} + \Pi^{\mathrm{HMH}}_{32} + \Pi^{\mathrm{HMH}}_{33} &= -\tfrac{1}{2} - \tfrac{1}{2} + 1 = 0\end{aligned} \tag{5.28}$$

when the Voigt notation for the Huber–von Mises matrix is used.

The structural 4th rank tensor of plastic anisotropy in Eq. (5.22) must be symmetric: $\Pi_{ijkl} = \Pi_{klij} = \Pi_{jikl} = \Pi_{ijlk}$, if stress tensor symmetry is assumed. Hence, in case if none other symmetry properties are implied, the von Mises plastic anisotropy tensor is defined by 21 modules. However, due to its invariance of the tensorial transformation rule, number of independent anisotropy modules is reduced to 18. Finally, the general *anisotropic von Mises criterion* can be furnished as

$$\begin{aligned}&\Pi_{xxxx}\sigma_x^2 + \Pi_{yyyy}\sigma_y^2 + \Pi_{zzzz}\sigma_z^2 +\\ &2\Pi_{xxyy}\sigma_x\sigma_y + 2\Pi_{yyzz}\sigma_y\sigma_z + 2\Pi_{zzxx}\sigma_z\sigma_x +\\ &4\Pi_{xxyz}\sigma_x\tau_{yz} + 4\Pi_{xxzx}\sigma_x\tau_{zx} + 4\Pi_{xxxy}\sigma_x\tau_{xy} +\\ &4\Pi_{yyyz}\sigma_y\tau_{yz} + 4\Pi_{yyzx}\sigma_y\tau_{zx} + 4\Pi_{yyxy}\sigma_y\tau_{xy} +\\ &4\Pi_{zzyz}\sigma_z\tau_{yz} + 4\Pi_{zzzx}\sigma_z\tau_{zx} + 4\Pi_{zzxy}\sigma_z\tau_{xy} +\\ &8\Pi_{xyyz}\tau_{xy}\tau_{yz} + 8\Pi_{yzzx}\tau_{yz}\tau_{zx} + 8\Pi_{zxxy}\tau_{zx}\tau_{xy} +\\ &4\Pi_{yzyz}\tau_{yz}^2 + 4\Pi_{zxzx}\tau_{zx}^2 + 4\Pi_{xyxy}\tau_{xy}^2 = 1\end{aligned} \tag{5.29}$$

where Π_{ijkl} denote 21 components of the von Mises plastic anisotropy tensor.

The *von Mises* 6×6 *matrix of plastic anisotropy*, being symmetric and fully populated matrix representation of the 4th rank anisotropy tensor Π_{ijkl} shown in (5.22), is furnished as follows:

$$[\overset{<4>}{\mathbb{III}}] = \left[\begin{array}{ccc|ccc}\Pi_{11} & \Pi_{12} & \Pi_{13} & \Pi_{14} & \Pi_{15} & \Pi_{16}\\ & \Pi_{22} & \Pi_{23} & \Pi_{24} & \Pi_{25} & \Pi_{26}\\ & & \Pi_{33} & \Pi_{34} & \Pi_{35} & \Pi_{36}\\ \hline & & & \Pi_{44} & \Pi_{45} & \Pi_{46}\\ & & & & \Pi_{55} & \Pi_{56}\\ & & & & & \Pi_{66}\end{array}\right] \quad \left[\begin{array}{ccc|ccc}\bullet & \bullet & \bullet & \bullet & \bullet & \bullet\\ & \bullet & \bullet & \bullet & \bullet & \bullet\\ & & \bullet & \bullet & \bullet & \bullet\\ \hline & & & \bullet & \bullet & \bullet\\ & & & & \bullet & \bullet\\ & & & & & \bullet\end{array}\right] \tag{5.30}$$

if engineering vectorial representation of the stress tensor $\{\boldsymbol{\sigma}\}$ is chosen as

$$\{\boldsymbol{\sigma}\} = \begin{Bmatrix} \sigma_1 \\ \sigma_2 \\ \sigma_3 \\ \sigma_4 \\ \sigma_5 \\ \sigma_6 \end{Bmatrix} = \begin{Bmatrix} \sigma_x \\ \sigma_y \\ \sigma_z \\ \tau_{yz} \\ \tau_{zx} \\ \tau_{xy} \end{Bmatrix} \tag{5.31}$$

When the matrix coordinates Π_{ij} (5.30) are consistently defined by the tensorial coordinates Π_{ijkl}

$$\begin{array}{lll} \Pi_{11} = \Pi_{xxxx} & \Pi_{22} = \Pi_{yyyy} & \Pi_{33} = \Pi_{zzzz} \\ \Pi_{12} = \Pi_{xxyy} & \Pi_{13} = \Pi_{xxzz} & \Pi_{23} = \Pi_{yyzz} \\ \Pi_{14} = 2\Pi_{xxyz} & \Pi_{15} = 2\Pi_{xxzx} & \Pi_{16} = 2\Pi_{xxxy} \ \ldots \\ \Pi_{44} = 4\Pi_{yzyz} & \Pi_{55} = 4\Pi_{zxzx} & \Pi_{66} = 4\Pi_{xyxy} \\ \Pi_{45} = 4\Pi_{yzzx} & \Pi_{46} = 4\Pi_{xyyz} & \Pi_{56} = 4\Pi_{zxxy} \end{array} \tag{5.32}$$

we arrive at the general anisotropic von Mises equation equivalent to (5.29)

$$\begin{array}{l} \Pi_{11}\sigma_x^2 + \Pi_{22}\sigma_y^2 + \Pi_{33}\sigma_z^2 + \\ 2(\Pi_{12}\sigma_x\sigma_y + \Pi_{23}\sigma_y\sigma_z + \Pi_{31}\sigma_z\sigma_x + \\ \Pi_{14}\sigma_x\tau_{yz} + \Pi_{15}\sigma_x\tau_{zx} + \Pi_{16}\sigma_x\tau_{xy} + \\ \Pi_{24}\sigma_y\tau_{yz} + \Pi_{25}\sigma_y\tau_{zx} + \Pi_{26}\sigma_y\tau_{xy} + \\ \Pi_{34}\sigma_z\tau_{yz} + \Pi_{35}\sigma_z\tau_{zx} + \Pi_{36}\sigma_z\tau_{xy} + \\ \Pi_{45}\tau_{yz}\tau_{zx} + \Pi_{46}\tau_{xy}\tau_{yz} + \Pi_{56}\tau_{zx}\tau_{xy}) + \\ \Pi_{44}\tau_{yz}^2 + \Pi_{55}\tau_{zx}^2 + \Pi_{66}\tau_{xy}^2 = 1 \end{array} \tag{5.33}$$

Representation of the anisotropic von Mises condition (5.23) in deviatoric form is not trivial. The von Mises equation in the vector–matrix notation depends on both the deviatoric $\mathbf{s}$ and the hydrostatic part $\sigma_{\mathrm{h}}\mathbf{1}$, when stress decomposition $\boldsymbol{\sigma} = \mathbf{s} + \sigma_{\mathrm{h}}\mathbf{1}$ is applied, namely

$$\{\mathbf{s}\}^{\mathrm{T}} [\overset{<4>}{\mathbb{III}}] \{\mathbf{s}\} + \underline{\left(2\{\mathbf{s}\}^{\mathrm{T}} + \sigma_{\mathrm{h}}\{\mathbf{1}\}^{\mathrm{T}}\right)\left([\overset{<4>}{\mathbb{III}}]\{\mathbf{1}\}\sigma_{\mathrm{h}}\right)} - 1 = 0 \tag{5.34}$$

The tensorial *von Mises* equation (5.34) can further be reduced to the *deviatoric form* independent of the hydrostatic pressure as follows:

$$\{\mathbf{s}\}^{\mathrm{T}} [{}_{\mathrm{dev}}\mathbb{III}] \{\mathbf{s}\} - 1 = 0 \tag{5.35}$$

only if the constraint

$$[\overset{<4>}{\mathbb{III}}]\{\mathbf{1}\} = 0 \tag{5.36}$$

is consistently applied. The constraint (5.36) guarantees the deviatoric von Mises equation (5.35) to be represented in the reduced six-dimensional stress space by a cylindrical surface defined by 15 independent anisotropy modules, when six constraints are satisfied

$$
\begin{aligned}
\Pi_{11} + \Pi_{12} + \Pi_{13} &= 0 \\
\Pi_{12} + \Pi_{22} + \Pi_{23} &= 0 \\
\Pi_{13} + \Pi_{23} + \Pi_{33} &= 0 \\
\Pi_{14} + \Pi_{24} + \Pi_{34} &= 0 \\
\Pi_{15} + \Pi_{25} + \Pi_{35} &= 0 \\
\Pi_{16} + \Pi_{26} + \Pi_{36} &= 0
\end{aligned}
\tag{5.37}
$$

However, the final matrix representation (5.30) with (5.37) employed depends on a choice of independent elements. Two of such representations are of special importance.

In the first case, the elements of matrix (5.30) considered as independent are: $\Pi_{12}, \Pi_{13}, \Pi_{23}; \Pi_{15}, \Pi_{16}, \Pi_{24}, \Pi_{26}, \Pi_{34}, \Pi_{35}$ and $\Pi_{44}, \Pi_{55}, \Pi_{66}; \Pi_{45}, \Pi_{46}, \Pi_{56}$, such that the following *first* representation for the *deviatoric von Mises matrix* is furnished

$$
[{}_{\mathrm{dev}}\mathbb{III}] = \left[\begin{array}{ccc|ccc}
-\Pi_{12} - \Pi_{13} & \Pi_{12} & \Pi_{13} & -\Pi_{24} - \Pi_{34} & \Pi_{15} & \Pi_{16} \\
 & -\Pi_{12} - \Pi_{23} & \Pi_{23} & \Pi_{24} & -\Pi_{15} - \Pi_{35} & \Pi_{26} \\
 & & -\Pi_{13} - \Pi_{23} & \Pi_{34} & \Pi_{35} & -\Pi_{16} - \Pi_{26} \\
\hline
 & & & \Pi_{44} & \Pi_{45} & \Pi_{46} \\
 & & & & \Pi_{55} & \Pi_{56} \\
 & & & & & \Pi_{66}
\end{array}\right]
\left[\begin{array}{ccc|ccc}
\circ & \bullet & \bullet & \circ & \bullet & \bullet \\
 & \circ & \bullet & \bullet & \circ & \bullet \\
 & & \circ & \bullet & \bullet & \circ \\
\hline
 & & & \bullet & \bullet & \bullet \\
 & & & & \bullet & \bullet \\
 & & & & & \bullet
\end{array}\right]
\tag{5.38}
$$

if constraints (5.37) are applied as follows

$$
\begin{aligned}
\Pi_{11} &= -\Pi_{12} - \Pi_{13}, & \Pi_{14} &= -\Pi_{24} - \Pi_{34} \\
\Pi_{22} &= -\Pi_{12} - \Pi_{23}, & \Pi_{25} &= -\Pi_{15} - \Pi_{35} \\
\Pi_{33} &= -\Pi_{13} - \Pi_{23}, & \Pi_{36} &= -\Pi_{16} - \Pi_{26}
\end{aligned}
\tag{5.39}
$$

In the second case, the elements of matrix (5.30) chosen as independent are: $\Pi_{11}, \Pi_{22}, \Pi_{33}; \Pi_{15}, \Pi_{16}, \Pi_{24}, \Pi_{26}, \Pi_{34}, \Pi_{35}$ and $\Pi_{44}, \Pi_{55}, \Pi_{66}; \Pi_{45}, \Pi_{46}, \Pi_{56}$, hence we arrive at the *second* representation of the *deviatoric von Mises matrix* as follows:

$$[{}_{\mathrm{dev}}\mathbb{III}] = \left[\begin{array}{ccc|ccc} \Pi_{11} & \frac{1}{2}(\Pi_{33}-\Pi_{11}-\Pi_{22}) & \frac{1}{2}(\Pi_{22}-\Pi_{11}-\Pi_{33}) & -\Pi_{24}-\Pi_{34} & \Pi_{15} & \Pi_{16} \\ & \Pi_{22} & \frac{1}{2}(\Pi_{11}-\Pi_{22}-\Pi_{33}) & \Pi_{24} & -\Pi_{15}-\Pi_{35} & \Pi_{26} \\ & & \Pi_{33} & \Pi_{34} & \Pi_{35} & -\Pi_{16}-\Pi_{26} \\ \hline & & & \Pi_{44} & \Pi_{45} & \Pi_{46} \\ & & & & \Pi_{55} & \Pi_{56} \\ & & & & & \Pi_{66} \end{array}\right] \left[\begin{array}{ccc|ccc} \bullet & \circ & \circ & \circ & \bullet & \bullet \\ & \bullet & \circ & \bullet & \circ & \bullet \\ & & \bullet & \bullet & \bullet & \circ \\ \hline & & & \bullet & \bullet & \bullet \\ & & & & \bullet & \bullet \\ & & & & & \bullet \end{array}\right] \tag{5.40}$$

if, instead of (5.39), other substitution is used

$$\begin{aligned} \Pi_{12} &= \tfrac{1}{2}(\Pi_{33}-\Pi_{11}-\Pi_{22}) \\ \Pi_{13} &= \tfrac{1}{2}(\Pi_{22}-\Pi_{11}-\Pi_{33}) \\ \Pi_{23} &= \tfrac{1}{2}(\Pi_{11}-\Pi_{22}-\Pi_{33}) \\ \Pi_{14} &= -\Pi_{24}-\Pi_{34} \\ \Pi_{25} &= -\Pi_{15}-\Pi_{35} \\ \Pi_{36} &= -\Pi_{16}-\Pi_{26} \end{aligned} \tag{5.41}$$

A choice of 15 elements in the von Mises matrix (5.30) considered as independent is not a unique procedure and can result in the different deviatoric von Mises equation forms. In particular, when a more convenient representation (5.38) is substituted for $[{}_{\mathrm{dev}}\mathbb{III}]$ in (5.35) we arrive at the following von Mises equation expressed in the *deviatoric stress space*

$$\begin{aligned} &\Pi_{12}\left(s_x-s_y\right)^2-\Pi_{13}\left(s_x-s_z\right)^2-\Pi_{23}\left(s_y-s_z\right)^2+ \\ &2\left\{\tau_{yz}\left[\Pi_{24}\left(s_y-s_x\right)+\Pi_{34}\left(s_z-s_x\right)\right]+\right. \\ &\tau_{zx}\left[\Pi_{15}\left(s_x-s_y\right)+\Pi_{35}\left(s_z-s_y\right)\right]+ \\ &\tau_{xy}\left[\Pi_{16}\left(s_x-s_z\right)+\Pi_{26}\left(s_y-s_z\right)\right]+ \\ &\left.\Pi_{45}\tau_{yz}\tau_{zx}+\Pi_{46}\tau_{xy}\tau_{yz}+\Pi_{56}\tau_{zx}\tau_{xy}\right\}+ \\ &\Pi_{44}\tau_{yz}^2+\Pi_{55}\tau_{zx}^2+\Pi_{66}\tau_{xy}^2=1 \end{aligned} \tag{5.42}$$

It is visible that above equation owns the clear deviatoric structure hence, when the tensorial stress space is used instead of the deviatoric one, the analogous equivalent to (5.42) representation of the *deviatoric von Mises equation* is also true in terms of stress components (cf. Szczepiński [58])

$$
\begin{aligned}
&-\Pi_{12}\left(\sigma_x-\sigma_y\right)^2-\Pi_{13}\left(\sigma_x-\sigma_z\right)^2-\Pi_{23}\left(\sigma_y-\sigma_z\right)^2+\\
&\underline{2\left\{\tau_{yz}\left[\Pi_{24}\left(\sigma_y-\sigma_x\right)+\Pi_{34}\left(\sigma_z-\sigma_x\right)\right]+\right.}\\
&\quad\underline{\tau_{zx}\left[\Pi_{15}\left(\sigma_x-\sigma_y\right)+\Pi_{35}\left(\sigma_z-\sigma_y\right)\right]+}\\
&\quad\underline{\tau_{xy}\left[\Pi_{16}\left(\sigma_x-\sigma_z\right)+\Pi_{26}\left(\sigma_y-\sigma_z\right)\right]+}\\
&\quad\underline{\left.\Pi_{45}\tau_{yz}\tau_{zx}+\Pi_{46}\tau_{xy}\tau_{yz}+\Pi_{56}\tau_{zx}\tau_{xy}\right\}+}\\
&\Pi_{44}\tau_{yz}^2+\Pi_{55}\tau_{zx}^2+\Pi_{66}\tau_{xy}^2=1
\end{aligned}
\tag{5.43}
$$

Note, that Eqs. (5.42) or (5.43) are defined by 15 elements Π_{ij}. However, the underlined terms are sensitive to change of sign of shear stresses, e.g., $\tau_{yz}(\sigma_y-\sigma_x)$ etc., which is physically questionable and, finally, such terms are consequently omitted in some cases (cf., e.g., Malinin and Rżysko [42]). Nevertheless, the full representation (5.43) might occur useful when the von Mises–Tsai–Wu extension to the brittle-like material is sought for (cf. Tsai and Wu [61]).

5.3 Orthotropic Initial Yield Criteria—The von Mises Orthotropic Criterion, the Hill Deviatoric Criterion

General form of the 21-parameter anisotropic von Mises criterion (5.33) involves none material symmetry property. In a particular case if plastic orthotropy is assumed for the initial yield criterion (5.23), when represented in principal orthotropy axes, the 9-parameter *orthotropic von Mises matrix* (5.30) takes the form

$$
[{}_{\mathrm{ort}}\mathbb{\Pi}] = \left[\begin{array}{ccc|ccc}
\Pi_{11} & \Pi_{12} & \Pi_{13} & 0 & 0 & 0\\
 & \Pi_{22} & \Pi_{23} & 0 & 0 & 0\\
 & & \Pi_{33} & 0 & 0 & 0\\
\hline
 & & & \Pi_{44} & 0 & 0\\
 & & & & \Pi_{55} & 0\\
 & & & & & \Pi_{66}
\end{array}\right]
\left[\begin{array}{ccc|ccc}
\bullet & \bullet & \bullet & & & \\
 & \bullet & \bullet & & & \\
 & & \bullet & & & \\
\hline
 & & & \bullet & & \\
 & & & & \bullet & \\
 & & & & & \bullet
\end{array}\right]
\tag{5.44}
$$

In such a case the general anisotropic von Mises equation (5.33) is reduced to the narrower 9-parameter *orthotropic von Mises criterion*

$$
\begin{aligned}
&\Pi_{11}\sigma_x^2+\Pi_{22}\sigma_y^2+\Pi_{33}\sigma_z^2+\\
&2(\Pi_{12}\sigma_x\sigma_y+\Pi_{23}\sigma_y\sigma_z+\Pi_{31}\sigma_z\sigma_x)+\\
&\Pi_{44}\tau_{yz}^2+\Pi_{55}\tau_{zx}^2+\Pi_{66}\tau_{xy}^2=1
\end{aligned}
\tag{5.45}
$$

When the *Voigt notation* is used, the 9-parameter orthotropic von Mises criterion takes the form

$$
\{\boldsymbol{\sigma}\}^{\mathrm{T}}[{}_{\mathrm{ort}}\mathbb{\Pi}]\{\boldsymbol{\sigma}\}-1=0 \tag{5.46}
$$

that involves definition (5.44). Note that equation (5.46) belongs to the class of *hydrostatic pressure sensitive criteria* (cf. item D8 in Table 6.3 Khan et al. [35, 36]).

In order to achieve pressure insensitive orthotropic criterion we apply a procedure described in Sect. 5.2. If we decompose again the stress tensor into deviatoric and volumetric parts $\boldsymbol{\sigma} = \boldsymbol{s} + \sigma_{\mathrm{h}}\mathbf{1}$ in the orthotropic von Mises equation (5.46) we arrive at the equation analogous to (5.34)

$$\{\boldsymbol{s}\}^{\mathrm{T}}\,[{}_{\mathrm{ort}}\mathbb{\Pi}]\,\{\boldsymbol{s}\} + \underline{\left(2\,\{\boldsymbol{s}\}^{\mathrm{T}} + \sigma_{\mathrm{h}}\,\{\mathbf{1}\}^{\mathrm{T}}\right)\left([{}_{\mathrm{ort}}\mathbb{\Pi}]\,\{\mathbf{1}\}\,\sigma_{\mathrm{h}}\right)} - 1 = 0 \tag{5.47}$$

Assuming further hydrostatic pressure insensitive form the following holds

$$[{}_{\mathrm{ort}}\mathbb{\Pi}]\,\{\mathbf{1}\} = 0 \tag{5.48}$$

which leads to three constraints instead of six in general case of von Mises anisotropic Eq. (5.37)

$$\begin{aligned} \Pi_{11} + \Pi_{12} + \Pi_{13} &= 0 \\ \Pi_{12} + \Pi_{22} + \Pi_{23} &= 0 \\ \Pi_{13} + \Pi_{23} + \Pi_{33} &= 0 \end{aligned} \tag{5.49}$$

In this way the *orthotropic von Mises criterion* (5.46) reduces to the *pressure insensitive criterion* called *Hill's criterion* [25, 26] that contains six independent modules

$$\{\boldsymbol{s}\}^{\mathrm{T}}\,[\mathbb{\Pi}^{\mathrm{H}}]\,\{\boldsymbol{s}\} - 1 = 0 \tag{5.50}$$

Hill's matrix $[\mathbb{\Pi}^{\mathrm{H}}]$ appearing in Eq. (5.50) contains six independent modules. A choice of the three independent modules form six involved in Eq. (5.49) is not unique. In what follows two of them are discussed (see two aforementioned forms (5.38) and (5.40)).

In this way we arrive at the following *Hill's matrices*

$$[\mathbb{\Pi}^{\mathrm{H}}] = \left[\begin{array}{ccc|ccc} -\Pi_{12}-\Pi_{13} & \Pi_{12} & \Pi_{13} & & & \\ & -\Pi_{12}-\Pi_{23} & \Pi_{23} & & & \\ & & -\Pi_{13}-\Pi_{23} & & & \\ \hline & & & \Pi_{44} & & \\ & & & & \Pi_{55} & \\ & & & & & \Pi_{66} \end{array}\right]$$

$$\left[\begin{array}{ccc|ccc} \circ & \bullet & \bullet & & & \\ & \circ & \bullet & & & \\ & & \circ & & & \\ \hline & & & \bullet & & \\ & & & & \bullet & \\ & & & & & \bullet \end{array}\right] \tag{5.51}$$

or

$$
[\mathbb{H}^{\mathrm{H}}] = \left[\begin{array}{ccc|ccc}
\Pi_{11} & \frac{\Pi_{33}-\Pi_{11}-\Pi_{22}}{2} & \frac{\Pi_{22}-\Pi_{11}-\Pi_{33}}{2} & & & \\
 & \Pi_{22} & \frac{\Pi_{11}-\Pi_{22}-\Pi_{33}}{2} & & & \\
 & & \Pi_{33} & & & \\
\hline
 & & & \Pi_{44} & & \\
 & & & & \Pi_{55} & \\
 & & & & & \Pi_{66}
\end{array}\right]
\quad
\left[\begin{array}{ccc|ccc}
\bullet & \circ & \circ & & & \\
 & \bullet & \circ & & & \\
 & & \bullet & & & \\
\hline
 & & & \bullet & & \\
 & & & & \bullet & \\
 & & & & & \bullet
\end{array}\right]
\tag{5.52}
$$

When the engineering notation is used, corresponding representations of the *Hill's criterion* are

$$
\begin{aligned}
-\left[\Pi_{23}\left(\sigma_y-\sigma_z\right)^2+\Pi_{13}\left(\sigma_z-\sigma_x\right)^2+\Pi_{12}\left(\sigma_x-\sigma_y\right)^2\right]+ \\
\Pi_{44}\tau_{yz}^2+\Pi_{55}\tau_{zx}^2+\Pi_{66}\tau_{xy}^2=1
\end{aligned}
\tag{5.53}
$$

or

$$
\begin{aligned}
\Pi_{11}\sigma_x^2+\Pi_{22}\sigma_y^2+\Pi_{33}\sigma_z^2+\left(\Pi_{33}-\Pi_{11}-\Pi_{22}\right)\sigma_x\sigma_y+ \\
\left(\Pi_{22}-\Pi_{11}-\Pi_{33}\right)\sigma_x\sigma_z+\left(\Pi_{11}-\Pi_{22}-\Pi_{33}\right)\sigma_y\sigma_z+ \\
\Pi_{44}\tau_{yz}^2+\Pi_{55}\tau_{zx}^2+\Pi_{66}\tau_{xy}^2=1
\end{aligned}
\tag{5.54}
$$

Both representations (5.53) or (5.54) describe the same Hill's limit surface, but applying two different choices of six independent elements of the Hill matrices (5.51) or (5.52). In order to calibrate Hill's criterion in the form (5.53) or (5.54) three

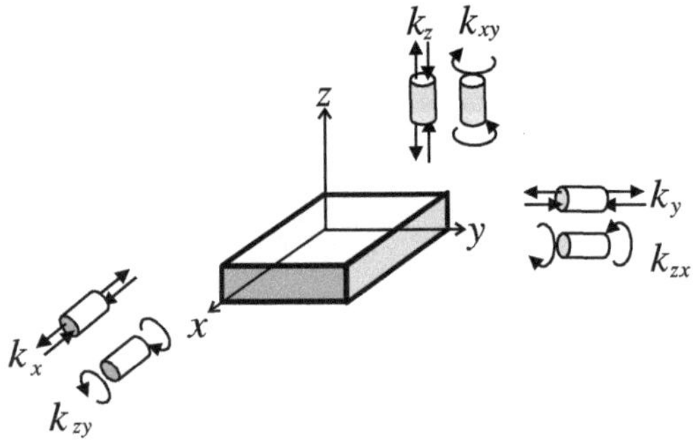

Fig. 5.9 Six tests for Hill's criterion calibration

tests of uniaxial tension $\sigma_x = k_x, \sigma_y = k_y, \sigma_z = k_z$ and three tests of pure shear $\tau_{xy} = k_{xy}, \tau_{yz} = k_{yz}, \tau_{zx} = k_{zx}$, in directions and planes of material orthotropy (Fig. 5.9), must be performed. These tests allow to express six modules of material orthotropy in Eqs. (5.53) and (5.54) in terms of 3 independent *plastic tension limits* k_x, k_y, k_z (in directions of orthotropy), and 3 independent *plastic shear limits* k_{yz}, k_{zx}, k_{xy} (in planes of material orthotropy). Hence,

$$\begin{aligned} -\Pi_{23} &= \frac{1}{2}\left(\frac{1}{k_y^2} + \frac{1}{k_z^2} - \frac{1}{k_x^2}\right), \quad \Pi_{44} = \frac{1}{k_{yz}^2} \\ -\Pi_{13} &= \frac{1}{2}\left(\frac{1}{k_z^2} + \frac{1}{k_x^2} - \frac{1}{k_y^2}\right), \quad \Pi_{55} = \frac{1}{k_{zx}^2} \\ -\Pi_{12} &= \frac{1}{2}\left(\frac{1}{k_x^2} + \frac{1}{k_y^2} - \frac{1}{k_z^2}\right), \quad \Pi_{66} = \frac{1}{k_{xy}^2} \end{aligned} \tag{5.55}$$

such that *orthotropic Hill's criteria* equivalent to (5.53) or (5.54) can be furnished in terms of plastic anisotropy limits as follows:

$$\begin{aligned} &\frac{1}{2}\left(\frac{1}{k_y^2} + \frac{1}{k_z^2} - \frac{1}{k_x^2}\right)(\sigma_y - \sigma_z)^2 + \frac{1}{2}\left(\frac{1}{k_z^2} + \frac{1}{k_x^2} - \frac{1}{k_y^2}\right)(\sigma_z - \sigma_x)^2 + \\ &\frac{1}{2}\left(\frac{1}{k_x^2} + \frac{1}{k_y^2} - \frac{1}{k_z^2}\right)(\sigma_x - \sigma_y)^2 + \left(\frac{\tau_{yz}}{k_{yz}}\right)^2 + \left(\frac{\tau_{zx}}{k_{zx}}\right)^2 + \left(\frac{\tau_{xy}}{k_{xy}}\right)^2 = 1 \end{aligned} \tag{5.56}$$

or

$$\begin{aligned} &\left(\frac{\sigma_x}{k_x}\right)^2 + \left(\frac{\sigma_y}{k_y}\right)^2 + \left(\frac{\sigma_z}{k_z}\right)^2 - \left(\frac{1}{k_x^2} + \frac{1}{k_y^2} - \frac{1}{k_z^2}\right)\sigma_x\sigma_y - \\ &\left(\frac{1}{k_y^2} + \frac{1}{k_z^2} - \frac{1}{k_x^2}\right)\sigma_y\sigma_z - \left(\frac{1}{k_z^2} + \frac{1}{k_x^2} - \frac{1}{k_y^2}\right)\sigma_z\sigma_x + \\ &\left(\frac{\tau_{yz}}{k_{yz}}\right)^2 + \left(\frac{\tau_{zx}}{k_{zx}}\right)^2 + \left(\frac{\tau_{xy}}{k_{xy}}\right)^2 = 1 \end{aligned} \tag{5.57}$$

Note that under a particular *plane stress condition*, e.g., in the x, y plane, when $\sigma_z = \tau_{zx} = \tau_{yz} = 0$, both formulas (5.56) and (5.57) reduce to the 4-parameter orthotropic Hill's condition

$$\frac{\sigma_x^2}{k_x^2} + \frac{\sigma_y^2}{k_y^2} - \left(\frac{1}{k_x^2} + \frac{1}{k_y^2} - \frac{1}{k_z^2}\right)\sigma_x\sigma_y + \frac{\tau_{xy}^2}{k_{xy}^2} = 1 \tag{5.58}$$

where initiation of plastic flow in the x, y plane is controlled not only by the in-plane limits k_x, k_y, and k_{xy}, but also by the out-of-plane limit k_z, which may finally lead to inadmissible loss of convexity by the yield surface. This will be discussed in detail in the next section.

Note that in case when $\Pi_{23} = \Pi_{13} = \Pi_{12} = -1/2k^2$ and $\Pi_{44} = \Pi_{55} = \Pi_{66} = 3/k^2$ the orthotropic Hill criterion (5.53) reduces to the *isotropic Huber–von Mises criterion*

$$\left(\sigma_y - \sigma_z\right)^2 + \left(\sigma_z - \sigma_x\right)^2 + \left(\sigma_x - \sigma_y\right)^2 + 6\left(\tau_{yz}^2 + \tau_{zx}^2 + \tau_{xy}^2\right) = 2k^2 \tag{5.59}$$

The Hill criterion (5.53) is formulated in the space of principal material directions of orthotropy which in general do not coincide with directions of principal stresses. In the particular case when the coaxiality holds $\sigma_x = \sigma_1$, $\sigma_y = \sigma_2$, $\sigma_z = \sigma_3$, $\tau_{xy} = \tau_{yz} = \tau_{zx} = 0$ we arrive at simplified

$$-\Pi_{23}\left(\sigma_2 - \sigma_3\right)^2 - \Pi_{13}\left(\sigma_3 - \sigma_1\right)^2 - \Pi_{12}\left(\sigma_1 - \sigma_2\right)^2 = 1 \tag{5.60}$$

or when calibration (5.55) is used the explicit form of (5.60) is finally furnished

$$\begin{gathered}\frac{1}{2}\left(\frac{1}{k_2^2} + \frac{1}{k_3^2} - \frac{1}{k_1^2}\right)\left(\sigma_2 - \sigma_3\right)^2 + \\ \frac{1}{2}\left(\frac{1}{k_3^2} + \frac{1}{k_1^2} - \frac{1}{k_2^2}\right)\left(\sigma_3 - \sigma_1\right)^2 + \\ \frac{1}{2}\left(\frac{1}{k_1^2} + \frac{1}{k_2^2} - \frac{1}{k_3^2}\right)\left(\sigma_1 - \sigma_2\right)^2 = 1\end{gathered} \tag{5.61}$$

Hill's condition (5.61) represents cylindrical elliptic surface whose axis coincides with the hydrostatic axis. Nevertheless in some cases, the limit surface looses closed form for high othotropy degree which may occur when one of following expressions $\frac{1}{k_2^2} + \frac{1}{k_3^2} - \frac{1}{k_1^2}$ elsewhere $\frac{1}{k_3^2} + \frac{1}{k_1^2} - \frac{1}{k_2^2}$ or $\frac{1}{k_1^2} + \frac{1}{k_2^2} - \frac{1}{k_3^2}$ changes the sign. Such behavior is not admissible and a way how to overcome it will be presented in the next section.

It is convenient to express Hill's limit surface by use of the *Haigh–Westergaard coordinates* (cf. Ganczarski and Lenczowski [15])

$$\begin{Bmatrix} \sigma_1 \\ \sigma_2 \\ \sigma_3 \end{Bmatrix} = \frac{\xi}{\sqrt{3}} \begin{Bmatrix} 1 \\ 1 \\ 1 \end{Bmatrix} + \sqrt{\frac{2}{3}}\rho(\theta) \begin{Bmatrix} \cos\theta \\ \cos(\theta - \frac{2\pi}{3}) \\ \cos(\theta + \frac{2\pi}{3}) \end{Bmatrix} \tag{5.62}$$

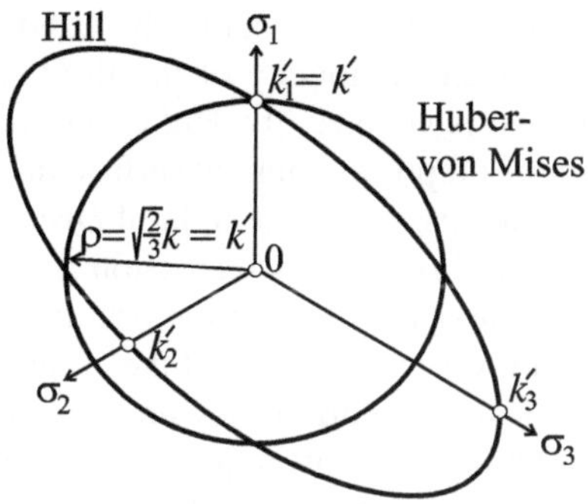

Fig. 5.10 Comparison of the Huber–von Mises and the Hill criteria in deviatoric plane applying the Haigh–Westergaard coordinates $\rho(\theta)$ ($k_1 = k$, $k_2 = 0.8k$, $k_3 = 1.5k$)

to finally obtain Hill's criterion in form $\rho(\theta)$

$$\rho(\theta) = \left[\frac{2}{\left(\frac{1}{k_2^2} + \frac{1}{k_3^2} - \frac{1}{k_1^2}\right)\sin^2\left(\theta + \frac{\pi}{3}\right) + \left(\frac{1}{k_3^2} + \frac{1}{k_1^2} - \frac{1}{k_2^2}\right)\sin^2\left(\theta - \frac{\pi}{3}\right) + \left(\frac{1}{k_1^2} + \frac{1}{k_2^2} - \frac{1}{k_3^2}\right)\sin^2\theta} \right]^{1/2} \tag{5.63}$$

Note that in case if $k_1 = k_2 = k_3 = k$ the *Huber–von Mises circular cylinder* is recovered Fig. 5.10

$$\rho = \sqrt{\frac{2}{3}}k = \text{const} \tag{5.64}$$

5.4 Hill's Criterion Versus Hu–Marin's Concept in Case of Strong Orthotropy

Classical orthotropic *Hill's criterion* [25], despite obvious advantages and wide technical applications, is limited however by some *constraints* of applicability, which are discussed in the present section following [18].

First limitation of applicability range of the classical Hill criterion is established through the inequality bounding the magnitudes of the engineering orthotropy constants k_1, k_2, and k_3 in order to avoid *ellipticity loss* of the limit surface in the stress space when the coordinate axes are aligned with the material axes of orthotropy (see, e.g., Ottosen and Ristinmaa [47], Ganczarski and Skrzypek [17, 18]). Such limit bounds put upon the orthotropy limits usually hold in case if the degree of material orthotropy is moderate. For example, if the material ensures the *transverse isotropy symmetry*, it is shown that the orthotropy degree bounded by the inequality $\frac{k_{max}}{k_{min}} < 2$ guarantees ellipticity of the limit surface to be saved. However, if the orthotropy bound is violated, the Hill criterion becomes useless when a possible degeneration of the elliptic cylindrical surface into two concave hyperbolic cylinders occurs, what is inadmissible in the light of *Drucker's* or *Sylvester's stability postulates*.

To illustrate this restriction, we consider two types of true materials for which the classical Hill criterion occurs to be: either useful, if material orthotropy degree is not very high such that the ellipticity property of the limit surface is preserved, or useless if the orthotropy degree is as high as the described limit surface no longer holds the ellipticity requirement. Other words, a physically inadmissible degeneration of the single convex and simply connected elliptical limit surface into two concave hyperbolic surfaces occurs.

The following inequality bounds the *range of applicability* for *Hill's criterion* (cf., e.g., Ottosen and Ristinmaa [47])

$$\frac{2}{k_1^2 k_2^2} + \frac{2}{k_2^2 k_3^2} + \frac{2}{k_3^2 k_1^2} > \frac{1}{k_1^4} + \frac{1}{k_2^4} + \frac{1}{k_3^4} \tag{5.65}$$

For simplicity, a coincidence of the principal stress axes with the material orthotropy axes is assumed in (5.65). In the narrower case of *transverse isotropy* $k_1 = k_2$, condition (5.65) reduces to the simple form

$$\frac{1}{k_3^2}\left(\frac{4}{k_1^2} - \frac{1}{k_3^2}\right) > 0 \tag{5.66}$$

Substitution of the dimensionless parameter $R = 2(\frac{k_3}{k_1})^2 - 1$, after Hosford and Backhofen [27], leads to the simplified restriction

$$R > -0.5 \tag{5.67}$$

If the above inequalities (5.65)–(5.67) do not hold, elliptic cross sections of the limit surface degenerate into two hyperbolic branches and the lack of convexity occurs. To illustrate this limitation, the yield curves in two planes: the transverse isotropy (σ_1, σ_2) and the orthotropy plane (σ_1, σ_3) for various R–values, are sketched in Fig. 5.11a, b, respectively. It is observed that when R starting from $R = 3$ approaches the limit $R = -0.5$, the curves change from closed ellipses to two parallel lines, whereas for $R < -0.5$ concave hyperbolas appear.

As example of orthotropic engineering material for which classical Hill's criterion can correctly predict the limit surface, consider first the *OTCz Titanium Alloy*, the mechanical orthotropic properties of which are given in Table 5.1 (cf. Malinin and Rżysko [42]). Note that, for the OTCz Titanium Alloy, yield limits in the plane of weak orthotropy 1,2 differ not so much, but the 3 axis is the dominant orthotropy axis. As a consequence, in the plane of weak orthotropy 1,2 Hill's ellipse is slightly rotated towards 2–axis ($\alpha_{12} \approx 45°$), in contrast to the plane of strong orthotropy

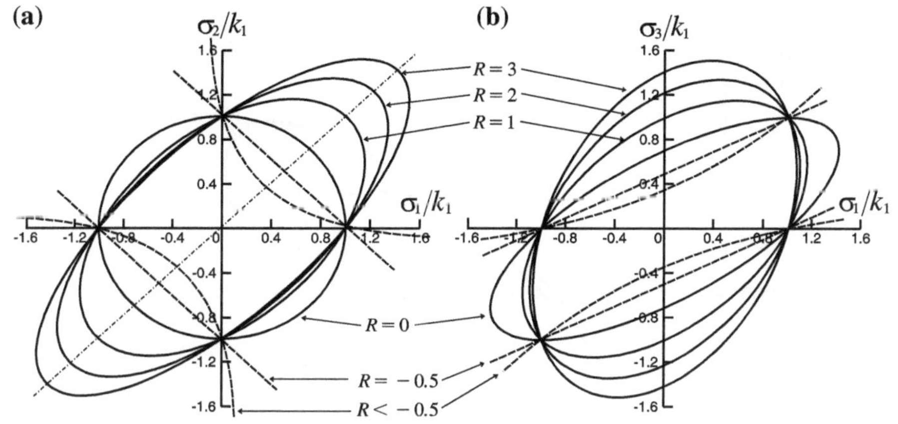

Fig. 5.11 Degeneration of the Hill's limit surface with the magnitude of the Hosford and Backhofen parameter R: **a** transverse isotropy plane, **b** orthotropy plane (after Ganczarski and Skrzypek [18])

Table 5.1 Mechanical properties of orthotropic OTCz Titanium Alloy after Malinin and Rżysko [42]

Yield limits	k_1 [MPa]	k_2 [MPa]	k_3 [MPa]
	490	520	800

1,3, where the rotation of the Hill ellipse is significant ($\alpha_{13} \approx 71°$), as shown in Fig. 5.12a, b, respectively.

In a case of *high orthotropy degree* (observed for majority of the long fiber reinforced composites, for instance: Boron/Al, SiC/Ti, Glass/Epoxy, Graphite/Epoxy, etc., e.g., Herakovich and Aboudi [23], Sun and Vaidya [57], and others), the concept other than Hill's is proposed. This new approach suggests formulation of limit criterion based on the 9-parameter von Mises condition, but enhanced by the Hu–Marin type biaxial orthotropic loading conditions (cf. Hu and Marin [30], Skrzypek and Ganczarski [54]). It will be demonstrated that, even in a case of arbitrarily strong orthotropy (for instance, $k_{max}/k_{min} \approx 9$, in case if brass Ł62 is tested) the property of ellipticity is saved.

In general case of strong orthotropy, when the ellipticity condition (5.65) does not hold, the deviatoric Hill criterion (5.56) or (5.57) becomes useless. Hence, in order to describe physically admissible closed and convex limit surface, the more general 9-parameter *orthotropic von Mises equation* (5.44) has to be recalled. In a narrower case of the principal stress axes coinciding with the material orthotropy axes the Eq. (5.45) reads as

$$\Pi_{11}\sigma_1^2 + \Pi_{22}\sigma_2^2 + \Pi_{33}\sigma_3^2 + 2(\Pi_{12}\sigma_1\sigma_2 + \Pi_{23}\sigma_2\sigma_3 + \Pi_{31}\sigma_3\sigma_1) = 1 \qquad (5.68)$$

The condition (5.68) is defined by six material parameters only, because $\tau_{23} \equiv \tau_{31} \equiv \tau_{12} \equiv 0$, hence its calibration requires six conditions:

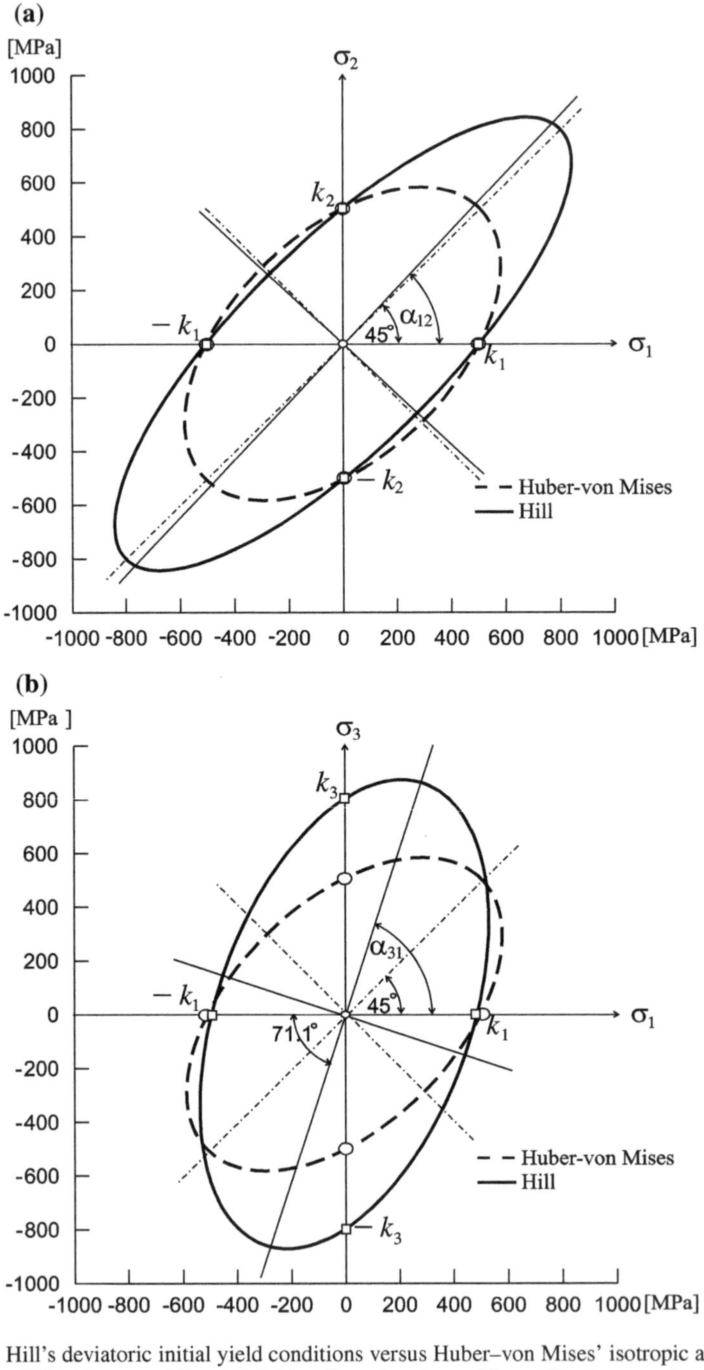

Fig. 5.12 Hill's deviatoric initial yield conditions versus Huber–von Mises' isotropic approximation for the OTCz Titanum Alloy (cf. Table 5.1): **a** the plane of "weak" orthotropy (σ_1, σ_2), **b** the plane of "strong" orthotropy (σ_1, σ_3) (after Ganczarski and Skrzypek [18])

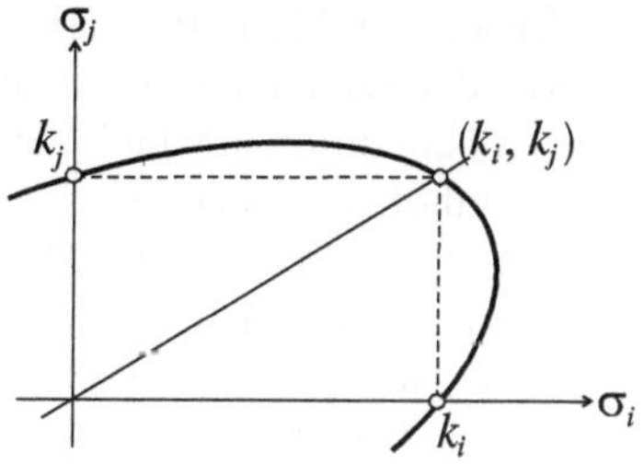

Fig. 5.13 Graphical illustration of biaxial loading conditions (5.70)

three tests of *uniaxial tension* along the orthotropy axes

$$\begin{aligned}\sigma_1 = k_1\ \sigma_2 = 0\ \sigma_3 = 0 &\longrightarrow \Pi_{11} = 1/k_1^2\\ \sigma_2 = k_2\ \sigma_1 = 0\ \sigma_3 = 0 &\longrightarrow \Pi_{22} = 1/k_2^2\\ \sigma_3 = k_3\ \sigma_1 = 0\ \sigma_2 = 0 &\longrightarrow \Pi_{33} = 1/k_3^2\end{aligned} \tag{5.69}$$

and three orthotropic *biaxial tension loading conditions* (k_i, k_j) cf. Fig. 5.13

$$\begin{aligned}\sigma_1 = k_1\ \sigma_2 = k_2\ \sigma_3 = 0 &\longrightarrow \Pi_{12} = -1/2k_1k_2\\ \sigma_1 = k_1\ \sigma_3 = k_3\ \sigma_2 = 0 &\longrightarrow \Pi_{13} = -1/2k_1k_3\\ \sigma_2 = k_2\ \sigma_3 = k_3\ \sigma_1 = 0 &\longrightarrow \Pi_{23} = -1/2k_2k_3\end{aligned} \tag{5.70}$$

The similar equibiaxial tension loading conditions are used, e.g., by Khan and Liu [35].

Calibration of the orthotropic von Mises criterion (5.68), performed with conditions (5.69) and (5.70) used, leads to the three-axial extension of the Hu–Marin type criterion (cf. Ganczarski and Skrzypek [16], Skrzypek and Ganczarski [54])

$$\left(\frac{\sigma_1}{k_1}\right)^2 - \frac{\sigma_1\sigma_2}{k_1k_2} + \left(\frac{\sigma_2}{k_2}\right)^2 - \frac{\sigma_2\sigma_3}{k_2k_3} + \left(\frac{\sigma_3}{k_3}\right)^2 - \frac{\sigma_1\sigma_3}{k_1k_3} = 1 \tag{5.71}$$

The *enhanced Mises–Hu–Marin type criterion* (5.71) is free from Hill's deficiency even in case of arbitrarily strong orthotropy degree, since it never violates the Drucker stability postulate, which is not guaranteed by Hill-type equations. The Hu–Marin-type Eq. (5.71) can easily be presented in the *"pseudo-deviatoric" format*

$$\left(\frac{\sigma_1}{k_1} - \frac{\sigma_2}{k_2}\right)^2 + \left(\frac{\sigma_2}{k_2} - \frac{\sigma_3}{k_3}\right)^2 + \left(\frac{\sigma_3}{k_3} - \frac{\sigma_1}{k_1}\right)^2 = 2 \tag{5.72}$$

Three orthotropy limit yield points k_1, k_2 and k_3 establish the proportional *stress/strength axis* of cylindrical Hu–Marin's surface. Note that this proportional stress/strength axis, which determines a position of the limit surface axis in the principal stress space, is different from the *hydrostatic axis*, but the condition of equal ratios $\sigma_i/k_i = \alpha$ holds at all points belonging to this axis. The extended von Mises–Hu–Marin type criteria (5.71–5.72) are always "unconditionally stable" criteria, that remain convex even for very strong orthotropy, by contrast to the classical Hill condition in which the possible loss of convexity can be met in the case of highly orthotropic materials. However, the fully deviatoric format of the Hill criteria (5.50–5.56) is lost in the Hu–Marin type format (5.72) where the *hydrostatic pressure insensitivity* is *relaxed.*

In the particular case of *plane stress state* $\sigma_3 = 0$ the three-parameter enhanced von Mises–Hu–Marin equation (5.71) is reduced to the two-parameter one, as proposed by Hu–Marin [30]

$$\left(\frac{\sigma_1}{k_1}\right)^2 - \frac{\sigma_1\sigma_2}{k_1k_2} + \left(\frac{\sigma_2}{k_2}\right)^2 = 1 \tag{5.73}$$

Comparison of the 2-parameter Hu–Marin plane stress equation (5.73) with the simplified 4-parameter plane stress Hill's equation (5.58) written for principal stress axes, leads to the 3-parameter form

$$\left(\frac{\sigma_1}{k_1}\right)^2 - \left(\frac{1}{k_1^2} + \frac{1}{k_2^2} - \frac{1}{k_3^2}\right)\sigma_1\sigma_2 + \left(\frac{\sigma_2}{k_2}\right)^2 = 1 \tag{5.74}$$

which becomes identical to the von Mises–Hu–Marin equation (5.73) only if following constraint holds

$$\frac{1}{k_3^2} = \frac{1}{k_1^2} + \frac{1}{k_2^2} - \frac{1}{k_1k_2} \tag{5.75}$$

which is usually not true.

In order to illustrate a suitability of the von Mises–Hu–Marin orthotropic Eq. (5.71), when compared to certain limitations of the Hill deviatoric Eq. (5.58), two engineering materials characterized by different degrees of orthotropy: *OTCz Titanium Alloy* ("weak" orthotropy) and *Ł62 brass* ("strong" orthotropy) are studied. The results are presented in Fig. 5.14a, b on the planes σ_1, σ_3 and σ_1, σ_2, respectively. In case of "weak" orthotropy both Hill's and Hu–Marin's ellipses differ not so much, and both concepts are recommended (Fig. 5.14a). However, in case of "strong" orthotropy, when the inequality (5.65) is not satisfied, following the Hill concept two concave hyperbolic cylinders are formed by opening of the elliptic cylinder towards the proportional stress/strength axis (Fig. 5.14b). On the other hand, the Hu–Marin type surface saves the ellipticity property regardless of the magnitude of orthotropy degree considered. In other words the *Hu–Marin surface* is "unconditionally stable" which remains convex for very strong orthotropy. It is possible due to three additional constraints (5.70) satisfied for the pairs of orthotropy yield limits (k_1, k_2), (k_2, k_3), and (k_3, k_1). But, it should be pointed out that the Hu–Marin cylindrical surface does not satisfy the condition of deviatoricity, hence this condition in fact should be classified as a specific representative of the hydrostatic pressure sensitive class of materials where the independence of the hydrostatic stress constraint is relaxed.

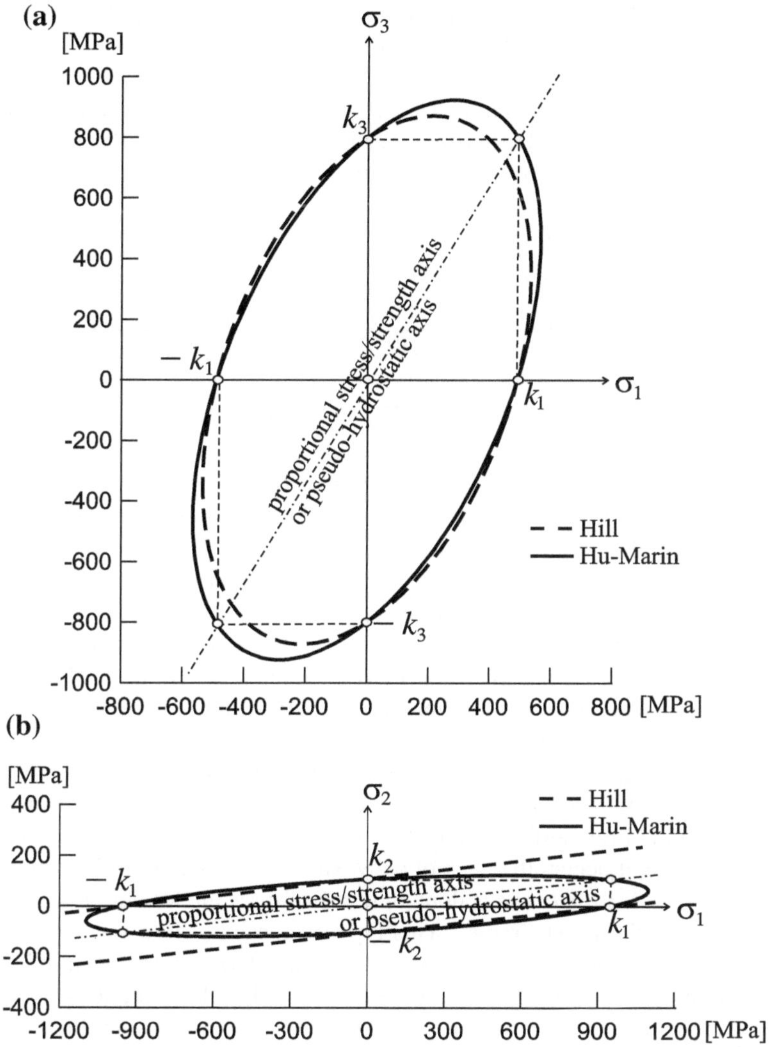

Fig. 5.14 Comparison of the Hill and the Hu–Marin plastic yield criteria for two orthotropic materials of different orthotropy degrees: **a** "weak" orthotropy in case of OTCz titanium alloy ($k_1 = 490\,\text{MPa}$, $k_2 = 520\,\text{MPa}$, $k_3 = 800\,\text{MPa}$), **b** "strong" orthotropy in case of Ł62 brass ($k_1 = 105\,\text{MPa}$, $k_2 = 120\,\text{MPa}$, $k_3 = 950\,\text{MPa}$) (after Ganczarski and Skrzypek [18])

The aforementioned possible *loss* of the *convexity* of classical *Hill's criterion* [25] (5.56) in case of highly orthotropic materials is even more pronounced when the *orthotropic* generalization of the isotropic *Hosford criterion* [27] (5.10) for higher (even) exponents is done

$$-\Pi_{23}|\sigma_y - \sigma_z|^m - \Pi_{13}|\sigma_z - \sigma_x|^m - \Pi_{12}|\sigma_x - \sigma_y|^m \\ + \Pi_{44}|\tau_{yz}|^m + \Pi_{55}|\tau_{zx}|^m + \Pi_{66}|\tau_{xy}|^m = 1 \tag{5.76}$$

Six generalized *orthotropy modules* $\Pi_{23}, \ldots, \Pi_{66}$ can be expressed in terms of six yield point stresses $k_x, \ldots, k_{xy}$ in analogous fashion as previously discussed manner of calibration for Hill's criterion (5.55), namely

$$\begin{aligned} -\Pi_{23} &= \frac{1}{2}\left(\frac{1}{|k_y|^m} + \frac{1}{|k_z|^m} - \frac{1}{|k_x|^m}\right), \quad \Pi_{44} = \frac{1}{|k_{yz}|^m} \\ -\Pi_{13} &= \frac{1}{2}\left(\frac{1}{|k_z|^m} + \frac{1}{|k_x|^m} - \frac{1}{|k_y|^m}\right), \quad \Pi_{55} = \frac{1}{|k_{zx}|^m} \\ -\Pi_{12} &= \frac{1}{2}\left(\frac{1}{|k_x|^m} + \frac{1}{|k_y|^m} - \frac{1}{|k_z|^m}\right), \quad \Pi_{66} = \frac{1}{|k_{xy}|^m} \end{aligned} \tag{5.77}$$

Note however that in this extended case (different from $m = 2$ and $m = 4$ when the orthotropic Hill is recovered) dimension of the orthotropy modules $\Pi_{23}, \ldots, \Pi_{66}$ depends on the value of power m and it is equal to MPa^{-m}.

Although the yield criterion defined by Eq. (5.76) with the calibration (5.77) used has been mathematically verified and its convexity has been proven in case of the *planar anisotropy* in the principal stress space if and only if $m \geq 1$ and the orthotropy modules $\Pi_{23}, \ldots, \Pi_{66}$ are positive constant coefficients (see Barlat and Lian [2], also Chu [10]), in case of the general orthogonal anisotropy in the six-dimensional stress space convexity is not obvious.

The more general case, when axes of material orthotropy are different from axes of principal stresses, was considered by Ganczarski and Lenczowski [15]. It was shown that, although the limit surface is closed and convex in space of *principal material orthotropy* frame, it occurs that lack of convexity is met when transformation to the space of *principal stress frame* is done in terms of three angles defining the mutual configuration of these two frames. This type of convexity loss was examined for the brass sheet Ł22 the six orthotropic yield points of which are given in Table 5.2 after Malinin and Rżysko [42] who gave three-axial yield point stresses whereas three shear yield point stresses were estimated in [15] using simplified formulas $k_{ij} = \sqrt{\frac{k_{ij}}{3}}$ for $m = 2$ and $k_{ij} = \frac{\sqrt{k_{ij}}}{2}$ for $m = 6, 8$. For simplicity the evolution of the generalized *orthotropic Hosford yield condition* ($m = 8$) with respect to only one of the Euler angles ϑ was considered. It represents a prism of the semi-hexagonal cross section with oval corners as presented in Fig. 5.15. The *loss of convexity* is observed for $18° \geq \vartheta \geq 26°$.

Table 5.2 Yield point stresses for brass Ł22 after Malinin and Rżysko [42]

m	k_x [MPa]	k_y [MPa]	k_z [MPa]	k_{zy} [MPa]	k_{zx} [MPa]	k_{xy} [MPa]
2	120	105	950	182	194	64.8
6 or 8	120	105	950	157	168	56.1

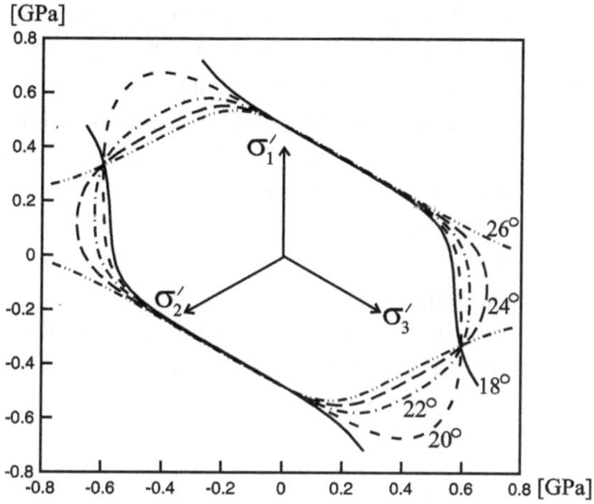

Fig. 5.15 Evolution of the generalized orthotropic Hosford yield condition ($m = 8$) versus the Euler angle $\vartheta = 18°, 20°, 22°, 24°$, and $26°$ for brass Ł22, after Ganczarski and Lenczowski [15]

It should be pointed out that the limit criteria considered throughout this section do not exhibit the strength differential effect such that they cannot be recommended as failure criteria for brittle materials where this effect is essential.

5.5 Transversely Isotropic Case—Hill-Type Tetragonal Symmetry Versus Hu–Marin-Type Hexagonal Symmetry Criteria

The second limitation of applicability range of classical Hill's criterion arises when the transverse isotropy property is considered. In this section it will be shown that, if reduction of Hill's criterion to the *transverse isotropy symmetry* is performed, the 4-parameter form that satisfies the *tetragonal symmetry class* is furnished (cf., e.g., Voyiadjis and Thiagarajan [62], Sun and Vaidya [57]). This type of symmetry is of particular importance in case of *unidirectional fiber reinforced composites*. In such a case moduli: k_x, k_y, k_z, and k_{xy} are considered as independent (z is the orthotropy axis), which makes impossible to reduce classical Hill's criterion to the isotropic von Mises condition in the plane of transverse isotropy.

To avoid this *irreducibility to isotropic von Mises*, the new *Hu–Marin-based transversely isotropic criterion* exhibiting *hexagonal symmetry* is proposed instead of *deviatoric transversely isotropic Hill's criterion* exhibiting *tetragonal symmetry*. It enables to achieve reducibility to the isotropic von Mises condition in the transverse isotropy plane, preserving cylindricity regardless of the magnitude of orthotropy degree.

Finally, it will be demonstrated that, for some composite materials it is necessary to further modify the 3-parameter Hu–Marin-type criterion to the new 4-parameter intermediate-type criterion between classical Hill's and hexagonal Hu–Marin's concepts, taking advantage of the bulge test. This new hybrid-type criterion differs essentially from both the Hu–Marin hexagonal type criterion and the isotropic von Mises criterion in the isotropy plane. *Bulge tests* have been performed and described, e.g., by Jackson et al. [33] with equipment used by Lankford et al. [39]. This new criterion is capable of properly describing the SiC/Ti long fiber reinforced composite examined by Herakovich and Aboudi [23].

Classical Hill's equation (5.53–5.54), which is expressed in terms of six independent plastic yield limits $k_x, k_y, k_z, k_{yz}, k_{zx}$, and k_{xy}, (5.56) is often too general for engineering applications. Orthotropic structural materials usually exhibit the transversely isotropic symmetry, basically due to either fabrication process or microstructure texture, as often observed in many long parallel fiber reinforced composites. In particular, if in elastic range the transversely isotropic symmetry group holds, it is expected that, also for the plastic yield initiation criterion such a narrower symmetry is true.

In what follows, a distinction between two symmetry classes of the *transverse isotropy–tetragonal* or *hexagonal type* has to be done. Such distinction is known, e.g., from definitions of Representative Unit Cell used in homogenization methods for *composite materials* (cf., e.g., Berryman [4], Sun and Vaidya [57], etc.).

Assume that the z-axis is the orthotropy axis, whereas x, y is the *transverse isotropy plane*. When applying Eq. (5.54) with calibrations (5.56) or (5.57) and additionally assuming $k_x = k_y \neq k_z$, $k_{zx} = k_{zy} \neq k_{xy}$, the number of independent limits in transversely isotropic Hill's equation reduces to four for instance: two axial yield limits k_x and k_z, and two shear yield limits k_{zx} and k_{xy} (Fig. 5.16).

In this way the following is furnished

$$\begin{aligned} -\Pi_{13} = -\Pi_{23} &= \frac{1}{2k_z^2}\,, & \Pi_{44} = \Pi_{55} &= \frac{1}{k_{zx}^2} \\ -\Pi_{12} &= \frac{1}{k_x^2} - \frac{1}{2k_z^2}\,, & \Pi_{66} &= \frac{1}{2k_{xy}^2} \end{aligned} \tag{5.78}$$

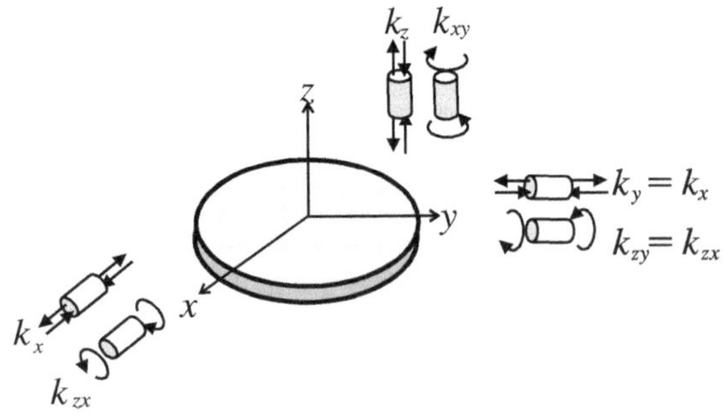

Fig. 5.16 Four independent tests for transversely isotropic Hill's criterion calibration

Substitution of (5.78) into (5.51) and (5.52) yields to *transversely isotropic Hill's matrices*

$$
\left[{}_{\text{tris}}\mathbb{H}^{\text{H}}\right] = \left[\begin{array}{ccc|ccc}
-\Pi_{12}-\Pi_{13} & \Pi_{12} & \Pi_{13} & & & \\
 & -\Pi_{12}-\Pi_{23} & \Pi_{13} & & & \\
 & & -2\Pi_{13} & & & \\
\hline
 & & & \Pi_{44} & & \\
 & & & & \Pi_{44} & \\
 & & & & & \Pi_{66}
\end{array}\right]
$$

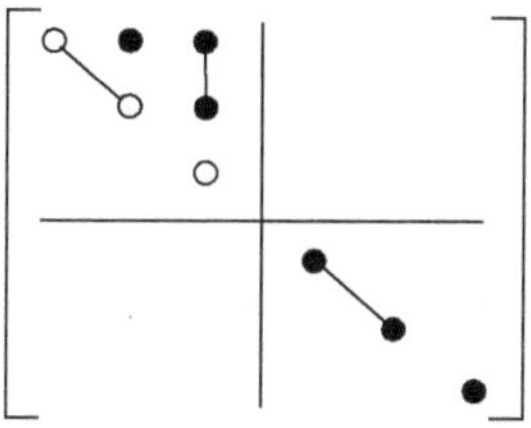

(5.79)

or

$$
\left[{}_{\text{tris}}\mathbb{H}^{\text{H}}\right] = \left[\begin{array}{ccc|ccc}
\Pi_{11} & \frac{\Pi_{33}-2\Pi_{11}}{2} & -\frac{\Pi_{33}}{2} & & & \\
 & \Pi_{11} & -\frac{\Pi_{33}}{2} & & & \\
 & & \Pi_{33} & & & \\
\hline
 & & & \Pi_{44} & & \\
 & & & & \Pi_{44} & \\
 & & & & & \Pi_{66}
\end{array}\right]
$$

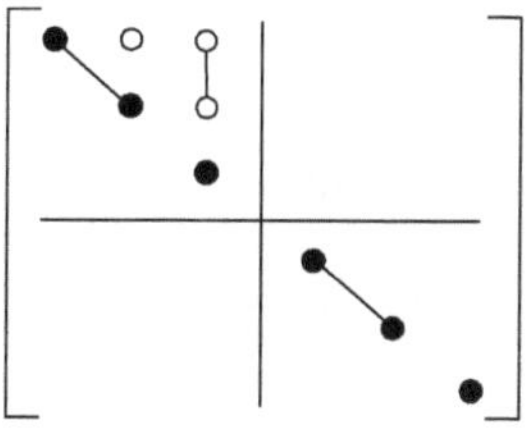

(5.80)

The transversely isotropic 4-parameter Hill criteria corresponding to *orthotropic Hill's criteria* (5.56) and (5.57) take the following representations

$$
\frac{(\sigma_y-\sigma_z)^2+(\sigma_z-\sigma_x)^2}{2k_z^2}+\underline{\left(\frac{1}{k_x^2}-\frac{1}{2k_z^2}\right)}(\sigma_x-\sigma_y)^2+\frac{\tau_{yz}^2+\tau_{zx}^2}{k_{zx}^2}+\frac{\tau_{xy}^2}{k_{xy}^2}=1 \tag{5.81}
$$

or equivalently

$$
\frac{\sigma_x^2+\sigma_y^2}{k_x^2}+\frac{\sigma_z^2}{k_z^2}-\underline{\left(\frac{2}{k_x^2}-\frac{1}{k_z^2}\right)}\sigma_x\sigma_y-\frac{\sigma_y\sigma_z+\sigma_z\sigma_x}{k_z^2}+\frac{\tau_{yz}^2+\tau_{zx}^2}{k_{zx}^2}+\frac{\tau_{xy}^2}{k_{xy}^2}=1 \tag{5.82}
$$

Both forms involve four plastic limits k_x, k_z, k_{zx}, and k_{xy}, considered as independent parameters. Underlined factor in (5.82) includes not only k_x but also k_z. The explicitly deviatoric form (5.81) exhibits the similar feature. The plastic state in the transverse isotropy plane x, y is controlled not only by the tensile yield limit in this plane k_x, but also by the out-of-plane tensile yield limit k_z. Concluding, *transversely isotropic Hill's criteria* (5.81) or (5.82) have to be classified as the *tetragonal symmetry format* (see Table 1.4). The assumption of tetragonal symmetry of the criteria (5.81–5.82) was also considered by Voyiadjis and Thiagarajan [62] in case of directionally reinforced metal matrix composites (Boron–Aluminum). Broader discussion that relates to distinction between the tetragonal versus hexagonal symmetry in the yield/failure criteria will be presented in Sect. 6.5, where additional constraint for case if $\Pi_{66} = -2(\Pi_{13} + 2\Pi_{12})$ is assumed, such that Π_{66} has to be considered as dependent plastic modulus. To this end, if aforementioned constraint postulated by Chen and Han [9] is applied, the equality holds

$$\Pi_{66} = \frac{4}{k_x^2} - \frac{1}{k_z^2} \tag{5.83}$$

instead of (5.78$_4$) and *transversely isotropic* 3-parameter *Hill's criteria* corresponding to (5.81) and (5.82) in following format

$$\frac{\frac{(\sigma_y - \sigma_z)^2 + (\sigma_z - \sigma_x)^2}{2k_z^2} + \left(\frac{1}{k_x^2} - \frac{1}{2k_z^2}\right)(\sigma_x - \sigma_y)^2 +}{\frac{\tau_{yz}^2 + \tau_{zx}^2}{k_{zx}^2} + \left(\frac{4}{k_x^2} - \frac{1}{k_z^2}\right)\tau_{xy}^2} = 1 \tag{5.84}$$

or equivalently

$$\frac{\frac{\sigma_x^2 + \sigma_y^2}{k_x^2} + \frac{\sigma_z^2}{k_z^2} - \left(\frac{2}{k_x^2} - \frac{1}{k_z^2}\right)\sigma_x\sigma_y - \frac{\sigma_y\sigma_z + \sigma_z\sigma_x}{k_z^2} +}{\frac{\tau_{yz}^2 + \tau_{zx}^2}{k_{zx}^2} + \left(\frac{4}{k_x^2} - \frac{1}{k_z^2}\right)\tau_{xy}^2} = 1 \tag{5.85}$$

can be written down.

In the particular case of plane stress state in the transverse isotropy plane (x, y) $\sigma_x, \sigma_y, \tau_{xy} \neq 0$ Eqs. (5.81) or (5.82) reduce to (5.58) with additional condition $k_x = k_y$

$$\frac{\sigma_x^2 + \sigma_y^2}{k_x^2} - \left(\frac{2}{k_x^2} - \frac{1}{k_z^2}\right)\sigma_x\sigma_y + \frac{\tau_{xy}^2}{k_{xy}^2} = 1 \tag{5.86}$$

The above form simply means that commonly used "transversely isotropic Hill's criterion" does not coincide in the "transverse isotropy plane" with the *isotropic Huber–von Mises equation*

$$\frac{\sigma_x^2 + \sigma_y^2}{k_x^2} - \frac{\sigma_x\sigma_y}{k_x^2} + 3\frac{\tau_{xy}^2}{k_{xy}^2} = 1 \tag{5.87}$$

In other words, when the new transversely isotropic yield criterion, that is free from inconsistencies between (5.86) and (5.87) is sought for, the material parameter preceding term $\sigma_x \sigma_y$ must be equal to $\Pi_{33} - 2\Pi_{11} = 1/k_x^2$ and not depend on k_z and, simultaneously, the material parameter $\Pi_{66} = 3/k_x^2$ must depend on k_x only.

In order to derive the transversely isotropic yield criterion reducible to coincidence with the Huber–von Mises criterion in the isotropy plane, the new *transversely isotropic hexagonal Hu–Marin equation* will be postulated. To obtain this criterion, the general *orthotropic von Mises equation* (5.45), which is not deviatoric, can be calibrated analogously to that presented in (5.69) and (5.70). Namely, when the constraints of transverse isotropy are imposed, we invoke:
the two *tensile tests* in the x- and the orthotropy z-axes and the *shear test* in the orthotropy zx-plane

$$
\begin{array}{ll}
\sigma_x = k_x,\ \ \sigma_y = \sigma_z = \tau_{xy} = \tau_{yz} = \tau_{zx} = 0 \longrightarrow \Pi_{11} = \Pi_{22} = 1/k_x^2 \\
\sigma_z = k_z,\ \ \sigma_x = \sigma_y = \tau_{xy} = \tau_{yz} = \tau_{zx} = 0 \longrightarrow \Pi_{33} = 1/k_z^2 \\
\tau_{zx} = k_{zx},\ \sigma_x = \sigma_y = \sigma_z = \tau_{xy} = \tau_{yz} = 0 \longrightarrow \Pi_{44} = \Pi_{55} = 1/k_{zx}^2
\end{array}
\tag{5.88}
$$

and the three *biaxial conditions* for coincidence of appropriate pairs of yield limits

$$
\begin{array}{ll}
\sigma_x = k_x, \sigma_y = k_x, \quad \sigma_z = \tau_{xy} = \tau_{yz} = \tau_{zx} = 0 \longrightarrow \Pi_{12} = -1/2k_x^2 \\
\sigma_x = k_x, \sigma_z = k_z, \quad \sigma_y = \tau_{xy} = \tau_{yz} = \tau_{zx} = 0 \longrightarrow \Pi_{13} = -1/2k_x k_z \\
\sigma_x = k_x, \tau_{xy} = k_x/\sqrt{3},\ \sigma_y = \sigma_z = \tau_{yz} = \tau_{zx} = 0 \longrightarrow \Pi_{66} = 3/k_x^2
\end{array}
\tag{5.89}
$$

Introduction of (5.88) and (5.89) into *orthotropic von Mises' criterion* (5.45) leads to *transversely isotropic* 3-parameter *hexagonal Hu–Marin's criterion* as follows:

$$
\frac{\sigma_x^2 + \sigma_y^2}{k_x^2} - \underline{\frac{\sigma_x \sigma_y}{k_x^2}} + \frac{\sigma_z^2}{k_z^2} - \frac{\sigma_y \sigma_z + \sigma_z \sigma_x}{k_z k_x} + \frac{\tau_{yz}^2 + \tau_{zx}^2}{k_{zx}^2} + \underline{\underline{3\frac{\tau_{xy}^2}{k_x^2}}} = 1 \tag{5.90}
$$

or

$$
\left(\frac{\sigma_x - \sigma_y}{k_x}\right)^2 + \left(\frac{\sigma_y}{k_x} - \frac{\sigma_z}{k_z}\right)^2 + \left(\frac{\sigma_z}{k_z} - \frac{\sigma_x}{k_x}\right)^2 + 3\frac{\tau_{yz}^2 + \tau_{zx}^2}{k_{zx}^2} + 6\frac{\tau_{xy}^2}{k_x^2} = 2 \tag{5.91}
$$

Note, that the above conditions correspond to generalized Hu–Marin's equations (5.71) or (5.72) with $k_1 = k_2$, but enhanced by the additional shear terms and referring to optional directions x, y, z. Equations (5.90) or (5.91) reduce to the Huber–von Mises equation (5.87) in case of *plane stress state* in the transverse isotropy plane (x, y), which means that this new criterion can finally be recognized as transversely isotropic hexagonal symmetry von Mises–Hu–Marin's based criterion.

Transversely isotropic *conditions—tetragonal Hill's* (5.71) or (5.72) and *hexagonal Hu–Marin's* (5.90) or (5.91), are examined for given orthotropy degrees $R = 2(\frac{k_z}{k_x})^2 - 1 = 2$,$k_{xy}/k_x = 0.8$, $k_{(xy)}/k_x = 0.9$, and $k_{zx}/k_x = 0.8$, for following stress states: biaxial normal stresses (σ_x, σ_y) and combined normal with shear stresses (σ_x, τ_{xy}) in the *transverse isotropy plane* (see Fig. 5.17a, b), as well as biaxial normal stresses (σ_x, σ_z) and combined normal with shear stresses (σ_x, τ_{zx})

in the *orthotropy plane* (see Fig. 5.18a, b). It is worth to mention that transversely isotropic Hill's condition of tetragonal symmetry (5.81) or (5.82) comprises four independent plastic yield limits: k_x, k_z, k_{zx}, and k_{xy}, because shear yield limit in isotropy plane k_{xy} is considered as independent.

Contrarily, transversely isotropic enhanced Hu–Marin-type condition, the symmetry class of which is hexagonal, is defined by three independent yield limits only: k_x, k_z, and k_{zx}, since in-plane shear yield limit k_{xy} must agree with the Huber–von Mises criterion in the isotropy plane $k_{xy} = \frac{k_x}{\sqrt{3}}$. Hence, representation of the *transversely isotropic hexagonal symmetry Hu–Marin-type* constitutive *matrix* of plasticity is as follows:

$$
\left[{}_{\rm tris}\mathbb{H}^{\rm HM}\right] = \left[\begin{array}{ccc|ccc}
\frac{1}{k_x^2} & -\frac{1}{2k_x^2} & -\frac{1}{2k_xk_z} & & & \\
 & \frac{1}{k_x^2} & -\frac{1}{2k_xk_z} & & & \\
 & & \frac{1}{k_z^2} & & & \\
\hline
 & & & \frac{1}{k_{zx}^2} & & \\
 & & & & \frac{1}{k_{zx}^2} & \\
 & & & & & \frac{3}{k_x^2}
\end{array}\right]
\tag{5.92}
$$

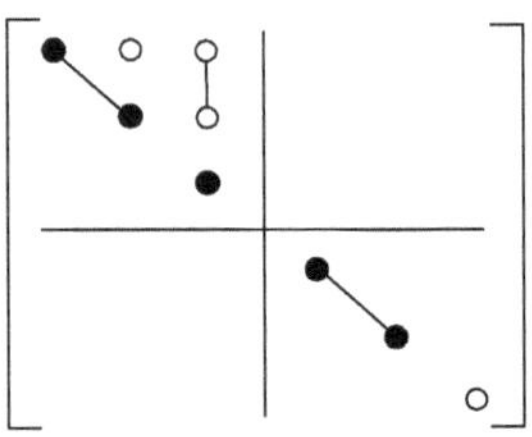

The general case of *transversely isotropic* 4-parameter *tetragonal symmetry Hu–Marin-type* yield criterion that preserves convexity but lost property of reducibility to the isotropic von Mises condition in the plane of transverse isotropy is considered by Voyiadjis and Thiagarajan [62]. The corresponding *matrix* of plasticity used by authors results from the general orthotropic matrix when four independent plastic onset limits are $k_1, k_2 = k_3, k_4 = k_5$, and k_6 (if original notation is saved 1 denotes fiber direction)

$$
\left[{}_{\rm tris}\mathbb{H}^{\rm VT}\right] =
\left[\begin{array}{ccc|ccc}
\frac{4}{9}k_1^2 & -\frac{2}{9}k_1k_2 & -\frac{2}{9}k_1k_2 & & & \\
 & \frac{4}{9}k_2^2 & -\frac{2}{9}k_1^2 & & & \\
 & & \frac{4}{9}k_2^2 & & & \\
\hline
 & & & \frac{4}{3}(k_1k_2+k_4^2) & & \\
 & & & & \frac{4}{3}(k_1k_2+k_4^2) & \\
 & & & & & \frac{4}{3}(k_2^2+k_6^2)
\end{array}\right]
\tag{5.93}
$$

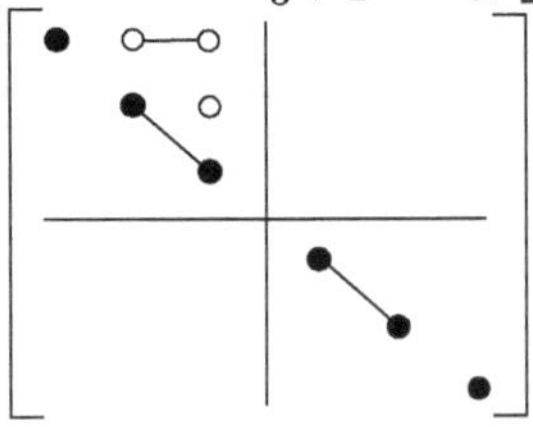

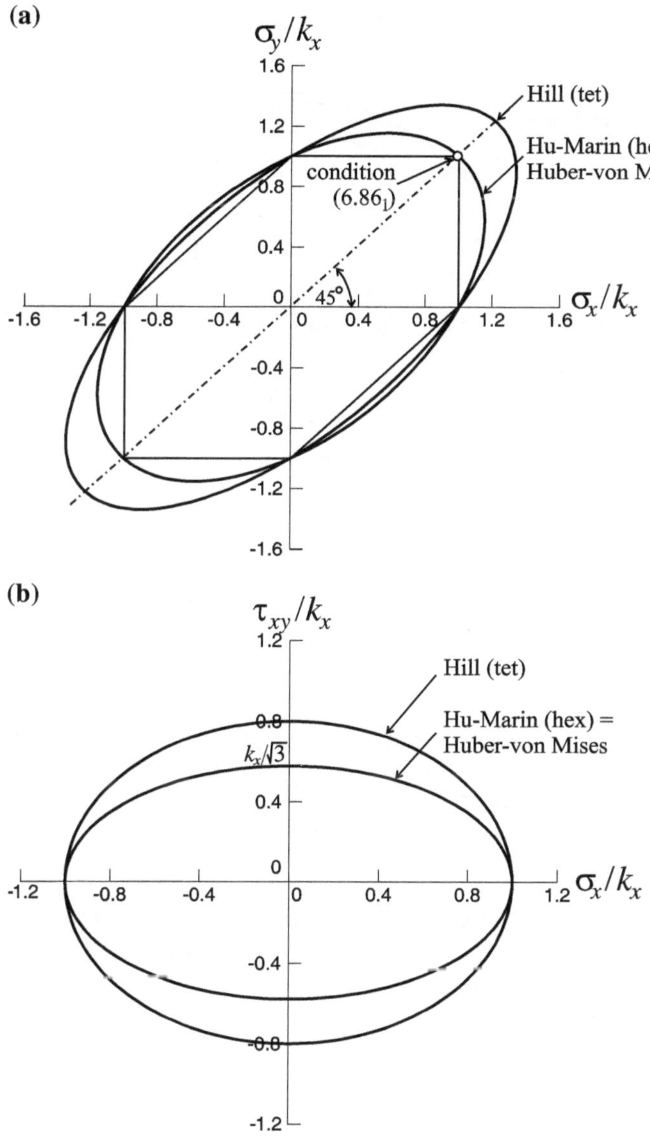

Fig. 5.17 Comparison of transversely isotropic criteria in the transverse isotropy planes: Hill's tetragonal (5.82), Hu–Marin's hexagonal (5.90) and Huber–von Mises' for given magnitudes of orthotropy ratios: $R = 2$, $k_{zx}/k_x = 0.8$, $k_{(xy)}/k_x = 0.9$ in case of 2D states of stress: **a** biaxial normal stresses (σ_x, σ_y) and **b** combined normal with shear stresses (σ_x, τ_{xy}) (after Ganczarski and Skrzypek [18])

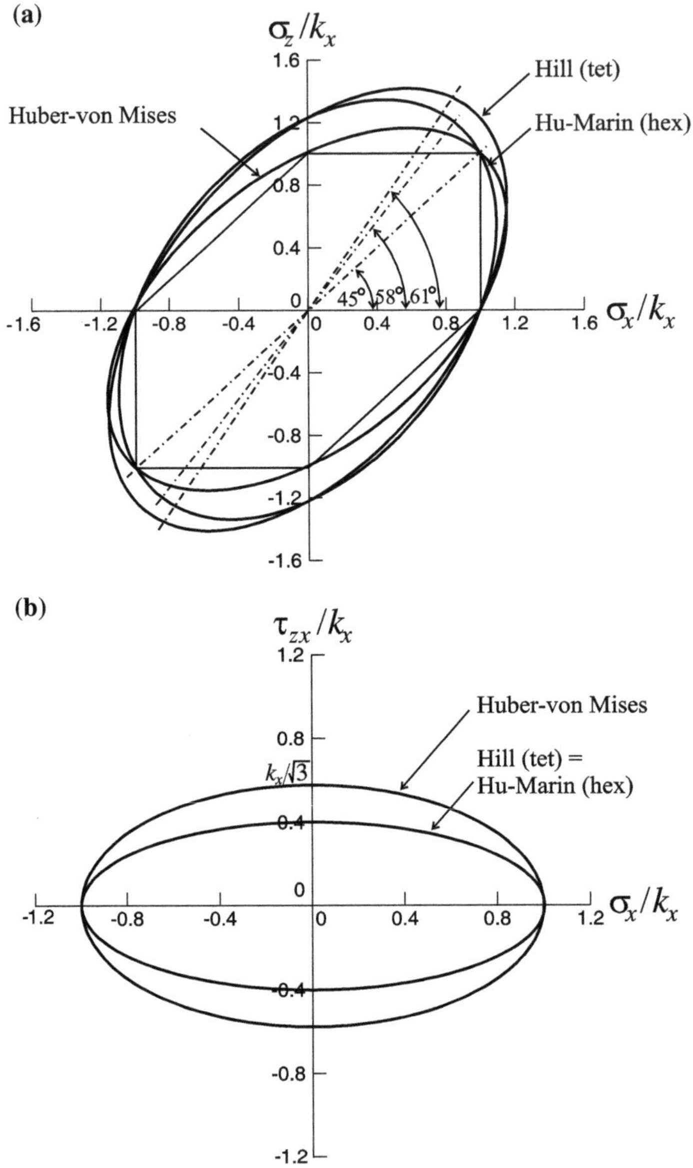

Fig. 5.18 Comparison of transversely isotropic criteria in the orthotropy plane: Hill's tetragonal (5.82), Hu–Marin's hexagonal (5.90), and Huber–von Mises for given magnitudes of orthotropy ratios: $R = 2$, $k_{zx}/k_x = 0.8$, $k_{(xy)}/k_x = 0.9$, in case of 2D states of stress: **a** biaxial normal stresses (σ_x, σ_z) and **b** combined normal with shear stresses (σ_x, τ_{zx}) (after Ganczarski and Skrzypek [18])

Introducing for k_1, k_2, k_4, and k_6 the following substitution $\frac{2}{9}k_1^2 = \frac{1}{k_z^2}$, $\frac{2}{9}k_2^2 = \frac{1}{k_x^2}$ $\frac{2}{3}(k_1k_2 + k_4^2) = \frac{1}{k_{zx}^2}$, $\frac{2}{3}(k_2^2 + k_6^2) = \frac{1}{k_{xy}^2}$ we end up with format of the *Voyiadjis and Thiagarajan condition* analogous to (5.92) however 4-parameter, where not only k_x, k_z, and k_{zx} but additionally k_{xy} are considered as independent (see doubly underlined terms in (5.90) and (5.94))

$$\frac{\sigma_x^2 + \sigma_y^2}{k_x^2} - \underline{\frac{\sigma_x\sigma_y}{k_x^2}} + \frac{\sigma_z^2}{k_z^2} - \frac{\sigma_y\sigma_z + \sigma_z\sigma_x}{k_zk_x} + \frac{\tau_{yz}^2 + \tau_{zx}^2}{k_{zx}^2} + \underline{\underline{\frac{\tau_{xy}^2}{k_{xy}^2}}} = 1 \tag{5.94}$$

Such criterion is irreducible to the isotropic von Mises type in the plane of isotropy, but it fits the experimental data for *Boron–Aluminum composite* tubular specimen having unidirectional lamina (Dvorak et al. [14] and Nigam et al. [45]).

Both *transversely isotropic criteria*: *Hill-type of tetragonal symmetry* (5.81) as well as *Hu–Marin-type of hexagonal symmetry* (5.90) describe cylindrical surfaces in space of principal stresses. However, Hill's type limit surface represents elliptical cylinder, the axis of which coincides with the hydrostatic axis, in contrast to enhanced Hu–Marin-type limit surface that represents elliptic cylinder, the axis of which forms a proportional stress/strength axis, different from the hydrostatic axis. It means that enhanced Hu–Marin's condition does not satisfy the deviatoricity property, which is a price for property of reducibility to the Huber–von Mises condition in the isotropy plane, with cylindricity ensured regardless of the magnitude of orthotropy degree.

A choice of appropriate transversely isotropic limit criterion, of either the tetragonal symmetry (5.81) or the hexagonal symmetry (5.90), depends on coincidence with experimental findings for real material. This may often lead to one of the two above considered symmetry classes, but sometimes material limit response is different even from both of them. Note that the shape of limit curves in the transverse isotropy plane is the key to appropriate classification of real transversely isotropic material as exhibiting *tetragonal symmetry* or *hexagonal* or *mixed symmetry* properties.

5.6 Hybrid Formulation of Enhanced Hu–Marin-Type Condition

In what follows a description of new limit criterion of the *hybrid symmetry property* between the tetragonal (5.81) or (5.82) and the hexagonal (5.90) or (5.91) symmetry classes, is proposed. The Hu–Marin type equation of pure hexagonal symmetry property (5.90) or (5.91) comprises three independent material constants k_x, k_z, and k_{zx}. However, real engineering materials of hybrid-type nature are frequently characterized by four independent material constants determined from four tests: two limits in uniaxial tensions k_x and k_z, shear limit in orthotropy plane k_{zx} (5.69) and additionally, in the *biaxial tension* test (bulge test) $k_{(xy)}$ instead of the first of condition

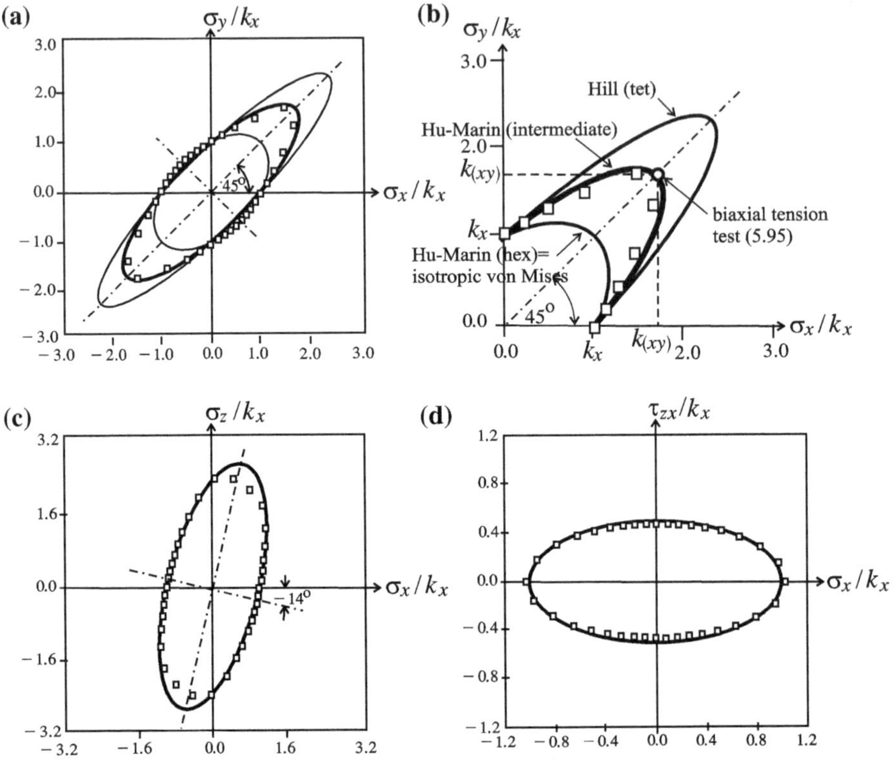

Fig. 5.19 Fitting of the initial yield surface of unidirectional SiC/Ti composite according to Herakovich and Aboudi findings [23] (symbol □) by the use of transversely isotropic Hu–Marin's hybrid-type criterion (5.90): **a**, **b** transverse isotropy plane (σ_x, σ_y), **c** orthotropy plane (σ_x, σ_z), **d** orthotropy shear plane (σ_x, τ_{zx}) (after Ganczarski and Skrzypek [18])

(5.70_1), namely

$$\sigma_x = \sigma_y = k_{(xy)} \; \sigma_z = \tau_{xy} = \tau_{yz} = \tau_{zx} = 0 \longrightarrow \Pi_{12} = -\frac{1}{2k^2_{(xy)}} \tag{5.95}$$

The above condition leads to the hybrid formulation of enhanced Hu–Marin's condition

$$\frac{\sigma_x^2 + \sigma_y^2}{k_x^2} - \underline{\frac{\sigma_x \sigma_y}{k^2_{(xy)}}} + \frac{\sigma_z^2}{k_z^2} - \frac{\sigma_y \sigma_z + \sigma_z \sigma_x}{k_z k_x} + \frac{\tau_{yz}^2 + \tau_{zx}^2}{k_{zx}^2} + 3\frac{\tau_{xy}^2}{k_x^2} = 1 \tag{5.96}$$

Equation (5.96) differs from the hexagonal form of Hu–Marin's condition (5.90) in the underlined term, where the fourth independent material constant $k_{(xy)}$ is taken from the *bulge test* (5.95), additionally to conditions ($5.89_{2,3}$). The hybrid formulation of 4-parameter *transversely isotropic Hu–Marin's* condition (5.96) has *matrix* representation as follows:

$$
[{}_{\text{hybr}}\mathbb{H}^{\text{HM}}] = \left[\begin{array}{ccc|ccc}
\frac{1}{k_x^2} & -\frac{1}{2k_{(xy)}^2} & -\frac{1}{2k_x k_z} & & & \\
 & \frac{1}{k_x^2} & -\frac{1}{2k_x k_z} & & & \\
 & & \frac{1}{k_z^2} & & & \\
\hline
 & & & \frac{1}{k_{zx}^2} & & \\
 & & & & \frac{1}{k_{zx}^2} & \\
 & & & & & \frac{3}{k_x^2}
\end{array}\right]
$$

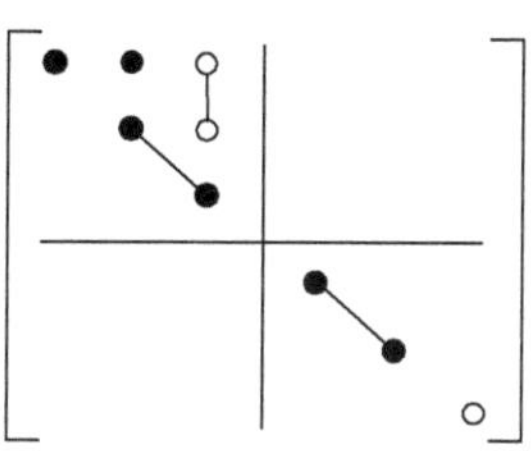

(5.97)

It is illustrated in Fig. 5.19a–d for the *SiC/Ti long fiber reinforced composite* by the use of thick solid line.

The hybrid-type enhanced Hu–Marin criterion is capable of capturing behavior of some *long fiber reinforced composite materials*, that in the transverse isotropy plane exhibit limit response different from both the Hill and the Huber–von Mises materials (cf., e.g., Herakovich and Aboudi [23]).

5.7 Comparison of Four Selected Transversely Isotropic Yield Criteria

Transition from the orthotropic yield criterion (von Mises or Hill) to the *transverse isotropy* is connected with the reduction of independent plastic modules (von Mises $9 \to 6$, Hill $6 \to 4$). These independent modules have to be identified by the use of appropriate number of tests and constraints. At present section the detailed discussion of the four selected yield criteria from Sect. 5.5 is performed. To this end we invoke following selected yield criteria, two of them based on the Hill origin and the other two based on the von Mises origin

$$
\begin{aligned}
&\frac{\sigma_x^2+\sigma_y^2}{k_x^2}+\frac{\sigma_z^2}{k_z^2}-\left(\frac{2}{k_x^2}-\frac{1}{k_z^2}\right)\sigma_x\sigma_y - \frac{\sigma_y\sigma_z+\sigma_z\sigma_x}{k_z^2} \\
&\qquad + \frac{\tau_{yz}^2+\tau_{zx}^2}{k_{zx}^2} + \frac{\tau_{xy}^2}{k_{xy}^2} = 1 \\
&\frac{\sigma_x^2+\sigma_y^2}{k_x^2}+\frac{\sigma_z^2}{k_z^2}-\left(\frac{2}{k_x^2}-\frac{1}{k_z^2}\right)\sigma_x\sigma_y - \frac{\sigma_y\sigma_z+\sigma_z\sigma_x}{k_z^2} \\
&\qquad + \frac{\tau_{yz}^2+\tau_{zx}^2}{k_{zx}^2} + \left(\frac{4}{k_x^2}-\frac{1}{k_z^2}\right)\tau_{xy}^2 = 1
\end{aligned}
\tag{5.98}
$$

$$\frac{\sigma_x^2+\sigma_y^2}{k_x^2}+\frac{\sigma_z^2}{k_z^2} \quad -\frac{\sigma_x\sigma_y}{k_x^2}-\frac{\sigma_y\sigma_z+\sigma_z\sigma_x}{k_zk_x} \\ +\frac{\tau_{yz}^2+\tau_{zx}^2}{k_{zx}^2} \quad +3\frac{\tau_{xy}^2}{k_x^2}=1$$

$$\frac{\sigma_x^2+\sigma_y^2}{k_x^2}+\frac{\sigma_z^2}{k_z^2} \quad -\frac{\sigma_x\sigma_y}{k_{(xy)}^2}-\frac{\sigma_y\sigma_z+\sigma_z\sigma_x}{k_zk_x} \\ +\frac{\tau_{yz}^2+\tau_{zx}^2}{k_{zx}^2} \quad +3\frac{\tau_{xy}^2}{k_x^2}=1$$

The above criteria have been derived in terms of different combinations of engineering modules based on appropriate identification procedures. All four criteria under consideration involve three common tests:

$$\sigma_x=k_x \qquad \sigma_z=k_z \qquad \tau_{zx}=k_{zx} \tag{5.99}$$

Additional conditions necessary for full identification take different forms.

In case of the classical Hill criterion (5.98_1) the additional fourth condition holds:

$$\tau_{xy}=k_{xy} \tag{5.100}$$

Aforementioned *Hill's* criterion contains four independent parameters and can be classified as *tetragonal symmetry form* (see Table 1.4d).

The second formulation (5.98_2) is also based on Hill's criterion however the additional constraint is imposed on the Π_{66} modulus (see Chen and Han [9])

$$\Pi_{66}=-2(\Pi_{13}+2\Pi_{12})=\frac{4}{k_x^2}-\frac{1}{k_z^2} \tag{5.101}$$

such that number of independent parameters is reduced to three k_x, k_z, and k_{zx} as a consequence this *Hill's criterion* exhibits property of *hexagonal symmetry*. However, it is irreducible to the Huber–von Mises criterion in transverse isotropy plane.

The third formulation (5.98_3) inherits the von Mises format hence, if reduction to transverse isotropy is performed, it requires identification of six plastic modules in terms of three independent plastic limits k_x, k_z, and k_{zx}. Therefore, except from three common conditions (5.99) the following additional three must be formulated

$$\sigma_x=\sigma_y=k_x \qquad \sigma_x=\sigma_y=k_x\wedge\sigma_z=k_z \qquad \tau_{xy}=k_x/\sqrt{3} \tag{5.102}$$

In other words, two biaxial conditions in planes (xy) and (xz) hold and additional condition imposed on Π_{66} exhibits Huber–von Mises reducibility property

$$\Pi_{66} = \frac{3}{k_x^2} \tag{5.103}$$

The fourth formulation (5.98_4) is of specific nature, namely it is based on the von Mises–Hu–Marin criterion and requires the following three additional conditions

$$\sigma_x = \sigma_y = k_{(xy)} \qquad \sigma_x = \sigma_y = k_x \wedge \sigma_z = k_z \qquad \tau_{xy} = k_x/\sqrt{3} \tag{5.104}$$

Hence, (5.104_1) essentially differs from (5.102_1) since it involves the new independent constant $k_{(xy)}$ established from the *bulge test* in the transverse isotropy plane (see Jackson et al. [33]).

In conclusion it is clear that only the *classical Hill* condition is of *tetragonal symmetry* whereas all three other proposals discussed above have to be classified as *hexagonal symmetry types* even though the reasons of reduction of independent parameters (4 or 3) are of different nature.

5.8 Implicit Formulation of Pressure Insensitive Anisotropic Initial Yield Criteria—Barlat's and Khan's Concepts

In this section another approach (implicit formulation) is discussed based on a series of papers developed by Barlat, Planckett, Cazacu, and Khan to mention some names only. The implicit formulation involves the linear transformation of the Cauchy stress tensor $\boldsymbol{\sigma}$ to the transformed stress $\boldsymbol{\Sigma} = \mathbb{L} : \boldsymbol{\sigma}$ by the use of transformation tensor $\mathbb{L}$ responsible for orthotropy. Such *linear transformation concept* of the stress tensor was first introduced by Sobotka [55] and Boehler and Sawczuk [6]

$$\widehat{\sigma}_{ij} = A_{ijkl}\sigma_{kl} \tag{5.105}$$

where A_{ijkl} stands for a certain dimensionless tensor of anisotropy that satisfies general symmetry conditions $A_{ijkl} = A_{jikl} = A_{ijlk} = A_{klij}$ and the well-known isotropic yield conditions to hold for anisotropic materials as well if σ_{ij} are replaced by $\widehat{\sigma}_{ij}$. This approach is not directly based on the theory of *common invariants* in the sense of Sayir, Goldenblat, Kopnov, Spencer, Boehler, Betten etc. formalism (explicit formulation). According to this *implicit approach* an extension of isotropic initial yield/failure criteria is performed to account for the *tension/compression asymmetry* property and to *material anisotropy frame* (usually orthotropy) by applying the linear transformation to the stress tensor and inserting this transformed stress tensor into the originally isotropic yield/failure criteria.

In the paper by Cazacu et al. [8] authors consider both the isotropic yield criterion for description of *asymmetric yielding*

$$f(J_{2s}, J_{3s}) = (|s_1| - \widehat{k}s_1)^a + (|s_2| - \widehat{k}s_2)^a + (|s_3| - \widehat{k}s_3)^a = 2k^a$$
$$\widehat{k} = \frac{1 - h(\frac{k_t}{k_c})}{1 + h(\frac{k_t}{k_c})} \qquad h\left(\frac{k_t}{k_c}\right) = \left[\frac{2^a - 2(\frac{k_t}{k_c})^a}{(2\frac{k_t}{k_c})^a - 2}\right]^{1/a} \tag{5.106}$$

where $s_i,\ i = 1, \ldots, 3$ are the principal values of the stress deviator and f gives the size of the yield locus (isotropic hardening), as well as its extension to include orthotropy by the use of linear *transformation of the Cauchy stress* deviator $\boldsymbol{\Sigma} = \mathbb{C} : \mathbf{s}$ through

$$\mathbb{C} = \left[\begin{array}{ccc|ccc} C_{11} & C_{12} & C_{13} & & & \\ C_{12} & C_{22} & C_{23} & & & \\ C_{13} & C_{23} & C_{33} & & & \\ \hline & & & C_{44} & & \\ & & & & C_{55} & \\ & & & & & C_{66} \end{array}\right] \tag{5.107}$$

which lead to following anisotropic equation

$$(|\Sigma_1| - \widehat{k}\Sigma_1)^a + (|\Sigma_2| - \widehat{k}\Sigma_2)^a + (|\Sigma_3| - \widehat{k}\Sigma_3)^a = 2k^a \tag{5.108}$$

Authors proved convexity of the isotropic yield form (5.106) as well as pressure insensitivity of its orthotropic form (5.108) obtained through the linear transformation to the transformed stress frame. However, the question of *convexity of the orthotropic form* (5.108) remains open in the light of discussion performed for Hill's (Fig. 5.11) and Hosford's (Fig. 5.15) extensions in case of a highly orthotropic materials.

The proposed yield function appears to be suitable for description of the strong asymmetry and anisotropy observed in *textured Mg-Th* and *Mg-Li* binary *alloy sheets* and for *titanium 4Al-1/4O*$_2$, see Cazacu et al. [8]. The orthotropic yield criterion proposed by Cazacu et al. [8] was also investigated in a series of multiaxial loading experiments on *Ti-6Al-4V titanium alloy* by Khan et al. [34].

Extension of *Drucker's isotropic yield criterion* (5.15) to anisotropy by use of common invariants J_2^0 and J_3^0 is due to Cazacu and Barlat [7], and investigated by Yoshida et al. [64]

$$(J_2^0)^{3/2} - cJ_3^0 - k^3 = 0 \tag{5.109}$$

The constant c in the Eq. (5.109) accounts for the *tension/compression asymmetry* defined as

$$c = \frac{3\sqrt{3}(k_t^3 - k_c^3)}{2(k_t^3 + k_c^3)} \tag{5.110}$$

and belongs to two ranges

$$c \in \begin{cases} \left(0, \frac{3\sqrt{3}}{2}\right) & \text{for } k_t > k_c > 0 \\ \left(-\frac{3\sqrt{3}}{2}, 0\right) & \text{for } 0 < k_t < k_c \end{cases} \tag{5.111}$$

The second and third *common invariants of orthotropy* are defined as

$$\begin{aligned} J_2^0 &= \tfrac{1}{6}\left[a_1(\sigma_x - \sigma_y)^2 + a_2(\sigma_y - \sigma_z)^2 + a_3(\sigma_z - \sigma_x)^2\right] \\ &\quad + a_4\tau_{xy}^2 + a_5\tau_{xz}^2 + a_6\tau_{zy}^2 \\ J_3^0 &= \tfrac{1}{27}\left\{(b_1 + b_2)\sigma_x^3 + (b_3 + b_4)\sigma_y^3 + [2(b_1 + b_4) - b_2 - b_3]\sigma_z^3\right\} \\ &\quad + 2b_{11}\tau_{xy}\tau_{yz}\tau_{zx} + \tfrac{1}{9}\left\{2(b_1 + b_2)\sigma_x\sigma_y\sigma_z - (b_1\sigma_y + b_2\sigma_z)\sigma_x^2\right. \\ &\quad \left. - (b_3\sigma_z + b_2\sigma_x)\sigma_y^2 - [(b_1 - b_2 + b_4)\sigma_x + (b_1 + b_3 + b_4)\sigma_y]\sigma_z^2\right\} \\ &\quad - \tfrac{1}{3}\left\{\tau_{yz}^2[(b_6 + b_7)\sigma_x - b_6\sigma_y - b_7\sigma_z]\right. \\ &\quad - \tau_{zx}^2[2b_9\sigma_y - b_8\sigma_z - (2b_9 - b_8)\sigma_x] \\ &\quad \left. - \tau_{xy}^2[2b_{10}\sigma_z - b_5\sigma_y - (2b_{10} - b_5)\sigma_x]\right\} \end{aligned} \tag{5.112}$$

The discussed anisotropic criterion was successfully verified for textured *magnesium Mg-Th* and *Mg-Li alloy sheets*. Authors proved convexity of the *enhanced isotropic yield criterion* only for $c(k_t/k_c)$ belonging to the range $[-\frac{3\sqrt{3}}{2}, -\frac{3\sqrt{3}}{2}]$. In case of the anisotropic form of *Cazacu and Barlat's criterion* (5.109) the general proof of convexity for the wide class of highly tension/compression asymmetric and anisotropic materials may not be possible.

More complete representation of J_2^0 and J_3^0 common invariants as well as the extended model (5.109) verification for high-purity *α-titanium* is done by Nixon et al. [46].

Korkolis and Kyriakides [38] applied *anisotropic extension of Hosford's* isotropic *criterion* (5.10) in terms of principal stress deviator s_1, s_2 in case of plane stress state

$$|s_1 - s_2|^n + |2s_1 + s_2|^n + |s_1 + 2s_2|^n = 2k^n \tag{5.113}$$

Folowing Barlat et al. [3] they introduced anisotropy by use of a concept of two linear transformations $\boldsymbol{S}' = \mathbb{L}' : \boldsymbol{s}$ and $\boldsymbol{S}'' = \mathbb{L}'' : \boldsymbol{s}$ where $\mathbb{L}'$ and $\mathbb{L}''$ are transformation tensors introducing anisotropy

$$|S_1' - S_2'|^n + |2S_1'' + S_2''|^n + |S_1'' + 2S_2''|^n = 2k^n \tag{5.114}$$

Experimental validation of (5.114) is due to Korkolis and Kyriakides [38] applied to *Al-6260-T4* as well as due to Dunand et al. [13], Luo et al. [41] applied to *AA6260-T6 alloys* under classical tensile and butterfly shear tests.

Comparison of two different approaches: *explicit formulation* based on common invariants and *implicit formulation* composed as extension of isotropic criteria to anisotropy and tension/compression asymmetry leads to the following characteristic features.

The implicit formulation is very advantageous and fruitful in order to build numerical models able to capture experimental evidence for broad class of innovative metallic materials (mainly metal-based alloys) that simultaneously exhibit *tension/compression asymmetry*, *anisotropy*, and *hydrostatic pressure insensitivity*. Apart from these advantages some open questions may be highlighted. Among them there might be mentioned not obvious physical interpretation for the extended criteria based on known isotropic forms enhanced through strength differential sensitivity and orthotropic linear transformation of stress. The general proof of convexity is rather cumbersome and not attached in a complete and convinced form. Although the isotropic equations are undestandable, have physical interpretations, and satisfy convexity requirements the transposition of these equations to the transformed stress frame may lead to the loss of convexity.

By contrast use of the explicit approach based on well-established theory of common invariants is more rigorous and so leads to more clear physical interpretation (energy) and convexity of quadratic or poly-quadratic forms. However, this consistent approach leads to major difficulties when numerical implementation and experimental validation are considered. Additional difficulties arise when implementing the explicit approach to more general cases if the material orthotropy frame does not coincide with the principal stress frame. Such more general problem was discussed by Ganczarski and Lenczowski [15] in case of Hill's and orthotropic Hosford's criteria. In such a case it is necessary to transform tensor of structural orthotropy to the frame of principal stress resulting in a possible loss of convexity and even degeneration of an initially closed surface into twofold surface (nonclosed).

5.9 Brief Survey of Commonly Used Pressure Insensitive Isotropic and Anisotropic Initial Yield Criteria

In this section a brief survey of the selected commonly used pressure insensitive initial yield criteria is presented. The survey is focused on following two aspects:

- isotropic versus anisotropic formulation,
- direct versus indirect dependence on the stress invariants or the common invariants.

Special attention is paid for invariant representation of invoked limit criteria. Chosen *isotropic yield criteria* are collected in Table 5.3. All cited criteria depend on the second deviatoric invariant and additionally they may depend on the third deviatoric invariant. Criteria A1, A2, and A3 are written down in the format directly

Table 5.3 Survey of pressure insensitive isotropic yield criteria

A.	Author(s)	Limit criterion
A1	Raniecki and Mróz [48] Eq. (5.13)	$(J_{2s})^{3n/2} - c(J_{3s})^n = k^{3n}$
A2	Cazacu and Barlat [7] Eq. (5.14)	$(J_{2s})^{3/2} - cJ_{3s} = k^3$
A3	Drucker [12] Eq. (5.15)	$(J_{2s})^3 - c(J_{3s})^2 = k^6$
A4	Huber [31], von Mises [43] Eq. (5.4)	$3J_{2s} = k^2$
A5	Tresca [60], Guest [20] Eq. (5.8) Reuss [49] Eq. (5.9)	$\max(\lvert\sigma_1 - \sigma_2\rvert, \lvert\sigma_2 - \sigma_3\rvert, \lvert\sigma_3 - \sigma_1\rvert) = k$ $4J_{2s}^3 - 27J_{3s}^2 - 9k^2J_{2s}^2 + 6k^4J_{2s} = k^6$ $\lvert\sigma_1 - \sigma_2\rvert + \lvert\sigma_2 - \sigma_3\rvert + \lvert\sigma_3 - \sigma_1\rvert = 2k$
A6	Schmidt [53], Ishlinsky [32] and Hill [26] Eq. (5.11)	$\max(\lvert\sigma_1 - \sigma_h\rvert, \lvert\sigma_2 - \sigma_h\rvert, \lvert\sigma_3 - \sigma_h\rvert) = \frac{2}{3}k$
A7	Hershey [24], Davies [11] and Hosford [29] Eq. (5.10)	$\lvert\sigma_1 - \sigma_2\rvert^m + \lvert\sigma_2 - \sigma_3\rvert^m + \lvert\sigma_3 - \sigma_1\rvert^m = 2k^m$
A8	Cazacu et al. [8] Eq. (5.106)	$(\lvert s_1\rvert - \widehat{k}s_1)^a + (\lvert s_2\rvert - \widehat{k}s_2)^a + (\lvert s_3\rvert - \widehat{k}s_3)^a = 2k^a$

dependent on both invariants. The existence of the third invariant being argument of different power functions enables to capture various asymmetry of the initial yield curve in the deviatoric plane. The particular case of aforementioned *Drucker-like criteria* when $c = 0$ is the classical *Huber–von Mises criterion* A4 in which influence of the third stress invariant is ignored. Another classical *Tresca's criterion* A5 is written down in the three equivalent formats: the form suggested by Tresca [60] and experimentally validated by Guest [20], explicitly invariant Reuss' form and the Cazacu and Barlat [7] form being a particular case of *Hosford's criterion* A7 when $m = 1$. The Tresca criterion represents the regular hexagonal prism in the Haigh–Westergaard space inscribed into the Huber–von Mises circular cylinder (see Fig. 5.4). The *maximal deviatoric stress*-based *criterion* A6 formulated by Schmidt [53], Ishlinsky [32], and Hill [25] also represents the regular hexagonal prism in the Haigh–Westergaard space, however circumscribed onto the Huber–von Mises circular cylinder (see Fig. 5.4). The Tresca and Schmidt–Ishlinsky–Hill criteria are useful as the *inner and outer bounds* for all isotropic third stress invariant insensitive criteria, however the existence of corners on initial yield surfaces is physically questionable because the uniqueness of plastic strain increment is lost (see Fig. 5.5). The direct generalization of the Tresca criterion A5 by the use of power form that eliminates corners is due to Hershey [24], Davies [11], and Hosford [27]. The exponent m that ensures convexity has to be taken from the range $1 \leq m < \infty$, see Fig. 5.6. The Tresca-like criteria A5, A6, and A7 do not account for the tension/compression asymmetry effect. Another original criterion proposed by Cazacu et al. [8] A8 is relevant to the Drucker criterion A3 in such a sense that it is a homogeneous function of degree a in stresses, the cross section of which represents a "triangle" with rounded corners, see Cazacu et al. [8]. The *strength differential effect* is included and controlled by a parameter $\widehat{k}(\frac{k_t}{k_c})$. The existence of absolute values in the criterion proposed results from a reversible shear mechanisms such as slip, since yielding depends only on the magnitude but not direction of the shear stress, yield criterion

$f(\boldsymbol{s}) = f(-\boldsymbol{s})$. Other yield criteria accounting for different representations of the second and the third invariants due to Sayir that exhibit discrete 120°-symmetry are discussed by Altenbach et al. [1].

Chosen *anisotropic yield criteria* are collected in Table 5.4. In the item B1 two examples of implementation of *implicit anisotropic extension of* the isotropic *Drucker yield criterion* (dependent on the second and the third deviatoric stress invariants) referring to works by Cazacu and Barlat [7] and Nixon et al. [46] are presented. The original notation used by the authors is given in Table 5.3. By contrast to original notation in item B1 of Table 5.4 the criterion is rewritten in a frame of transformed stress $\boldsymbol{\Sigma} = \mathbb{L} : \boldsymbol{\sigma}$ instead of the Cauchy stress frame $\boldsymbol{\sigma}$. Due to this concept the second J_2^0 and the third J_3^0 *transformed invariants* are expressed in terms of only one fourth-rank *transformation tensor* $\mathbb{L}$ instead of the second-rank $\boldsymbol{s} : {}_{\mathrm{dev}}\overset{<4>}{\mathbb{I}} : \boldsymbol{s}$ and the third-rank common invariants $\boldsymbol{s} : {}_{\mathrm{dev}}\overset{<6>}{\mathbb{I}} : \boldsymbol{s} : \boldsymbol{s}$ necessary to be implemented when the *Goldenblat–Kopnov explicit formulation* would be used. The discussed implicit formulation shows essential reduction of the number of material constants that have to be identified in order to capture experimental data (see discussion in Sect. 5.2). Note that the transformation tensor $\mathbb{L}$ exhibits format of the Hill orthotropy matrix however it is dimensionless. When comparing items B2 and B3 corresponding to the deviatoric von Mises criterion (5.43) written in the form suggested by Szczepiński [58] and to the Hill criterion (5.53) [25, 26] different population of corresponding plastic matrices is applied. In case of Hill's format the terms which are sensitive to

Table 5.4 Survey of pressure insensitive anisotropic yield criteria

B.	Author(s)	Limit criterion
B1	Cazacu and Barlat [7] and Nixon et al. [46] Eq. (5.109)	$\left\{\frac{1}{2}\mathrm{tr}\,[(\mathbb{L}:\boldsymbol{\sigma})\cdot(\mathbb{L}:\boldsymbol{\sigma})]\right\}^{3/2}$ $-c\frac{1}{3}\mathrm{tr}\,[(\mathbb{L}:\boldsymbol{\sigma})\cdot(\mathbb{L}:\boldsymbol{\sigma})\cdot(\mathbb{L}:\boldsymbol{\sigma})] = k^3$
B2	Szczepiński [58] Eq. (5.43)	$\boldsymbol{s} : {}_{\mathrm{dev}}\overset{<4>}{\mathbb{I}} : \boldsymbol{s} = 1$
B3	Hill [25, 26] Eq. (5.53)	$\boldsymbol{s} : \overset{<4>}{\mathbb{I}}{}^{\mathrm{H}} : \boldsymbol{s} = 1$
B4	Voyiadjis and Thiagarajan [62] Eq. (5.94)	$\boldsymbol{\sigma} : {}_{\mathrm{tris}}\overset{<4>}{\mathbb{I}}{}^{\mathrm{VT}} : \boldsymbol{\sigma} = 1$
B5	Skrzypek and Ganczarski [18, 54] Eq. (5.90)	$\boldsymbol{\sigma} : {}_{\mathrm{tris}}\overset{<4>}{\mathbb{I}}{}^{\mathrm{HM}} : \boldsymbol{\sigma} = 1$
B6	Ganczarski and Skrzypek [18] Eq. (5.96)	$\boldsymbol{\sigma} : {}^{\mathrm{hybr}}_{\mathrm{tris}}\overset{<4>}{\mathbb{I}} : \boldsymbol{\sigma} = 1$
B7	Cazacu et al. [8] and Khan et al. [34] Eq. (5.108)	$(\lvert\Sigma_1\rvert - \widehat{k}\Sigma_1)^a + (\lvert\Sigma_2\rvert - \widehat{k}\Sigma_2)^a +$ $(\lvert\Sigma_3\rvert - \widehat{k}\Sigma_3)^a = 2k^a$
B8	Ganczarski and Lenczowski [15] Eq. (5.76)	$a_1\lvert\sigma_y - \sigma_z\rvert^m + a_2\lvert\sigma_z - \sigma_x\rvert^m +$ $a_3\lvert\sigma_x - \sigma_y\rvert^m + a_4\lvert\tau_{yz}\rvert^m +$ $a_5\lvert\tau_{zx}\rvert^m + a_6\lvert\tau_{xy}\rvert^m = 1$
B9	Korkolis and Kyriakides [38] Eq. (5.114)	$\lvert S_1' - S_2'\rvert^n + \lvert 2S_1'' + S_2''\rvert^n$ $+\lvert S_1'' + 2S_2''\rvert^n = 2k^n$

change of sign of shear stresses, for instance $\tau_{yz}(s_y - s_z), \ldots, \tau_{yz}\tau_{zx}, \ldots$ are omitted. It is equivalent to the reduction of a number of independent plastic modules from 15 to 6.

Items B4, B5, and B6 refer to the *transversely isotropic criteria* of initial yield/failure in unidirectionally reinforced Boron–Aluminum fibrous composites. Voyiadjis and Thiagarajan [62] used generally *transversely isotropic tetragonal symmetry* form of the yield criterion. However the experimental data used for calibration based on Dvorak et al. [14] and Nigam et al. [45] were limited to narrower case in which only plane stress state in the orthotropy plane was considered without distinction between the tetragonal and hexagonal symmetries. All three formulas B4, B5, and B6 describe cylindrical limit surfaces in stress space, the axis of which does not coincide with the hydrostatic axis.

The key difference between the Voyiadjis and Thiagarajan formulation B4 and the Skrzypek and Ganczarski approach B5 both related to the transversely isotropic materials, lies in the format of doubly underlined terms in Eqs. (5.94) and (5.90), respectively. Namely when Eq. (5.94) is used the fourth constant k_{xy} is independent and determined from experiment, whereas in Eq. (5.90) the fourth constant is dependent and equals to $k_{xy} = \frac{k_x}{\sqrt{3}}$. In other words, the *Voyiadjis and Thiagarajan criterion* (5.94) is irreducible to the Huber–von Mises criterion in the transverse isotropy plane, whereas the *Skrzypek and Ganczarski criterion* is reducible. This means that the Voyiadjis and Thiagarajan criterion possesses *tetragonal symmetry* whereas the Skrzypek and Ganczarski criterion exhibits *hexagonal symmetry*.

The full reducibility requirement in Eq. (5.90) may occur too restrictive when some composite materials are experimentally tested. In such a case the *hybrid formulation* (5.96) is proposed where $k_{(xy)}$ taken from the bulge test in the transverse isotropy plane is independent leading to 4-parameter *tetragonal format* (k_x, k_z, k_{zx}, and $k_{(xy)}$). The considered criteria B4, B5, and B6 are in fact secondary *pressure* sensitive, however this *sensitivity* property is *inquired* due to preserved cylindricity. The property of cylindricity is predominant and justifies the appearance of criteria B4, B5, and B6 in this section.

To describe both the *asymmetry* between *tension and compression* and the anisotropy observed in hexagonal closed packed metal sheets, Cazacu et al. [8] and Khan et al. [34] proposed extension of isotropic criterion (5.106) to the case of orthotropy represented by item B7. It consists in application of fourth-order *linear transformation operator* on the Cauchy stress tensor expressed by its principal values. The proposed anisotropic criterion was successively applied to the description of the anisotropy and asymmetry of the yield loci of textured *polycrystalline magnesium* and binary *Mg–Th, Mg–Li alloys* and *α titanium*.

Orthotropic generalization of the Hosford criterion (item A7) in which principal axes of material orthotropy do not coincide with principal stress axes was proposed by Ganczarski and Lenczowski [15] in the form of item B8. Next the convexity check of the yield condition was performed in case of the brass sheet of Russian commercial symbol Ł22, that is material of strong orthotropy slightly different from transverse isotropy.

The last criterion (item B9) is another anisotropic *generalization of Hosford's* isotropic *criterion* (item A7) done by Korkolis and Kyriakides [38] and addressed to Al-6260-T4 tubes inflated under combined internal pressure and axial load.

References

1. Altenbach, H., Bolchoun, A., Kolupaev, V.A.: Phenomenological yield and failure criteria. In: Altenbach, H., Öchsner, A. (eds.) Plasticity of Pressure-Sensitive Materials. Springer, Heidelberg (2014)
2. Barlat, F., Lian, J.: Plastic behavior and stretchability of sheet metals. Int. J. Plast. **5**(1), 51 (1989)
3. Barlat, F., Brem, J.C., Yoon, J.W., Chung, K., Dick, R.E., Lege, D.J., Pourboghrat, F., Choi, S.-H., Chu, E.: Plane stress function for aluminium alloy sheets—part I: theory. Int. J. Plast. **19**, 1297–1319 (2003)
4. Berryman, J.G.: Bounds and self-consistent estimates for elastic constants of random polycrystals with hexagonal, trigonal, and tetragonal symmetries. J. Mech. Phys. Solids **53**, 2141–2173 (2005)
5. Betten, J.: Applications of tensor functions to the formulation of yield criteria for anisotropic materials. Int. J. Plast. **4**, 29–46 (1988)
6. Boehler, J.P., Sawczuk, A.: Equilibre limite des sols anisotropes. J. Mécanique **9**, 5–33 (1970)
7. Cazacu, O., Barlat, F.: A criterion for description of anisotropy and yield differential effects in pressure-insensitive materials. Int. J. Plast. **20**, 2027–2045 (2004)
8. Cazacu, O., Planckett, B., Barlat, F.: Orthotropic yield criterion for hexagonal close packed metals. Int. J. Plast. **22**, 1171–1194 (2006)
9. Chen, W.F., Han, D.J.: Plasticity for Structural Engineers. Springer, Berlin (1995)
10. Chu, E.: Generalization of Hill's 1979 anisotropic yield criteria. In: Proceedings of the NUMISHEETS'89, pp. 199–208 (1989)
11. Davies, E.A.: The Bailey flow rule and associated yield surface. Trans. ASME **E28**(2), 310 (1961)
12. Drucker, D.C.: Relation of experiments to mathematical theories of plasticity. J. Appl. Mech. **16**, 349–357 (1949)
13. Dunand, M., Maertens, A.P., Luo, M., Mohr, D.: Experiments and modeling of anisotropic aluminum extrusions under multi-axial loading—part I: plasticity. Int. J. Plast. **36**, 34–49 (2012)
14. Dvorak, G.J., Bahei-El-Din, Y.A., Macheret, Y., Liu, C.H.: An experimental study of elastic-plastic behavior of a fibrous boron-aluminum composite. Int. J. Mech. Phys. Solids **36**, 655–687 (1988)
15. Ganczarski, A., Lenczowski, J.: On the convexity of the Goldenblatt-Kopnov yield condition. Arch. Mech. **49**(3), 461–475 (1997)
16. Ganczarski, A., Skrzypek, J.: Modeling of limit surfaces for transversely isotropic composite SCS-6/Ti-15-3. Acta Mechanica et Automatica **5**(3), 24–30 (2011) (in Polish)
17. Ganczarski, A., Skrzypek, J.: Mechanics of Novel Materials (in Polish). Wydawnictwo Politechniki Krakowskiej, Kraków (2013)
18. Ganczarski, A., Skrzypek, J.: Constraints on the applicability range of Hill's criterion: strong orthotropy or transverse isotropy. Acta Mech. **225**, 2568–2582 (2014)
19. Goldenblat, I.I., Kopnov, V.A.: Obobshchennaya teoriya plasticheskogo techeniya anizotropnyh sred, pp. 307–319. Sbornik Stroitelnaya Mehanika, Stroĭizdat, Moskva (1966)
20. Guest, J.J.: On the strength of ductile materials under combined stress. Philos. Mag. **50**, 69–132 (1900)
21. Haigh, B.F.: The strain-energy function and the elastic limit. Eng. Lond. **109**, 158–160 (1920)
22. Hencky, H.: Zur Theorie plastischer Deformationen und der hierdurch im Material hervorgerufen Nach-Spannungen. ZAMM **4**, 323–334 (1924)

23. Herakovich, C.T., Aboudi, J.: Thermal effects in composites. In: Hetnarski, R.B. (ed.) Thermal Stresses V, pp. 1–142. Lastran Corporation Publishing Division, Rochester (1999)
24. Hershey, A.V.: The plasticity of an isotropic aggregate of anisotropic face—centred cubic crystals. J. Appl. Mech. **21**(3), 241–249 (1954)
25. Hill, R.: A theory of the yielding and plastic flow of anisotropic metals. Proc. R. Soc. Lond. **A193**, 281–297 (1948)
26. Hill, R.: The Mathematical Theory of Plasticity. Oxford University Press, Oxford (1950)
27. Hosford, W.F., Backhofen, W.A.: Strength and plasticity of textured metals. In: Backhofen, W.A., Burke, J., Coffin, L., Reed, N., Weisse, V. (eds.) Fundamentals of Deformation Processing, pp. 259–298. Syracuse University Press, Syracuse (1964)
28. Hosford, W.F.: Texture Strengthening. Met. Eng. Q. **6**, 13–19 (1966)
29. Hosford, W.F.: A generalized isotropic yield criterion. Trans. ASME **E39**(2), 607–609 (1972)
30. Hu, Z.W., Marin, J.: Anisotropic loading functions for combined stresses in the plastic range. J. Appl. Mech. **22**, 1 (1956)
31. Huber, M.T.: Właściwa praca odkształcenia jako miara wytężenia materiału, Czas. Techn. 22, 34–40, 49–50, 61–62, 80–81, Lwów, Pisma, Vol. II, PWN, Warszawa 1956, 3–20 (1904)
32. Ishlinskiĭ, A.Yu.: Gipoteza prochnosti formoizmeneniya, p. 46. University, Mekh, Uchebnye Zapiski Mosk (1940)
33. Jackson, L.R., Smith, K.F., Lankford, W.T.: Plastic flow in anisotropic sheet steel. Am. Inst. Min. Metall. Eng. **2440**, 1–15 (1948)
34. Khan, A.S., Kazmi, R., Farrokh, B.: Multiaxial and non-proportional loading responses, anisotropy and modeling of Ti-6Al-4V titanium alloy over wide ranges of strain rates and temperatures. Int. J. Plast. **23**, 931–950 (2007)
35. Khan, A.S., Liu, H.: Strain rate and temperature dependent fracture criteria for isotropic and anisotropic metals. Int. J. Plast. **37**, 1–15 (2012)
36. Khan, A.S., Yu, S., Liu, H.: Deformation enhanced anisotropic responses of Ti-6Al-4V alloy, part II: a stress rate and temperature dependent anisotropic yield criterion. Int. J. Plast. **38**, 14–26 (2012)
37. Kowalsky, U.K., Ahrens, H., Dinkler, D.: Distorted yield surfaces—modeling by higher order anisotropic hardening tensors. Comput. Math. Sci. **16**, 81–88 (1999)
38. Korkolis, Y.P., Kyriakides, S.: Inflation and burst of aluminum tubes. part II: an advanced yield function including deformation-induced anisotropy. Int. J. Plast. **24**, 1625–1637 (2008)
39. Lankford, W.T., Low, J.R., Gensamer, M.: The plastic flow of aluminium alloy sheet under combined loads. Trans. AIME 171, 574; TP 2238, Met. Techn., Aug. 1947
40. Lode, W.: Der Einfluss der mittleren Hauptspannung auf der Fliessen der Metalle, Forschungs arbeiten auf dem Gebiete des Ingenieurusesen, 303 (1928)
41. Luo, M., Dunand, M., Mohr, D.: Experiments and modeling of anisotropic aluminum extrusions under multi-axial loading—part II: ductile fracture. Int. J. Plast. **32–33**, 36–58 (2012)
42. Malinin, N.N., Rżysko, J.: Mechanika Materiałów. PWN, Warszawa (1981)
43. von Mises, R.: Mechanik der festen Körper im plastisch deformablen Zustand, Götingen Nachrichten. Math. Phys. **4**(1), 582–592 (1913)
44. von Mises, R.: Mechanik der plastischen Formänderung von Kristallen. ZAMM **8**(13), 161–185 (1928)
45. Nigam, H., Dvorak, G.J., Bahei-El-Din, Y.A.: An experimental investigation of elastic-plastic behavior of a fibrous Boron-Aluminum composite. I. Matrix-dominated mode. Int. J. Plast. **10**, 23–48 (1933)
46. Nixon, M.E., Cazacu, O., Lebensohn, R.A.: Anisotropic response of high-purity α-titanium: experimental characterization and constitutive modeling. Int. J. Plast. **26**, 516–532 (2010)
47. Ottosen, N.S., Ristinmaa, M.: The Mechanics of Constitutive Modeling. Elsevier, Amsterdam (2005)
48. Raniecki, B., Mróz, Z.: Yield or martensitic phase transformation conditions and dissipative functions for isotropic, pressure-insensitive alloys exhibiting SD effect. Acta Mech. **195**, 81–102 (2008)

49. Reuss, A.: Vereifachte Berechnung der plastischen Formänderungen in der Plastizitätstheorie. ZAMM **10**(3), 266–274 (1933)
50. Rogers, T.G.: Yield criteria, flow rules, and hardening in anisotropic plasticity. In: Boehler, J.P. (ed.) Yielding, Damage and Failure of Anisotropic Solids, pp. 53–79. Mechanical Engineering Publications, London (1990)
51. Rymarz, Cz.: Continuum Mechanics (in Polish). PWN, Warszawa (1993)
52. Sayir, M.: Zur Fließbedingung der Plastizitätstheorie. Ingenierurarchiv **39**, 414–432 (1970)
53. Schmidt, R.: Über den Zusammenhang von Spannungen und Formänderungen im Vestigungsgebiet. Ing.-Arch. **3**, 215–235 (1932)
54. Skrzypek, J., Ganczarski, A.: Anisotropic initial yield and failure criteria including temperature effect. In: Hetnarski, R.B. (ed.) Encyclopedia of Thermal Stresses, vol. A, pp. 146–159. Springer, Dordrecht (2014)
55. Sobotka, Z.: Theorie des plastischen Fliessens von anisotropen Körpern. Z. Angew. Math. Mechanik **49**, 25–32 (1969)
56. Spencer, A.J.M.: Theory of invariants. In: Eringen, C. (ed.) Continuum Physics, pp. 239–353. Academic Press, New York (1971)
57. Sun, C.T., Vaidya, R.S.: Prediction of composite properties from a representative volume element. Compos. Sci. Technol. **56**, 171–179 (1996)
58. Szczepiński, W.: On deformation-induced plastic anisotropy of sheet metals. Arch. Mech. **45**(1), 3–38 (1993)
59. Tamma, K.K., Avila, A.F.: An integrated micro/macro modelling and computational methodology for high temperature composites. In: Hetnarski, R.B. (ed.) Thermal Stresses V, pp. 143–256. Lastran Corporation Publishing Division, Rochester (1999)
60. Tresca, H.: Mémoire sur l'écoulement des corps solids soumis á de fortes pressions. Comptes Rendus de l'Académie des Sciences **59**, 754–758 (1864)
61. Tsai, S.T., Wu, E.M.: A general theory of strength for anisotropic materials. Int. J. Numer. Methods Eng. **38**, 2083–2088 (1971)
62. Voyiadjis, G.Z., Thiagarajan, G.: An anisotropic yield surface model for directionally reinforced metal-matrix composites. Int. J. Plast. **11**, 867–894 (1995)
63. Westergaard, H.M.: On the resistance of ductile materials to combined stresses in two and three directions perpendicular to one another. J. Frankl. Inst. **189**, 627–640 (1920)
64. Yoshida, F., Hamasaki, H.M., Uemori, T.: A user-friendly 3D yield function to describe anisotropy of steel sheets. Int. J. Plast. **45**, 119–139 (2013)
65. Życzkowski, M.: Anisotropic yield conditions. In: Lemaitre, J. (ed.) Handbook of Materials Behavior Models, pp. 155–165. Academic Press, San Diego (2001)

Chapter 6
Termination of Elastic Range of Pressure Sensitive Materials—Isotropic and Anisotropic Initial Yield/Failure Criteria

Jacek J. Skrzypek and Artur W. Ganczarski

Abstract Yield/failure initiation criteria discussed in this chapter account for the three following effects: the hydrostatic pressure dependence, tension/compression asymmetry, and isotropic or anisotropic material response. For isotropic materials, the criteria accounting for pressure/compression asymmetry (strength differential effect) must include all three stress invariants (Iyer, Gao, Yoon, Coulomb–Mohr criteria). In a narrower case, when only pressure sensitivity is accounted for, rotationally symmetric surfaces independent of the third invariant are considered and broadly discussed (Burzyński, Drucker–Prager criteria). For anisotropic materials, the explicit formulation based on either all three common invariants (Goldenblat–Kopnov, Kowalsky) or the first and second common invariants (von Mises–Tsai–Wu) is addressed, especially in case of transverse isotropy when the difference between tetragonal and hexagonal symmetries is highlighted. A mixed way to formulate pressure sensitive tension/compression asymmetric initial failure criteria capable of describing fully distorted limit surfaces, which are based on all stress invariants and also the second common invariant (Khan, Liu) alone, are received and particularly addressed to orthotropic materials where fourth-order linear transformation tensors are used to achieve extension of the isotropic criterion.

Keywords Pressure sensitive failure criteria · Strength differential effect · Distorted surfaces · Explicit, implicit or mixed formulations · von Mises–Tsai–Wu criteria · Transverse isotropy—tetragonal or hexagonal

J.J. Skrzypek (✉) · A.W. Ganczarski
Solid Mechanics Division, Institute of Applied Mechanics,
Cracow University of Technology, al. Jana Pawła II 37, 31-864 Kraków, Poland
e-mail: Jacek.Skrzypek@pk.edu.pl

A.W. Ganczarski
e-mail: Artur.Ganczarski@pk.edu.pl

J.J. Skrzypek and A.W. Ganczarski (eds.), *Mechanics of Anisotropic Materials*, Engineering Materials, DOI 10.1007/978-3-319-17160-9_6

6.1 Nature of Yield/Failure Initiation in Pressure Sensitive Materials

Initial yield conditions discussed in Chap. 5 are applicable for ductile materials in which it is justified to ignore both the tension/compression asymmetry and the hydrostatic pressure sensitivity. In a majority of metallic polycrystalline materials, termination of the elastic range corresponds to initiation of plastic microslips. In case of *brittle materials*, such as the majority of *ceramic materials*, *rocks*, *concrete*, *ceramic matrix composites* CMC, *columnar ice*, etc., *material failure* is initiated not by the plastic slips but *microcracks* (damage), which by way of evolution and aggregation processes may lead to initiation and formation of macrocracks (failure).

The general *anisotropic* form of *failure initiation criterion* in brittle materials may be assumed in the analogous form as that used as an initial yield criterion proposed by Goldenblat and Kopnov [16] and others in Eq. (5.16), namely

$$f^{\mathrm{d}} = f^{\mathrm{d}}\left(\Pi_i, \Pi_{ij}\sigma_{ij}, \Pi_{ijkl}\sigma_{ij}\sigma_{kl}, \Pi_{ijklmn}\sigma_{ij}\sigma_{kl}\sigma_{mn}, \ldots\right) = 0 \qquad (6.1)$$

Symbols $\Pi_i, \Pi_{ij}, \Pi_{ijkl}, \Pi_{ijklmn}, \ldots$ stand for the material constants Π_i and the *structural tensors of material anisotropy*. It is commonly assumed that damage is not associated with plastic slip evolution. However, recently developed theories introduce a distinction between *elastic damage* and *plastic damage evolution*, see Abu Al-Rub and Voyiadjis [1], Egner [10].

Assuming by analogy to Eq. (5.17) the tensorially polynomial format of the criterion (6.1), the general *anisotropic Goldenblat–Kopnov criterion of failure initiation* is furnished as

$$\left(\Pi_{ij}\sigma_{ij}\right)^{\alpha} + \left(\Pi_{ijkl}\sigma_{ij}\sigma_{kl}\right)^{\beta} + \left(\Pi_{ijklmn}\sigma_{ij}\sigma_{kl}\sigma_{mn}\right)^{\gamma} + \cdots - 1 = 0 \qquad (6.2)$$

Although the general format of failure initiation criterion (6.2) is analogous to the previously introduced criterion of yield initiation (5.17), further reduction of this equation to be applicable for brittle materials has to be performed applying different assumptions from that used for ductile materials. Plastic yield initiation in metallic polycrystalline materials is traditionally characterized by the following features (compare Chap. 5):

- Yield initiation is usually hydrostatic pressure insensitive (independent of the first common invariant).
- The criterion of yield initiation exhibits symmetry with respect to tension and compression (no strength differential effect).
- Initial yield surface has to be convex (Drucker's postulate).

On the other hand, failure initiation in some metallic or nonmetallic materials is commonly described including the more complex behavior:

- Failure initiation is hydrostatic pressure sensitive.

- The criterion of failure initiation exhibits asymmetry with respect to tension and compression (also called strength differential effect), since for the majority of brittle materials strength resistance for compression is frequently much higher than for tension (essential strength differential effect).
- Initial failure surface has to be convex (positive definiteness of tangent stiffness matrix in Sylvester's sense).

The above-mentioned "sharp" distinction between ductile and brittle materials is often not exactly justified, especially when more advanced structural material behaviors are considered. Numerous experimental findings referred to novel materials suggest another classification with respect to distinct mechanical responses, among which three are of particular importance:

- *Hydrostatic pressure sensitivity.*
- *Tension versus compression asymmetry.*
- *Material anisotropy.*

The above features are briefly discussed in Sect. 4.2. In light of experimental observation of metal alloys, distinction between plastic mechanism and brittle mechanism at failure initiation may not be true and justified.

Indeed, a majority of pressure insensitive metallic materials show rather plastic yield initiation mechanism of either isotropic nature (NiTi shape memory alloys, Raniecki and Mróz [31]; Mg, Mg–Th or Mg–Li alloys, Cazacu and Barlat [5]; 4Al–$\frac{1}{4}O_2$ Titanium alloy, Cazacu et al. [6]) or anisotropic nature (Al 6061–T6511 alloy, Cazacu and Barlat [5]; Ti–6Al-4V Titanium alloy, Khan et al. [19]; Al 6260–T4 alloy, Korkolis and Kyriakides [22]).

However, some experimental evidence for *pressure sensitive metallic alloys* show *combined plastic/fracture mechanism* with pronounced tension/compression asymmetry effect of either isotropic nature *Nickel-base Inconel 718*, Iyer [17], Pęcherski et al. [29]; *5083 Aluminum alloy*, Gao et al. [15]; or anisotropic response (*Ti–6Al–4V Titanium alloy*, Khan et al. [21], Khan and Liu [20]; *AA2008–T4 Aluminum alloy*; *AZ31 Magnesium alloy*, Yoon et al. [39]).

Nevertheless, it should be pointed out that in all described cases of physically different coupled mechanisms, the *limit surface* of *yield and/or failure initiation* must definitely satisfy the *convexity* requirement in the sense of either *Drucker's* or *Sylvester's* material *stability postulates*.

Criteria of yield initiation in polycrystalline materials are discussed in Sect. 5.1. These materials commonly called ductile (majority of metals, steels, alloys, etc.) do not exhibit strong sensitivity of limit surface to hydrostatic pressure. It was shown that in such a case limit surfaces are independent of the first stress invariant ($J_{1\sigma}$) or equivalently of the first Haigh–Westergaard coordinate (ξ). Initial yield surfaces are therefore represented by cylindrical surfaces whose axis is the hydrostatic axis. Hence, it is not possible to distinguish limit states of failure initiation at tension k_t and compression k_c, because $k_t = k_c$ must hold.

There exists a broad class of *isotropic materials* for which the above limitation does not obey (concrete, rocks, soils, cast iron, particle reinforced composites).

Therefore, it is necessary to include the *first stress invariant* $J_{1\sigma}$ or the *first Haigh–Westergaard coordinate* ξ in the criterion of failure initiation. Limit surfaces represented in the space of principal stress lose the cylindrical form, which allows to distinguish failure strength at tension and compression $k_{\rm t} \neq k_{\rm c}$ commonly called *strength differential effect*. This is the direct consequence of the first stress invariant existence that changes the size of the limit surface cross-section along the hydrostatic axis.

In this section we confine our considerations to the isotropic limit surfaces associated with microslip systems in plastic metals. In the general case of yield initiation in isotropic materials the equation of limit surface in terms of three *basic stress invariants* $J_{1\sigma}$, $J_{2\sigma}$, $J_{3\sigma}$ takes the form

$$f\left(J_{1\sigma}, J_{2\sigma}, J_{3\sigma}; k_i\right) = 0 \tag{6.3}$$

where definitions (1.12) are applied. This equation describes the general class of isotropic hydrostatic pressure sensitive materials, see items C1–C5 in Table 6.2.

In the narrower case of failure of isotropic materials the equation of limit surface is commonly written in terms of the following three invariants: the *first stress tensor invariant* $J_{1\sigma}$, the *second*, and the *third stress deviator invariants* J_{2s} and J_{3s}

$$f\left(J_{1\sigma}, J_{2s}, J_{3s}; k_i\right) = 0 \tag{6.4}$$

or alternatively in terms of the three *Haigh–Westergaard coordinates* (5.2)

$$f\left(\xi, \rho, \theta; k_i\right) = 0 \tag{6.5}$$

The advantage of Eq. (6.4) is that it separates the hydrostatic stress $J_{1\sigma}$ from the influence of deviatoric stresses expressed by J_{2s} and J_{3s}. On the other hand Eq. (6.5) has a geometrical interpretation showing dependence of the limit surface cross-section on the position at the hydrostatic axis ξ. Such surface is no longer cylindrical, hence it naturally exhibits *tension/compression asymmetry*. All limit surfaces belonging to the class considered exhibit certain sectorial symmetry with respect to hydrostatic axis (see discussion in Chap. 5), however, only in a particular case of independence of the third invariant J_{3s} (or θ) it is fully rotational symmetry.

6.2 Isotropic Initial Yield/Failure Criteria Accounting for Pressure Sensitivity and Strength Differential Effect

As discussed in Sect. 6.1, there exists a wide class of materials, metallic and nonmetallic, in which both a hydrostatic pressure dependence and the strength differential effect are pronounced and have to be incorporated into the threshold criteria describing *onset of yield* or *fracture* as well as *phase change*. In case of isotropic materials such criteria have to be expressed in terms of all three invariants $J_{1\sigma}$, $J_{2\sigma}$, and $J_{3\sigma}$ (6.3) or $J_{1\sigma}$, J_{2s}, and J_{3s} (6.4).

The Ottosen and Ristinmaa [28] mixed format separates the influence of hydrostatic pressure $J_{1\sigma}$ from the deviatoric stress represented by J_{2s} and $\cos 3\theta$

$$f\,(J_{1\sigma}, J_{2s}, \cos 3\theta; k_i) = 0 \tag{6.6}$$

where

$$\cos(3\theta) = \frac{3\sqrt{3}}{2}\frac{J_{3s}}{(J_{2s})^{3/2}} \quad J_{2s} = \frac{1}{2}s_{ij}s_{ji} \quad J_{3s} = \frac{1}{3}s_{ij}s_{jk}s_{ki} \tag{6.7}$$

The invariants $J_{1\sigma}$, J_{2s}, and $\cos 3\theta$ have clear interpretation: $J_{1\sigma}$ tells about the *hydrostatic pressure sensitivity* (noncylindrical limit surface), J_{2s} represents the distance of the point at the deviatoric limit curve (the magnitude of the deviatoric stress), and $\cos 3\theta$ informs about influence of the direction of deviatoric stress (nonrotationally symmetric *limit surfaces asymmetry* in so-called *strength differential materials*), cf. Ottosen and Ristinmaa [28]. Applying consistently system of the first stress invariant and the second and the third stress deviator invariants, Eq. (6.6) can also be written as (6.4).

The general form of the power threshold function applicable to advanced metals having not only three threshold parameters a, b, and c but also including two independent powers p and r in the yield/failure criterion for isotropic materials can be furnished as follows:

$$\left(aJ_{1\sigma}^{2p} + bJ_{2s}^{p} + cJ_{3s}^{2p/3}\right)^{1/r} - 1 = 0 \tag{6.8}$$

It will be shown below that a majority of criteria met in the literature to predict onset of yield, failure, or even phase transformation (Iyer [17], Pęcherski et al. [29], Gao et al. [15], Iyer and Lissenden [18], Brünig et al. [3], Raniecki and Mróz [31]) can be captured as specific cases of this general format (6.8).

Assuming $r = p$ in Eq. (6.8) the Iyer [17] *yield/failure onset criterion* in isotropic materials is recovered as

$$\left(aJ_{1\sigma}^{2p} + bJ_{2s}^{p} + cJ_{3s}^{2p/3}\right)^{1/p} - 1 = 0 \tag{6.9}$$

Equation (6.9) is successfully used to describe the Nickel-base alloy Inconel 718 at elevated temperature 650 °C, e.g., by Iyer and Lissenden [18]. If $r = p = 1$ the specific format used by Iyer and Lissenden [18], Pęcherski et al. [29] is obtained as

$$aJ_{1\sigma}^{2} + bJ_{2s} + cJ_{3s}^{2/3} - 1 = 0 \tag{6.10}$$

Equation of this type that represents asymmetric either paraboloidal or ellipsoidal surface was calibrated for *Inconel 718* by Pęcherski et al. [29]. Note that the Huber–von Mises $f(J_{2s})$, the Drucker–Prager $f(J_{1\sigma}, J_{2s})$, and the Drucker $f(J_{2s}, J_{3s})$

yield functions can be obtained as special cases. Assuming another combination of powers $p/r = 1/2$ and $r = 6$ we arrive at the Gao et al. [15] yield function

$$c_1 \left(a_1 J_{1\sigma}^6 + 27 J_{2s}^3 + b_1 J_{3s}^2 \right)^{1/6} - k = 0 \tag{6.11}$$

calibrated and verified for the *5083 Aluminum alloy*. The constant c_1 can be found by introducing the uniaxial condition onto Eq. (6.11)

$$c_1 = \left(a_1 + \frac{4}{729} b_1 + 1 \right)^{-1/6} \tag{6.12}$$

Another special case of Eq. (6.8) when $p/r = 1/2$ and $r = 1$ was considered by Brünig et al. [3]

$$a J_{1\sigma} + \sqrt{J_{2s}} + b \sqrt[3]{J_{3s}} = c \tag{6.13}$$

who confirmed its applicability for failure initiation in *Aluminum alloys* and *high-strength steels*. Equation of this type represents asymmetric cone and is capable of capturing yield onset in the high-strength *4310* and *4330 steels*.

Direct extension of the Cazacu and Barlat [5] asymmetric yield function (5.14) to *pressure sensitive materials* is due to Yoon et al. [39]

$$a[b J_{1\sigma} + (J_{2s}^{3/2} - c J_{3s})^{1/3}] - 1 = 0 \tag{6.14}$$

From among three material constants a, b, and c in (6.14) only two are independent since the third a has to capture uniaxial tensile test, namely

$$a = \frac{1}{b + \left(\frac{1}{3\sqrt{3}} - \frac{c}{27} \right)^{1/3}} \tag{6.15}$$

Additionally, the convexity of the proposed yield function is satisfied if $c \in [\frac{-3\sqrt{3}}{4}, \frac{3\sqrt{3}}{4}]$. This isotropic equation is also extended to material anisotropy in order to properly describe various metals of *AA 2008-T4*, high-purity *α-Titanium*, and *AZ31 Magnesium alloy*.

The aforementioned criteria (6.8)–(6.14) that include the effect of all three invariants are essentially applicable to advanced metals and metal alloys. However, there exists a broad class of conventional *nonmetallic materials* (concrete, soils, ceramics, etc.) that account for both *hydrostatic pressure sensitivity* and *strength differential effect*. The useful criterion for description of *soils*, later generalized for description of *concrete* by Mohr [27] in 1900, was originally proposed by Coulomb [8] in 1776. As shown later, this criterion can be considered as extension of both the Tresca condition and the Drucker–Prager criterion. The original form proposed by Coulomb is well known in the literature dealing with *soil mechanics*, namely

$$|\tau| = c - \sigma \tan\phi \tag{6.16}$$

Two material constants c and ϕ stand for *cohesion* and the *angle of internal friction*, respectively. According to Mohr's circles concept, Eq. (6.16) means that failure of material occurs if the radius of the largest principal circle is tangent to an envelope formed by two straight lines inclined by angle 2ϕ as shown in Fig. 6.1a. In the special case of *frictionless material* $\phi = 0$ criterion (6.16) reduces to the Tresca criterion when the cohesion is identical to the yield stress under pure shear $c = k_s$.

Equation (6.16) defining traditional format of *Coulomb–Mohr criterion* can be rewritten in terms of principal stresses. Assuming order of the principal stresses $\sigma_1 \geq \sigma_2 \geq \sigma_3$ and considering arbitrary stresses τ and σ as shown in Fig. 6.1b

$$\begin{aligned}\tau &= r\cos\phi = \frac{1}{2}(\sigma_1 - \sigma_3)\cos\phi \\ \sigma &= \frac{1}{2}(\sigma_1 + \sigma_3) + r\sin\phi = \frac{1}{2}(\sigma_1 + \sigma_3) + \frac{1}{2}(\sigma_1 - \sigma_3)\sin\phi\end{aligned} \tag{6.17}$$

we arrive at

$$\frac{1}{2}(\sigma_1 - \sigma_3)\cos\phi = c - \left[\frac{1}{2}(\sigma_1 + \sigma_3) + \frac{\sigma_1 - \sigma_3}{2}\sin\phi\right]\tan\phi \tag{6.18}$$

The above equation can also be written in the abbreviated form

$$\sigma_1 \frac{1 + \sin\phi}{2c\cos\phi} - \sigma_3 \frac{1 - \sin\phi}{2c\cos\phi} = 1 \tag{6.19}$$

When the different *failure strengths* for *compression* k_c and *tension* k_t, $k_c > k_t$ are defined as follows:

$$k_c = \frac{2c\cos\phi}{1 - \sin\phi} \qquad k_t = \frac{2c\cos\phi}{1 + \sin\phi} \tag{6.20}$$

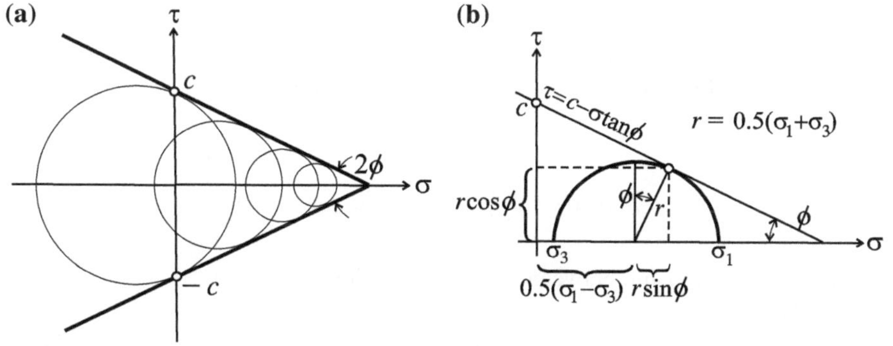

Fig. 6.1 Coulomb–Mohr failure criterion in-plane $\sigma - \tau$: **a** graphical representation as failure envelope, **b** Mohr's transformation of arbitrary state of stress to the principal state of stress

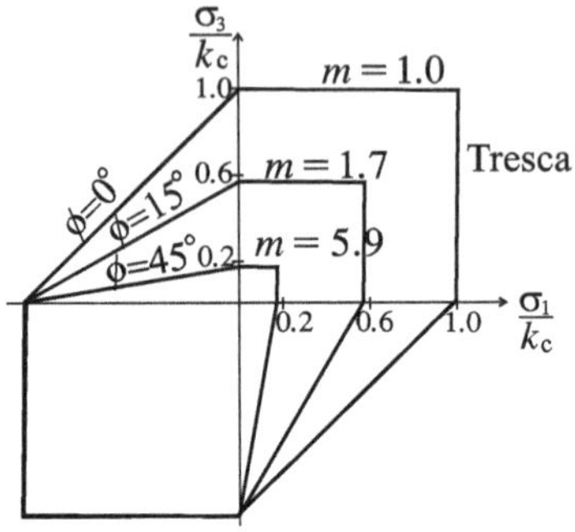

Fig. 6.2 Failure loci of Coulomb–Mohr criterion in plane $\sigma_2 = 0$ for several values of ratio $m = k_c/k_t$

the *extension of Tresca's* format is recovered from Eq. (6.19)

$$m\sigma_1 - \sigma_3 = k_c \tag{6.21}$$

where m stands for the compressive to tensile strengths ratio $m = k_c/k_t$. The Coulomb–Mohr equation in the above format (6.21) in the *plane stress state* assuming intermediate stress $\sigma_2 = 0$, with the convention $\sigma_1 > \sigma_2 > \sigma_3$ used, can graphically be represented by deformed irregular hexagons following the magnitude ratio $m \geq 1$ in Eq. (6.21), as shown in Fig. 6.2. In case of $k_c = k_t$ ($m = 1$) the classical *Tresca condition* is recovered, however, when Coulomb–Mohr equation is considered for $k_c > k_t$ ($m > 1$) a reduction of the admissible field in quarters I, II, and IV is observed whereas the quarter III remains unchanged.

When using the *mixed invariant system* $(J_{1\sigma}, J_{2s}, \theta)$ the *implicit invariant format* of the *Coulomb–Mohr criterion* (6.21) is reached as

$$\begin{aligned} f(J_{1\sigma}, J_{2s}, \theta; c, \phi) &= \frac{1}{3}J_{1\sigma}\sin\phi + \sqrt{J_{2s}}\sin\left(\theta + \frac{\pi}{3}\right) \\ &+ \sqrt{\frac{J_{2s}}{3}}\cos\left(\theta + \frac{\pi}{3}\right)\sin\phi - c\cos\phi = 0 \end{aligned} \tag{6.22}$$

where $\cos(3\theta) = \frac{3\sqrt{3}}{2}\frac{J_{3s}}{(J_{2s})^{3/2}}$, and $0 \leq \theta \leq \pi/3$.

Alternatively, in terms of the *Haigh–Westergaard coordinates* ξ, ρ, θ the following *explicit format* of *Coulomb–Mohr criterion* is furnished as

$$\begin{aligned} f(\xi, \rho, \theta; c, \phi) &= \sqrt{2}\xi\sin\phi + \sqrt{3}\rho\sin\left(\theta + \frac{\pi}{3}\right) \\ &+ \rho\cos\left(\theta + \frac{\pi}{3}\right)\sin\phi - \sqrt{6}c\cos\phi = 0 \end{aligned} \tag{6.23}$$

where a new relationship between c, k_c, and ϕ holds as

$$c = k_c\frac{1 - \sin\phi}{\cos\phi} \tag{6.24}$$

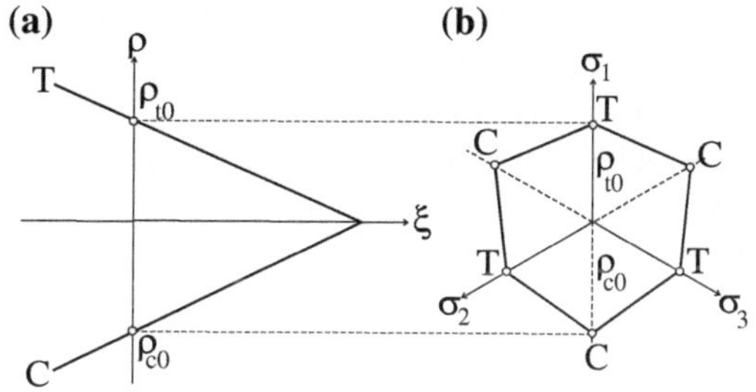

Fig. 6.3 Cross-sections of Coulomb–Mohr failure surface in: **a** plane $\xi - \rho$, **b** deviatoric plane

In the *Haigh–Westergaard space* the Coulomb–Mohr criterion represents *irregular hexagonal pyramid* the *tensile* T and the *compressive* C *meridians* whose straight lines are unequally inclined to the ξ axis (Fig. 6.3a)

$$\rho_{t0} = \frac{2\sqrt{6}c\cos\phi}{3+\sin\phi} = \frac{\sqrt{6}k_c(1-\sin\phi)}{3+\sin\phi}$$
$$\rho_{c0} = \frac{2\sqrt{6}c\cos\phi}{3-\sin\phi} = \frac{\sqrt{6}k_c(1-\sin\phi)}{3-\sin\phi} \quad (6.25)$$

In the deviatoric plane the Coulomb–Mohr surface cross-section has the form of irregular hexagon shown in Fig. 6.3b.

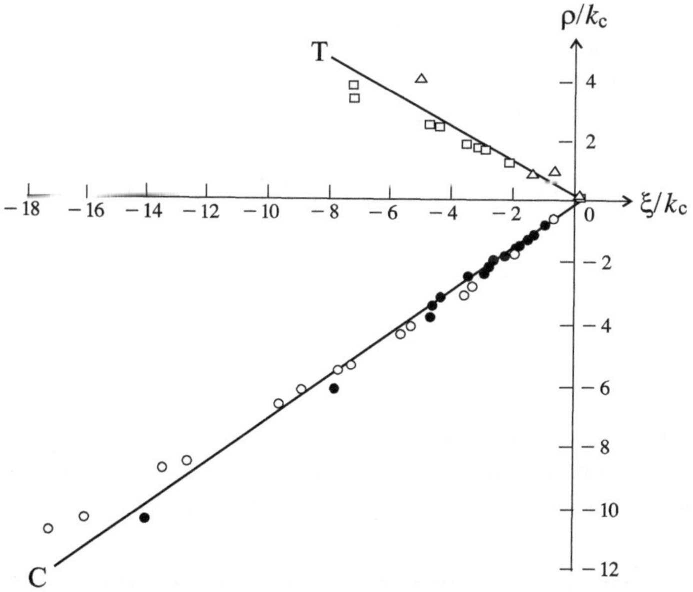

Fig. 6.4 Experimental verification of Coulomb–Mohr failure criterion in the $\xi - \rho$ plane for concrete specimens, according to (□, ○) Richart et al. [32], (△) Kupfer et al. [25], (•) Balmer [2]

The Coulomb–Mohr criterion is widely used for concrete. The experimental findings by Richart et al. [32], Kupfer et al. [25] and Balmer [2] show applicability of the *Coulomb–Mohr failure criterion* of *concrete* Fig. 6.4. Note however that the admissible tensile field bounded by meridians T and C for $\xi > 0$ reduces nearly to zeroth since the tensile strength k_{t} is almost equal to zero for the considered class of materials.

6.3 Isotropic Rotationally Symmetric Initial Failure Surfaces—Hydrostatic Pressure σ_{h} and J_{2s} Sensitivity but J_{3s} Insensitivity: Burzyński and Drucker–Prager Criteria

Let us limit a geometrical symmetry of considered isotropic limit surfaces to the case of complete rotational symmetry with respect to the hydrostatic axis characterized by the condition $\sigma_1 = \sigma_2 = \sigma_3$. The rotationally symmetric isotropic surfaces have to be dependent neither on the third deviatoric stress invariant J_{3s} nor on the third Haigh–Westergaard coordinate θ, equivalently.

Examine a universal form of the three-parameter isotropic *rotationally symmetric initial yield/failure surface* originally introduced by Burzyński [4] (item C6 in Table 6.3) as follows:

$$A\sigma_{\mathrm{eq}}^2 + B\sigma_{\mathrm{h}}^2 + C\sigma_{\mathrm{h}} - 1 = 0 \tag{6.26}$$

where A, B, and C are the material constants, while σ_{eq} and σ_{h} are equivalent and hydrostatic stresses, respectively. The *Burzyński criterion* (6.26) can also be written in the equivalent invariant fashion in terms of the first stress invariant $J_{1\sigma}$ and the second deviatoric stress J_{2s} invariant

$$A3J_{2s} + B\left(\frac{J_{1\sigma}}{3}\right)^2 + C\left(\frac{J_{1\sigma}}{3}\right) - 1 = 0 \tag{6.27}$$

Alternatively, in the Haigh–Westergaard space the Burzyński criterion takes the form

$$A\frac{3}{2}\rho^2 + B\left(\frac{\xi}{\sqrt{3}}\right)^2 + C\frac{\xi}{\sqrt{3}} - 1 = 0 \tag{6.28}$$

Three constants in the Burzyński criterion A, B, and C are determined based on three tests: the uniaxial tension (k_{t}), uniaxial compression (k_{c}), and simple shear (k_{s}). These calibrations lead to the *three-parameter Burzyński criterion* general form

$$\frac{k_{\mathrm{t}}k_{\mathrm{c}}}{3k_{\mathrm{s}}^2}\sigma_{\mathrm{eq}}^2 + \left(9 - \frac{3k_{\mathrm{t}}k_{\mathrm{c}}}{k_{\mathrm{s}}^2}\right)\sigma_{\mathrm{h}}^2 + 3\left(k_{\mathrm{c}} - k_{\mathrm{t}}\right)\sigma_{\mathrm{h}} - k_{\mathrm{t}}k_{\mathrm{c}} = 0 \tag{6.29}$$

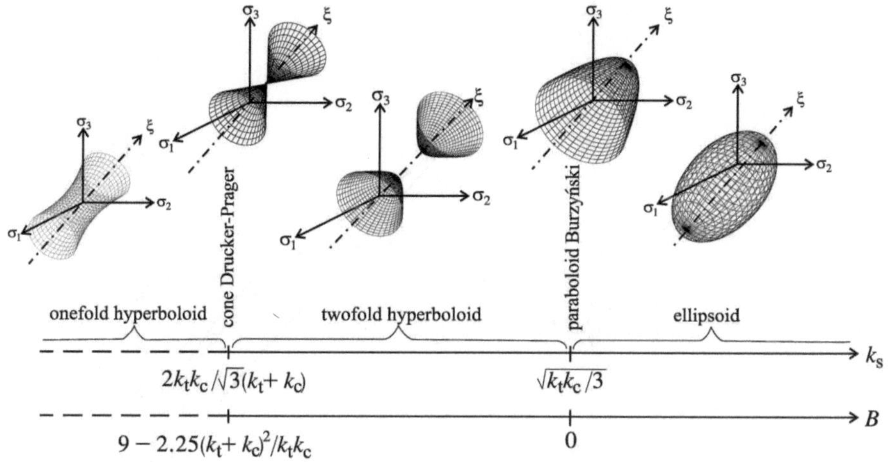

Fig. 6.5 Different types of Burzyński's rotationally symmetric yield/failure surface versus mutual relationships between k_t, k_c, and k_s or magnitude of parameter B in Eq. (6.29)

Such formulation, although frequently forgotten, presumes not only *tension/compression asymmetry* $k_t \neq k_c$ but also the third *shear limit point* k_s considered as independent. Equation (6.29) represents different types of Burzyński's rotationally symmetric surface depending on mutual relationships between k_t, k_c, and k_s. However, all of them are independent of the third Haigh–Westergaard coordinate θ or the third deviatoric stress invariant J_{3s}, alternatively. In case the shear yield/failure point stress k_s is greater than $\sqrt{\frac{k_tk_c}{3}}$ Eq. (6.29) represents *rotationally symmetric ellipsoid*, if the shear yield point stress is equal to $\sqrt{\frac{k_tk_c}{3}}$ it represents *rotationally symmetric paraboloid*, whereas if the shear yield point stress is less than $\sqrt{\frac{k_tk_c}{3}}$ it represents twofold *rotationally symmetric hyperboloid*. If k_s reaches its lower admissible bound $\frac{2k_tk_c}{\sqrt{3}(k_t+k_c)}$ the hyperboloid transforms to the *Drucker–Prager cone*. Shear yield stress points less than $\frac{2k_tk_c}{\sqrt{3}(k_t+k_c)}$ are inadmissible in sense of Drucker's postulate since in such case Eq. (6.29) represents onefold concave hyperboloid (Fig. 6.5):

- $k_s > \sqrt{\frac{k_tk_c}{3}}$ ellipsoid
- $k_s = \sqrt{\frac{k_tk_c}{3}}$ paraboloid
- $k_s < \sqrt{\frac{k_tk_c}{3}}$ twofold hyperboloid
- $k_s = \frac{2k_tk_c}{\sqrt{3}(k_t+k_c)}$ Drucker–Prager cone
- $k_s < \frac{2k_tk_c}{\sqrt{3}(k_t+k_c)}$ onefold concave hyperboloid

In case of a twofold hyperboloid only one fold that includes stress origin has physical sense.

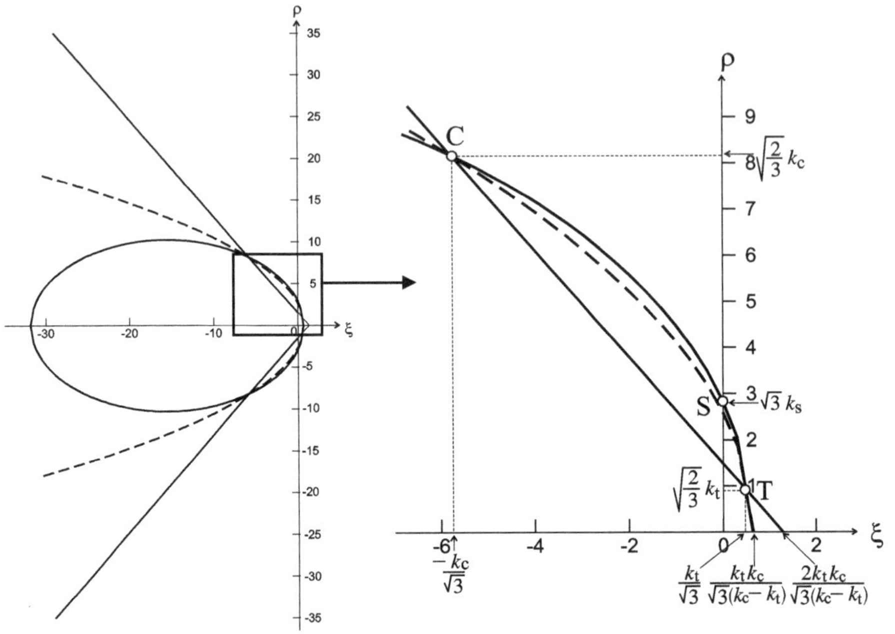

Fig. 6.6 Cross-sections of Burzyński's yield surface: elliptic, parabolic and straight-line at the Haigh–Westergaard plane $\xi - \rho$ in case of material characterized by ratio $k_c/k_t = 10$

In *nonmetallic materials* (for instance *concrete*) the ratio k_c/k_t may be much higher than one reaching the magnitude of 10. Then volume bounded by Burzyński's surface: ellipsoid, paraboloid or cone is basically limited to the compression zone, only slightly entering into the tensile zone ($\xi > 0$) as sketched in Fig. 6.6. The three-parameter rotationally symmetric format of Burzyński's criterion (6.29) guarantees that the appropriate ellipsoid includes exactly three *meridians of uniaxial tension* T(k_t), *uniaxial compression* C(k_c), and the *simple shear* S(k_s). It is a consequence of the fact that the three constants k_t, k_c, and k_s are independent. Contrarily, *two-parameter Burzyński's approximations*, paraboloidal or conical include only two of the three meridians, tensile T(k_t) and compressive C(k_c), which means that the shear point stress k_s is treated as dependent $k_s(k_t, k_c)$.

The two-parameter rotationally symmetric *paraboloidal approximation* of *Burzyński's surface* ($k_s = \sqrt{\frac{k_t k_c}{3}}$) can be written as follows:

$$\sigma_{eq}^2 + 3\,(k_c - k_t)\,\sigma_h - k_t k_c = 0 \tag{6.30}$$

or

$$3J_{2s} + (k_c - k_t)\,J_{1\sigma} - k_t k_c = 0 \tag{6.31}$$

or

$$\frac{3}{2}\rho^2 + (k_c - k_t)\sqrt{3}\xi - k_t k_c = 0 \tag{6.32}$$

It is experimentally verified for metallic alloys; for instance, material constants for *Inconel 718* (*Ni-base* like *alloy* including Cr) cited by Pęcherski et al. [29] are following:

$$k_t = 779 \text{ MPa}, \qquad k_c = 878 \text{ MPa}, \qquad k_s = 473 \text{ MPa} \tag{6.33}$$

Note that yield/failure stress points fulfil the condition $k_s = \sqrt{\frac{k_t k_c}{3}}$ which means that Burzyński's paraboloid is guaranteed.

In the other particular case if

$$k_s = \frac{2k_t k_c}{\sqrt{3}(k_t + k_c)} \tag{6.34}$$

the two-parameter *conical approximation* of *Burzyński's surface* is furnished

$$\sigma_{eq} + 3\frac{k_c - k_t}{k_t + k_c}\sigma_h - 2\frac{k_t k_c}{k_t + k_c} = 0 \tag{6.35}$$

or

$$\sqrt{3J_{2s}} + \frac{k_c - k_t}{k_t + k_c}J_{1\sigma} - 2\frac{k_t k_c}{k_t + k_c} = 0 \tag{6.36}$$

or

$$\rho + \frac{k_c - k_t}{k_t + k_c}\sqrt{2}\xi - 2\sqrt{\frac{2}{3}}\frac{k_t k_c}{k_t + k_c} = 0 \tag{6.37}$$

The above condition can be reduced to the *Drucker–Prager condition* commonly met in literature. Note however that in the light of above discussion, the Drucker–Prager condition can be considered as the limit case of the applicability range of Burzyński's criterion. Behind this limit when $k_s < \frac{2k_t k_c}{\sqrt{3}(k_t + k_c)}$, the conical surface deforms into the concave onefold hyperboloid which is inadmissible following *Drucker's stability postulate*.

Finally, in order to explicitly show reduction of Burzyński's conical criterion (6.35–6.37) to Drucker–Prager's condition [9] the following new parameters are defined

$$\alpha = \frac{1}{\sqrt{3}}\frac{k_c - k_t}{k_t + k_c}, \qquad k = \frac{2}{\sqrt{3}}\frac{k_t k_c}{k_t + k_c} \tag{6.38}$$

When the Drucker–Prager notation is used, we end up with the format explicitly written down in terms of the first stress invariant and the second deviatoric stress invariant (item C7 in Table 6.3), cf. (6.36)

$$f(J_{1\sigma}, J_{2s}; k) = \alpha J_{1\sigma} + \sqrt{J_{2s}} - k = 0 \tag{6.39}$$

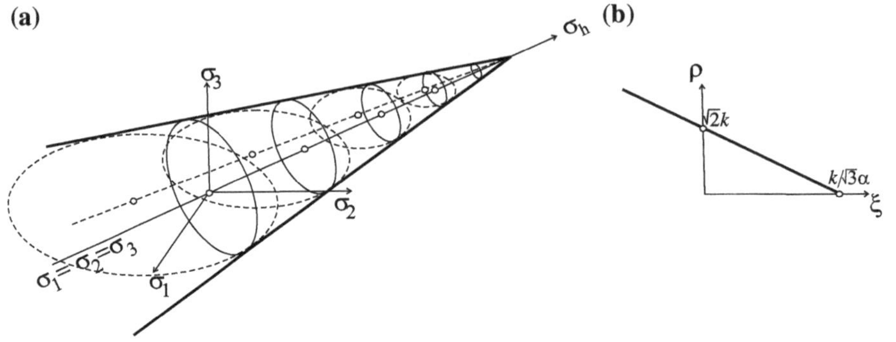

Fig. 6.7 Drucker–Prager's yield/failure surface in: **a** principal stress space, **b** Haigh–Westergaard's coordinates

Analogously, when the *Haigh–Westergaard coordinates* are employed we arrive at the form

$$f\left(\xi, \rho; k\right) = \alpha\sqrt{3}\xi + \frac{\rho}{\sqrt{2}} - k = 0 \tag{6.40}$$

The Drucker–Prager condition (6.39–6.40) describes conical circular surface the axis of which is the hydrostatic axis (equally inclined to principal stress axes σ_1, σ_2 and σ_3) as shown in Fig. 6.7. Cross-sections of the Drucker–Prager cone are subject to isotropic contraction under hydrostatic tension ($\sigma_h > 0$) contrary to isotropic extension under hydrostatic compression ($\sigma_h < 0$). In the particular case $\alpha = 0$, the Drucker–Prager cone transforms to the Huber–von Mises cylinder. Note however that the Drucker–Prager cone is bounded by inequality $0 \leq \xi < \frac{k}{\sqrt{3}\alpha}$ at the tensile side of the hydrostatic axis whereas it is not possible to reach failure $-\infty < \xi \leq 0$ on the compressive side of the hydrostatic axis. The Drucker–Prager criterion is used in the description of variety of metallic engineering materials such as Aluminium, steel AISI 4330, Fe-based, Ti-based alloys etc. for which hydrostatic pressure effect has to be included.

When (6.39)–(6.40) format of Drucker–Prager's criterion is used, we arrive at the expanded equation

$$\alpha\left(\sigma_1 + \sigma_2 + \sigma_3\right) + \sqrt{\frac{\sigma_1^2 - \sigma_1\sigma_2 + \sigma_2^2 - \sigma_2\sigma_3 + \sigma_3^2 - \sigma_3\sigma_1}{3}} - k = 0 \tag{6.41}$$

In a particular case of the *plane stress state* $\sigma_3 = 0$, the above equation yields the following one

$$\alpha\left(\sigma_1 + \sigma_2\right) + \sqrt{\frac{\sigma_1^2 - \sigma_1\sigma_2 + \sigma_2^2}{3}} - k = 0 \tag{6.42}$$

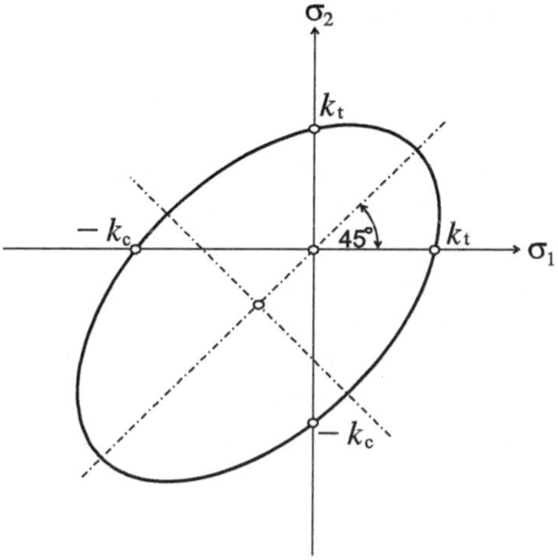

Fig. 6.8 Exemplary cross-section of Drucker–Prager's limit surface by plane $\sigma_3 = 0$

Equation (6.42) represents an off-center ellipse in the σ_1, σ_2 plane (Fig. 6.8), the major axis of which is inclined at 45° to σ_1, σ_2 axes and which intersects stress axes at different ordinates k_t and k_c

$$k_t = \frac{\sqrt{3}k}{1+\sqrt{3}\alpha}, \qquad k_c = \frac{\sqrt{3}k}{1-\sqrt{3}\alpha} \tag{6.43}$$

that stand for the uniaxial tensile and uniaxial compressive yield/failure onset points $k_c > k_t$, hence the *strength differential effect* is captured.

6.4 Anisotropic Yield/Failure Criteria for Hydrostatic Pressure Sensitive Materials—von Mises–Tsai–Wu Type Criteria (Explicit Formulation)

Based on the discussion performed in Sect. 6.1, the conclusion can be drawn that the linear term $\Pi_{ij}\sigma_{ij}$ in the *Goldenblat–Kopnov criterion* (6.2) plays an essential role and cannot be omitted ($\alpha \neq 0$) when the pressure sensitive materials are considered.

As a rule it is convenient to reduce the general Goldenblat–Kopnov criterion (6.2) to the narrower format which exhibits dimensional homogeneity assuming $\alpha = 1, \beta = 1/2, \gamma = 1/3$ as follows:

$$\Pi_{ij}\sigma_{ij} + \left(\Pi_{ijkl}\sigma_{ij}\sigma_{kl}\right)^{1/2} + \left(\Pi_{ijklmn}\sigma_{ij}\sigma_{kl}\sigma_{mn}\right)^{1/3} - 1 = 0 \tag{6.44}$$

Limiting ourselves to the linear and the quadratic terms in the Eq. (6.44), in other words neglecting the third common invariant responsible for a distortion, we arrive at the particular sub-case of the Goldenblat–Kopnov criterion, cf. Życzkowski [41]

$$\Pi_{ij}\sigma_{ij} + \sqrt{\Pi_{ijkl}\sigma_{ij}\sigma_{kl}} - 1 = 0 \tag{6.45}$$

which, on the other hand, can be treated as an *extension of isotropic Drucker–Prager's failure criterion* (6.39) *to* the case of *anisotropy*. Note however that due to material isotropy, the Drucker–Prager criterion contains only stress invariants $J_{1\sigma}$ and J_{2s}, whereas in Eq. (6.45) to describe anisotropy the common invariants of stress and the structural anisotropy tensors Π_{ij} and Π_{ijkl} must be used (cf. Chap. 5). By contrast to the cases of the Huber–von Mises and the Hill yield conditions which represent the circular and the elliptic cylinders, respectively (Fig. 6.10), in the considered case of the Drucker–Prager criterion and its anisotropic generalization (6.45), the respective failure surfaces can be recognized as the circular and the elliptic cones, respectively.

Another special sub-case of the Goldenblat–Kopnov criterion (6.2) is obtained if $\alpha = 1, \beta = 1$ and consecutive limitation of this format to the linear and the quadratic terms hold such that at the *anisotropic extension of Burzyński's paraboloid* (6.30–6.32) is met, cf. Ganczarski and Lenczowski [11]

$$\Pi_{ij}\sigma_{ij} + \Pi_{ijkl}\sigma_{ij}\sigma_{kl} - 1 = 0 \tag{6.46}$$

Note that the quadratic term $\Pi_{ijkl}\sigma_{ij}\sigma_{kl}$ in Eq. (6.45) appears under square root whereas in Eq. (6.46) does not; hence, the condition (6.45) can be interpreted as a noncircular cone whereas the condition (6.46) as a noncircular paraboloid. Compare also the relevant discussion referring to the isotropic subcases of Burzyński's the circular cone (6.37) and the circular paraboloid (6.32) discussed in Sect. 6.3.

Structural tensors of the second Π_{ij} and fourth Π_{ijkl} orders appearing in Eqs. (6.45)–(6.46) stand for two independent yield/failure anisotropy tensors the identification of which has to be performed on the basis of respective yield/failure tests in analogous way as that discussed in Chap. 5. However, in the present case, two anisotropy tensors have to be calibrated; hence, the appropriate number of tests increases such that the difference between the tension and the compression uniaxial tests can be captured. By substituting for convenience Voigt's vector-matrix notation, both the above tensors can be represented as follows:

$$[\boldsymbol{\pi}] = \begin{bmatrix} \pi_{11} & \pi_{12} & \pi_{13} \\ & \pi_{22} & \pi_{23} \\ & & \pi_{33} \end{bmatrix} \begin{bmatrix} \bullet & \bullet & \bullet \\ & \bullet & \bullet \\ & & \bullet \end{bmatrix} \tag{6.47}$$

or

$$\{\boldsymbol{\pi}\}^{\mathrm{T}} = \left\{ \pi_{11}\ \pi_{22}\ \pi_{33} \middle| \pi_{23}\ \pi_{13}\ \pi_{12} \right\}^{\mathrm{T}} \left\{ \bullet\ \bullet\ \bullet \middle| \bullet\ \bullet\ \bullet \right\}^{\mathrm{T}} \tag{6.48}$$

and

$$
[\mathbb{III}] = \left[\begin{array}{ccc|ccc} \Pi_{11} & \Pi_{12} & \Pi_{13} & \Pi_{14} & \Pi_{15} & \Pi_{16} \\ & \Pi_{22} & \Pi_{23} & \Pi_{24} & \Pi_{25} & \Pi_{26} \\ & & \Pi_{33} & \Pi_{34} & \Pi_{35} & \Pi_{36} \\ \hline & & & \Pi_{44} & \Pi_{45} & \Pi_{46} \\ & & & & \Pi_{55} & \Pi_{56} \\ & & & & & \Pi_{66} \end{array}\right] \quad \left[\begin{array}{ccc|ccc} \bullet & \bullet & \bullet & \bullet & \bullet & \bullet \\ & \bullet & \bullet & \bullet & \bullet & \bullet \\ & & \bullet & \bullet & \bullet & \bullet \\ \hline & & & \bullet & \bullet & \bullet \\ & & & & \bullet & \bullet \\ & & & & & \bullet \end{array}\right] \tag{6.49}
$$

where yield/failure loci are determined by two *yield/failure characteristic matrices* $[\pi]$ of the dimension (3×3) and $[\mathbb{III}]$ of the dimension (6×6). Hence in the considered case of general anisotropy, the number of modules defining yield/failure initiation is equal to $27 = 6 + 21$.

The condition of *yield/failure initiation in anisotropic materials* (6.46) takes in Voigt's notation the equivalent format

$$
\{\pi\}\{\sigma\} + \{\sigma\}^{\mathrm{T}}[\mathbb{III}]\{\sigma\} - 1 = 0 \tag{6.50}
$$

where

$$
\{\sigma\} = \begin{Bmatrix} \sigma_x \\ \sigma_y \\ \sigma_z \\ \tau_{yz} \\ \tau_{zx} \\ \tau_{xy} \end{Bmatrix} \tag{6.51}
$$

being an extension of anisotropic von Mises' criterion of plastic materials (5.23), however enriched by the additional term. Among the general anisotropy number of modules 27, only $24 = 6 + 18$ are truly independent. However, in practical application, this number of 24 material modules can further be reduced by assuming a certain symmetry group. For instance, in case of *orthotropy*, a hypothesis that anisotropy exhibited in the elastic range is inherited also by the limit criterion (initiation of yield/failure), may be formulated, as shown in Table 1.4.

In what follows an *extension of anisotropic von Mises' criterion*, (5.22) enhanced by linear terms (6.46) is considered. Assume the deviatoric form of von Mises criterion (5.43) but enhanced by including the linear terms

$$
\begin{aligned}
&-\Pi_{12}\left(\sigma_x - \sigma_y\right)^2 - \Pi_{13}\left(\sigma_x - \sigma_z\right)^2 - \Pi_{23}\left(\sigma_y - \sigma_z\right)^2 + \\
&2\left\{\tau_{yz}\left[\Pi_{24}\left(\sigma_y - \sigma_x\right) + \Pi_{34}\left(\sigma_z - \sigma_x\right)\right] + \right. \\
&\tau_{zx}\left[\Pi_{15}\left(\sigma_x - \sigma_y\right) + \Pi_{35}\left(\sigma_z - \sigma_y\right)\right] + \\
&\tau_{xy}\left[\Pi_{16}\left(\sigma_x - \sigma_z\right) + \Pi_{26}\left(\sigma_y - \sigma_z\right)\right] + \\
&\left.\Pi_{45}\tau_{yz}\tau_{zx} + \Pi_{46}\tau_{xy}\tau_{yz} + \Pi_{56}\tau_{zx}\tau_{xy}\right\} +
\end{aligned} \tag{6.52}
$$

$$\Pi_{44}\tau_{yz}^2+\Pi_{55}\tau_{zx}^2+\Pi_{66}\tau_{xy}^2+$$
$$\pi_{11}\sigma_x+\pi_{22}\sigma_y+\pi_{33}\sigma_z+\pi_{12}\tau_{xy}+\pi_{13}\tau_{zx}+\pi_{23}\tau_{yz}=1$$

Note that this form is strictly pressure insensitive only in the quadratic terms but it is pressure sensitive as far as the linear terms are concerned. In order to obtain the form of Eq. (6.52), fully pressure insensitive the following additional constraint has to be satisfied

$$\pi_{11}+\pi_{22}+\pi_{33}=0 \tag{6.53}$$

The condition (6.53) can be understood in one of the three following ways:

$$\pi_{11}=-(\pi_{22}+\pi_{33}) \quad \text{or} \quad \pi_{22}=-(\pi_{11}+\pi_{33}) \quad \text{or} \quad \pi_{33}=-(\pi_{11}+\pi_{22}) \tag{6.54}$$

For instance, substituting the first of conditions (6.54) in the Eq. (6.52), we arrive at the first of three *deviatoric* forms of the *von Mises–Tsai–Wu criterion*, which is hydrostatic pressure insensitive

$$\begin{aligned}
&-\Pi_{12}\left(\sigma_x-\sigma_y\right)^2-\Pi_{13}\left(\sigma_x-\sigma_z\right)^2-\Pi_{23}\left(\sigma_y-\sigma_z\right)^2+\\
&2\left\{\tau_{yz}\left[\Pi_{24}\left(\sigma_y-\sigma_x\right)+\Pi_{34}\left(\sigma_z-\sigma_x\right)\right]+\right.\\
&\quad\tau_{zx}\left[\Pi_{15}\left(\sigma_x-\sigma_y\right)+\Pi_{35}\left(\sigma_z-\sigma_y\right)\right]+\\
&\quad\tau_{xy}\left[\Pi_{16}\left(\sigma_x-\sigma_z\right)+\Pi_{26}\left(\sigma_y-\sigma_z\right)\right]+\\
&\quad\left.\Pi_{45}\tau_{yz}\tau_{zx}+\Pi_{46}\tau_{xy}\tau_{yz}+\Pi_{56}\tau_{zx}\tau_{xy}\right\}+\\
&\Pi_{44}\tau_{yz}^2+\Pi_{55}\tau_{zx}^2+\Pi_{66}\tau_{xy}^2+\\
&\pi_{22}\left(\sigma_y-\sigma_x\right)+\pi_{33}\left(\sigma_z-\sigma_x\right)+\pi_{12}\tau_{xy}+\pi_{13}\tau_{zx}+\pi_{23}\tau_{yz}=1
\end{aligned} \tag{6.55}$$

The form analogous to (6.55) was considered by Szczepiński [34] where the first of constraints $(6.54)_1$ was chosen when calibrating anisotropic modules $\Pi_{12}, \Pi_{13}, \Pi_{23}, \ldots, \Pi_{66}$ (15 modules) and $\pi_{22}, \pi_{33}, \ldots, \pi_{23}$ (5 modules).

Experimental verification of Eq. (6.55) was done by Kowalewski and Śliwowski [23] where the *low carbon steel 18G2A* specimens were used. In this experiment, the cross-sections of limit surface (6.55) in the plane σ_x, τ_{xy}

$$-\left(\Pi_{12}+\Pi_{13}\right)\sigma_x^2+2\Pi_{16}\tau_{xy}\sigma_x+\Pi_{66}\tau_{xy}^2-\left(\pi_{22}+\pi_{33}\right)\sigma_x+\pi_{12}\tau_{xy}=1 \tag{6.56}$$

was considered. The Eq. (6.56) represents ellipse the center of which is shifted from the origin (σ_x, τ_{xy}), the axes are rotated with longer to shorter axes ratio slightly different from that in isotropic material (Huber–von Mises' criterion), Fig. 6.9.

Limiting further considerations to orthotropic materials, both the characteristic matrices $[{}_{\mathrm{ort}}\pi]$ and $[{}_{\mathrm{ort}}\Pi]$ (6.49) and (6.50) take following forms valid for principal directions of orthotropy

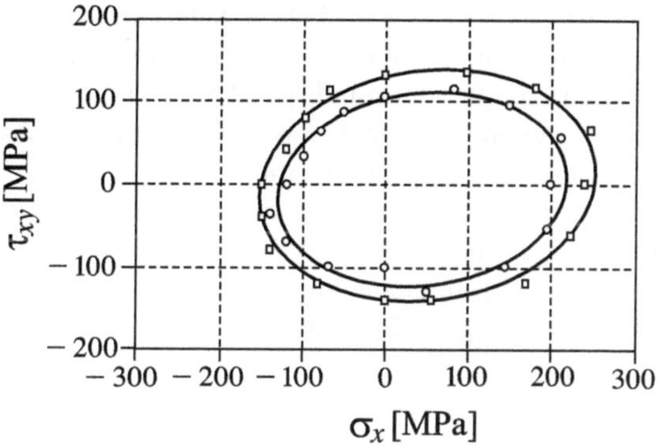

Fig. 6.9 Experimental verification of Eq. (6.56) in case of low carbon steel 18G2A specimen subjected to monotonic prestrain $\circ - \varepsilon_{\text{off}} = 1 \times 10^{-5}$, $\square - \varepsilon_{\text{off}} = 5 \times 10^{-5}$, after Kowalewski and Śliwowski [23]

$$
\left[{}_{\text{ort}}\boldsymbol{\pi}\right] = \begin{bmatrix} \pi_{11} & 0 & 0 \\ & \pi_{22} & 0 \\ & & \pi_{33} \end{bmatrix} \begin{bmatrix} \bullet & & \\ & \bullet & \\ & & \bullet \end{bmatrix} \{\bullet\bullet\bullet|\ \}^{\mathrm{T}}
$$

$$
\left[{}_{\text{ort}}\mathbb{\Pi}\right] = \left[\begin{array}{ccc|ccc} \Pi_{11} & \Pi_{12} & \Pi_{13} & 0 & 0 & 0 \\ & \Pi_{22} & \Pi_{23} & 0 & 0 & 0 \\ & & \Pi_{33} & 0 & 0 & 0 \\ \hline & & & \Pi_{44} & 0 & 0 \\ & & & & \Pi_{55} & 0 \\ & & & & & \Pi_{66} \end{array}\right] \left[\begin{array}{ccc|ccc} \bullet & \bullet & \bullet & & & \\ & \bullet & \bullet & & & \\ & & \bullet & & & \\ \hline & & & \bullet & & \\ & & & & \bullet & \\ & & & & & \bullet \end{array}\right] \tag{6.57}
$$

The second rank matrix $[{}_{\text{ort}}\boldsymbol{\pi}]$ is of the dimension 3×3 whereas the fourth rank matrix $[{}_{\text{ort}}\mathbb{\Pi}]$ dimension is 6×6. The matrix $[{}_{\text{ort}}\boldsymbol{\pi}]$ has diagonal form and the matrix $[{}_{\text{ort}}\mathbb{\Pi}]$ is of the identical symmetry as the *von Mises plastic orthotropy matrix* (5.44). Both matrices (6.57) are defined by $12 = 3 + 9$ modules.

Therefore the condition of *yield/failure initiation* for *anisotropic materials* (6.50) takes a form typical for the rotationally symmetric group

$$
\{{}_{\text{ort}}\boldsymbol{\pi}\}\{\boldsymbol{\sigma}\} + \{\boldsymbol{\sigma}\}^{\mathrm{T}}[{}_{\text{ort}}\mathbb{\Pi}]\{\boldsymbol{\sigma}\} - 1 = 0 \tag{6.58}
$$

being an extension of the von Mises orthotropic yield condition (5.46) for pressure sensitive materials. The *von Mises orthotropic yield/failure* initiation *criterion* (6.58) can be written down in the following extended form

$$
\begin{aligned}
&\Pi_{11}\sigma_x^2 + \Pi_{22}\sigma_y^2 + \Pi_{33}\sigma_z^2 + \\
&2(\Pi_{12}\sigma_x\sigma_y + \Pi_{23}\sigma_y\sigma_z + \Pi_{31}\sigma_z\sigma_x) + \\
&\Pi_{44}\tau_{yz}^2 + \Pi_{55}\tau_{zx}^2 + \Pi_{66}\tau_{xy}^2 + \\
&\pi_{11}\sigma_x + \pi_{22}\sigma_y + \pi_{33}\sigma_z - 1 = 0
\end{aligned} \tag{6.59}
$$

Note that above equation represents fully tensorial form of the orthotropic yield/failure criterion contrary to the deviatoric form which is characteristic for the *Hill yield criterion*. This means that $12 = 3 + 9$ material modules defining yield/failure material characteristic tensors ${}_{\mathrm{ort}}\boldsymbol{\pi}$ and ${}_{\mathrm{ort}}\mathbb{\Pi}$ are required for its identification. The first term in Eq. (6.58) refers to the *strength differential effect* whereas the second one represents a von Mises-type surface the axis of which generally does not coincide with the hydrostatic axis.

Consider now reduction of criterion (6.58) to a narrower form known in the literature as the *Tsai–Wu orthotropic criterion of failure*. The Tsai–Wu criterion is characterized simultaneously by *strength differential effect* and *pressure insensitivity* of $[{}_{\mathrm{ort}}\mathbb{\Pi}]$ in Eq. (6.57) such that $[{}_{\mathrm{ort}}\mathbb{\Pi}] \longrightarrow [\mathbb{\Pi}^{\mathrm{TW}}]$ (see (6.52))

$$\left\{\boldsymbol{\pi}^{\mathrm{TW}}\right\}\{\boldsymbol{\sigma}\} + \{\boldsymbol{s}\}^{\mathrm{T}}\left[\mathbb{\Pi}^{\mathrm{TW}}\right]\{\boldsymbol{s}\} - 1 = 0 \tag{6.60}$$

where

$$\{\boldsymbol{\sigma}\} = \begin{Bmatrix} \sigma_x \\ \sigma_y \\ \sigma_z \\ \tau_{yz} \\ \tau_{zx} \\ \tau_{xy} \end{Bmatrix} \qquad \{\boldsymbol{s}\} = \begin{Bmatrix} \sigma_x - \sigma_{\mathrm{h}} \\ \sigma_y - \sigma_{\mathrm{h}} \\ \sigma_z - \sigma_{\mathrm{h}} \\ \tau_{yz} \\ \tau_{zx} \\ \tau_{xy} \end{Bmatrix} \tag{6.61}$$

This leads to the following representation of both characteristic matrices

$$\left[\boldsymbol{\pi}^{\mathrm{TW}}\right] = \begin{bmatrix} \pi_{11} & 0 & 0 \\ & \pi_{22} & 0 \\ & & \pi_{33} \end{bmatrix} \begin{bmatrix} \bullet & & \\ & \bullet & \\ & & \bullet \end{bmatrix} \Rightarrow \{\bullet\,\bullet\,\bullet|\ \ \}^{\mathrm{T}} \tag{6.62}$$

$$\left[\mathbb{\Pi}^{\mathrm{TW}}\right] = \left[\begin{array}{ccc|ccc} -\Pi_{12} - \Pi_{13} & \Pi_{12} & \Pi_{13} & 0 & 0 & 0 \\ & -\Pi_{12} - \Pi_{23} & \Pi_{23} & 0 & 0 & 0 \\ & & -\Pi_{13} - \Pi_{23} & 0 & 0 & 0 \\ \hline & & & \Pi_{44} & 0 & 0 \\ & & & & \Pi_{55} & 0 \\ & & & & & \Pi_{66} \end{array}\right] \left[\begin{array}{ccc|ccc} \circ & \bullet & \bullet & & & \\ & \circ & \bullet & & & \\ & & \circ & & & \\ \hline & & & \bullet & & \\ & & & & \bullet & \\ & & & & & \bullet \end{array}\right] \tag{6.63}$$

In what above the Nye graphics is adopted in order to distinguish the independent $\bullet$ from dependent $\circ$ 4th-rank matrix elements.

Equation (6.60) and $\mathbb{\Pi}^{\mathrm{TW}}$ representation (6.63) reflect "hybrid notation" in the following sense: the first term represents the *linear common invariant* of the stress tensor $\boldsymbol{\sigma}$ and the structural tensor $\boldsymbol{\pi}^{\mathrm{TW}}$ (analogy to the pressure sensitivity in case of isotropic material) whereas the second term represents *quadratic common invariant* of the stress deviator $\boldsymbol{s}$ and the structural tensor $\mathbb{\Pi}^{\mathrm{TW}}$ (defining shape and orientation of surface in the stress space). The criterion (6.60) takes therefore explicit form of 9–parameter *Tsai–Wu's criterion* [36]

$$
\begin{aligned}
-\left[\Pi_{23}\left(\sigma_y-\sigma_z\right)^2+\Pi_{13}\left(\sigma_z-\sigma_x\right)^2+\Pi_{12}\left(\sigma_x-\sigma_y\right)^2\right]+\Pi_{44}\tau_{yz}^2+ \\
\Pi_{55}\tau_{zx}^2+\Pi_{66}\tau_{xy}^2+\pi_{11}\sigma_x+\pi_{22}\sigma_y+\pi_{33}\sigma_z-1=0
\end{aligned}
\tag{6.64}
$$

As a matter of fact, any addition of a hydrostatic pressure to all normal stresses $\sigma_x \rightarrow \sigma_x \pm \sigma_{\mathrm{h}}$ does not change the magnitude of quadratic terms in condition (6.64) but simultaneously causes the linear terms still dependent on σ_{h}. Hence, finally the Tsai–Wu criterion in the format given by (6.64) remains the *pressure sensitive* one through the linear terms.

6.5 Transversely Isotropic Case Tsai–Wu Type Tetragonal Versus Hexagonal Symmetry Criteria

Similar to Sect. 5.5, a reduction of 9–parameter yield/failure orthotropic Tsai–Wu's criterion (6.64) to narrower case of the *transverse isotropy* requires precise distinction between the *tetragonal* and *hexagonal symmetry classes*. Assuming after Chen and Han [7] plane of transverse isotropy xy, the 4th-rank orthotropy matrix $[\mathbb{\Pi}^{\mathrm{TW}}]$ (6.63) reduces to the transversely isotropic format $[{}_{\mathrm{tris}}\mathbb{\Pi}^{\mathrm{TW}}]$ analogously to the transversely isotropic Hill criterion (5.79) or (5.80) possessing only four independent material constants whereas the 2nd-rank transversely isotropic matrix $[{}_{\mathrm{tris}}\boldsymbol{\pi}^{\mathrm{TW}}]$ reduces to a form possessing only two independent material constants. Finally, assuming $\Pi_{23} = \Pi_{13}$, $\Pi_{44} = \Pi_{55}$, $\pi_{11} = \pi_{22}$ in (6.62) and (6.63) instead of the 9–parameter form (6.64), we arrive at two 6–parameter forms of the *transversely isotropic yield/failure criterion* of *tetragonal symmetry* that directly refer to formulations (5.79) or (5.80), namely

$$
\left[{}_{\mathrm{tris}}\boldsymbol{\pi}^{\mathrm{TW}}\right]=\begin{bmatrix}\pi_{11} & 0 & 0\\ & \pi_{11} & 0\\ & & \pi_{33}\end{bmatrix}
$$

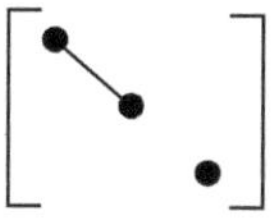

(6.65)

and

$$
\left[{}_{\mathrm{tris}}\mathbb{III}^{\mathrm{TW}}\right] = \left[\begin{array}{ccc|ccc}
-\Pi_{12}-\Pi_{13} & \Pi_{12} & \Pi_{13} & 0 & 0 & 0 \\
 & -\Pi_{12}-\Pi_{13} & \Pi_{13} & 0 & 0 & 0 \\
 & & -2\Pi_{13} & 0 & 0 & 0 \\
\hline
 & & & \Pi_{44} & 0 & 0 \\
 & & & & \Pi_{44} & 0 \\
 & & & & & \Pi_{66}
\end{array}\right]
$$

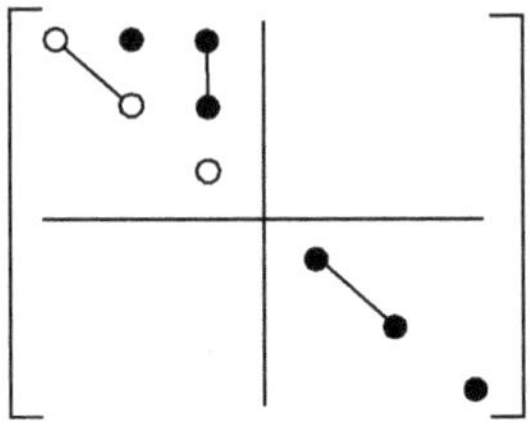

(6.66)

or alternatively

$$
\left[{}_{\mathrm{tris}}\boldsymbol{\pi}^{\mathrm{TW}}\right] = \left[\begin{array}{ccc}
\pi_{11} & 0 & 0 \\
 & \pi_{11} & 0 \\
 & & \pi_{33}
\end{array}\right]
$$

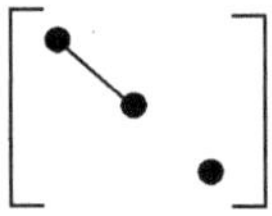

(6.67)

and

$$
\left[{}_{\mathrm{tris}}\mathbb{III}^{\mathrm{TW}}\right] = \left[\begin{array}{ccc|ccc}
\Pi_{11} & \frac{\Pi_{33}-2\Pi_{11}}{2} & -\frac{\Pi_{33}}{2} & 0 & 0 & 0 \\
 & \Pi_{11} & -\frac{\Pi_{33}}{2} & 0 & 0 & 0 \\
 & & \Pi_{33} & 0 & 0 & 0 \\
\hline
 & & & \Pi_{44} & 0 & 0 \\
 & & & & \Pi_{44} & 0 \\
 & & & & & \Pi_{66}
\end{array}\right]
$$

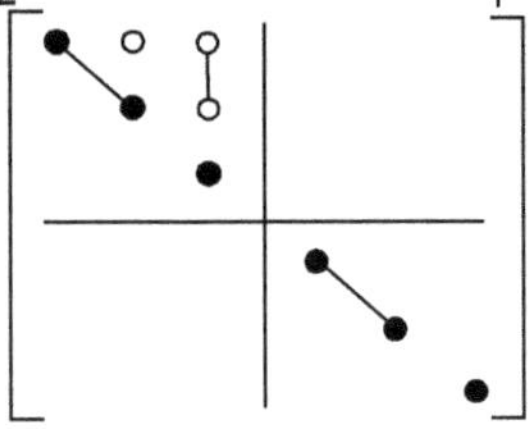

(6.68)

In case of transversely isotropic symmetries of both the 2nd-rank and the 4th-rank anisotropic matrices defining yield/failure onset, the pairs ●—● or ○—○ stand for identical matrix elements considered as independent or dependent pairs, respectively. In case when non abbreviated notation is used, the 6–parameter *transversely isotropic Tsai–Wu yield/failure criterion* of *tetragonal symmetry* takes the following form

$$-\Pi_{13}\left[(\sigma_y-\sigma_z)^2+(\sigma_z-\sigma_x)^2\right]-\Pi_{12}\left(\sigma_x-\sigma_y\right)^2+\Pi_{44}\left(\tau_{yz}^2+\tau_{zx}^2\right) \\ +\Pi_{66}\tau_{xy}^2+\pi_{11}\left(\sigma_x+\sigma_y\right)+\pi_{33}\sigma_z-1=0 \tag{6.69}$$

or alternatively

$$\Pi_{11}\left(\sigma_x^2+\sigma_y^2\right)+\Pi_{33}\sigma_z^2+(\Pi_{33}-2\Pi_{11})\,\sigma_x\sigma_y-\Pi_{33}\left(\sigma_x\sigma_z+\sigma_y\sigma_z\right) \\ +\Pi_{44}\left(\tau_{yz}^2+\tau_{zx}^2\right)+\Pi_{66}\tau_{xy}^2+\pi_{11}\left(\sigma_x+\sigma_y\right)+\pi_{33}\sigma_z-1=0 \tag{6.70}$$

The above transversely isotropic limit equations are expressed in terms of six material anisotropy modules: $\Pi_{12}, \Pi_{13}, \Pi_{44}, \Pi_{66}, \pi_{11}, \pi_{33}$ or $\Pi_{11}, \Pi_{33}, \Pi_{44}, \Pi_{66}, \pi_{11}, \pi_{33}$ if corresponding matrix representations (6.65)–(6.66) or (6.67)–(6.68) are implemented.

However, the number of independent modules can further be reduced to five, since the sixth diagonal modulus Π_{66} has to satisfy the relationships (cf. Chen and Han [7], Ganczarski and Skrzypek [12])

$$\Pi_{66}=-2(\Pi_{13}+2\Pi_{12}) \qquad \text{or} \qquad \Pi_{66}=4\Pi_{11}-\Pi_{33} \tag{6.71}$$

if the corresponding formats (6.66) or (6.68) are used. The conditions (6.71) satisfy the reducibility of the criteria (6.66) or (6.68) to the forms invariant with respect to two equivalent stress states $\tau_{xy}=\sigma$ and $\sigma_x=\sigma, \sigma_y=-\sigma$ in the transverse isotropy plane.

Taking above conditions into account, equations (6.69) and (6.70) contain only five independent material coefficients referring to appropriate tensile and compressive strengths $k_{\text{tx}}, k_{\text{cx}}, k_{\text{tz}}, k_{\text{cz}}$ and shear strength k_{zx}. Hence, in order to calibrate them, the following tests have to be performed if, for instance, the format (6.69) is used:

$$\begin{aligned}
&\sigma_x=k_{\text{tx}}, \quad \sigma_y=\ldots=\tau_{zx}=0 \longrightarrow (-\Pi_{13}-\Pi_{12})\,k_{\text{tx}}^2+\pi_{11}k_{\text{tx}}=1\\
&\sigma_x=-k_{\text{cx}}, \sigma_y=\ldots=\tau_{zx}=0 \longrightarrow (-\Pi_{13}-\Pi_{12})\,k_{\text{cx}}^2-\pi_{11}k_{\text{cx}}=1\\
&\sigma_z=k_{\text{tz}}, \quad \sigma_x=\ldots=\tau_{zx}=0 \longrightarrow -2\Pi_{13}k_{\text{tz}}^2+\pi_{33}k_{\text{tz}}=1\\
&\sigma_z=-k_{\text{cz}}, \sigma_x=\ldots=\tau_{zx}=0 \longrightarrow -2\Pi_{13}k_{\text{cz}}^2-\pi_{33}k_{\text{cz}}=1\\
&\tau_{zx}=k_{zx}, \quad \sigma_x=\ldots=\tau_{yz}=0 \longrightarrow \Pi_{44}k_{zx}^2=1
\end{aligned} \tag{6.72}$$

Solution of Eq. (6.72) with respect to $\Pi_{13}, \Pi_{12}, \Pi_{44}, \pi_{11}$ and π_{33} takes the form

$$\begin{aligned}
&-\Pi_{13}=\frac{1}{2k_{\text{tz}}k_{\text{cz}}}, \ -\Pi_{12}=\frac{1}{k_{\text{tx}}k_{\text{cx}}}-\frac{1}{2k_{\text{tz}}k_{\text{cz}}}, \ \Pi_{44}=\frac{1}{2k_{zx}^2}\\
&\pi_{11}=\frac{1}{k_{\text{tx}}}-\frac{1}{k_{\text{cx}}}, \ \pi_{33}=\frac{1}{k_{\text{tz}}}-\frac{1}{k_{\text{cz}}},
\end{aligned} \tag{6.73}$$

Magnitude of material modulus Π_{66}, referring to shear strength in the plane of transverse isotropy is not independent but given by Eq. (6.71), hence

$$\Pi_{66} = \frac{4}{k_{\mathrm{t}x} k_{\mathrm{c}x}} - \frac{1}{k_{\mathrm{t}z} k_{\mathrm{c}z}} \tag{6.74}$$

Note that both formats in Eq. (6.71) lead to the same calibration for Π_{66} (6.74). Hence, after substitution of Eqs. (6.73)–(6.74) to Eq. (6.69) one can get the final form of the *hexagonal transversely isotropic Tsai–Wu criterion* in terms of five independent constants $k_{\mathrm{t}x}$, $k_{\mathrm{c}x}$, $k_{\mathrm{t}z}$, $k_{\mathrm{c}z}$ and k_{zx}

$$\begin{aligned}\frac{\sigma_x^2+\sigma_y^2}{k_{\mathrm{t}x} k_{\mathrm{c}x}} + \frac{\sigma_z^2}{k_{\mathrm{t}z} k_{\mathrm{c}z}} - \left(\frac{2}{k_{\mathrm{t}x} k_{\mathrm{c}x}} - \frac{1}{k_{\mathrm{t}z} k_{\mathrm{c}z}}\right)\sigma_x\sigma_y - \frac{\sigma_y\sigma_z+\sigma_x\sigma_z}{k_{\mathrm{t}z} k_{\mathrm{c}z}} \\ + \frac{\tau_{yz}^2+\tau_{zx}^2}{k_{zx}^2} + \underline{\left(\frac{4}{k_{\mathrm{t}x} k_{\mathrm{c}x}} - \frac{1}{k_{\mathrm{t}z} k_{\mathrm{c}z}}\right)\tau_{xy}^2} \\ + \left(\frac{1}{k_{\mathrm{t}x}} - \frac{1}{k_{\mathrm{c}x}}\right)(\sigma_x+\sigma_y) + \left(\frac{1}{k_{\mathrm{t}z}} - \frac{1}{k_{\mathrm{c}z}}\right)\sigma_z = 1\end{aligned} \tag{6.75}$$

Inspection of the transversely isotropic format of the Tsai–Wu criterion (6.75) reveals that underlined coefficient preceding τ_{xy} differs in format from the analogous term in the *transversely isotropic Hill criterion* (5.82) since independent shear limit in the transverse isotropy plane k_{xy} is used. Obviously, the transition from the Tsai–Wu criterion (6.75) to the Hill criterion (5.82) requires to ignore the tension/compression asymmetry effect $k_{\mathrm{t}x} = k_{\mathrm{c}x}$ and $k_{\mathrm{t}z} = k_{\mathrm{c}z}$ which leads simultaneously to vanishing of linear terms. In other words, in this case, the Tsai–Wu transversely isotropic criterion reducible to the Hill criterion becomes pressure insensitive, by contrast to the Eq. (6.75) in which *pressure sensitivity* $k_{\mathrm{t}x} \neq k_{\mathrm{c}x}$ and $k_{\mathrm{t}z} \neq k_{\mathrm{c}z}$ plays essential role.

It is seen that material coefficients in the xy-plane of transverse isotropy that precede the terms $\sigma_x\sigma_y$ and τ_{xy}^2 are not fully independent since they contain not only the in-plane tensile and compressive limits $k_{\mathrm{t}x}$, $k_{\mathrm{c}x}$ but also the out-of-plane tensile and compressive limits $k_{\mathrm{t}z}$ and $k_{\mathrm{c}z}$. Consequently, Eq. (6.75) can be classified as the *hexagonal transversely isotropic Tsai–Wu criterion* of initial yield/failure.

Applicability range of the Tsai–Wu orthotropic criterion (6.75) to properly describe initiation of failure in some engineering materials that exhibit *high orthotropy degree*, is bounded by a possible *ellipticity loss of the limit surface*, see Ganczarski and Adamski [14]. In other words, a physically inadmissible degeneration of a single convex and simply connected elliptic limit surface into two concave hyperbolic surfaces occurs.

The following inequality bounds the range of applicability of the transversely isotropic Tsai–Wu criterion to ensure convexity

$$\frac{1}{k_{tz}k_{cz}}\left(\frac{4}{k_{tx}k_{cx}}-\frac{1}{k_{tz}k_{cz}}\right)>0 \tag{6.76}$$

which can easily be recognized as an extension of the relevant bounding inequality for Hill's criterion (5.66). Substitution of the dimensionless parameter $\overline{R}=2(\frac{k_{tz}k_{cz}}{k_{tx}k_{cx}})-1$, (extension of the Hosford and Backhofen parameter), leads to the simplified restriction

$$\overline{R}>-0.5 \tag{6.77}$$

If the above inequalities (6.76)–(6.77) do not hold, elliptic cross-sections of the limit surface degenerate to two hyperbolic branches and the lack of convexity occurs. To illustrate this limitation, the yield curves in two planes:
the transverse isotropy plane (σ_x, σ_y)

$$\sigma_x^2-\frac{2\overline{R}}{1+\overline{R}}\sigma_x\sigma_y+\sigma_y^2+(k_{cx}-k_{tx})\left(\sigma_x+\sigma_y\right)=k_{tx}k_{cx} \tag{6.78}$$

and the orthotropy plane (σ_x, σ_z)

$$\sigma_x^2-\frac{2}{1+\overline{R}}\sigma_x\sigma_z+\frac{2}{1+\overline{R}}\sigma_z^2+(k_{cx}-k_{tx})\,\sigma_x+k_{tz}k_{cz}\sigma_z=k_{tx}k_{cx} \tag{6.79}$$

for various $\overline{R}$-values, are sketched in Fig. 6.10a, b, respectively. It is observed that when $\overline{R}$, starting from $\overline{R}=3$, approaches the limit $\overline{R}=-0.5$, the limit curves change from closed ellipses to two parallel lines, whereas for $\overline{R}<-0.5$, concave hyperbolas appear.

Except the hexagonal transversely isotropic Tsai–Wu criterion Eq. (6.75), one can introduce the other hexagonal transversely isotropic Tsai–Wu failure criterion, see Ganczarski and Adamski [14]. In order to do this, let us consider the more general transverse isotropic von Mises–Tsai–Wu criterion of the format

$$\begin{aligned}&\Pi_{11}\left(\sigma_x^2+\sigma_y^2\right)+\Pi_{33}\sigma_z^2+2\Pi_{12}\sigma_x\sigma_y+2\Pi_{13}\left(\sigma_x+\sigma_y\right)\sigma_z\\&+\Pi_{44}\left(\tau_{yz}^2+\tau_{zx}^2\right)+\Pi_{66}\tau_{xy}^2+\pi_{11}\left(\sigma_x+\sigma_y\right)+\pi_{33}\sigma_z=1\end{aligned} \tag{6.80}$$

Equation (6.80) contains $8=6+2$ independent modules and it is straightforward simplification of the orthotropic von Mises–Tsai–Wu criterion (6.59) by introducing obvious symmetry conditions $\Pi_{11}=\Pi_{22}$, $\Pi_{23}=\Pi_{31}$, $\Pi_{44}=\Pi_{55}$ and $\pi_{11}=\pi_{22}$. For calibration of it following tests are to be performed:
six uniaxial tension/compression and shear conditions

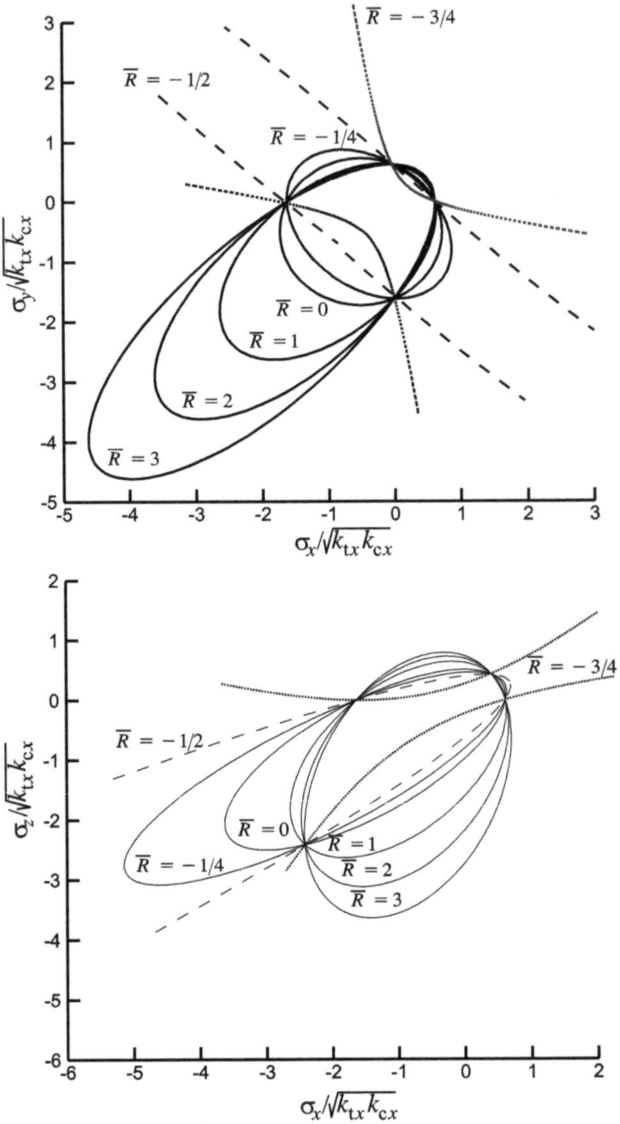

Fig. 6.10 Degeneration of the Tsai–Wu limit surface with the magnitude of the generalized Hosford and Backhofen parameter $\overline{R}$: **a** transverse isotropy plane, **b** orthotropy plane, after Ganczarski and Adamski [14]

$$\begin{array}{lll} \sigma_x = k_{tx}, & \sigma_y = \ldots = \tau_{zx} = 0 \longrightarrow \Pi_{11} k_{tx}^2 + \pi_{11} k_{tx} = 1 \\ \sigma_x = -k_{cx}, & \sigma_y = \ldots = \tau_{zx} = 0 \longrightarrow \Pi_{11} k_{cx}^2 - \pi_{11} k_{cx} = 1 \\ \sigma_z = k_{tz}, & \sigma_x = \ldots = \tau_{zx} = 0 \longrightarrow \Pi_{33} k_{tz}^2 + \pi_{33} k_{tz} = 1 \\ \sigma_z = -k_{cz}, & \sigma_x = \ldots = \tau_{zx} = 0 \longrightarrow \Pi_{33} k_{cz}^2 - \pi_{33} k_{cz} = 1 \\ \tau_{zx} = k_{zx}, & \sigma_x = \ldots = \tau_{yz} = 0 \longrightarrow \Pi_{44} k_{zx}^2 = 1 \\ \tau_{xy} = \sqrt{k_{tx} k_{cx}/3}, & \sigma_x = \ldots = \tau_{yz} = 0 \longrightarrow \Pi_{66} k_{tx} k_{cx}/3 = 1 \end{array} \tag{6.81}$$

and two biaxial conditions that allows to capture magnitudes of Π_{12} and Π_{13}

$$\begin{array}{l} \sigma_x = \sigma_y = k_{(xy)} = -(k_{cx} - k_{tx}) \mp \sqrt{\Delta_1}, \sigma_z = \ldots = \tau_{yz} = 0 \\ \longrightarrow \dfrac{2k_{(xy)}^2}{k_{tx} k_{cx}} - 2\Pi_{12} k_{(xy)}^2 + \dfrac{k_{cx} - k_{tx}}{k_{tx} k_{cx}} k_{(xy)} = 1 \\ \sigma_x = \sigma_z = k_{(xz)} = -\dfrac{1}{2} \left[(k_{cx} - k_{tx}) + (k_{cz} - k_{tz}) \dfrac{k_{tx} k_{cx}}{k_{tz} k_{cz}} \pm \sqrt{\Delta_2} \right], \\ \sigma_y = \ldots = \tau_{yz} = 0 \longrightarrow \dfrac{k_{(xz)}^2}{k_{tx} k_{cx}} + \dfrac{k_{(xz)}^2}{k_{tz} k_{cz}} - 2\Pi_{13} k_{(xz)}^2 \\ + \left(\dfrac{1}{k_{tx}} - \dfrac{1}{k_{cx}} \right) k_{(xz)} + \left(\dfrac{1}{k_{tz}} - \dfrac{1}{k_{cz}} \right) k_{(xz)} = 1 \end{array} \tag{6.82}$$

Symbols Δ_1 and Δ_2 used for brevity denote: $\Delta_1 = (k_{cx} - k_{tx})^2 + k_{tx} k_{cx}$ and $\Delta_2 = [(k_{cx} - k_{tx}) + (k_{cz} - k_{tz}) \frac{k_{tx} k_{cx}}{k_{tz} k_{cz}}]^2 + 4k_{tx} k_{cx}$. Solution of Eqs. (6.81–6.82) with respect to Π_{11}, Π_{12}, Π_{13}, Π_{33}, Π_{44}, Π_{66}, π_{11} and π_{33} yields

$$\begin{array}{lll} \Pi_{11} = 1/k_{tx} k_{cx} & \Pi_{12} = -1/2k_{tx} k_{cx} & \Pi_{13} = -1/2k_{tz} k_{cz} \\ \Pi_{33} = 1/k_{tz} k_{cz} & \Pi_{44} = 1/k_{zx}^2, & \Pi_{66} = 3/k_{tx} k_{cx} \\ \pi_{11} = 1/k_{tx} - 1/k_{cx} & \pi_{33} = 1/k_{tz} - 1/k_{cz} & \end{array} \tag{6.83}$$

which finally leads to the new *hexagonal transversely isotropic von Mises–Tsai–Wu failure criterion* also in terms of five independent constants k_{tx}, k_{cx}, k_{tz}, k_{cz} and k_{zx}, but different from (6.75)

$$\begin{array}{l} \dfrac{\sigma_x^2 + \sigma_y^2}{k_{tx} k_{cx}} + \dfrac{\sigma_z^2}{k_{tz} k_{cz}} - \underline{\dfrac{\sigma_x \sigma_y}{k_{tx} k_{cx}}} - \dfrac{\sigma_y \sigma_z + \sigma_x \sigma_z}{k_{tz} k_{cz}} + \dfrac{\tau_{yz}^2 + \tau_{zx}^2}{k_{zx}^2} \\ + \underline{\dfrac{3}{k_{tx} k_{cx}}} \tau_{xy}^2 + \left(\dfrac{1}{k_{tx}} - \dfrac{1}{k_{cx}} \right) (\sigma_x + \sigma_y) + \left(\dfrac{1}{k_{tz}} - \dfrac{1}{k_{cz}} \right) \sigma_z = 1 \end{array} \tag{6.84}$$

Note that the coefficients preceding $\sigma_x \sigma_y$ and τ_{xy}^2, underlined terms in (6.84) are always positive by contrast to analogous terms in (6.75) that can change sign. These prevent elliptic form of failure curves from loss of ellipticity and reduce Eq. (6.84) to the "shifted" Huber–von Mises ellipse from the origin of coordinate system in case of transverse isotropy plane. In other words, this new hexagonal format of

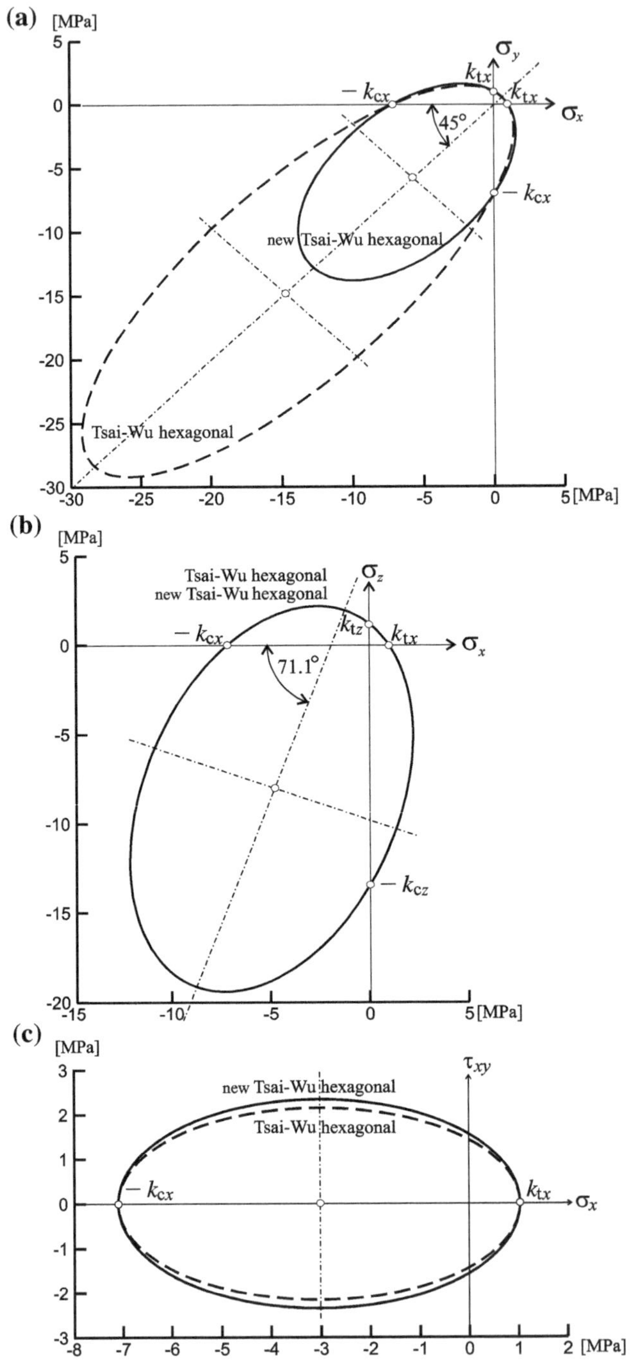

Fig. 6.11 Comparison of transversely isotropic Tsai–Wu's initial failure criteria of hexagonal and new hexagonal types for columnar ice: **a** plane of transverse isotropy (σ_x, σ_y), **b** plane of orthotropy (σ_x, σ_z), **c** shear plane (σ_x, τ_{xy}), after Ganczarski and Adamski [14]

Table 6.1 Experimental data for columnar ice after Ralston [30]

Tensile strength		Compressive strength	
k_{tx}	1.01 MPa	k_{cx}	7.11 MPa
k_{tz}	1.21 MPa	k_{cz}	13.5 MPa

Tsai–Wu's failure criterion is unconditionally stable and preserves reducibility to isotropic Huber–von Mises ellipse but shifted in the isotropy plane.

Both the Tsai–Wu transversely isotropic initial failure criteria, hexagonal Eq. (6.75), and new hexagonal type Eq. (6.84) are compared for columnar ice, the experimental data of which was established by Ralston [30] in Table 6.1 in plane of transverse isotropy (σ_x, σ_y), shear plane (σ_x, τ_{xy}), and in plane of orthotropy (σ_x, σ_z), see Fig. 6.11. Subsequent cross-sections of the limit surface are ellipses that exhibit strong oblateness in case of tetragonal symmetry; the centers of which are shifted outside the origin of coordinate system toward the quarter referring to compressive stresses. In case of cross-section by plane of transverse isotropy (see Fig. 6.11a), the symmetry axis has obviously inclination equal 45° to the axes of coordinate system; in other words, it overlaps projection of hydrostatic axis at the transverse isotropy plane (σ_x, σ_y), contrary to the cross-section by plane of orthotropy (see Fig. 6.11b) the main semi-axis of ellipse is inclined by 71.1°. It has to be emphasized that in case of columnar ice compressive strength along othotropy axis, k_{cz} is over 10 times greater than tensile strength k_{tz}; whereas, analogous ratio k_{cx}/k_{tx} is approximately equal to 7 in case of transverse isotropy plane. Moreover, ratio of semi-axes for Tsai–Wu tetragonal ellipse in (σ_x, σ_y) plane essentially exceeds analogous ratio for Huber–von Mises ellipse, contrary to the case of Tsai–Wu hexagonal ellipse. It is also worth to emphasize that although the hexagonal transversely isotropic Tsai–Wu failure criterion Eq. (6.75) and the new hexagonal transversely isotropic Tsai–Wu failure criterion Eq. (6.84) contain the same number of five independent strengths $k_{tx}, k_{cx}, k_{tz}, k_{cz}$, and k_{zx}, only criterion (6.84) is free from convexity loss and simultaneously truly transversely isotropic in sense of hexagonal class of symmetry.

6.6 Implicit Formulation of Pressure Sensitive Anisotropic Initial Failure Criteria—Khan's Concept

In the Sect. 5.7, representative papers based on the implicit approach to anisotropic yield criteria not accounting for pressure sensitivity were discussed. In what follows selected examples of implementation of the *implicit approach to* the broader class accounting for *anisotropy*, *tension/compression asymmetry*, and *pressure sensitivity* are thoroughly considered.

Khan and Liu [20] applied the following extension of the nine-parameter orthotropic von Mises criterion (5.45) to describe the ductile fracture of the Ti-6Al-4V alloy accounting for hydrostatic pressure sensitivity, anisotropy, and significant tension/compression asymmetry effect

$$\sqrt{\exp[C(\zeta+1)]\left(F\sigma_1^2+G\sigma_2^2+H\sigma_3^2+L\sigma_1\sigma_2+M\sigma_2\sigma_3+N\sigma_1\sigma_3 + P\sigma_{12}^2+Q\sigma_{13}^2+R\sigma_{23}^2\right)} = \exp\left(c_1\frac{I_1}{\sqrt{3}}\right) \tag{6.85}$$

Both the hydrostatic pressure dependence I_1 and the tension/compression asymmetry J_3 are included in an implicit fashion as arguments of two exponential functions appearing as multipliers at the right- and the left-hand sides of orthotropic von Mises' equation. According to authors, interpretation of the main advantage of such formulation is that the anisotropy and tension/compression asymmetry are uncoupled into separate multiplicative terms which allow the anisotropic parameters and tension/compression asymmetry coefficient to be determined independently. The following definitions hold F, G, H, L, M, N, P, Q, and R are anisotropic parameters; C is the tension/compression asymmetry coefficient, ζ denotes the Lode parameter $\zeta = \cos 3\theta = \frac{27}{2}\frac{J_3}{(\sqrt{3J_2})^3}$, where θ is the Lode angle, I_1 is the first stress invariant, whereas J_2 and J_3 are the second and the third invariants of deviatoric stress tensor. Although the general form of limit criterion (6.85) accounts for all three features, anisotropy, tension/compression asymmetry, and hydrostatic pressure dependence, in fact its calibration performed by authors leads the form capturing only the tension/compression asymmetry and hydrostatic pressure dependence. As a consequence, the limit curve of *Al2024-T351 alloy* exhibits only one axis of symmetry which means that this corresponds to the case of *partly distorted limit surface*. By the use of above formula, authors succeeded with fitting experimental data in rolling direction (RD), transverse to rolling direction (TD), and the thickness direction (ND) Fig. 6.12. However, hydrostatic pressure dependence introduced by the use of right-hand side exponential function leads to *loss of convexity* of the *fracture surface* along the meridian direction in the Haigh–Westergaard space as it was shown by Khan and Liu [20]. The convexity loss discussed in this case is significant only from theoretical point of view, because in such a case Drucker postulate is violated. However, for the data cited by authors the *concave meridian effect* is very small such that it can probably be ignored from engineering point of view for the considered material data. Nevertheless, in spite of possible convexity loss along meridian no convexity loss along circumference is observed; although there exists second exponential function dependent on J_2 and J_3 being a multiplier of the Hill form on the left-hand side of Eq. (6.85).

In another paper by Khan et al. [21], the direct hydrostatic pressure dependence (through I_1) is dropped; however, both significant anisotropy (fully *anisotropic calibration* of all material constants $F, G, \ldots, R$) and *tension/ compression asymmetry* are saved

$$\exp[-C(\zeta+1)]\left(F\sigma_1^2+G\sigma_2^2+H\sigma_3^2+L\sigma_1\sigma_2+M\sigma_2\sigma_3+N\sigma_1\sigma_3 + P\sigma_{12}^2+Q\sigma_{13}^2+R\sigma_{23}^2\right) = 1 \tag{6.86}$$

Although the general form of limit criterion (6.86) accounts for nine independent anisotropy parameters, in example considered by authors in [21], due to calibration

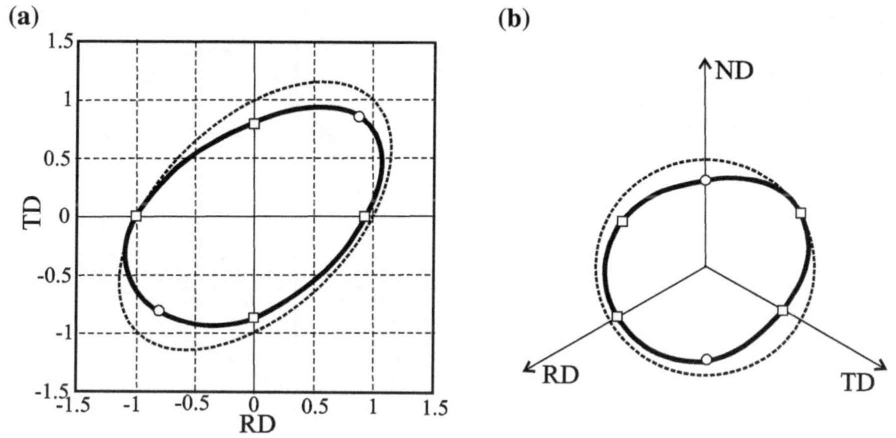

Fig. 6.12 Correlation of the 0.2 % yield loci of Ti-6Al-4V alloy (○—experimental data points, □—points calculated from ND experimental data) with the yield function proposed by Khan et al. [21] (*solid line*) and Huber–von Mises criterion (*dashed line*): **a** comparison in RD-TD plane, **b** projection on deviatoric plane

the material constant G is determined from the equi-biaxial compression test; so it depends on three compression limits like in case of Hill criterion. Under assumption of plane stress state, it reduces to four-parameter *orthotropic Hill condition* (5.58). Fitting of experimental data for *Ti-6Al-4V alloy* at different strain rates and temperatures shows excellent coincidence between the experimental findings and simulation. By contrast to the previous formulation (6.85), the symmetry of the limit curve is lost completely (nonaxis of symmetry exists) as shown in Fig. 6.12.

Orthotropic yield criterion proposed by Yoon et al. [39] being anisotropic extension of the isotropic criterion (6.14)

$$\widetilde{I}_1 + \left(J_2'^{3/2} - J_3''\right)^{1/3} = 1 \tag{6.87}$$

where

$$\begin{aligned} \widetilde{I}_1 &= h_x\sigma_{xx} + h_y\sigma_{yy} + h_z\sigma_{zz} \\ J_2' &= \tfrac{1}{2}\mathbf{s}' : \mathbf{s}' \\ J_3'' &= \det(\mathbf{s}'') \end{aligned} \tag{6.88}$$

The stress tensors $\mathbf{s}'$ and $\mathbf{s}''$ (6.88) are transformed from the stress tensor $\boldsymbol{\sigma}$ to the transformed space by two *fourth-order linear transformation tensors* $\mathbb{L}'$ and $\mathbb{L}''$ as follows $\mathbf{s}' = \mathbb{L}' : \boldsymbol{\sigma}$ and $\mathbf{s}'' = \mathbb{L}'' : \boldsymbol{\sigma}$ with

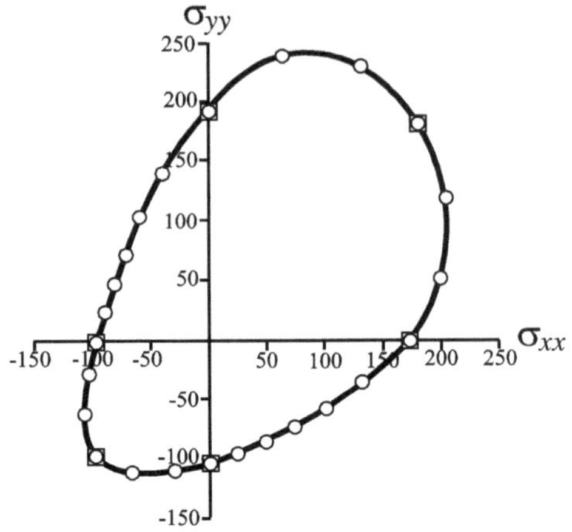

Fig. 6.13 Comparison of the yield surface of AZ31 Magnesium alloy: o—experimental data points, □—yield function proposed by Yoon et al. [39] (unit MPa)

$$\mathbb{L}^{(i)} = \begin{bmatrix} (C_2^{(i)}+C_3^{(i)})/3 & -C_3^{(i)}/3 & -C_2^{(i)}/3 & & & \\ -C_3^{(i)}/3 & (C_3^{(i)}+C_1^{(i)})/3 & -C_1^{(i)}/3 & & & \\ -C_1^{(i)}/3 & -C_1^{(i)}/3 & (C_1^{(i)}+C_2^{(i)})/3 & & & \\ & & & C_4^{(i)} & & \\ & & & & C_5^{(i)} & \\ & & & & & C_6^{(i)} \end{bmatrix} \tag{6.89}$$

where superscript $(i) = \prime$ or $\prime\prime$.

This highly extended yield criterion is capable of capturing all three features: *anisotropy*, *tension/compression asymmetry* and *hydrostatic pressure sensitivity* of various metals like *AA 2008-T4*, high-purity *α-Titanium*, and *AZ31 Magnesium alloy*. Excellent fitting of proposed yield criterion and experimental data of AZ31 is shown in Fig. 6.13. This sufficiently general form can unconditionally be recommended as very effective and specially addressed to model *totally distorted response* of cold-rolled metals.

6.7 Review of Isotropic and Explicit or Implicit Anisotropic Initial Failure Criteria

In this section, a brief review of the selected *pressure sensitive initial yield/failure criteria* is demonstrated. Contrary to the survey given in Sect. 5.9, in the case considered here, the review of pressure sensitive criteria has to account for three characteristic properties:

- the first stress $J_{1\sigma}$ or the first common $\Pi_{ij}\sigma_{ij}$ invariants have to be present in the yield/failure criterion,

- isotropic versus anisotropic formulation,
- direct versus indirect dependence on the stress invariants or the common invariants.

In case of *isotropic pressure sensitive criteria*, the attention is paid to invariant representation of invoked criteria. Selected isotropic yield/failure criteria are collected in Table 6.2. All cited criteria depend on both the first stress invariant and the second deviatoric invariant but additionally they may also depend on the third deviatoric invariant. Criteria C1 Iyer [17], Gao et al. [15] and C3 Iyer and Lissenden [18], Pęcherski et al. [29] are special cases of the general criterion C4 Eq. (6.8), see Sect. 6.2. Criterion C2 by Yoon et al. [39] has slightly different format and cannot be derived form the general criterion C4 as a particular case but it can be considered as the extension of the Cazacu and Barlat [5] pressure insensitive yield criterion A2 in Table 5.3 to the case of *hydrostatic pressure sensitivity*. Chronologically first *yield/failure Coulomb–Mohr's criterion* C5 has been presented in the three equivalent formats: the original Coulomb format (6.16), the Mohr format (6.18) explicitly expressed in terms of principal stresses, and the mixed invariant format (6.22) in which definition $\cos(3\theta) = \frac{3\sqrt{3}}{2}\frac{J_{3s}}{(J_{2s})^{3/2}}$ holds and explicit dependence on the third stress deviator invariant is visible, see Chen and Han [7]. All above criteria C1–C5 represent in the *High–Westergaard space asymmetric yield/failure surfaces* hence tensile and compressive meridians stay in different distance from hydrostatic axes. Next criterion C6 originated by Burzyński [4] represents in the High-Westergaard space *rotationally symmetric surface* of various shapes: *ellipsoidal, paraboloidal, hyperboloidal*, or *conical*, see Fig. 6.5. Hypothetically possible onefold hyperboloidal surface has to be excluded on the base of *Drucker's convex-*

Table 6.2 Review of pressure sensitive isotropic yield/failure criteria

C	Author(s)	Stress invariants
C1	Iyer [17], Gao et al. [15] Eq. (6.9)	$\left(aJ_{1\sigma}^{2p} + bJ_{2s}^{p} + cJ_{3s}^{2p/3}\right)^{1/p} = 1$
C2	Yoon et al. [39] Eq. (6.14)	$a[bJ_{1\sigma} + (J_{2s}^{3/2} - cJ_{3s})^{1/3}] = 1$
C3	Iyer and Lissenden [18], Pęcherski et al. [29] Eq. (6.10)	$aJ_{1\sigma}^{2} + bJ_{2s} + cJ_{3s}^{2/3} = 1$
C4	Extension of C1 and C3 formats, Eq. (6.8)	$\left(aJ_{1\sigma}^{2p} + bJ_{2s}^{p} + cJ_{3s}^{2p/3}\right)^{1/r} - 1 = 0$
C5	Coulomb [8], Mohr [27] Eq. (6.16) principal stress format Eq. (6.18) mixed invariant format Eq. (6.22)	$\lvert\tau\rvert = c - \sigma\tan\phi$ $\frac{1}{2}(\sigma_1 - \sigma_3)\cos\phi = c - \left[\frac{1}{2}(\sigma_1 + \sigma_3) + \frac{\sigma_1 - \sigma_3}{2}\sin\phi\right]\tan\phi$ $\frac{1}{3}J_{1\sigma}\sin\phi + \sqrt{J_{2s}}\sin\left(\theta + \frac{\pi}{3}\right) + \sqrt{\frac{1}{3}J_{2s}}\cos\left(\theta + \frac{\pi}{3}\right)\sin\phi - c\cos\phi = 0$
C6	Burzyński [4] Eq. (6.27)	$A3J_{2s} + B\left(\frac{J_{1\sigma}}{3}\right)^2 + C\left(\frac{J_{1\sigma}}{3}\right) = 1$
C7	Drucker and Prager [9] Eq. (6.39)	$\alpha J_{1\sigma} + \sqrt{J_{2s}} = k$

ity postulate. The limit format of the Burzyński criterion which satisfies Drucker's convexity postulate is the Drucker–Prager [9] criterion C7 that represents the conical surface. Note that both *Burzyński's* and *Drucker–Prager's criteria* degenerate to *Huber–von Mises'* cylindrical *surface* in case when dependence on hydrostatic pressure is neglected.

Selected pressure sensitive *anisotropic yield/failure criteria* are written in Table 6.3. Most of the criteria presented in this table, namely items D1–D13, deal with *explicit formulation* of the anisotropic yield/failure criteria being consistently formulated in the frame of common stress and structural tensors $\Pi_{ij}\sigma_{ij}$, $\Pi_{ijkl}\sigma_{ij}\sigma_{kl}$ and $\Pi_{ijklmn}\sigma_{ij}\sigma_{kl}\sigma_{mn}$. On the other hand, the last two items D14 and D15 comprise

Table 6.3 Review of pressure sensitive anisotropic yield/failure criteria

D	Author(s)	Common invaritants
D1	Goldenblat and Kopnov [16], Sayir [33] Eq. (6.2)	$(\Pi_{ij}\sigma_{ij})^{\alpha} + \left(\Pi_{ijkl}\sigma_{ij}\sigma_{kl}\right)^{\beta} + \left(\Pi_{ijklmn}\sigma_{ij}\sigma_{kl}\sigma_{mn}\right)^{\gamma} + \ldots = 1$
D2	Życzkowski [40] Eq. (6.44)	$\Pi_{ij}\sigma_{ij} + \left(\Pi_{ijkl}\sigma_{ij}\sigma_{kl}\right)^{1/2} + \left(\Pi_{ijklmn}\sigma_{ij}\sigma_{kl}\sigma_{mn}\right)^{1/3} = 1$
D3	Kowalsky et al. [24] Eq. (5.21)	$h^{(0)} + h^{(1)}_{ij}s_{ij} + s_{ij}h^{(2)}_{ijkl}s_{kl} + s_{ij}s_{kl}h^{(3)}_{ijklmn}s_{mn} = 0$
D4	Życzkowski [41] Eq. (6.44)	$\Pi_{ij}\sigma_{ij} + \sqrt{\Pi_{ijkl}\sigma_{ij}\sigma_{kl}} = 1$
D5	Ganczarski and Lenczowski [11], Ganczarski and Skrzypek [13] Eq. (6.46)	$\Pi_{ij}\sigma_{ij} + \Pi_{ijkl}\sigma_{ij}\sigma_{kl} = 1$
D6	Ganczarski and Skrzypek [13] Eq. (6.52)	$\Pi_{ij}\sigma_{ij} + \Pi_{ijkl}s_{ij}s_{kl} = 1$
D7	Orthotropic von Mises–Tsai–Wu Eq. (6.58)	${}_{\mathrm{ort}}\Pi_{ij}\sigma_{ij} + {}_{\mathrm{ort}}\Pi_{ijkl}\sigma_{ij}\sigma_{kl} = 1$
D8	von Mises [37, 38]	$\Pi_{ijkl}\sigma_{ij}\sigma_{kl} = 1$
D9	Khan et al. [21]	${}_{\mathrm{ort}}\Pi_{ijkl}\sigma_{ij}\sigma_{kl} = 1$
D10	Theocaris [35], Liu et al. [26]	${}_{\mathrm{ort}}\Pi_{ij}\sigma_{ij} + \sqrt{\Pi^{\mathrm{H}}_{ijkl}s_{ij}s_{kl}} = 1$
D11	Tsai and Wu [36] Eq. (6.64)	${}_{\mathrm{ort}}\Pi_{ij}\sigma_{ij} + \Pi^{\mathrm{H}}_{ijkl}s_{ij}s_{kl} = 1$
D12	Tetragonal transversely isotropic Tsai–Wu Eq. (6.75)	${}_{\mathrm{tris}}\Pi^{\mathrm{TW}}_{ij}\sigma_{ij} + {}^{\mathrm{tet}}_{\mathrm{tris}}\Pi^{\mathrm{TW}}_{ijkl}s_{ij}s_{kl} = 1$ (6 matrial constants)
D13	Hexagonal transversely isotropic Tsai–Wu, Ganczarski and Adamski [14] Eq. (6.84)	${}_{\mathrm{tris}}\Pi^{\mathrm{TW}}_{ij}\sigma_{ij} + {}^{\mathrm{hex}}_{\mathrm{tris}}\Pi^{\mathrm{TW}}_{ijkl}s_{ij}s_{kl} = 1$ (5 material constants)
D14	Khan and Liu [20] Eq. (6.85), Khan et al. [21] Eq. (6.86)	$\sqrt{\exp[C(\zeta+1)]_{\mathrm{ort}}\Pi_{ijkl}\sigma_{ij}\sigma_{kl}} = \exp(c_1\frac{I_1}{3})$ $\exp[-C(\zeta+1)]_{\mathrm{ort}}\Pi_{ijkl}\sigma_{ij}\sigma_{kl} = 1$
D15	Yoon et al. [39] Eq. (6.87)	$\widetilde{I}_1 + (J_2'^{3/2} - J_3'')^{1/3} = 1$

exemplary *anisotropic yield/failure criteria* based on *implicit formulation* where anisotropy is introduced by linear transformation imposed on the stress tensor, and next generalization of the known pressure sensitive isotropic criteria are done by replacing stresses or stress invariants by transformed ones. All aforementioned criteria include first and second common or transformed invariants by definition (presence of the first invariant is necessary in order to account for hydrostatic pressure sensitivity and the second invariant ensures energy based interpretation of the limit criterion), whereas appearance of the third common or transformed invariant is optional.

The most general form D1 originated by *Goldenblat and Kopnov* [16], Sayir [33] is written in a *polynomial format* where the exponents $\alpha, \beta, \gamma, \ldots$ are arbitrary constants and number of terms is arbitrarily chosen, but usually limited to the first three terms. Two particular cases of the criterion D1 are of special interest. Assuming $\alpha = 1, \beta = 1/2, \gamma = 1/3$ the homogeneity of the polynomial function on the left-hand side is assured, e.g., Życzkowski [40] D2. On the other hand, the criterion D3 used by Kowalsky et al. [24] does not satisfy homogeneity requirement where $\alpha = \beta = \gamma = 1$ holds. In the criteria D1–D2, all three common invariants are saved; hence, the total number of *independent material constants* corresponding to the first $\Pi_{ij}\sigma_{ij}$ the second $\Pi_{ijkl}\sigma_{ij}\sigma_{kl}$ and the third $\Pi_{ijklmn}\sigma_{ij}\sigma_{kl}\sigma_{mn}$ common invariants is equal to $6+21+56 = 83$. Both criteria D1–D2 are formulated in the space of stress tensor components. However, when the majority of metallic materials is considered the stress deviator space is more adequate to formulate limit criteria. Criterion D3 described in this space, having reduced total number of independent material constants, was proposed by Kowalsky et al. [24]. Engineering application of the full format including all three common invariants is very complicated because it requires identification of large number of modules of the third common invariant (up to 56 in general case). The *third common invariant* is responsible for *distortion of limit surface* hence in all cases where distortion is not very significant it is reasonable to neglect the third common invariant. Items D4–D13 take advantage of aforementioned simplification, what means that only first two common invariants are saved, which drastically reduces number of independent material constants down to $6 + 21 = 27$. Both items D4 and D5 are consequently written in the stress space. However, item D4 represents the conical-type limit surface, being *anisotropic generalization of* the isotropic *Drucker–Prager cone*; whereas, item D5 represents paraboloidal-type limit surface, being *anisotropic generalization of* isotropic rotationally symmetric *Burzyński's paraboloid.* Of course due to anisotropy, both discussed criteria do not satisfy the rotational symmetry property. In some cases, it is justified to use deviatoric format of the second common invariant only that leads to some reduction of number of independent material constants $6 + 15 = 21$. The representative of such limit criterion is item D6. If fully deviatoric format of both the first and the second common invariants is used, we arrive at the *hydrostatic pressure independent criterion* considered by Szczepiński [34] (6.55) that has $5 + 15 = 20$ independent material constants. This criterion does not appear in Table 6.3 since it is pressure independent; however, its simplified form that has not the first deviatoric common invariant (5.43) was discussed in Sect. 5.2. The essential difference between both criteria (5.43) and (6.55) is that Eq. (5.43) represents limit surface that axis coincides with the hydrostatic axis whereas Eq. (6.55) represents limit surface that axis is shifted from the hydrostatic axis. The appropriate limit surfaces degenerate to not rotationally symmetric cylinders.

Criterion D7 may be considered as an extension of the orthotropic von Mises criterion (5.45) by use of the first common invariant described $3+9=12$ independent material constants. Criteria presented in items D8 and D9 do not contain the first common invariant and they are written down in the stress space. This means that both discussed criteria are pressure sensitive ones. The general von Mises criterion D8 is described by 21 independent material constants whereas the criterion D9, suggested by Khan et al. [21] contains only 9 independent material constants since it describes material orthotropy. Next two items, namely D10 and D11, can be considered as narrower formats of items D4 and D5. Both criteria are determined by $3+6=9$ independent material constants. This reduction is furnished by simultaneous use of two substitutions: first, substitution of *Hill's structural tensor* (6 independent material constants), instead of Mises' tensor (21 independent material constants); and second, substitution of stress deviator by stress tensor in the second common invariant.

Criteria D12 and D13 describe yield/failure surfaces in case of *transverse isotropy of tetragonal* Eq. (6.69) and new *hexagonal* Eq. (6.84) *symmetry* however they differ each from the other in this sense that format Eq. (6.69) is described by $2+4=6$ independent material constants by contrast to the format Eq. (6.84) in which $2+3=5$ independent material constants is present. In such a way the narrower hexagonal form assures its reducibility to the shifted *Huber–von Mises* type *ellipse* in the plane of transverse isotropy.

Criteria D14 and D15 belong to separate type of limit criteria in that sense that they are neither the common invariant-based explicit equations nor linear transformation-based implicit generalization of chosen isotropic criteria. These original mixed concepts are difficult to be classified in sense of either implicit or explicit approaches because involved simultaneously all three invariants. In D14 criterion suggested by Khan and Liu [20], Khan et al. [21], the second common invariant $\Pi^{\text{ort}}_{ijkl}\sigma_{ij}\sigma_{kl}$ and all stress invariants I_1, J_2, J_3 are involved. In D15 criterion proposed by Yoon et al. [39], the first common invariant $\Pi_{ij}\sigma_{ij}$ together with the second and the third transformed invariants J_2', J_3' are used. The format with J_2', J_3' turns out to be anisotropic extension of Drucker's criterion which appears in power $1/3$ due to assure dimension homogeneity with the first common invariant $\widetilde{I}_1$.

References

1. Abu Al-Rub, R.K., Voyiadjis, G.Z.: On the coupling of anisotropic damage and plasticity models for ductile materials. Int. J. Solids Struct. **40**, 2611–2643 (2003)
2. Balmer, G.G.: Shearing strength of concrete under high triaxial stress—computation of Mohr's envelope as a curve. Structural Research Laboratory Report SP-23. Denver, Colorado (1949)
3. Brünig, M., Berger, S., Obrecht, H.: Numerical simulation of the localization behavior of hydrostatic-stress-sensitive metals. Int. J. Mech. Sci. **42**, 2147 (2000)

4. Burzyński, W.: Study on Strength Hypotheses (in Polish). Akademia Nauk Technicznych, Lwów (1928)
5. Cazacu, O., Barlat, F.: A criterion for description of anisotropy and yield differential effects in pressure-insensitive materials. Int. J. Plast. **20**, 2027–2045 (2004)
6. Cazacu, O., Planckett, B., Barlat, F.: Orthotropic yield criterion for hexagonal close packed metals. Int. J. Plast. **22**, 1171–1194 (2006)
7. Chen, W.F., Han, D.J.: Plasticity for Structural Engineers. Springer, Berlin-Heidelberg (1995)
8. Coulomb, C.: Essai sur une application des régles maximis et minimis á quelques problémes de statique rélatifs á l'architecture. Mémoires de l'Acad. Roy. des Sci. **7**, 43 (1776)
9. Drucker, D.C., Prager, W.: Solid mechanics and plastic analysis of limit design. Q. Appl. Math. **10**, 157–165 (1952)
10. Egner, H.: On the full coupling between thermo-plasticity and thermo-damage in thermodynamic modeling of dissipative materials. Int. J. Solids Struct. **49**, 279–288 (2012)
11. Ganczarski, A., Lenczowski, J.: On the convexity of the Goldenblatt-Kopnov yield condition. Arch. Mech. **49**(3), 461–475 (1997)
12. Ganczarski, A., Skrzypek, J.: Plasticity of Engineering Materials (in Polish). Wydawnictwo Politechniki Krakowskiej (2009)
13. Ganczarski, A., Skrzypek, J.: Mechanics of novel materials (in Polish). Wydawnictwo Politechniki Krakowskiej (2013)
14. Ganczarski, A., Adamski, M.: Tetragonal or hexagonal symmetry in modeling of yield criteria for transversely isotropic materials. Acta Mechanica et Automatica **8**(3), 125–128 (2014)
15. Gao, X., Zhang, T., Zhou, J., Graham, S.M., Hayden, M., Roe, C.: On stress-state dependent plasticity modeling significance at the hydrostatic stress, the third invariant of stress deviator and the non-associated flow rule. Int. J. Plast. **27**, 217–231 (2011)
16. Goldenblat, I.I., Kopnov V.A.: Obobshchennaya teoriya plasticheskogo techeniya anizotropnyh sred, Sbornik Stroitelnaya Mehanika, pp. 307–319. Stroĭizdat, Moskva (1966)
17. Iyer, S.K.: Viscoplastic model development to account for strength differential: application to aged Inconel 718 at elevated temperature. Ph.D thesis, The Pennsylvania State University (2000)
18. Iyer, S.K., Lissenden, C.J.: Multiaxial constitutive model accounting for the strength-differential in Inconel 718. Int. J. Plast. **19**, 2055–2081 (2003)
19. Khan, A.S., Kazmi, R., Farrokh, B.: Multiaxial and non-proportional loading responses, anisotropy and modeling of Ti-6Al-4V titanium alloy over wide ranges of strain rates and temperatures. Int. J. Plast. **23**, 931–950 (2007)
20. Khan, A.S., Liu, H.: Strain rate and temperature dependent fracture criteria for isotropic and anisotropic metals. Int. J. Plast. **37**, 1–15 (2012)
21. Khan, A.S., Yu, S., Liu, H.: Deformation induced anisotropic responses of Ti-6Al-4V alloy, part II: a strain rate and temperature dependent anisotropic yield criterion. Int. J. Plast. **38**, 14–26 (2012)
22. Korkolis, Y.P., Kyriakides, S.: Inflation and burst of aluminum tubes, part II: an advanced yield function including deformation-induced anisotropy. Int. J. Plast. **24**, 1625–1637 (2008)
23. Kowalewski, Z.L., Śliwowski, M.: Effect of cyclic loading on the yield surface evolution of 18G2A low-alloy steel. Int. J. Mech. Sci. **39**(1), 51–68 (1997)
24. Kowalsky, U.K., Ahrens, H., Dinkler, D.: Distorted yield surfaces—modeling by higher order anisotropic hardening tensors. Comput. Mater. Sci. **16**, 81–88 (1999)
25. Kupfer, H., Hilsdorf, H.K., Rusch, H.: Behavior of concrete under biaxial stresses. ACI J. **66**(8), 656–666 (1969)
26. Liu, C., Huang, Y., Stout, M.G.: On the asymmetric yield surface of plastically orthotropic materials: a phenomenological study. Acta Mater. **45**, 2397–2406 (1997)
27. Mohr, O.: Welche Umstände bedingen die Elastizitätsgrenze und den Bruch eines Materials? Verein Deuts. Ing. Zeit. **44**(1524–1530), 1572–1577 (1900)
28. Ottosen, N.S., Ristinmaa, M.: The Mechanics of Constitutive Modeling. Elsevier, Amsterdam (2005)

29. Pęcherski, R.B., Szeptyński, P., Nowak, M.: An extension of Burzyński hypothesis of material effort accounting for the third invariant of stress tensor. Arch. Metall. Mater. **56**, 503–508 (2011)
30. Ralston, T.D.: Yield and plastic deformation in ice crushing failure. In: ICSI.AIDJEX Symposium on Sea Ice-Processes and Models. Seattle, Washington (1977)
31. Raniecki, B., Mróz, Z.: Yield or martensitic phase transformation conditions and dissipative functions for isotropic, pressure-insensitive alloys exhibiting SD effect. Acta Mech. **195**, 81–102 (2008)
32. Richart, F.E., Brandtzaeg, A., Brown, R.L.: A study of the failure of concrete under combined compressive stresses. University of Illinois Engineering Experimental Station Bulletin, 185 (1928)
33. Sayir, M.: Zur Fließbedingung der Plastiztätstheorie. Ingenierarchiv **39**, 414–432 (1970)
34. Szczepiński, W.: On deformation-induced plastic anisotropy of sheet metals. Arch. Mech. **45**(1), 3–38 (1993)
35. Theocaris, P.S.: Weighting failure tensor polynomial criteria for composites. Int. J. Damage Mech. **1**, 4–46 (1992)
36. Tsai, S.T., Wu, E.M.: A general theory of strength for anisotropic materials. Int. J. Numer. Methods Eng. **38**, 2083–2088 (1971)
37. von Mises, R.: Mechanik der festen Körper im plastisch deformablen Zustand, Nachrichten der Geselschaft der Wissenschaften zu Götingen (1913)
38. von Mises, R.: Mechanik der plastischen Formänderung von Kristallen. ZAMM **8**(13), 161–185 (1928)
39. Yoon, J.W., Lou, Y., Yoon, J., Glazoff, M.V.: Asymmetric yield function based on stress invariants for pressure sensitive metals. Int. J. Plast. (2014) (in press)
40. Życzkowski, M.: Combined Loadings in the Theory of Plasticity. Polish Scientific Publishers, Warszawa (1981)
41. Życzkowski, M.: Anisotropic yield conditions. In: Lemaitre, J. (ed.) Handbook of Materials Behavior Models, pp. 155–165. Academic Press, San Diego (2001)

Chapter 7
Classification of Constitutive Equations for Dissipative Materials—General Review

H. Egner and W. Egner

Abstract In the present chapter the general features of thermodynamically based constitutive modeling are described. In such approach a basic hypothesis is that the state of a material is entirely determined by certain values of some independent variables, called variables of state. This type of constitutive modeling is particularly well adapted to the formulation of constitutive equations for deformable solids with several dissipative phenomena. A common three-stage procedure in the definition of a constitutive model is discussed: (1) choice of the state variables, (2) definition of the state potential from which the state relations (between strain-like variables and their dual conjugated forces) are derived, and (3) choice of the dissipation potential from which the rate equations of state variables are derived. The classification of constitutive equations is then presented for elastic-damage, elastic-plastic, thermo-elastic-(visco)plastic, and elastic-plastic-damage materials. Damage-induced anisotropy and unilateral damage effect are accounted. When plasticity is considered, an alternative multiscale approach, based on polycrystalline calculations for the description of yielding anisotropy and its evolution with accumulated deformation, is also discussed. As an example of thermoplastic coupling, the fatigue behavior of martensitic hot work tool steel in nonisothermal conditions is analyzed. In this example two cases are compared: (1) partial coupling, when changing temperature is accounted only in changing material parameters, and (2) full coupling, when additional terms proportional to temperature rate are added in the kinetic equations of thermodynamic conjugate forces. Numerical simulations are performed, which indicate the significant influence of temperature rate on the response of constitutive model when cyclic thermomechanical loading is considered.

H. Egner (✉)
Solid Mechanics Division, Institute of Applied Mechanics,
Cracow University of Technology, al. Jana Pawła II 37, 31-864 Kraków, Poland
e-mail: Halina.Egner@pk.edu.pl

W. Egner
Division of Technical Mechanics, Institute of Applied Mechanics,
Cracow University of Technology, al. Jana Pawła II 37, 31-864 Kraków, Poland
e-mail: Wladyslaw.Egner@pk.edu.p

J.J. Skrzypek and A.W. Ganczarski (eds.), *Mechanics of Anisotropic Materials*,
Engineering Materials, DOI 10.1007/978-3-319-17160-9_7

Keywords Constitutive modeling · Dissipative material · Thermoplastic coupling · Damage-induced anisotropy

7.1 Coupled Dissipative Phenomena

We consider a specific portion of the physical universe, called a system. A system is closed, which means that there is no exchange of matter between it and its surroundings. To describe the state of a system we need a set of macroscopic quantities, which are characteristics of a system, and which can be scalars, vectors, or tensors (matrices), such as temperature or a strain tensor. Such quantities are called *state variables*, and functions of state variables are *state functions*. State functions depend only on the state of the system and not on the manner in which this state is achieved. A system subjected to loading undergoes the process of deformation. If both the system and its surroundings can be brought back to their initial conditions, the process of deformation is called reversible. If the restoration of a system to its initial conditions requires changing of the conditions of its surroundings, the process is called irreversible. For *reversible processes*, the material after unloading returns to its initial state, and its characteristics do not change. In the case of *irreversible process* the material after unloading does not return to its initial state, but to some residual state, characterized by residual strains and stresses, and changed material properties. Irreversible phenomena are accompanied by the *dissipation of energy* introduced to the material in the course of deformation. For this reason they are called *dissipative phenomena*. In particular, among reversible phenomena we may indicate *linear* or *nonlinear elasticity*, while *plasticity*, *creep*, and/or *damage* are irreversible and lead to various rearrangements of a material microstructure. Elastic response is independent of the load history, so that the response for loading and unloading follows the same path (path independent). In accordance, the stresses are uniquely given by the strains through the constitutive relation.

Reversible deformation is limited when irreversible *rearrangements* of a *material microstructure* are initiated. Most often the material degradation connected with slip rearrangements of crystallographic planes through dislocation motion is observed (ductile materials) and/or the development of microcracks and microvoids takes place (brittle materials). Such behavior may be illustrated on the example of uniaxial loading/unloading for three typical groups of engineering materials: *(visco)plastic* (or ductile—most of metals), *brittle* (some CMC-type composites, ceramic materials, concrete, rocks) and mixed *(visco)plastic/brittle* (most of MMC composites, cast iron). Figure 7.1a illustrates the response of ductile (visco)plastic material. After reaching *threshold stress* k^{vp} the initially linear stress–strain diagram becomes nonlinear due to the development of (visco)plastic strains. On the basis of experimental observations it can be assumed that during this hardening process the elastic modulus remains unchanged, $E^{\mathrm{vp}} = E^{\mathrm{e}}$. The areas $\mathcal{W}^{\mathrm{e}}$, $\mathcal{W}^{\mathrm{vp}}$, and $\mathcal{W}^{\mathrm{svp}}$ in Fig. 7.1a illustrate, respectively: the *elastic strain energy* that can be retrieved during unloading, the *dissipated energy* (mainly in the form of heat), and the *energy* used *for rearrangements*

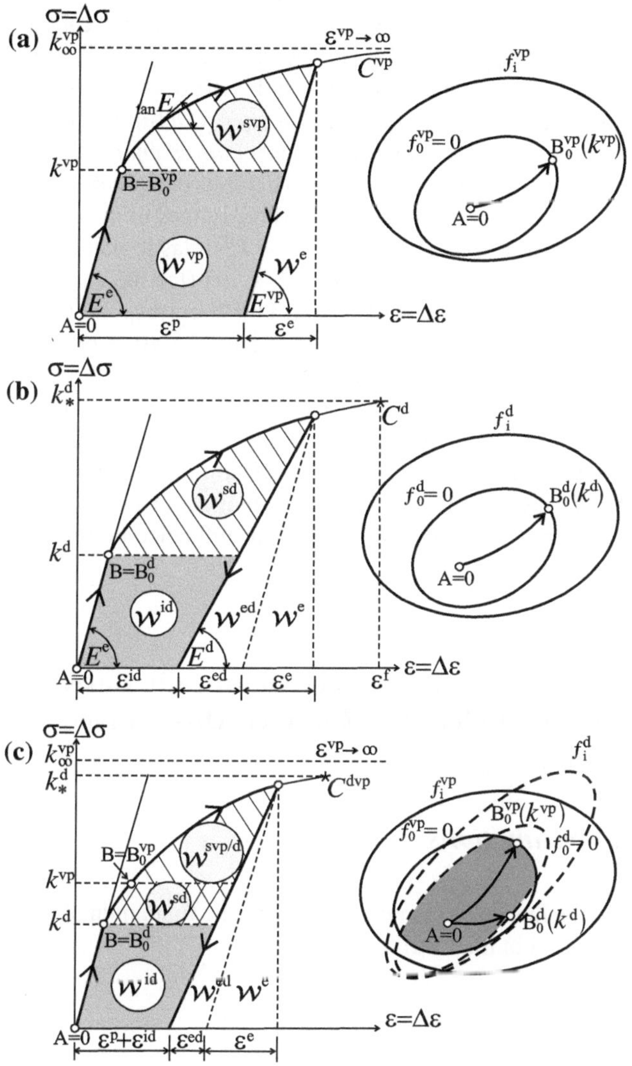

Fig. 7.1 Different dissipative phenomena observed during uniaxial loading/unloading: **a** (visco)plasticity; **b** damage; **c** plasticity with damage, after Ganczarski and Skrzypek [32]

of material *microstructure* (stored energy). Stress level k_{∞}^{vp} is an asymptotic value, to which the stress tends when the viscoplastic strain tends to infinity, $\varepsilon^{\mathrm{vp}} \rightarrow \infty$ (asymptotic hardening).

Brittle materials exhibit quite different behavior. After reaching threshold value k^{d} the initiations of microcracks and/or microvoids are observed. Further increase of the load causes the development of microdamage, and finally leads to *brittle fracture* when fracture strain ε^{f} is reached (see Fig. 7.1b). Microdamage development causes material degradation (damage softening) observed on macroscale as a progressive drop of elastic modulus, $E^{\mathrm{d}} < E^{\mathrm{e}}$. The areas $\mathcal{W}^{\mathrm{e}}$, $\mathcal{W}^{\mathrm{ed}}$, $\mathcal{W}^{\mathrm{id}}$, and $\mathcal{W}^{\mathrm{sd}}$ illustrate, respectively, the *elastic strain energy*, reversible part of the energy related to partial *closure of microcracks* during unloading, the *energy dissipated by damage*, and the energy used for microstructural rearrangements (*stored energy*).

In a more complex case, when the material exhibits both ductile and brittle features (for example, spheroidal graphite cast iron) three ranges of the deformation process can be distinguished (see Fig. 7.1c): elastic ($\sigma < k^{\mathrm{d}}$), elastic-damage ($k^{\mathrm{d}} < \sigma < k^{\mathrm{vp}}$), and mixed *(visco)plastic-damage range* ($k^{\mathrm{vp}} < \sigma$) (for the case when $k^{\mathrm{d}} < k^{\mathrm{vp}}$). Respective areas: $\mathcal{W}^{\mathrm{e}}$, $\mathcal{W}^{\mathrm{ed}}$, $\mathcal{W}^{\mathrm{id}}$, $\mathcal{W}^{\mathrm{sd}}$, and $\mathcal{W}^{\mathrm{svpd}}$ correspond to the *elastic strain energy*, reversible part of the *energy* related to partial *closure of microcracks* during unloading, the *energy dissipated by damage*, the *energy stored by damage hardening*, and the energy stored by simultaneous development of microdamage and plastic strains. The elastic range in such complex case is limited by a surface which is a common part of two other surfaces, related to *plastic slip initiation*, f^{vp}, and *damage initiation*, f^{d}. During coupled process of damage and (visco)plastic dissipation, plastic strains, as well as damage development, influence both plastic and damage surfaces (cf. Egner [26]).

7.2 General Features of Thermodynamically Based Constitutive Modeling of Coupled Dissipative Phenomena

7.2.1 Basic Assumptions

The motions of the thermodynamic system obey the fundamental laws of *continuum mechanics* expressed in the local form:

- conservation of mass

$$\dot{\rho} + \rho v_{i|i} = 0 \tag{7.1}$$

- conservation of linear momentum

$$\sigma_{ij|j} + \rho b_i = \rho \dot{v}_i \tag{7.2}$$

- conservation of angular momentum

$$\sigma_{ij} = \sigma_{ji} \tag{7.3}$$

- the first law of thermodynamics

$$\rho \dot{u} - \dot{\varepsilon}_{ij}\sigma_{ij} - r + q_{i|i} = 0 \tag{7.4}$$

- the second law of thermodynamics

$$\rho\dot{s} - \frac{r}{T} + \left(\frac{q_i}{T}\right)_{|i} \geq 0 \tag{7.5}$$

The following notation is used: ρ is the mass density per unit volume; v_i is the material velocity; σ_{ij} is the stress tensor; b_i is the body force density per unit mass; u is the internal energy per unit mass; ε_{ij} is the strain tensor; r is the distributed heat source per unit volume; q_i is the outward heat flux; s stands for the internal entropy production per unit mass; T is the absolute temperature. Depending on the scale, different approaches may be used in order to describe an overall structural response of a dissipative structure on the macroscale. In general, *micromechanical models* relate the macroproperties and the macroresponse of a structure to its microstructure. In such approach the rearrangements of microstructure are discrete and stochastic phenomena induced by a number of weakly or strongly interacting microchanges that influence the overall structural response. The micromechanical models have the advantage of being able to sustain heterogeneous structural details on the microscale and mesoscale, and to allow a micromechanical formulation of the evolution equations based on the accurate microchange growth processes involved (cf. Chaboche et al. [17]; Boudifa et al. [11], Aboudi [1]).

Continuum Mechanics approach discussed in the present chapter provides the constitutive modeling in the framework of thermodynamics of irreversible processes with internal state variables. This approach is based on a concept of the *effective quasi-continuum* (see Fig. 7.2c). The material heterogeneity (on the micro and mesoscale) is smeared out over the Representative Volume Element (RVE) of the piecewise discontinuous material. The true state of a material within RVE, represented by the topology, size, orientation, and number of microchanges, is mapped to a material point of the effective quasi-continuum. The true distribution of microchanges within the RVE, and the correlation between them are measured by the change of the effective constitutive properties. The microstructural rearrangements are defined by the set of *state variables* of the scalar, vectorial, or tensorial nature (cf. Murakami and Ohno [54]; Litewka [49]; Chaboche [18]; Skrzypek and Ganczarski [69]; Skrzypek et al. [70]; Ganczarski et al. [30]). The constitutive tensors for the dissipative material are defined by the use of even-rank effect tensors (*damage effect tensor, phase transformation effect tensor*, etc.) that map thermodynamic forces from the physical (discontinuous) to the fictitious (pseudocontinuous) configurations, see Fig. 7.2.

7.2.2 State Potential and State Equations

In constitutive modeling, the well-known formalism of *thermodynamics of irreversible processes* with *internal state variables*, and the *local state method* are often adopted. In this approach a central hypothesis is that the state of a material is entirely determined by certain values of some independent variables, called variables of

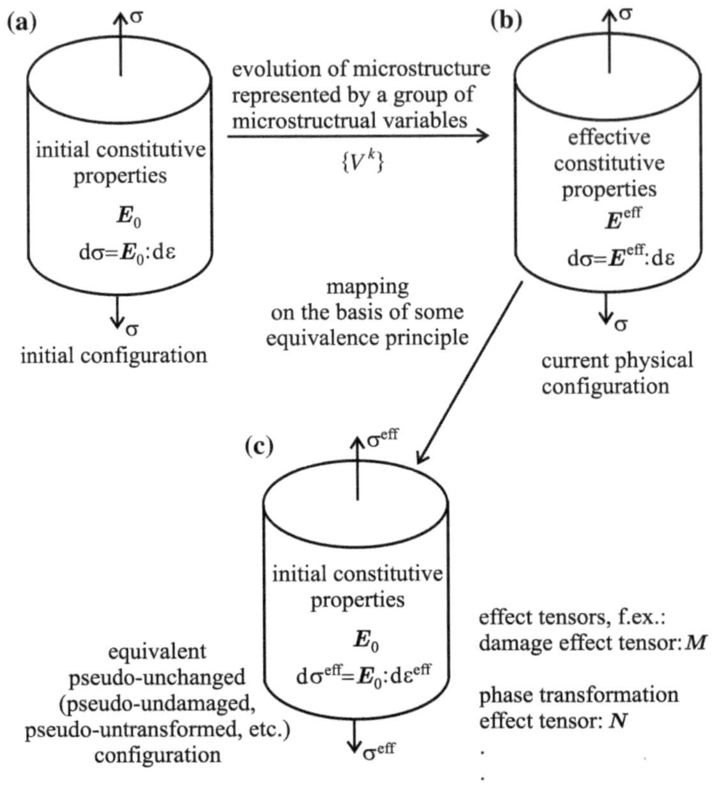

Fig. 7.2 **a** Virgin (initial), **b** physical (changed), and **c** equivalent pseudo-unchanged continuum; the equivalence principles are used in order to smear out the true microchanges distribution over the RVE to yield the effective constitutive modules for dissipative material

state (see Chap. 1). This type of approach is particularly well suitable to the formulation of constitutive equations for deformable solids with several dissipative phenomena. The constitutive behavior is defined by the specification of two potentials: an energy (state) potential and a dissipation potential. A *state potential* is a closed, convex, and scalar-valued function of the overall state variables. Usually, Helmholtz's free energy density is adopted, decomposed into elastic ($\rho\psi^{\mathrm{e}}$), plastic ($\rho\psi^{\mathrm{p}}$), damage ($\rho\psi^{\mathrm{d}}$), phase change ($\rho\psi^{\mathrm{ph}}$), etc. terms (cf. Lemaitre and Chaboche [48], Abu Al-Rub and Voyiadjis [3]):

$$\rho\psi = \sum_{k=1}^{n} \rho\psi^{k} = \rho\psi^{\mathrm{e}} + \rho\psi^{\mathrm{p}} + \rho\psi^{\mathrm{d}} + \rho\psi^{\mathrm{ph}} + \cdots \tag{7.6}$$

By eliminating all the reversible processes from the Clausius–Duhem inequality (7.5) the following *state equations* which express the thermodynamic forces conjugated to the observable state variables (see Chap. 1) are obtained:

$$\sigma_{ij} = \frac{\partial(\rho\psi)}{\partial\varepsilon^{\mathrm{e}}_{ij}} = \left.\frac{\partial(\rho\psi)}{\partial\varepsilon^{\mathrm{e}}_{pq}}\right|_{\varepsilon^{*}_{ij}=\mathrm{const}} + \frac{\partial(\rho\psi)}{\partial\varepsilon^{*}_{kl}}\frac{\partial\varepsilon^{*}_{kl}}{\partial\varepsilon^{\mathrm{e}}_{ij}}$$
$$s = -\frac{\partial\psi}{\partial T} \tag{7.7}$$

The second term in (7.7_1) introduces the fourth-order tensor $D^{(\varepsilon)}_{ijkl} = \frac{\partial\varepsilon^{*}_{kl}}{\partial\varepsilon^{e}_{ij}}$ which accounts for the unilateral damage effect. The definition of tensor ε^{*}_{ij} is given by Eq. (1.58) in Chap. 1 of this book. In addition, the pairs of forces conjugated to other microstructural or *hidden state variables* are postulated in a similar form to (7.7_1) and (7.7_2) (cf. Chaboche [18]):

$$-Y^k = \frac{\partial(\rho\psi)}{\partial V^k}, \quad H^k = \frac{\partial(\rho\psi)}{\partial h^k} \tag{7.8}$$

In the above equations Y^k stand for *thermodynamic forces* conjugated to *microstructural state variables* V^k Eq. (1.54), whereas *hardening forces* H^k are conjugated to *hidden state variables* h^k. In particular, when hardening variables (1.56) are used, thermodynamic forces $H^k = \left\{R^k, X^k_{ij}, L^k_{ijkl}, G^k_{ijklmn}\right\}$, conjugate of corresponding internal variables h^k (1.56) are defined for each dissipation mechanism ($k = \mathrm{p}, \mathrm{d}, \mathrm{ph}, \ldots$) (see Table 7.1):

$$R^k = \frac{\partial(\rho\psi)}{\partial r^k}, \; X^k_{ij} = \frac{\partial(\rho\psi)}{\partial\alpha^k_{ij}}, \; L^k_{ijpq} = \frac{\partial(\rho\psi)}{\partial l^k_{ijpq}}, \; G^k_{ijpqrs} = \frac{\partial(\rho\psi)}{\partial g^k_{ijpqrs}} \tag{7.9}$$

If we now define the *thermodynamic conjugate force vector* $\boldsymbol{J}$ and the *flux vector* $\dot{\boldsymbol{P}}$ as:

$$\boldsymbol{J} = \left[\sigma_{ij}, \frac{T_{|i}}{T}; -Y^k, H^k\right], \quad \dot{\boldsymbol{P}} = \left[\dot{\varepsilon}^{I}_{ij}, -q_i; -\dot{V}^k, -\dot{h}^k\right]^T \tag{7.10}$$

then the *dissipation inequality* (7.5) can be expressed as the scalar product of $\boldsymbol{J}$ and $\dot{\boldsymbol{P}}$ as follows (Krajcinovic [44]; Ottosen and Ristinmaa [60]):

$$\sqcap = \boldsymbol{J}\cdot\dot{\boldsymbol{P}} \geq 0 \tag{7.11}$$

where $\sqcap$ is the dissipation function,

$$\sqcap = \underbrace{\sigma_{ij}\dot{\varepsilon}^{I}_{ij} + Y^k\dot{V}^k - H^k\dot{h}^k}_{\sqcap^{\mathrm{mech}}} - \underbrace{q_i T_{|i}/T}_{\sqcap^{\mathrm{T}}} \tag{7.12}$$

that can be fartherly decomposed into mechanical $\sqcap^{\mathrm{mech}}$ and thermal $\sqcap^{\mathrm{T}}$ terms.

The state relations (7.7)–(7.9) are deduced from the fundamental *Clausius–Duhem inequality*. The kinetic equations of force-like variables are then obtained by taking

Table 7.1 State variables and corresponding thermodynamic conjugate forces for the general thermo-elastic-plastic-damage two phase material

Phenomenon	State variables	Conjugate forces
Mechanical variables	Observable state variables: ε_{ij}, T	σ_{ij}, s
Plastic flow	$\varepsilon^{\mathrm{e}}_{ij}$ (or $\varepsilon^{\mathrm{p}}_{ij}$)	σ_{ij} (or $-\sigma_{ij}$)
Plastic hardening	r^{p} (isotropic)	R^{p}
	α^{p}_{ij} (kinematic)	X^{p}_{ij}
	l^{p}_{ijkl} (anisotropic)	L^{p}_{ijkl}
	g^{p}_{ijklmn} (distortional)	G^{p}_{ijkl}
Damage	$V^{\mathrm{d}}_{ij} = D_{ij}$	$-Y^{\mathrm{d}}_{ij}$
Damage hardening	r^{d} (isotropic)	R^{d}
	α^{d}_{ij} (kinematic)	X^{d}_{ij}
	l^{d}_{ijkl} (anisotropic)	L^{d}_{ijkl}
	g^{d}_{ijklmn} (distortional)	G^{d}_{ijkl}
Phase transformation	$V^{\mathrm{ph}}_{ij} = \xi_{ij}$	$-Y^{\mathrm{d}}_{ij}$
Phase transformation hardening	r^{ph} (isotropic)	R^{ph}
	$\alpha^{\mathrm{ph}}_{ij}$ (kinematic)	X^{ph}_{ij}
	l^{ph}_{ijkl} (anisotropic)	L^{ph}_{ijkl}
	g^{ph}_{ijklmn} (distortional)	G^{ph}_{ijkl}

time derivatives. Various coupling terms appear in the kinetic equations, that are necessary for proper description of a material behavior, especially when cyclic loading is considered [27, 28].

7.2.3 Dissipation Potential and Rate Equations

Dissipation Potential

Once the force-like variables are known from the state relations, it remains to define the flux variables so that the volumic dissipation $\sqcap$ (7.12) is always nonnegative. In order to define the evolution equations to dissipative phenomena, the existence of several *dissipation potentials* F^k may be assumed, corresponding to each microstructural rearrangement (due to plastic flow p, damage growth d, phase change ph etc.), and defined independently but partly coupled (*weak dissipation coupling*, Chaboche [18]). Dissipation potentials are assumed in the form of positive, convex, closed, and scalar-valued functions of force-like variables, with the *associated state variables* able to serve as parameters.

Dissipation functions in general can be expressed in the following nonassociated form:

$$F^k = f^k + g^k_{\text{iso}}(R^k) + g^k_{\text{kin}}(X^k_{ij}) + g^k_{\text{aniso}}(L^k_{ijpq}) + g^k_{\text{dist}}(G^k_{ijpqrs}) \tag{7.13}$$

where g^k_{iso}, g^k_{kin}, g^k_{aniso}, and g^k_{dist} are functions corresponding to recovery effects of partial progressive return to the original microstructure (Kuo and Lin [45]; Mirzakhani et al. [52]). Only the first two terms, related to isotropic and kinematic effects, are used in the majority of existing models. Usually, the recovery functions are defined as quadratic functions of thermodynamic forces conjugated to hardening variables, R^k, X^k_{ij}, L^k_{ijpq}, G^k_{ijpqrs}.

Loading Functions

In Eq. (7.13) f^k stands for a loading surface related to the kth dissipative phenomenon. A first classical approach to the definition of the loading/failure criteria, usually in relation to the phenomenon of plastic flow, is based on a concept of *common invariants* of the stress and structural anisotropy tensors (see Chap. 4). Loading functions f^k, described by relevant thermodynamic forces which are tensors of different order, can be listed in a polynomial hierarchy with increasing complexity and hardening properties, see Kowalsky et al. [43]. For example, the plastic yield function f^{p} of third degree with *distortional hardening* is given by

$$f^{\text{p}} = \sqrt{\frac{3}{2} s^{\text{ef}}_{ij} L^{p}_{ijkl} s^{\text{ef}}_{kl} + s^{\text{ef}}_{ij} s^{\text{ef}}_{kl} G^{\text{p}}_{ijklmn} s^{\text{ef}}_{mn}} - (R^{\text{p}}_0 + R^{\text{p}}) = 0 \tag{7.14}$$

where $s^{\text{ef}}_{ij} = \sigma^{\text{ef}}_{ij} - \frac{1}{3}\sigma^{\text{ef}}_{kk}\delta_{ij}$ and $\sigma^{\text{ef}}_{ij} = \sigma_{ij} - X^{\text{p}}_{ij}$. Other loading functions of third degree (damage, phase transformation, etc.) may be defined as

$$\begin{aligned} f^{\text{d}} &= \sqrt{Y^{\text{def}}_{ij} L^{\text{d}}_{ijkl} Y^{\text{def}}_{kl} + Y^{\text{def}}_{ij} Y^{\text{def}}_{kl} G^{\text{d}}_{ijklmn} Y^{\text{def}}_{mn}} - (R^{\text{d}}_0 + R^{\text{d}}) = 0 \\ f^{\text{ph}} &= \sqrt{Y^{\text{phef}}_{ij} L^{\text{ph}}_{ijkl} Y^{\text{phef}}_{kl} + Y^{\text{phef}}_{ij} Y^{\text{phef}}_{kl} G^{\text{ph}}_{ijklmn} Y^{\text{phef}}_{mn}} - (R^{\text{ph}}_0 + R^{\text{ph}}) = 0 \end{aligned} \tag{7.15}$$

where $Y^{\text{def}}_{ij} = Y^{\text{d}}_{ij} - X^{\text{d}}_{ij}$ and $Y^{\text{phef}}_{ij} = Y^{\text{ph}}_{ij} - X^{\text{ph}}_{ij}$. Investigations by Streilein [71] have shown that such polynomial yield functions, including hardening tensors up to the sixth order are best covering the experimental data. To account for the unilateral damage effect, the damage loading function f^{d} may be expressed in terms of the modified damage-conjugated thermodynamic force $Y^{\text{def}*}_{ij}$ (see Challamel et al. [22]):

$$f^{\text{d}} = \sqrt{Y^{\text{def}*}_{ij} L^{\text{d}}_{ijkl} Y^{\text{def}*}_{kl} + Y^{\text{def}*}_{ij} Y^{\text{def}*}_{kl} G^{\text{d}}_{ijklmn} Y^{\text{def}*}_{mn}} - (R^{\text{d}}_0 + R^{\text{d}}) = 0 \tag{7.16}$$

where (see Eq. (1.58))

$$Y_{ij}^{\mathrm{def}*} = \sum_{I=1}^{3} \kappa(Y_{kl}^{\mathrm{def}}) n_{iI}^{(Y)} n_{jI}^{(Y)} n_{Ik}^{(Y)} n_{Il}^{(Y)} Y_{kl}^{\mathrm{def}} = B_{ijkl}^{(Y)} Y_{kl}^{\mathrm{def}} \tag{7.17}$$

Second approach, developed recently for example by Barlat et al. [4–6], Cazacu and Barlat [14], Plunkett et al. [65], Yoon et al. [78] is of mixed type: classical isotropic yield criteria were extended to anisotropy by the application of *linear transformation of* the *stress tensor*. The proposed anisotropic yield function represents with great accuracy both the tension/compression anisotropy and the strength differential effect.

A new yield criterion to describe the *anisotropic yield* behavior and *tension/compression asymmetry* was also proposed by Khan et al. [42], and Khan and Liu [41]. Both effects were uncoupled into multiplicative terms, which allowed to determine the anisotropic coefficients and tension/compression asymmetry parameters independently. Additionally, by introducing a shape-dependent term different types of *yield surfaces* can be predicted: *quadratic*—Hill type, *non quadratic*—Tresca type, or *intermediate*.

Rate Equations

A constitutive model that fulfills the *Clausius–Duhem inequality*, fulfills all formal requirements. However, this does not guarantee that the model provides a good approximation of the real material behavior. If the internal state variables (1.57) chosen in modeling are not identified with underlying physical mechanisms, responsible for dissipation, the theory may be physically empty. There are various approaches for establishment of the rate laws, so that the dissipation inequality is fulfilled:

- *Direct approach*. On the basis of experimental observations some *evolution laws* for the components of flux vector $\dot{P}$ Eq. (7.10) are postulated. Then a posteriori check is performed that the dissipation inequality (7.11) is fulfilled, and this must be done for each material model.
- *Onsager's approach*. Rate laws are established on the basis of the Onsager reciprocal relations (cf. Onsager [58, 59]):

$$\dot{P}_{\Theta} = T_{\Theta Y} J_Y \tag{7.18}$$

where coefficients $T_{\Theta Y}$ create symmetric and positive definite coefficient matrix. Insertion of (7.18) into (7.11) always provides:

$$\sqcap = J \cdot \dot{P} = J_{\Theta} T_{\Theta Y} J_Y \geq 0 \tag{7.19}$$

- *Potential approach*. The existence of a *dissipation potential* $F(J_{\Theta}, Z_K)$ is assumed, which is a closed, convex, and scalar-valued function of the thermodynamic forces J_{Θ}, and some other possible variables Z_K. If function F fulfills the condition $F(J_{\Theta}, Z_K) - F(0, Z_K) \geq 0$, then the evolution laws:

$$\dot{P}_\Theta = \dot{\lambda}\frac{\partial F}{\partial J_\Theta} \qquad \dot{\lambda} \geq 0 \tag{7.20}$$

fulfill the dissipation inequality (7.11). This approach was pioneered by Halphen and Nguyen [36] and is known in literature as a *generalized normality rule.*

- *Postulate of maximum dissipation* (cf. Ziegler [84]). According to this postulate, from all admissible fluxes these are taken which maximize the mechanical dissipation $\sqcap^{\mathrm{mech}}$ (7.12) under the constraint that $f^k \leq 0$ (7.13). This problem is a Lagrange minimization problem with a constraint in terms of inequality:

$$L = -\sqcap^{\mathrm{mech}} + \dot{\lambda}^k f^k = -\sigma_{ij}\dot{\varepsilon}^I_{ij} - Y^k \dot{V}^k + H^k \dot{h}^k + \dot{\lambda}^k f^k \tag{7.21}$$

Necessary conditions of extremum lead to the following rate laws:

$$\begin{aligned}
\dot{\varepsilon}^{\mathrm{I}}_{ij} &= \sum_{k=1}^{n} \dot{\lambda}^k \frac{\partial f^k}{\partial \sigma_{ij}} = \underbrace{\dot{\lambda}^{\mathrm{p}}\frac{\partial f^{\mathrm{p}}}{\partial \sigma_{ij}}}_{\dot{\varepsilon}^{\mathrm{p}}_{ij}} + \underbrace{\dot{\lambda}^{\mathrm{d}}\frac{\partial f^{\mathrm{d}}}{\partial \sigma_{ij}}}_{\dot{\varepsilon}^{\mathrm{d}}_{ij}} + \underbrace{\dot{\lambda}^{\mathrm{ph}}\frac{\partial f^{\mathrm{ph}}}{\partial \sigma_{ij}}}_{\dot{\varepsilon}^{\mathrm{ph}}_{ij}} + \cdots \\
\dot{V}^k &= \sum_{i=1}^{n} \dot{\lambda}^i \frac{\partial f^i}{\partial Y^k} = \underbrace{\dot{\lambda}^{\mathrm{p}}\frac{\partial f^{\mathrm{p}}}{\partial Y^k}}_{\dot{V}^{\mathrm{pk}}} + \underbrace{\dot{\lambda}^{\mathrm{d}}\frac{\partial f^{\mathrm{d}}}{\partial Y^k}}_{\dot{V}^{\mathrm{dk}}} + \underbrace{\dot{\lambda}^{\mathrm{ph}}\frac{\partial f^{\mathrm{ph}}}{\partial Y^k}}_{\dot{V}^{\mathrm{phk}}} + \cdots \\
-\dot{h}^k &= \sum_{i=1}^{n} \dot{\lambda}^i \frac{\partial f^i}{\partial H^k} = -\left(\underbrace{\dot{\lambda}^{\mathrm{p}}\frac{\partial f^{\mathrm{p}}}{\partial H^k}}_{\dot{h}^{\mathrm{pk}}} + \underbrace{\dot{\lambda}^{\mathrm{d}}\frac{\partial f^{\mathrm{d}}}{\partial H^k}}_{\dot{h}^{\mathrm{dk}}} + \underbrace{\dot{\lambda}^{\mathrm{ph}}\frac{\partial f^{\mathrm{ph}}}{\partial H^k}}_{\dot{h}^{\mathrm{phk}}} + \cdots\right)
\end{aligned} \tag{7.22}$$

where $\dot{\lambda}^k$ are nonnegative consistency multipliers and k is a number of a dissipative phenomenon, like plastic flow, damage growth, phased change, etc., taking place in the material. The postulate of maximum dissipation leads to associated theories, since the loading criterion f^k is associated with the potential function.

Both potential approach and approach based on the *postulate of maximum dissipation* may be generalized into the following evolution rules (cf. Chaboche [18], Egner [26]).

$$\begin{aligned}
\dot{\varepsilon}^{\mathrm{I}}_{ij} &= \sum_{k=1}^{n} \dot{\lambda}^k \frac{\partial F^k}{\partial \sigma_{ij}} = \dot{\lambda}^{\mathrm{p}}\frac{\partial F^{\mathrm{p}}}{\partial \sigma_{ij}} + \dot{\lambda}^{\mathrm{d}}\frac{\partial F^{\mathrm{d}}}{\partial \sigma_{ij}} + \dot{\lambda}^{\mathrm{ph}}\frac{\partial F^{\mathrm{ph}}}{\partial \sigma_{ij}} + \cdots \\
\dot{V}^k &= \sum_{i=1}^{n} \dot{\lambda}^i \frac{\partial F^i}{\partial Y^k} = \underbrace{\dot{\lambda}^{\mathrm{p}}\frac{\partial F^{\mathrm{p}}}{\partial Y^k}}_{\dot{V}^{\mathrm{pk}}} + \underbrace{\dot{\lambda}^{\mathrm{d}}\frac{\partial F^{\mathrm{d}}}{\partial Y^k}}_{\dot{V}^{\mathrm{dk}}} + \underbrace{\dot{\lambda}^{\mathrm{ph}}\frac{\partial F^{\mathrm{ph}}}{\partial Y^k}}_{\dot{V}^{\mathrm{phk}}} + \cdots \\
-\dot{h}^k &= \sum_{i=1}^{n} \dot{\lambda}^i \frac{\partial F^i}{\partial H^k} = -\left(\underbrace{\dot{\lambda}^{\mathrm{p}}\frac{\partial F^{\mathrm{p}}}{\partial H^k}}_{\dot{h}^{pk}} + \underbrace{\dot{\lambda}^{\mathrm{d}}\frac{\partial F^{\mathrm{d}}}{\partial H^k}}_{\dot{h}^{\mathrm{dk}}} + \underbrace{\dot{\lambda}^{\mathrm{ph}}\frac{\partial F^{ph}}{\partial H^k}}_{\dot{h}^{\mathrm{phk}}} + \cdots\right)
\end{aligned} \tag{7.23}$$

For *rate-independent problems* the consistency multipliers may be calculated from the consistency conditions:

$$\dot{f}^k = 0, \quad \Longrightarrow \quad \begin{cases} \dot{f}^{\mathrm{p}} = 0 \\ \dot{f}^{\mathrm{d}} = 0 \\ \dot{f}^{\mathrm{ph}} = 0 \\ \ldots \end{cases} \tag{7.24}$$

The parameters $\dot{\lambda}^k$ are assumed to obey the classical *Kuhn–Tucker loading/unloading conditions*:

$$f^k \leq 0 \quad \text{and} \quad \dot{f}^k \begin{cases} < 0 \text{ and } \dot{\lambda}^k = 0 \Longrightarrow \text{passive loading} \\ = 0 \text{ and } \dot{\lambda}^k = 0 \Longrightarrow \text{neutral loading} \\ = 0 \text{ and } \dot{\lambda}^k > 0 \Longrightarrow \text{active loading} \end{cases} \tag{7.25}$$

Note that if the damage loading function is expressed in terms of the modified thermodynamic force $Y_{ij}^{\mathrm{def}*}$, see Eqs. (7.16) and (7.17), then another fourth-rank tensor, $D_{ijkl}^{(Y)} = \partial Y_{kl}^{\mathrm{def}*} / \partial Y_{ij}^{\mathrm{def}}$, which accounts for different damage evolution under tension and compression, appears in expression (7.23$_2$) written for $k = \mathrm{d}$:

$$\dot{V}_{ij}^{\mathrm{d}} = \dot{D}_{ij} = \sum_{m=1}^{n} \dot{\lambda}^m \frac{\partial F^m}{\partial Y_{ij}^{\mathrm{d}}} = \sum_{m=1}^{n} \dot{\lambda}^m \frac{\partial F^m}{\partial Y_{kl}^{\mathrm{def}*}} \frac{\partial Y_{kl}^{\mathrm{def}*}}{\partial Y_{ij}^{\mathrm{d}}} \tag{7.26}$$

The kinetic laws for thermodynamic conjugate forces (R^k, X_{ij}^k, L_{ijkl}^k, $G_{ijklmn}^k, \ldots$) are obtained by taking time derivatives of the state laws (7.8) and (7.9). By the use of tensorial forces L_{ijkl}^k and G_{ijklmn}^k conjugated to hardening variables l_{ijkl}^k and g_{ijklmn}^k a significantly better agreement with experimental data may be obtained, as proved by Streilein [71]. The kinetic laws for the overstress model by Chaboche and Rousselier [21], extended to rotational and distortional hardening up to the sixth order, were given by Kowalsky et al. [43]. The deviation between the experimental and numerical results decreased from 11.8 % for the original model to 5.1 % for the model extended to rotational and distortional hardening, see Fig. 7.3.

When the classical approach based on the *normality rule* is used, the rate of a given state variable is derived from one dissipation function, related to dissipative phenomenon represented by this variable. On the other hand, if another approach, based on the *postulate of maximum dissipation* (cf. for example, Abu Al-Rub and Voyiadjis [3]) is applied, coupling between dissipation phenomena is possible to represent in evolution equations, however only associated theories are then described, since the kinetic laws result from side conditions of a minimization Lagrange problem, which are imposed on the loading functions and not dissipation functions. Note that Eq. (7.23) describe both coupling between dissipation phenomena (so that all

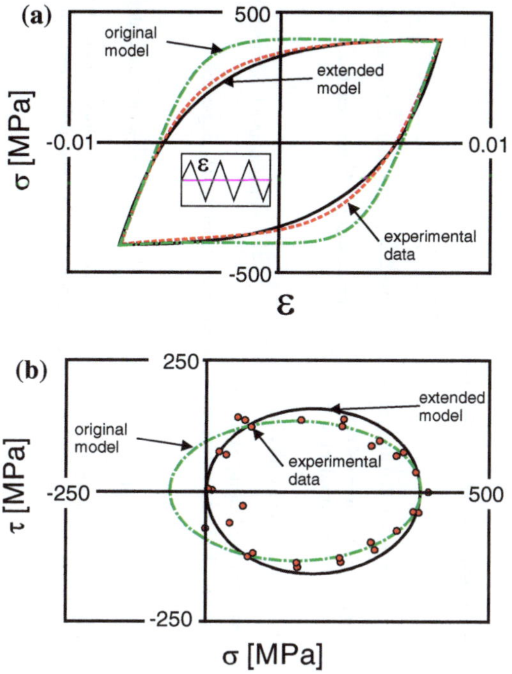

Fig. 7.3 Hysteresis loops and yield surfaces for the uniaxial tensile test, after Kowalsky et al. [43]

dissipation functions may appear in each kinetic law) and *nonassociated rules*. Consequently, the inelastic strain rate consists not only of the plastic strain rate, but also of strain rates related to other dissipative phenomena. As well, the rates of microstructural state variables, $\dot{V}^k$ include terms resulting from coupling of kth dissipative phenomenon with other dissipative phenomena. At the same time the description of nonassociated theories is possible. Therefore, the evolution rules (7.23) may be considered as the generalization of classical normality rules and approaches based on the postulate of maximum dissipation: for associated theories, when the considered dissipation function is equal to the loading function, $F^k = f^k$, or for nonassociated theories in which the recovery terms are independent of thermodynamic forces associated with the microstructural rearrangements, Eq. (7.23) become equivalent to the approaches based on the principle of maximum dissipation (cf. for example Abu Al-Rub and Voyiadjis [3]). On the other hand, if coupling between individual dissipation potentials is neglected, Eq. (7.23) reduce to the classical normality rule (cf. for example Ganczarski and Skrzypek [31]). The comparison between kinetic equation (7.23) presented in Egner [26] and approaches presented in Abu Al-Rub and Voyiadjis [3] and in Chaboche [18] for *elastic-plastic-damage material* is presented in Table 7.2.

Table 7.2 Comparison between evolution rules proposed here and in Abu Al-Rub and Voyiadjis [3] and in Chaboche [18]

	Egner [26]	Abu Al-Rub and Voyiadjis [3]	Chaboche [18]
(a)	$\dot{\varepsilon}^{\mathrm{I}}_{ij} = \dot{\lambda}^{\mathrm{p}} \frac{\partial F^{\mathrm{p}}}{\partial \sigma_{ij}} + \dot{\lambda}^{\mathrm{d}} \frac{\partial F^{\mathrm{d}}}{\partial \sigma_{ij}} = \dot{\varepsilon}^{\mathrm{p}}_{ij} + \dot{\varepsilon}^{\mathrm{id}}_{ij}$ nonassociated and coupled with damage flow rule for inelastic strain $\varepsilon^{\mathrm{I}}_{ij}$	$\dot{\varepsilon}^{\mathrm{I}}_{ij} = \dot{\lambda}^{\mathrm{p}} \frac{\partial f^{\mathrm{p}}}{\partial \sigma_{ij}} + \dot{\lambda}^{\mathrm{d}} \frac{\partial f^{\mathrm{d}}}{\partial \sigma_{ij}} = \dot{\varepsilon}^{\mathrm{p}}_{ij} + \dot{\varepsilon}^{\mathrm{id}}_{ij}$ coupled with damage but associated with plastic and damage loading surfaces; equivalent to (a) if recovery terms in (7.13) are independent of σ_{ij}	$\dot{\varepsilon}^{\mathrm{p}}_{ij} = \dot{\lambda}^{\mathrm{p}} \frac{\partial F^{\mathrm{p}}}{\partial \sigma_{ij}}$ nonassociated, but uncoupled from damage, $\dot{\varepsilon}^{\mathrm{id}}_{ij}$ neglected
(b)	$\dot{D}_{ij} = \dot{\lambda}^{\mathrm{p}} \frac{\partial F^{\mathrm{p}}}{\partial Y^{\mathrm{d}}_{ij}} + \dot{\lambda}^{\mathrm{d}} \frac{\partial F^{\mathrm{d}}}{\partial Y^{\mathrm{d}}_{ij}} = \dot{D}^{\mathrm{p}}_{ij} + \dot{D}^{\mathrm{d}}_{ij}$ nonassociated and coupled with plasticity flow rule for damage variable D_{ij}	$\dot{D}_{ij} = \dot{\lambda}^{\mathrm{p}} \frac{\partial f^{\mathrm{p}}}{\partial Y^{\mathrm{d}}_{ij}} + \dot{\lambda}^{\mathrm{d}} \frac{\partial f^{\mathrm{d}}}{\partial Y^{\mathrm{d}}_{ij}} = \dot{D}^{\mathrm{p}}_{ij} + \dot{D}^{\mathrm{d}}_{ij}$ coupled with plasticity but associated with plastic and damage loading surfaces; equivalent to (b) if recovery terms in (7.13) are independent of Y^{d}_{ij}	$\dot{D}_{ij} = \dot{\lambda}^{\mathrm{d}} \frac{\partial F^{\mathrm{d}}}{\partial Y^{\mathrm{d}}_{ij}}$ nonassociated, but uncoupled from plasticity, $\dot{D}^{\mathrm{d}}_{ij}$ neglected
(c)	$\dot{\alpha}^{\mathrm{p}}_{ij} = -\dot{\lambda}^{\mathrm{p}} \frac{\partial F^{\mathrm{p}}}{\partial X^{\mathrm{p}}_{ij}} - \dot{\lambda}^{\mathrm{d}} \frac{\partial F^{\mathrm{d}}}{\partial X^{\mathrm{p}}_{ij}}$ nonassociated and coupled with damage flow rule for plastic kinematic hardening variable α^{p}_{ij}	$\dot{\alpha}^{\mathrm{p}}_{ij} = -\dot{\lambda}^{\mathrm{p}} \frac{\partial F^{\mathrm{p}}}{\partial X^{\mathrm{p}}_{ij}}$ nonassociated, but uncoupled from damage; equivalent to (c) if damage dissipation potential does not depend on plastic hardening variable	$\dot{\alpha}^{\mathrm{p}}_{ij} = -\dot{\lambda}^{\mathrm{p}} \frac{\partial F^{\mathrm{p}}}{\partial X^{\mathrm{p}}_{ij}} - \dot{\lambda}^{\mathrm{s}} \frac{\partial F^{\mathrm{s}}}{\partial X^{\mathrm{p}}_{ij}}$ nonassociated, F^{s} is additional, static microstructural evolution potential, but uncoupled from damage
(d)	$\dot{r}^{\mathrm{p}} = -\dot{\lambda}^{\mathrm{p}} \frac{\partial F^{\mathrm{p}}}{\partial R^{\mathrm{p}}} - \dot{\lambda}^{\mathrm{d}} \frac{\partial F^{\mathrm{d}}}{\partial R^{\mathrm{p}}}$ nonassociated and coupled with damage flow rule for plastic isotropic hardening variable r^{p}	$\dot{r}^{\mathrm{p}} = -\dot{\lambda}^{\mathrm{p}} \frac{\partial F^{\mathrm{p}}}{\partial R^{\mathrm{p}}}$ nonassociated, but uncoupled from damage; in a general case inconsistent with (a) (cf. Egner [26])	$\dot{r}^{\mathrm{p}} = -\dot{\lambda}^{\mathrm{p}} \frac{\partial F^{\mathrm{p}}}{\partial R^{\mathrm{p}}} - \dot{\lambda}^{\mathrm{s}} \frac{\partial F^{\mathrm{s}}}{\partial R^{\mathrm{p}}}$ nonassociated, F^{s} is additional, static microstructural evolution potential, but uncoupled from damage
(e)	$\dot{\alpha}^{\mathrm{d}}_{ij} = -\dot{\lambda}^{\mathrm{p}} \frac{\partial F^{\mathrm{p}}}{\partial X^{\mathrm{d}}_{ij}} - \dot{\lambda}^{\mathrm{d}} \frac{\partial F^{\mathrm{d}}}{\partial X^{\mathrm{d}}_{ij}}$ nonassociated and coupled with plasticity flow rule for damage kinematic hardening variable α^{d}_{ij}	$\dot{\alpha}^{\mathrm{d}}_{ij} = -\dot{\lambda}^{\mathrm{d}} \frac{\partial F^{\mathrm{d}}}{\partial X^{\mathrm{d}}_{ij}}$ nonassociated, but uncoupled from plasticity; equivalent to (e) if plastic dissipation potential does not depend on damage hardening variable	$\dot{\alpha}^{\mathrm{d}}_{ij} = -\dot{\lambda}^{\mathrm{d}} \frac{\partial F^{\mathrm{d}}}{\partial X^{\mathrm{d}}_{ij}}$ nonassociated, but uncoupled from plasticity; equivalent to (e) if plastic dissipation potential does not depend on damage hardening variable
(f)	$\dot{r}^{\mathrm{d}} = -\dot{\lambda}^{\mathrm{p}} \frac{\partial F^{\mathrm{p}}}{\partial R^{\mathrm{d}}} - \dot{\lambda}^{\mathrm{d}} \frac{\partial F^{\mathrm{d}}}{\partial R^{\mathrm{d}}}$ nonassociated and coupled with plasticity flow rule for damage isotropic hardening variable r^{d}	$\dot{r}^{\mathrm{d}} = -\dot{\lambda}^{\mathrm{d}} \frac{\partial F^{\mathrm{d}}}{\partial R^{\mathrm{d}}}$ nonassociated, but uncoupled from plasticity; in a general case inconsistent with (a)	$\dot{r}^{\mathrm{d}} = -\dot{\lambda}^{\mathrm{d}} \frac{\partial F^{\mathrm{d}}}{\partial R^{\mathrm{d}}}$ nonassociated, but uncoupled from plasticity

7.3 Elastic-Damage Material

Linear elastic material behavior is based on the assumption that the elastic stiffness (or compliance) tensor is constant through the entire service time. In general, this assumption is not true, because of the appearance of several irreversible mechanical phenomena, e.g. plasticity and damage, etc. which all affect the elasticity tensors. Roughly speaking, due to these material degradation phenomena, a drop in stiffness and an increase in compliance is observed, such that the initially linear behavior becomes nonlinear. Neglecting, at this point, the plastic dissipation mechanism, we confine ourselves to the influence of the damage dissipation mechanism on the properties of nonlinear elastic-damage material. If the damage process is active, additional damage-induced strains, reversible $\varepsilon_{ij}^{\mathrm{ed}}$ and irreversible $\varepsilon_{ij}^{\mathrm{id}}$, are observed, see Eq. (1.38): $\varepsilon_{ij} = \varepsilon_{ij}^{\mathrm{e}} + \varepsilon_{ij}^{\mathrm{ed}} + \varepsilon_{ij}^{\mathrm{id}}$. In other words, nucleation of microcracks and microcavities, growth, and coalescence, as well as decohesion, grain boundary cracks, etc., are the source of nonlinearity. It may be described by the effective stress concept or, more generally, the effective variables concept [24]. According to this formalism, the effective (damage influenced) stress $\widetilde{\sigma}_{ij}$ is obtained from the Cauchy stress σ_{ij} through the linear tensorial transformation, by the use of the fourth-rank *damage effect tensor* $M_{ijkl}(D_{pq})$, the elements of which depend on the current components of the *second-rank damage tensor* D_{pq}. Hence, the following linear tensorial transformation rule holds:

$$\widetilde{\sigma}_{ij} = M_{ijkl}\sigma_{kl} \tag{7.27}$$

Matrix representation of the damage effect tensor is complicated and not unique, as shown by Chen and Chow [23], see Chap. 1. Summing up, when the effect of damage growth is taken into account, the initially linear elasticity equations become nonlinear following the *stiffness deterioration* (and compliance increase) *due to damage*:

$$\sigma_{ij} = M_{ipjq}E_{pqrs}M_{rksl}\varepsilon_{kl}^{\mathrm{e}} \tag{7.28}$$

Usually, the initial damage threshold is observed, the exceeding of which matches the active damage growth. Below the *damage threshold* processes are purely elastic, but on reverse loading the additional effect of damage hardening may occur. In spite of mechanical properties damage also influences thermal expansion and thermal conductivity. In order to derive the *damage affected thermal expansion tensor* [29] application of the stress equivalence principle gives:

$$[\widetilde{\alpha}] = \begin{bmatrix} \alpha_1(1-D_1) & & \\ & \alpha_2(1-D_2) & \\ & & \alpha_3(1-D_3) \end{bmatrix} \tag{7.29}$$

In an analogous way, postulating the entropy equivalence principle one may introduce a *damage affected thermal conductivity tensor* [69]:

$$[\widetilde{\lambda}] = \begin{bmatrix} \lambda_1(1-D_1) & & \\ & \lambda_2(1-D_2) & \\ & & \lambda_3(1-D_3) \end{bmatrix} \tag{7.30}$$

7.4 Elastic-Plastic Material

A second basic dissipative mechanical phenomenon that causes a loss of material linearity is material degradation connected with plasticity. In the case of the plasticity dissipation mechanism (as observed in the majority of ductile metals) the irreversible plastic strain $\varepsilon_{ij}^{\mathrm{p}}$ in a loaded/unloaded specimen remains in the stress-free state. Assumption of the class of symmetry of material in the elastic range (orthotropy, transverse isotropy, isotropy) is the key point for proper definition of the yield criterion. If material is isotropic in the elastic range, it deforms approximately isotropically also at the initial phase of plastic flow. In the case of deep plastic deformation, however, a specifically ordered material texture is formed during fabrication process, such as metal forming, rolling, deep drawing, plastic penetration, etc. On the other hand, materials that are anisotropic in the elastic range, either virgin or damage acquired (e.g. long fiber reinforced composites, thin ceramic layers deposited by different techniques, bones, etc.) retain anisotropy also in the plastic range. The most general criterion of the transition of anisotropic material from the elastic to the elastic-plastic range, based on the von Mises concept, is known as the von Mises criterion:

$$\sigma_{ij}\Pi_{ijkl}\sigma_{kl} = 1 \tag{7.31}$$

The *von Mises anisotropy tensor* Π_{ijkl} is characterized by 21 independent moduli, see broader discussion in Chap. 5. However, this criterion is difficult to practical applications due to expensive material tests. In the case of isotropy, the *Huber–von Mises isotropic yield condition* is often applied:

$$\sigma_{\mathrm{eq}}^{\mathrm{HMH}} = \frac{1}{\sqrt{2}}\left[\left(\sigma_x-\sigma_y\right)^2+\left(\sigma_y-\sigma_z\right)^2+(\sigma_z-\sigma_x)^2+ 6\left(\tau_{xy}^2+\tau_{yz}^2+\tau_{zx}^2\right)\right]^{1/2} = k \tag{7.32}$$

7.4.1 Classical Isotropic Plastic Hardening

In the case of J_2-type mixed isotropic/kinematic hardening model of the isotropic material, the hardening effect can be decomposed into the isotropic growth of the diameter of the plastic dissipation surface and a rigid movement of the surface center, cf. Fig. 7.4. Corresponding equations for the J_2-type isotropic, kinematic, or mixed plasticity hardening functions, also called loading functions, are:

$$
\begin{aligned}
f^{\mathrm{piso}} &= \sqrt{\frac{3}{2}s_{ij}s_{ij}} - k\,(T) - R^{\mathrm{p}}\,(p) = 0 \\
f^{\mathrm{pkin}} &= \sqrt{\frac{3}{2}(s_{ij} - X'^{\mathrm{p}}_{ij})(s_{ij} - X'^{\mathrm{p}}_{ij})} - k\,(T) = 0 \\
f^{\mathrm{pmix}} &= \sqrt{\frac{3}{2}(s_{ij} - X'^{\mathrm{p}}_{ij})(s_{ij} - X'^{\mathrm{p}}_{ij})} - k\,(T) - R^{\mathrm{p}}\,(p) = 0
\end{aligned}
\tag{7.33}
$$

In the above equations, the scalar variable $R^{\mathrm{p}}(p)$ represents the isotropic term, or drag stress, the tensorial variable $X^{\mathrm{p}}_{ij}(\varepsilon^{\mathrm{p}}_{kl}, p)$ is the translation tensor, or the back stress tensor. The scalar variable $p(\varepsilon^{\mathrm{p}}_{ij})$ is the accumulated plastic strain, $\mathrm{d}p = \sqrt{\frac{2}{3}\mathrm{d}\varepsilon^{\mathrm{p}}_{ij}\mathrm{d}\varepsilon^{\mathrm{p}}_{ij}}$, function $k(T)$ is the initial temperature-dependent yield point stress, whereas X'^{p}_{ij} stands for the deviator of X^{p}_{ij}. Isotropic and kinematic plastic hardening functions, $X^{\mathrm{p}}_{ij}(\varepsilon^{\mathrm{p}}_{kl}, p)$ and $R^{\mathrm{p}}(p)$ are, in general, given by the nonlinear relationships. When the *Armstrong–Frederick mixed hardening model* is used, the following nonlinear evolution rules can be applied:

$$
\begin{aligned}
f^{\mathrm{pA-F}} &= J_2(\sigma_{ij} - X^{\mathrm{p}}_{ij}) - k\,(T) - R^{\mathrm{p}}\,(p) = 0 \\
J_2(\sigma_{ij} - X^{\mathrm{p}}_{ij}) &= \sqrt{\frac{3}{2}(s_{ij} - X'^{\mathrm{p}}_{ij})(s_{ij} - X'^{\mathrm{p}}_{ij})} \\
\mathrm{d}R^{\mathrm{p}} &= b^{\mathrm{p}}(R^{\mathrm{p}}_{\infty} - R^{\mathrm{p}})\mathrm{d}p \\
\mathrm{d}X^{\mathrm{p}}_{ij} &= \frac{2}{3}C^{\mathrm{p}}(p)\mathrm{d}\varepsilon^{\mathrm{p}}_{ij} - \gamma^{\mathrm{p}}(p)X^{\mathrm{p}}_{ij}\mathrm{d}p
\end{aligned}
\tag{7.34}
$$

Parameters $C^{\mathrm{p}}(p)$ and $\gamma^{\mathrm{p}}(p)$ are known scalar functions of the cumulative plastic strain. For the uniaxial stress state both the drag stress $R^{\mathrm{p}}(\varepsilon^{p})$ and the back stress $X^{\mathrm{p}}(\varepsilon^{\mathrm{p}})$ are described by the scalar functions, namely:

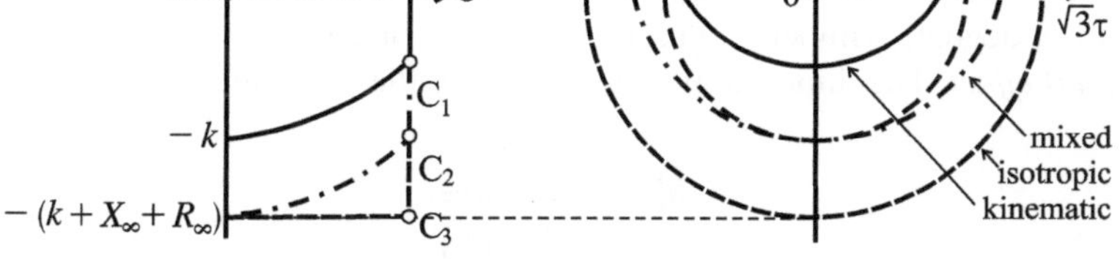

Fig. 7.4 Illustration of the Armstrong–Frederick nonlinear isotropic/kinematic hardening, after Ganczarski et al. [30]

$$
\begin{aligned}
f^{\mathrm{pA-F}} &= |\sigma - X^{\mathrm{p}}| - k\,(T) - R^{\mathrm{p}}\,(\varepsilon^{\mathrm{p}}) = 0 \\
\mathrm{d}R^{\mathrm{p}} &= b^{\mathrm{p}}(R^{p}_{\infty} - R^{\mathrm{p}})\,|\mathrm{d}\varepsilon^{\mathrm{p}}| \\
\mathrm{d}X^{\mathrm{p}} &= \frac{2}{3}C^{\mathrm{p}}(\varepsilon^{\mathrm{p}})\mathrm{d}\varepsilon^{\mathrm{p}} - \gamma^{\mathrm{p}}(\varepsilon^{\mathrm{p}})X^{\mathrm{p}}\,|\mathrm{d}\varepsilon^{\mathrm{p}}|
\end{aligned}
\tag{7.35}
$$

For some applications, Chaboche and Rousselier [21] proposed multi-kinematic hardening model:

$$
\begin{aligned}
f^{\mathrm{p}} &= J_2(\sigma_{ij} - X^{\mathrm{p}}_{ij}) - k\,(T) = 0 \\
X^{\mathrm{p}}_{ij} &= \sum_{n=1}^{3} X^{\mathrm{pn}}_{ij} \\
\mathrm{d}X^{\mathrm{pn}}_{ij} &= \frac{2}{3}C^{\mathrm{pn}}(p)\mathrm{d}\varepsilon^{\mathrm{p}}_{ij} - \gamma^{\mathrm{pn}}(p)X^{\mathrm{pn}}_{ij}\mathrm{d}p
\end{aligned}
\tag{7.36}
$$

Note that all aforementioned *mixed, isotropic/kinematic* J_2-type *plastic hardening* models assume that plastic surfaces follow the isotropy condition. This means that subsequent yield surfaces are similar to one another, and no distortion effects are considered. In general, this is not true, as shown, for example, in experiments by Phillips and Tang [61].

7.4.2 Homogeneous Yield Function-Based Anisotropic Hardening

Asymmetric Yielding

An alternative to kinematic hardening in classical plasticity is the asymmetric yielding model, presented by Barlat et al. [6]. The approach is based on homogeneous yield functions/plastic potentials, combining a stable, isotropic, or anisotropic hardening-type component $f(s_{ij})$ and a fluctuation component $f_{\mathrm{h}}(s_{ij})$:

$$
\begin{aligned}
f^{\mathrm{pB}} &= \left\{\left[f(s_{ij})\right]^q + \left[f_{\mathrm{h}}(s_{ij})\right]^q\right\}^{\frac{1}{q}} - \bar{k}(\bar{\varepsilon}) \\
&= \{[f(s_{ij})]^q + |p_1|^q\,|\widehat{h}^{\mathrm{s}}_{ij}s_{ij} - |\widehat{h}^{\mathrm{s}}_{ij}s_{ij}||^q \\
&\quad + |p_2|^q\,|\widehat{h}^{\mathrm{s}}_{ij}s_{ij} + |\widehat{h}^{\mathrm{s}}_{ij}s_{ij}||^q]^{\frac{1}{q}} - \bar{k}(\bar{\varepsilon}) = 0
\end{aligned}
\tag{7.37}
$$

where p_1, p_2, q are coefficients, while $\bar{k}$ and $\bar{\varepsilon}$ denote equivalent stress and strain, respectively. Stable function $f(s_{ij})$ may be isotropic or anisotropic, homogeneous of an arbitrary degree, symmetric with respect to the origin, or capturing the *strength differential effect*. The dimensionless tensor $\widehat{\boldsymbol{h}}^{\mathrm{s}}$, of the components:

$$
\widehat{h}^{\mathrm{s}}_{ij} = \frac{h^{\mathrm{s}}_{ij}}{\sqrt{\frac{8}{3}h^{\mathrm{s}}_{kl}h^{\mathrm{s}}_{kl}}}
\tag{7.38}
$$

is a *structural tensor* called the *microstructure deviator*. As an example, if $\widehat{\boldsymbol{h}}^{\mathrm{s}}$ corresponds to uniaxial tension, it is represented in its matrix form by

$$\left[\widehat{\boldsymbol{h}}^{s}\right] = \begin{bmatrix} \frac{1}{2} & 0 & 0 \\ 0 & -\frac{1}{4} & 0 \\ 0 & 0 & -\frac{1}{4} \end{bmatrix} \tag{7.39}$$

Note that in the particular case of $p_1 = p_2 = 0$ the yield function f^{pB} reduces to the traditional isotropic or anisotropic yield function $f^{pB} = f(s_{ij}) = \bar{k}(\bar{\varepsilon})$. If only $p_2 = 0$ expression (7.37) becomes:

$$f^{pB} = \left\{\left[f(s_{ij})\right]^{q} + |p_1|^{q}\left|\widehat{h}^{s}_{ij}s_{ij} - |\widehat{h}^{s}_{ij}s_{ij}|\right|^{q}\right\}^{\frac{1}{q}} - \bar{k}(\bar{\varepsilon}) = 0 \tag{7.40}$$

Therefore, different yield stress values are obtained depending on the sign of loading. For example, if a uniaxial tension is considered, then the stress deviator is:

$$[s_t] = k_t \begin{bmatrix} \frac{2}{3} & 0 & 0 \\ 0 & -\frac{1}{3} & 0 \\ 0 & 0 & -\frac{1}{3} \end{bmatrix}, \qquad \widehat{h}^{s}_{ij}s_{ij} = \frac{1}{2}k_t \tag{7.41}$$

and (7.40) becomes

$$f^{pB} = k_t - \bar{k}(\bar{\varepsilon}) = 0 \tag{7.42}$$

On the other hand, the uniaxial compression gives:

$$[s_c] = k_c \begin{bmatrix} -\frac{2}{3} & 0 & 0 \\ 0 & \frac{1}{3} & 0 \\ 0 & 0 & \frac{1}{3} \end{bmatrix}, \qquad \widehat{h}^{s}_{ij}s_{ij} = -\frac{1}{2}k_c \tag{7.43}$$

leading to

$$f^{pB} = k_c[1 + (p_1)^{q}]^{\frac{1}{q}} - \bar{k}(\bar{\varepsilon}) = 0 \tag{7.44}$$

From (7.42) and (7.44) the ratio of compression to tension yield stress can be obtained:

$$\frac{k_c}{k_t} = [1 + (p_1)^{q}]^{-\frac{1}{q}} \tag{7.45}$$

A yield surface example in the π- plane for $q = 2$ and $p_1 = 4/3$ (then $k_c/k_t = 3/5$) is shown in Fig. 7.5a.

Evolution of the Yield Surface: Forward Loading

In Barlat et al. [6] the flow stress is a function of the plastic work:

$$\frac{k_c}{k_t} = \frac{k_c}{\bar{k}} = g_1(\bar{\varepsilon}) \tag{7.46}$$

and it is assumed that this stress decreases according to:

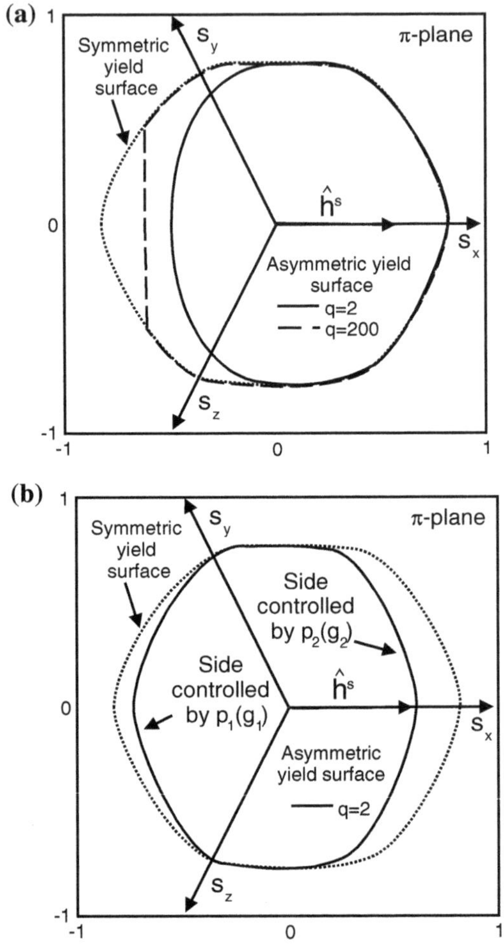

Fig. 7.5 **a** Isotropic, asymmetric, and truncated yield loci in π—plane with $p_1 = 4/3$, $p_2 = 0$, and corresponding microstructure deviator $\widehat{h}^s_{ij}$; **b** Isotropic and asymmetric yield loci after stress reversal (with $p_1 \neq 0$ and $p_2 \neq 0$) (after Barlat et al. [6])

$$\frac{dg_1}{d\bar{\varepsilon}} = c_2 \left(c_3 \frac{\bar{k}_0}{\bar{k}} - g_1 \right) \tag{7.47}$$

In Eq. (7.47) $\bar{k} = \bar{k}(\bar{\varepsilon})$ represents the reference stress–strain curve, and $\bar{k}_0 = \bar{k}$ $(\bar{\varepsilon} = 0)$. From (7.47) and (7.44) the evolution of coefficient p_1 is obtained as:

$$p_1 = \left[(g_1)^{-q} - 1 \right]^{\frac{1}{q}} \tag{7.48}$$

The relationship (7.48) allows to describe the progressive flattening of the yield surface on the opposite side of the loading direction, without affecting the shape of the yield surface at locations near the current loading state.

Table 7.3 Exemplary set of parameters for homogeneous yield function-based anisotropic hardening model

a	Coefficient of the stable yield function of Hershey [38] type: $f(s_{ij}) = \{\frac{1}{2}(\lvert s_1 - s_2\rvert^a + \lvert s_2 - s_3\rvert^a + \lvert s_1 - s_3\rvert^a)\}^{\frac{1}{a}} - \bar{k}(\bar{\varepsilon}) = 0$
C, n	Coefficients of the classical hardening curve (Swift [72] behavior is assumed) $\bar{k}(\bar{\varepsilon}) = C(\varepsilon_0 + \bar{\varepsilon})^n$
q	Parameter affecting the flatness of the back of the yield surface with respect to the loading direction
c_1, c_2, c_3	Parameters affecting the new flow stress and hardening rate after unloading and reloading
c	Parameter associated to the rotation rate of the microstructure history deviator
c_4, c_5	Additional parameters controlling the flow stress rate if permanent softening is considered

Evolution of the Yield Surface: Reverse Loading

If the stress is reversed, the new yield stress in compression, k_c, is different from the flow stress in tension just before unloading (k_t), as can be concluded from (7.45) and (7.48). It means that the Barlat model is capable of capturing the Bauschinger effect. As observed experimentally, the flow stress under compression increases rapidly, and approximately recovers the tensile flow stress (from (7.46) it follows then that $g_1(\bar{\varepsilon}) \to 1$). A simple evolution for g_1 proposed in [6] is:

$$\frac{\mathrm{d}g_1}{\mathrm{d}\bar{\varepsilon}} = c_1 \frac{1 - g_1}{g_1} \tag{7.49}$$

Similar to p_1, the parameter p_2 can be expressed as a function of the variable $g_2(\bar{\varepsilon}) = k_c/\bar{k}$, so that p_2 starts to increase for reverse loading in the same way as p_1 increased for forward loading. This allows the yield surface to be distorted in two opposite directions (Fig. 7.1b). If the flow stress after reversal does not recover the level of the monotonic curve (permanent softening), two additional coefficients, c_4 and c_5 and two additional state functions, g_3 and g_4 may be introduced to capture the effect (see Tables 7.3 and 7.4).

General Deformation Paths

When loading changes direction, the rotation of the microstructure history deviator $\widehat{h}^s_{ij}$ takes place. As long as $\widehat{h}^s_{ij} s_{ij}$ remains positive (forward loading), tensor $\widehat{h}^s_{ij}$ rotates towards the direction of the current stress tensor, according to ([6]):

$$\frac{\mathrm{d}\widehat{h}^s_{ij}}{\mathrm{d}\bar{\varepsilon}} = c\left[\widehat{s}_{ij} - \frac{8}{3}\widehat{h}^s_{ij}(\widehat{s}_{kl}\widehat{h}^s_{kl})\right] \tag{7.50}$$

where $\widehat{s}_{ij}$ is the normalized stress deviator, in the same manner as (7.38).

Table 7.4 Summary of hardening equations for homogeneous yield function-based anisotropic hardening model

Loading case	Hardening equations
Forward loading: $\widehat{h}^s_{ij} s_{ij} \geq 0$	$\frac{\mathrm{d}g_1}{\mathrm{d}\bar{\varepsilon}} = c_2\left(c_3 \frac{\bar{k}_0}{\bar{k}} - g_1\right), \frac{\mathrm{d}g_2}{\mathrm{d}\bar{\varepsilon}} = c_1 \frac{g_3 - g_2}{g_2}$
	$\frac{\mathrm{d}g_4}{\mathrm{d}\bar{\varepsilon}} = c_5(c_4 - g_4), \frac{\mathrm{d}\widehat{h}^s_{ij}}{\mathrm{d}\bar{\varepsilon}} = c\left[\widehat{s}_{ij} - \frac{8}{3}\widehat{h}^s_{ij}(\widehat{s}_{kl}\widehat{h}^s_{kl})\right]$
Reverse loading: $\widehat{h}^s_{ij} s_{ij} < 0$	$\frac{\mathrm{d}g_1}{\mathrm{d}\bar{\varepsilon}} = c_1 \frac{g_4 - g_1}{g_1}, \frac{\mathrm{d}g_2}{\mathrm{d}\bar{\varepsilon}} = c_2\left(c_3 \frac{\bar{k}_0}{\bar{k}} - g_2\right)$
	$\frac{\mathrm{d}g_3}{\mathrm{d}\bar{\varepsilon}} = c_5(c_4 - g_3), \frac{\mathrm{d}\widehat{h}^s_{ij}}{\mathrm{d}\bar{\varepsilon}} = c\left[-\widehat{s}_{ij} + \frac{8}{3}\widehat{h}^s_{ij}(\widehat{s}_{kl}\widehat{h}^s_{kl})\right]$

When $\widehat{h}^s_{ij} s_{ij} < 0$ (reverse loading), the tensor $\widehat{h}^s_{ij}$ rotates towards the direction opposite to the current stress tensor:

$$\frac{\mathrm{d}\widehat{h}^s_{ij}}{\mathrm{d}\bar{\varepsilon}} = c\left[-\widehat{s}_{ij} + \frac{8}{3}\widehat{h}^s_{ij}(\widehat{s}_{kl}\widehat{h}^s_{kl})\right] \tag{7.51}$$

The approach presented above is capable of describing the Bauschinger effect, as well as *asymmetric yielding* and *anisotropic hardening*. The formulation is effective for the modeling of a number of materials, including low carbon, *dual phase* and *ferritic stainless steel* sheet samples (for more details see [6]).

7.5 Remarks on Implicit Multiscale Formulations of Hardening Descriptions

7.5.1 General Multiscale Procedure Based on Polycrystalline Calculations

When only plasticity is concerned as an irreversible phenomenon, the classical isotropic hardening models, reflecting the proportional expansion of the surface, are suitable for simulation of sheet forming operations of cubic metals (both fcc and bcc) [62]. Kinematic hardening, accounting for pure translation of the initial yield surface, is necessary to model more accurately the smooth elastic-plastic transition under reverse loading. For this purpose various nonlinear kinematic hardening models and multi-surface models have been developed. However, in many cases the evolution of the material texture influences so significantly the material properties even for the simplest monotonic loading paths, that traditional hardening laws cannot accurately model these phenomena.

The anisotropic initial yield criteria are extensively discussed in Chap. 5 of the present book. They involve a number of coefficients required for proper description of various anisotropic effects. The determination of analytic laws of variation for the anisotropy coefficients requires a large amount of data. For this reason a general *multiscale procedure* based on polycrystalline calculations for the description of *yielding anisotropy* and its evolution with accumulated deformation has been proposed by Plunkett et al. [62] and Plunkett and Cazacu [64]. The approach consists in determination of the flow stress in various stress directions either by experimental measurements or polycrystalline calculations, next an interpolation technique is used to construct subsequent yield surfaces. The *crystal plasticity model* allows for describing the deformation of a material by crystallographic slip. It also accounts for the reorientation of the crystal lattice. The initial texture, needed as an input of the polycrystal model, is obtained from the experiment, while the final texture is given by the *grain reorientations* associated with shears in the active deformation systems (slip and/or twinning) in the grains. When a viscoplastic self-consistent (VPSC) model is used (cf. [46]) to simulate the interaction of a grain with the surroundings, each grain is treated as an *ellipsoidal inclusion* embedded in a uniform matrix having unknown properties to be determined. The information about the evolution of the yield loci is then generated in numerical tests.

The procedure consists of the following steps: (I) fitting the parameters of a *polycrystal model* to reproduce the mechanical response of the material along a given deformation path; (II) probing numerically the pre-strained polycrystal along various directions in order to quantify the induced anisotropy; (III) determining coefficients of the macroscopic yield surface on the basis of the yield stresses calculated from polycrystal model (this procedure is repeated for various pre-strain levels); (IV) obtaining the *macroscopic yield surfaces* corresponding to any pre-strain level, using an interpolation technique (see Fig. 7.6).

The above *implicit multiscale procedure* was successfully used to model the mechanical response of *hexagonal closed packed metals* (high purity *Zirconium*, *Magnesium alloys*, Titanium, etc.) [55, 62, 63, 78] and *Aluminum alloy* sheets [76, 77]. Two examples of this approach applied to simulations of behavior of different materials are briefly presented below.

7.5.2 Cube-Textured Aluminum Alloy Sheets

Two *Aluminum alloy sheets*, 1050-O (strongly cube textured) and 6022-T4 (mildly cube textured) were investigated by Yoon et al. [76]. The constitutive model included the Yld2004-18p yield function (cf. [65]):

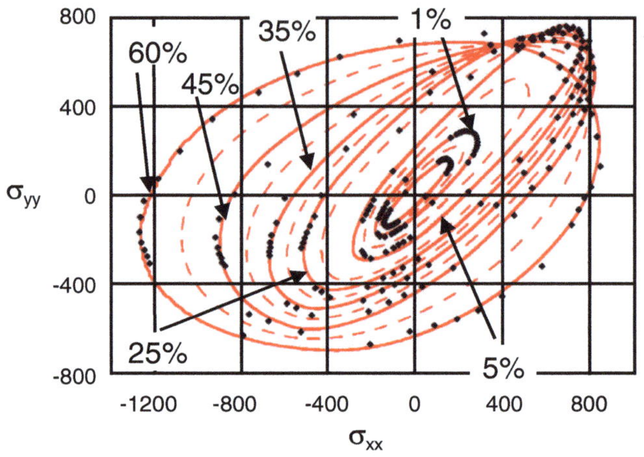

Fig. 7.6 Theoretical yield surface evolution for a zirconium clock-rolled plate during in-plane compression for various levels of pre-strain. *Solid lines* yield surfaces corresponding to fixed pre-strain levels, determined using the VPSC model (*symbols*). *Dashed lines* yield surfaces for intermediate pre-strain levels obtained by linear interpolation (after [63])

$$\begin{aligned} f^{\mathrm{p}}(\Sigma_{ij}) = &\left|\widetilde{S}_1' - \widetilde{S}_1''\right|^a + \left|\widetilde{S}_1' - \widetilde{S}_2''\right|^a + \left|\widetilde{S}_1' - \widetilde{S}_3''\right|^a \\ &+ \left|\widetilde{S}_2' - \widetilde{S}_1''\right|^a + \left|\widetilde{S}_2' - \widetilde{S}_2''\right|^a + \left|\widetilde{S}_2' - \widetilde{S}_3''\right|^a \\ &+ \left|\widetilde{S}_3' - \widetilde{S}_1''\right|^a + \left|\widetilde{S}_3' - \widetilde{S}_2''\right|^a + \left|\widetilde{S}_3' - \widetilde{S}_3''\right|^a = 4\bar{k}^a \end{aligned} \tag{7.52}$$

where $\widetilde{S}_1'$, $\widetilde{S}_2'$, $\widetilde{S}_3'$ and $\widetilde{S}_1''$, $\widetilde{S}_2''$, $\widetilde{S}_3''$ are principal components of the *linearly transformed stress tensors*:

$$\widetilde{S}_{ij}' = L_{ijkl}'\sigma_{kl} \quad \text{and} \quad \widetilde{S}_{ij}'' = L_{ijkl}''\sigma_{kl} \tag{7.53}$$

The Swift or the Voce isotropic strain hardening laws were used to model the strain hardening behavior of Aluminum alloy 1050-O and 6022-T, respectively, in uniaxial tension. The yield surfaces for both sheet samples are shown in Fig. 7.7. The overall results for mildly cube-textured material confirmed that the uniaxial flow curves in different directions did not exhibit the same stress level due to plastic anisotropy. However, the strain hardening rates were similar. For this case the yield function associated with the isotropic hardening rule is able to reflect the uniaxial anisotropic behavior sufficiently well. For strongly cube-textured sheet sample the uniaxial stress–strain curves are clearly different in different directions. The simple shear hardening curves in the investigated directions appeared to be drastically different. In this case crystal plasticity, which accounts for *crystallographic texture evolution*, allowed to fully explain the behavior of 1050-O.

Similar investigations are presented in [5] for *6111-T4* and *2090-T3 Aluminum alloy* sheet samples, and in [77] for *AA2090-T3 Aluminum alloy* sheet sample, subjected to deep drawing process. They demonstrated that the anisotropic properties of

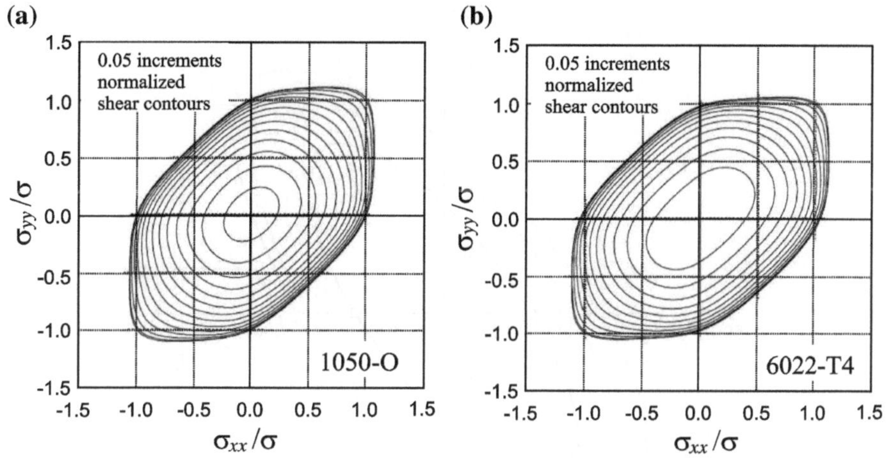

Fig. 7.7 Yield surfaces for: **a** 1050-O and **b** 6022-T4 sheet samples (after [76])

sheet materials subjected to uniaxial tension were described very accurately. Also, the predicted and experimental cup height profiles with six ears were shown to be in excellent agreement.

7.5.3 Hexagonal Close-Packed Metals

Hexagonal Close-Packed metals (HCP) exhibit a deformation behavior which is quite different from cubic crystalline structure materials. As a consequence, rolled or extruded products of these materials show a *strong anisotropy* and *compression–tension asymmetry* (*strength differential effect*). Recently, the anisotropic behavior of hexagonal close-packed metals was investigated by Plunkett et al. [62, 63], Yoon et al. [78], Cazacu et al. [15], and Nixon et al. [55], among others. In all these papers yielding is described using a criterion which can capture both anisotropy and strength differential effect. The expressions for the evolution laws were derived using the above-mentioned multiscale procedure (experimental investigations of uniaxial stress–strain curves and crystallographic texture, *crystal plasticity-based calculations*, and macroscopic scale interpolation techniques). The comparison between simulated and experimental results validated that the proposed methodology can provide good predictability of *anisotropic behavior* and *strength differential effect* of the considered materials (see Fig. 7.8).

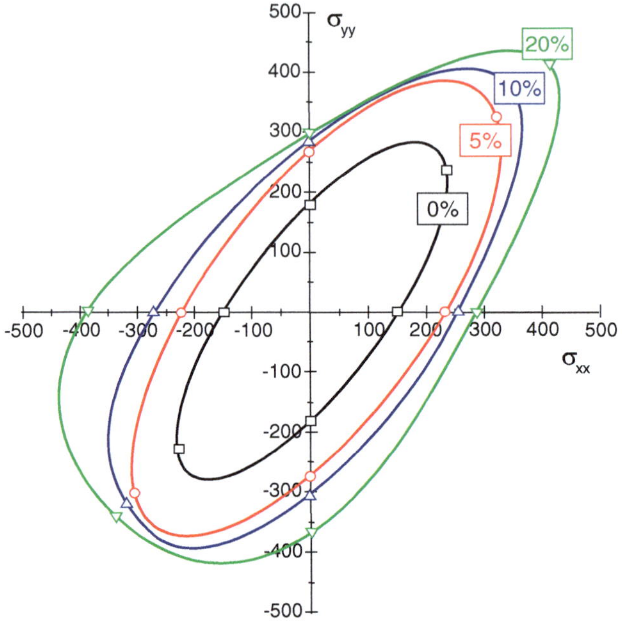

Fig. 7.8 Comparison of the simulated yield surfaces of high-purity α-titanium under plane stress with experimental results (after [78])

7.6 Thermoplastic Coupling

7.6.1 Introduction

When the elastic-plastic material is loaded so that not only inelastic strains develop, but also the temperature is changed, then *thermoelasticity* and *thermoplasticity* are encountered. The experimental results (Bednarek and Kamocka [7]) proved that not only the temperature itself but also the *heating rate* makes a significant impact on parameters that determine carrying capacity at elevated temperatures, and that heating rate should be accounted for in the strength analysis of structures exposed to high temperatures. Increasing the heating rate results in decreasing the slip along grain boundaries and leads to the creation of local empty spaces, which decrease the cross-section area and give reasons for more brittle cracking than in the case of a long-time low heating rate. The low heating rate creates a significant grain deformation within the pearlite-and-ferrite areas, accompanied by the ductile damage, while the high heating rate causes small grain deformation accompanied by the brittle damage (see Fig. 7.9). The need for the additional term, proportional to the temperature rate in the evolution equation for the back stress was already considered by Prager [66]. It was introduced in the description of hardening behavior under thermomechanical loading also by Ohno [56], McDowell [50], or by Chaboche [19] in the unified viscoplastic constitutive equations using the Armstrong–Frederic format. A thermomechanical development of plasticity presented in Lagrangian form with the use of *rate-type constitutive equations* in a strain-temperature space setting was included in Casey [13]. A formulation of elastic-plastic theory for rate-independent materials, based on the use of thermodynamic potentials with thermal effects included, is attributed to Houlsby and Puzrin [39]. They applied the four energy functions commonly used in thermodynamics to provide descriptions depending on which combinations of stress, strain, temperature, and entropy are taken as independent variables. A systematic presentation is made of 16 possible ways of formulating constitutive behavior within this framework. A general framework for rate-independent, small-strain, *thermoinelastic material behavior* is presented also in the paper by Benallal and Bigoni [8], which includes thermoplasticity as a particular case. Strain localization and the development of material instabilities are investigated to highlight the roles of thermal effects and *thermomechanical couplings*. Thermodynamic laws based on consistent Eulerian formulation of finite elastoplasticity with thermal effects were presented by Xiao et al. [75]. In Chaboche [16] the argument is made for the necessity of *temperature rate terms* in the context of hardening rules. The temperature rate terms of the back stresses are also considered by Yu et al. [79] in a thermo-viscoplastic constitutive model derived by the authors. A thermodynamic framework for constitutive modeling of time- and rate-dependent materials (viscoelastic, viscoplastic, viscodamage, and microdamage healing) was derived by Abu Al-Rub and Darabi [2]. The emphasis in their paper was placed on the decomposition of thermodynamic conjugate forces into energetic and dissipative components. It was shown that such decomposition is necessary for accurate estimation of the rate of energy dissipation. Thermomechanical coupling was also considered by Saanouni [67], summarizing the current most effective methods for modeling, simulating, and optimizing metal forming processes. Ganczarski and Skrzypek [31] and Ganczarski et al. [30] take into account the temperature dependence of all material functions that characterize plasticity and damage components, which results in *extended thermoplastic-damage equations*, with the additional temperature rate terms in all evolution equations of thermodynamic conjugate forces. More general case of the *nonassociated plasticity* and *nonassociated damage*, when not only *temperature softening* but also *damage softening* is taken into account is due to Egner [25, 26] and Egner and Egner [28]. In Egner and Egner [27] the influence of temperature rate is investigated quantitatively on the example of thermomechanic *low cycle fatigue* of *AISI L6 steel*.

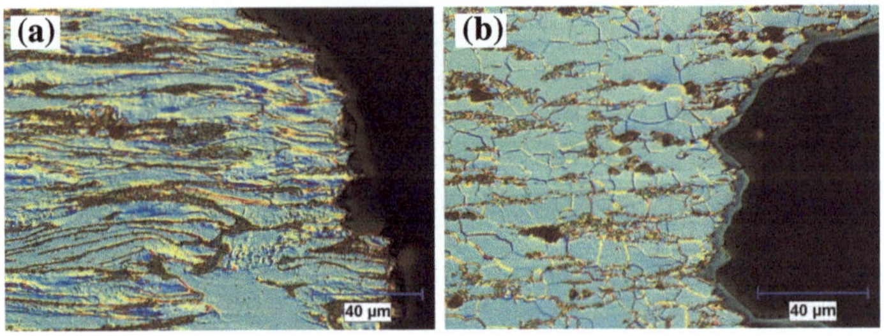

Fig. 7.9 S235JRG2 steel microstructure after mechanical tests: **a** 50 C/min heating rate; **b** 500 C/min heating rate, after Bednarek and Kamocka [7]

7.6.2 Example of Thermoplastic Coupling: Tempered Martensitic Hot Work Tool Steel Subjected to Cyclic Thermomechanical Loading

Low Cycle Fatigue of AISI L6 Steel

To investigate the influence of temperature rate on the response of the constitutive model of *thermo-elastic-(visco)plastic material*, the tempered martensitic hot work tool steel, widely used in the forging industry will be considered. The isothermal low cycle fatigue behavior of this steel is well described in the literature (cf. Zhang et al. [83], Velay et al. [74], Zhang et al. [80]). The steel undergoes cyclic softening, regardless of the testing temperature (see Fig. 7.10). During the initial few hundred cycles (for accumulated plastic strain less than one) rapid softening is observed, followed by a slow, quasilinear softening till rupture (cf. Mebarki et al. [51], Bernhart et al. [9]). The first stage is generally explained by the rapid change of dislocation density inherited from the quench treatment, while the second is related to the formation of dislocation substructure and carbide coarsening under the action of time, temperature, and cyclic load (cf. Gibbons and Dunn [33], Zhang et al. [81], Golański and Mroziński [34, 35]). The considered steel is not stable during fatigue (see Fig. 7.11). For test temperatures lower than tempering temperature (left part of Fig. 7.11, see Table 7.5) the maximum stress decreases linearly with the difference between both temperatures for each level of hardness with nearly the same slope. Above the tempering temperature the maximum stress decreases more violently (the slope changes) because of the interaction of temperature and stabilization of steel. This is related to the thermal aging effect and indicates that the microstructure can be modified by the thermal cycle when the steel is subjected to temperatures equal to or higher than the tempering temperatures.

Equations of 2M1C Constitutive Model

Basic Assumptions

The material behavior is described by the use of 2M1C constitutive model, derived by Cailletaud and Saï [12] and extended by Velay et al. [74]. This RVE-based model

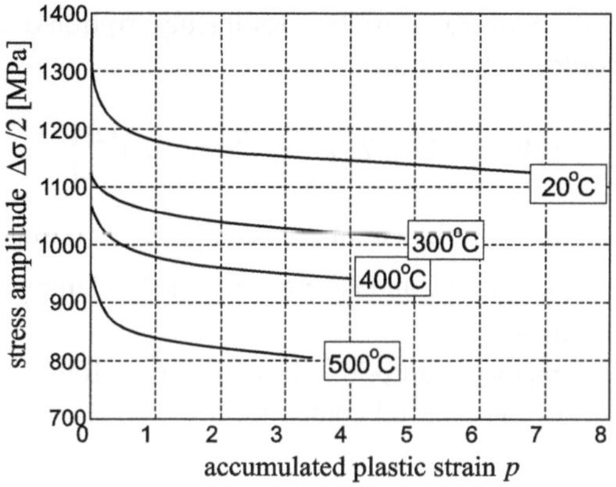

Fig. 7.10 Cyclic softening with a constant strain range ($\Delta\varepsilon/2 = \pm 0.8\,\%$), after Velay et al. [74]

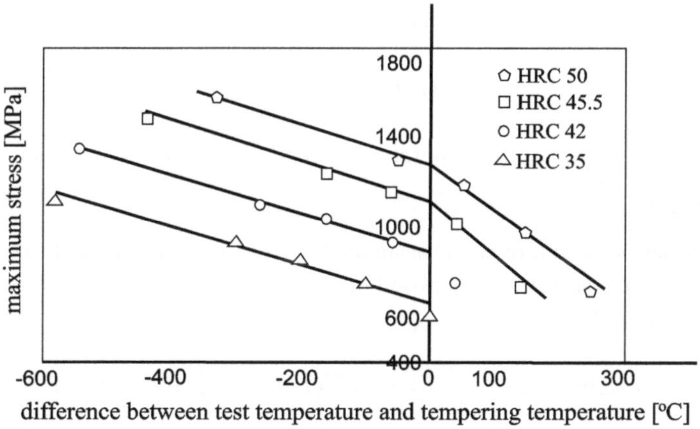

Fig. 7.11 Maximum stress versus difference between test temperature and tempering temperature (after [81])

Table 7.5 Quenching and tempering conditions of fatigue samples made of 55NiCrMoV7 hot work tool steel (after [82])

Tempering temperature (°C)	350	460	560	600
Tempering time (h)	2	2	2	2
Hardness (HRC)	50	45.5	42	35

involves the following set of state variables for the description of the current state of *thermo-elastic-viscoplastic material*:

$$\{V_\alpha\} = \left\{\varepsilon_{ij}^{\rm e}, T, r^{(1)}, r^{(2)}, \alpha_{ij}^{(1)}, \alpha_{ij}^{(2)}\right\} \tag{7.54}$$

where elastic strain $\varepsilon_{ij}^{\rm e}$ and absolute temperature T serve as observable variables and four internal variables: kinematic and isotropic plastic hardening variables $\alpha_{ij}^{(1)}, \alpha_{ij}^{(2)}$ and $r^{(1)}, r^{(2)}$ are introduced.

The model is based on the assumption of small strains. Total strain is decomposed into an elastic, inelastic, and thermal components:

$$\varepsilon_{ij} = \varepsilon_{ij}^{\rm e} + \varepsilon_{ij}^{\rm I} + \varepsilon_{ij}^{\rm T} \tag{7.55}$$

The inelastic strain is farther decomposed into two different strain mechanisms inducing the *cyclic softening of* the *tempered martensitic steels*, namely the decrease of the dislocation density inherited from the quench treatment, and the *carbide coarsening*:

$$\varepsilon_{ij}^{\rm I} = A_1(T)\varepsilon_{ij}^{(1)} + A_2(T)\varepsilon_{ij}^{(2)} \tag{7.56}$$

where $A_1(T)$ and $A_2(T)$ are temperature-dependent proportionality factors.

Equations of State

The state equations result from the assumed form of the state potential, which is here the *Helmholtz free energy*, decomposed into *thermoelastic* ($\rho\psi^{\rm te}$) and *thermoplastic* ($\rho\psi^{\rm tp}$) terms, after [74]:

$$\rho\psi(V_{\rm st}) = \rho\psi^{\rm te}(\varepsilon_{ij} - \varepsilon_{ij}^{\rm I}, T) + \rho\psi^{\rm tp}(T, r^{(k)}, \alpha_{ij}^{(k)}), \quad k = 1, 2 \tag{7.57}$$

where

$$\begin{aligned}
\rho\psi^{\rm te} &= \rho h(T) + \frac{1}{2}(\varepsilon_{ij} - \varepsilon_{ij}^{\rm I})E_{ijkl}(T)(\varepsilon_{kl} - \varepsilon_{kl}^{I}) \\
&\quad - \beta_{ij}(T)(\varepsilon_{ij} - \varepsilon_{ij}^{\rm I})(T - T_0), \quad \beta_{ij}(T) = E_{ijkl}(T)\alpha_{kl}^{\rm T}(T) \\
\rho\psi^{\rm tp} &= \frac{1}{3}[C_{11}(T)\alpha_{ij}^{(1)}\alpha_{ij}^{(1)} + 2C_{12}(T)\alpha_{ij}^{(1)}\alpha_{ij}^{(2)} + C_{22}(T)\alpha_{ij}^{(2)}\alpha_{ij}^{(2)}] \\
&\quad + \frac{1}{2}[b_1(T)Q_1(T)r^{(1)2} + b_2(T)Q_2(T)r^{(2)2}]
\end{aligned} \tag{7.58}$$

In Eq. (7.58) $\alpha_{ij}^{\rm T}(T)$ is the thermal expansion tensor; $h(T)$ is a function of temperature; $C_{11}(T)$, $C_{12}(T)$, $C_{22}(T)$, $b_1(T)$, $b_2(T)$, $Q_1(T)$, $Q_2(T)$ are material parameters, which in general may be temperature dependent. Symbol T_0 stands for the reference temperature at which no thermal strains exists. All coefficients appearing in the considered model are summarized in Table 7.6. In Eq. (7.58) partial

Table 7.6 2M1C model coefficients

Young modulus and yield stress	$E_0(T),\ R_0(T)$
Viscous coefficients	$K(T), n(T)$
Parameters of kinematic part	$C_{11}(T), C_{12}(T), C_{22}(T)$
Localisation coefficients of strain mechanisms	$A_1(T),\ A_2(T)$
Static recovery terms	$M_1(T),\ M_2(T), m_1(T), m_2(T)$
Parameters of isotropic part	$Q_{1\infty}(T), Q_2(T), b_1(T), b_2(T)$

Table 7.7 State equations of the 2M1C constitutive model (after [74])

State variable	Conjugated force
$\varepsilon^{\mathrm{e}}_{ij}$ (elastic strain)	$\sigma_{ij} = \dfrac{\partial(\rho\psi)}{\partial\varepsilon^{\mathrm{e}}_{ij}} = E_{ijkl}(T)\varepsilon^{\mathrm{e}}_{kl}$
T (temperature)	$s = -\dfrac{\partial\psi}{\partial T}$
$\alpha^{(1)}_{ij}$ (kinematic hardening)	$X^{(1)}_{ij} = \dfrac{\partial(\rho\psi)}{\partial\alpha^{(1)}_{ij}} = \dfrac{2}{3}[C_{11}(T)\alpha^{(1)}_{ij} + C_{12}(T)\alpha^{(2)}_{ij}]$
$\alpha^{(2)}_{ij}$ (kinematic hardening)	$X^{(2)}_{ij} = \dfrac{\partial(\rho\psi)}{\partial\alpha^{(2)}_{ij}} = \dfrac{2}{3}[C_{22}(T)\alpha^{(2)}_{ij} + C_{12}(T)\alpha^{(1)}_{ij}]$
$r^{(1)}$ (isotropic hardening)	$R^{(1)} = \dfrac{\partial(\rho\psi)}{\partial r^{(1)}} = b_1(T)Q_1(T)r^{(1)}$
$r^{(1)}$ (isotropic hardening)	$R^{(2)} = \dfrac{\partial(\rho\psi)}{\partial r^{(2)}} = b_2(T)Q_2(T)r^{(2)}$

kinematic–kinematic state coupling is introduced, while no coupling is considered for the isotropic hardening effects, cf. Blaj and Cailletaud [10]. The state equations for 2M1C model are given in Table 7.7. Isotropic hardening force $R^{(1)}$ (the fifth equation presented in Table 7.7) corresponds to the strong softening during the initial several hundred cycles until the accumulated plastic strain does not exceed unity (see Fig. 7.10). Experimental observations (cf. Zhang et al. [83]) confirm the softening dependence on the initial plastic strain level (so-called strain range memorization effect, cf. Chaboche [16]). Therefore, the additional internal variables need to be introduced, see Table 7.9.

Evolution of Strain-Like Variables

Potential of *dissipation* F, after Velay et al. [74], is assumed not equal to plastic yield surface (*nonassociated thermo-viscoplasticity*). This allows us to obtain *nonlinear plastic hardening rules*, which give a more realistic description of the material response:

$$F = F^{\mathrm{vp}}(f^{\mathrm{p}}) + g(X^{(k)}_{ij}) \tag{7.59}$$

with

$$F^{\mathrm{vp}} = \frac{K(T)}{n(T)+1}\left\langle \frac{f^{\mathrm{p}}}{K(T)} \right\rangle^{n(T)+1} \tag{7.60}$$

and

$$g = \sum_{k=1}^{2} \frac{M_k(T)}{m_k(T)+1}\left\langle \frac{J(X_{ij}^{(k)})}{M_k(T)} \right\rangle^{m_k(T)+1}, \quad J(X_{ij}^{(k)}) = \sqrt{\frac{3}{2} X_{ij}^{\prime(k)} X_{ij}^{\prime(k)}} \tag{7.61}$$

$$X_{ij}^{\prime(k)} = X_{ij}^{(k)} - \tfrac{1}{3}(X_{nn}^{(k)})\delta_{ij}$$

where $K(T)$ and $n(T)$ are viscous coefficients, and $M_1(T)$, $M_2(T)$, $m_1(T)$, $m_2(T)$ are static recovery terms (see Table 7.6). Symbol $f^{\mathrm{p}} = 0$ is the *J-type plastic yield surface*:

$$f^{\mathrm{p}} = \sqrt{J(\sigma_{ij}^{(1)} - X_{ij}^{(1)})^2 + J(\sigma_{ij}^{(2)} - X_{ij}^{(2)})^2} - (R_0(T) + R^{(1)} + R^{(2)}) = 0 \tag{7.62}$$

where $J(\sigma_{ij}) = \sqrt{3J_{2s}}$ (see Eq. (1.15) Chap. 1).

On the basis of the *generalized normality rule* (cf. Chaboche [18]) the rate of inelastic strains is given by the following relations:

$$\begin{aligned}
\dot{\varepsilon}_{ij}^{(1)} &= \dot{\lambda}\frac{\partial f^{\mathrm{p}}}{\partial \sigma_{ij}^{(1)}} = \frac{3}{2}\left\langle \frac{f^{\mathrm{p}}}{K(T)} \right\rangle^{n(T)} \frac{s_{ij}^{(1)} - X_{ij}^{(1)}}{\sqrt{J(\sigma_{ij}^{(1)} - X_{ij}^{(1)})^2 + J(\sigma_{ij}^{(2)} - X_{ij}^{(2)})^2}} = \dot{\lambda} n_{ij}^{(1)} \\
\dot{\varepsilon}_{ij}^{(2)} &= \dot{\lambda}\frac{\partial f^{\mathrm{p}}}{\partial \sigma_{ij}^{(2)}} = \frac{3}{2}\left\langle \frac{f^{\mathrm{p}}}{K(T)} \right\rangle^{n(T)} \frac{s_{ij}^{(2)} - X_{ij}^{(2)}}{\sqrt{J(\sigma_{ij}^{(1)} - X_{ij}^{(1)})^2 + J(\sigma_{ij}^{(2)} - X_{ij}^{(2)})^2}} = \dot{\lambda} n_{ij}^{(2)}
\end{aligned} \tag{7.63}$$

In the present formulation, which is based on the model derived by Cailletaud and Saï [12] and extended by Velay et al. [74], the model response depends on two mechanisms (2M) while only one criterion (1C) characterizing the elastic domain is defined. The equations of evolution resulting from the assumed potential (7.59), derived in [74], are summarized in Table 7.8. Experimental tests have shown that the asymptotic value Q of the isotropic hardening variable r may depend of the plastic strain range (cf. Chaboche [16], Jiang and Zhang [40], Saï [68]). The *strain memory effect* can be incorporated into all isotropic hardening variables used. For the sake of simplicity, the strain memory effect was here introduced into the first variable, $r^{(1)}$ only, after [73, 74]. In [74] the plastic strain range memorization effect is introduced through the asymptotic value Q_1, which is subjected to change when the *inelastic strain* exceeds a certain *threshold* f^* introduced in the plastic strain space. For this reason two additional internal state variables have been defined: the radius p^* and the center ε_{ij}^* of the *memory surface* f^* (see Table 7.9).

Table 7.8 Evolution equations of internal state variables

State variable	Evolution law
$\alpha_{ij}^{(1)}$ (kinematic hardening)	$\dot{\alpha}_{ij}^{(1)} = \dot{\varepsilon}_{ij}^{(1)} - \frac{3}{2}\frac{X_{ij}^{(1)}}{J(X_{ij}^{(1)})}\left\langle\frac{J(X_{ij}^{(1)})}{M_1}\right\rangle^{m_1}$
$\alpha_{ij}^{(2)}$ (kinematic hardening)	$\dot{\alpha}_{ij}^{(2)} = \dot{\varepsilon}_{ij}^{(2)} - \frac{3}{2}\frac{X_{ij}^{(2)}}{J(X_{ij}^{(2)})}\left\langle\frac{J(X_{ij}^{(2)})}{M_2}\right\rangle^{m_2}$
$r^{(1)}$ (isotropic hardening)	$\dot{\lambda}\left(1 - \frac{R^{(1)}}{Q_1}\right) = \dot{r}^{(1)}$
$r^{(1)}$ (isotropic hardening)	$\dot{\lambda}\left(1 - \frac{R^{(2)}}{Q_2}\right) = \dot{r}^{(2)}$

Table 7.9 Memory effect of the strain range

Parameter of isotropic hardening part related to initial rapid softening	$Q_1 = Q_{1\infty}(T)[1 - \exp(-2\mu p^*)]$
Memory surface	$f^* = \frac{2}{3}J(\varepsilon_{ij}^I - \varepsilon_{ij}^*) - p^*$
Evolution of additional internal state variable memorizing prior maximum plastic strain range	$\dot{p}^* = \eta H(f^*) < n_{ij}n_{ij}^* > \dot{p}$
Evolution of additional internal state variable memorizing prior maximum plastic strain range	$\dot{\varepsilon}_{ij}^* = \sqrt{\frac{3}{2}}(1-\eta)H(f^*) < n_{kl}n_{kl}^* > \dot{p}n_{ij}^*$
Unit normal to the memory surface $f^* = 0$	$n_{ij}^* = \frac{\partial f^*}{\partial \varepsilon_{ij}^I} / \left\| \frac{\partial f^*}{\partial \varepsilon_{ij}^I} \right\| = \sqrt{\frac{3}{2}}\frac{\varepsilon_{ij}^{\prime I} - \varepsilon_{ij}^{\prime *}}{J(\varepsilon_{ij}^I - \varepsilon_{ij}^*)}$
	$\varepsilon_{ij}^{\prime I} = \varepsilon_{ij}^I - \frac{1}{3}(\varepsilon_{kk}^{\prime I})\delta_{ij}$
	$\varepsilon_{ij}^{\prime *} = \varepsilon_{ij}^* - \frac{1}{3}(\varepsilon_{kk}^*)\delta_{ij}$
Unit normal to the yield surface $f^{\mathrm{p}} = 0$	$n_{ij} = \frac{\partial f^{\mathrm{p}}}{\partial \sigma_{ij}} / \left\| \frac{\partial f^{\mathrm{p}}}{\partial \sigma_{ij}} \right\|$

Evolution of Stress-Like Variables: Effects of Temperature Rate

The influence of temperature rates is often disregarded in kinetic equations. The need for such terms was discussed in many papers, for example by Moreno and Jordan [53], Hartman [37], Ohno et al. [57], Ohno [56], Lee and Krempl [47], Chaboche [16], Ganczarski and Skrzypek [31], Egner [26], Egner and Egner [27, 28]. To investigate the problem qualitatively and quantitatively the evolution equations for thermodynamic conjugate forces, extended with terms proportional to temperature rate were derived in Egner and Egner [28] accounting for full coupling with temperature (see Table 7.10).

Table 7.10 Evolution laws for thermodynamic conjugate forces

	Partial coupling	Full coupling with temperature
$\dot{\sigma}_{ij} =$	$E_{ijkl}(T)(\dot{\varepsilon}_{kl} - \dot{\varepsilon}^{\mathrm{I}}_{kl})$	$+\left[\dfrac{\partial E_{ijkl}(T)}{\partial T}(\varepsilon_{kl} - \varepsilon^{\mathrm{I}}_{kl}) - \dfrac{\partial \beta_{ij}(T)}{\partial T}(T - T_0) - \beta_{ij}(T)\right]\dot{T}$
$\dot{X}^{(1)}_{ij} =$	$\frac{2}{3}[C_{11}(T)\dot{\alpha}^{(1)}_{ij} + C_{12}(T)\dot{\alpha}^{(2)}_{ij}]$	$+\dfrac{2}{3}\left[\dfrac{\partial C_{11}(T)}{\partial T}\alpha^{(1)}_{ij} + \dfrac{\partial C_{12}(T)}{\partial T}\alpha^{(2)}_{ij}\right]\dot{T}$
$\dot{X}^{(2)}_{ij} =$	$\frac{2}{3}[C_{22}(T)\dot{\alpha}^{(2)}_{ij} + C_{12}(T)\dot{\alpha}^{(1)}_{ij}]$	$+\dfrac{2}{3}\left[\dfrac{\partial C_{22}(T)}{\partial T}\alpha^{(2)}_{ij} + \dfrac{\partial C_{12}(T)}{\partial T}\alpha^{(1)}_{ij}\right]\dot{T}$
$\dot{R}^{(1)} =$	$b_1(T)[Q_1(T)\dot{r}^{(1)} + \dot{Q}_1(T)r^{(1)}]$	$+\left[\dfrac{\partial b_1(T)}{\partial T}Q_{1\infty}(T) + b_1(T)\dfrac{\partial Q_{1\infty}(T)}{\partial T}\right] \times [1 - \exp(-2\mu p^*)]r^{(1)}\dot{T}$
$\dot{R}^{(2)} =$	$b_2(T)Q_2(T)\dot{r}^{(2)}$	$+\left[\dfrac{\partial b_2(T)}{\partial T}Q_2(T) + b_2(T)\dfrac{\partial Q_2(T)}{\partial T}\right]r^{(2)}\dot{T}$

Heat Balance Equation

In the case of *thermo-elastic-viscoplastic material*, for which the number of state variables is reduced to the set $\{V_\alpha\}$ given by relation (7.54), the *general coupled heat equation* takes the following form:

$$\begin{aligned}\rho c^{\mathrm{T}}_{\varepsilon}\dot{T} &= -q_{i,i} + \rho r_{\mathrm{ext}} + \rho T\frac{\partial^2\psi}{\partial\varepsilon_{ij}\partial T}(\dot{\varepsilon}_{ij} - \dot{\varepsilon}^{\mathrm{I}}_{ij}) + \rho\frac{\partial\psi}{\partial\varepsilon_{ij}}\dot{\varepsilon}^{\mathrm{I}}_{ij}\\ &+\left(\sigma_{ij} - \rho\frac{\partial\psi}{\partial\varepsilon_{ij}}\right)\dot{\varepsilon}_{ij} - \rho\left(\frac{\partial\psi}{\partial r^{(1)}} - T\frac{\partial^2\psi}{\partial T\partial r^{(1)}}\right)\dot{r}^{(1)}\\ &-\rho\left(\frac{\partial\psi}{\partial r^{(2)}} - T\frac{\partial^2\psi}{\partial T\partial r^{(2)}}\right)\dot{r}^{(2)} - \rho\left(\frac{\partial\psi}{\partial\alpha^{(1)}_{ij}} - T\frac{\partial^2\psi}{\partial T\partial\alpha^{(1)}_{ij}}\right)\dot{\alpha}^{(1)}_{ij}\\ &-\rho\left(\frac{\partial\psi}{\partial\alpha^{(2)}_{ij}} - T\frac{\partial^2\psi}{\partial T\partial\alpha^{(2)}_{ij}}\right)\dot{\alpha}^{(2)}_{ij}\end{aligned} \tag{7.64}$$

which is nonlinear and fully coupled to mechanical problem. In the above equation r_{ext} is the specific external heat gained by the body (e.g. through radiation), and heat flux q_i is given by the Fourier law:

$$q_i = -\lambda_{ij}T_{|j} \tag{7.65}$$

By the use of state equations presented in Table 7.7 and law (7.65), the equation of heat balance (7.64) can be transformed to the following form (cf. Abu Al-Rub and Darabi [2], Saanouni [67], Ottosen and Ristinmaa [60], Egner [26]):

$$
\begin{aligned}
\rho c_{\varepsilon}^{\mathrm{T}} \dot{T} &= (\lambda_{ij} T_{|j})_{|i} + \rho r_{\mathrm{ext}} + \sqcap^{\mathrm{mech}} \\
&+ \underbrace{T \left[\frac{\partial R^{(1)}}{\partial T} \dot{r}^{(1)} + \frac{\partial R^{(2)}}{\partial T} \dot{r}^{(2)} + \frac{\partial X_{ij}^{(1)}}{\partial T} \dot{\alpha}_{ij}^{(1)} + \frac{\partial X_{ij}^{(2)}}{\partial T} \dot{\alpha}_{ij}^{(2)} - P_{ij} (\dot{\varepsilon}_{ij} - \dot{\varepsilon}_{ij}^{\mathrm{I}}) \right]}_{\text{thermomechanical coupling term}}
\end{aligned}
\tag{7.66}
$$

where $\sqcap^{\mathrm{mech}}$ is the mechanical dissipation:

$$
\sqcap^{\mathrm{mech}} = \sigma_{ij} \dot{\varepsilon}_{ij}^{\mathrm{I}} - R^{(1)} \dot{r}^{(1)} - R^{(2)} \dot{r}^{(2)} - X_{ij}^{(1)} \dot{\alpha}_{ij}^{(1)} - X_{ij}^{(2)} \dot{\alpha}_{ij}^{(2)} \tag{7.67}
$$

and tensor P_{ij} is introduced in the form:

$$
P_{ij} = -\frac{\partial E_{ijkl}}{\partial T} (\varepsilon_{kl} - \varepsilon_{kl}^{\mathrm{I}}) + \frac{\partial \beta_{ij}}{\partial T} (T - T_0) + \beta_{ij} \tag{7.68}
$$

Material Data

The model coefficients, summarized in Table 7.6, were identified by Velay et al. [74] for martensitic *hot work tool steel AISI L6* in different test temperatures, below tempering temperature of the steel. The results of experimental identifications are presented in Table 7.11. To illustrate the variation with temperature of model parameters the chosen functions are plotted in Fig. 7.12. Following

Table 7.11 55NiCrMoV7 parameters (after [74])

	20 °C	300 °C	400 °C	500 °C
E (MPa)	206,580	188,940	176,580	156,935
R_0 (MPa)	790	525	455	410
K (MPa)	130	165	195	268
n	19.5	18	17	15
C_{11}	450,480	406,585	378,675	195,655
C_{22}	124,980	91,520	41,965	13,215
C_{12}	−149,925	−126,100	−84,843	−40,500
A_1	0.78	0.74	0.66	0.65
A_2	0.4	0.436	0.46	0.48
M_1	795	760	740	705
m_1	22	20	18	10.5
M_2	890	850	800	700
m_2	11.75	9.5	7	4.3
$Q_{1\infty}$	−295	−80	−68	−100
b_1	11	7	6	5.5
Q_2	−75	−75	−75	−75
b_1	0.2	0.2	0.2	0.2

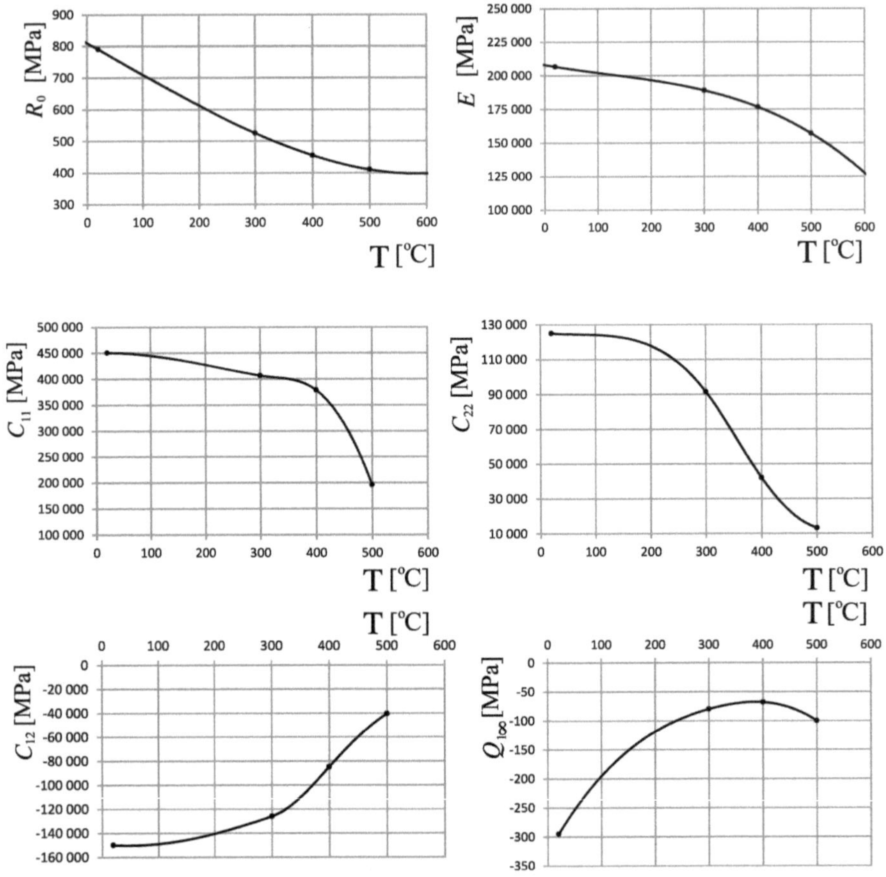

Fig. 7.12 The variation with temperature of chosen model parameters

Zhang et al. [83], the memory parameter μ is taken equal to 420 for all the temperature levels. Additionally, if only symmetrical strain controlled tests are considered, η equals 0.5 (*instantaneous memorization*); in the other cases, η equals 0.1 for *progressive memorization* (Chaboche [20]).

Numerical Results

Several *anisothermal fatigue tests* were subjected to numerical analysis, according to the strain and temperature controlling presented in Fig. 7.13.

Two cases are compared in each example: (1) temperature rate terms in kinetic equations presented in Table 7.10 (third column) are disregarded, and the influence of temperature changes is accounted for only by changing material characteristics according to the second column of Table 7.10 (case 1); (2) all temperature rate terms are included according to the second and third columns of Table 7.10 (case 2).

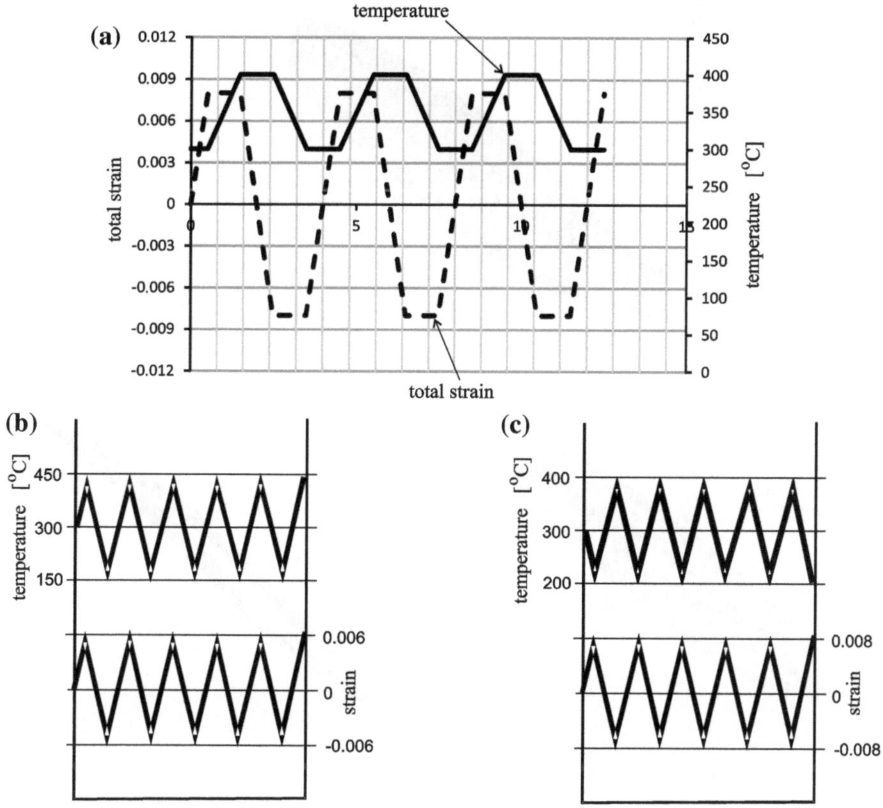

Fig. 7.13 Anisothermal fatigue tests: total strain and temperature changes. **a** Test I; **b** test II; **c** test III

The stress strain loops for all the tests are presented in Fig. 7.14. Qualitatively different results are obtained: without temperature rate terms the response exhibits unreasonable shift of hysteresis loops along the stress axis, while including additional temperature rate-dependent terms allows to preserve stable behavior. Such effect was already indicated by Chaboche [16], and is expected to be even more significant for materials exhibiting cyclic hardening (here the shift of stress–strain loops in test I and test III is mitigated by cyclic softening material effect).

The quantitative difference between both cases, which represents the influence of temperature rate on the response of the considered constitutive model, can be estimated from curves in Fig. 7.15 showing the evolution of maximal stress versus cumulated plastic strain. When full coupling with temperature is accounted, the numerical simulations of test I and III confirm cyclic softening behavior throughout

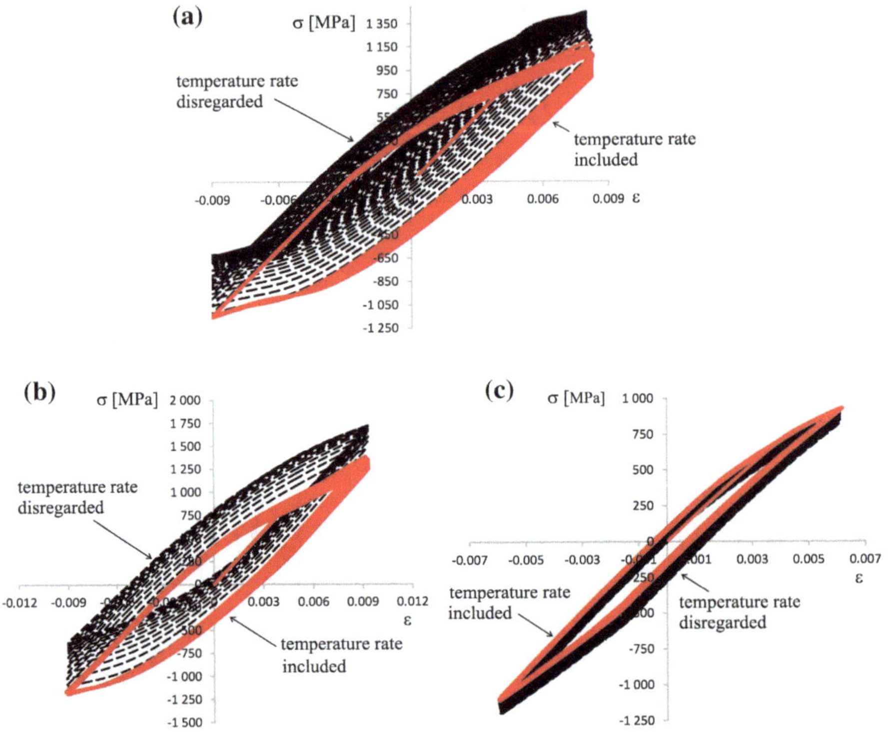

Fig. 7.14 Hysteresis loops for **a** test I; **b** test II; **c** test III

all the fatigue tests. If the influence of temperature rate is neglected, and change of temperature during the tests is accounted only in changing material parameters (partial coupling, case 1), the numerical simulations exhibit cyclic hardening of the material, which is in contradiction with the experimental observations (at least for isothermal conditions, see Fig. 7.10). The maximum observed difference between the values of maximum stress on cycle is here even as large as 30 %, and is observed during the first stage of fatigue test (rapid softening). The results presented here indicate that coupling between temperature and dissipative phenomena taking place in the material may have a significant influence on the response of a constitutive model. Disregarding the rate of temperature in the evolution of thermodynamic forces related to hardening effects may lead to erroneous results, especially when solving high temperature problems, such as fire conditions or thermal shock.

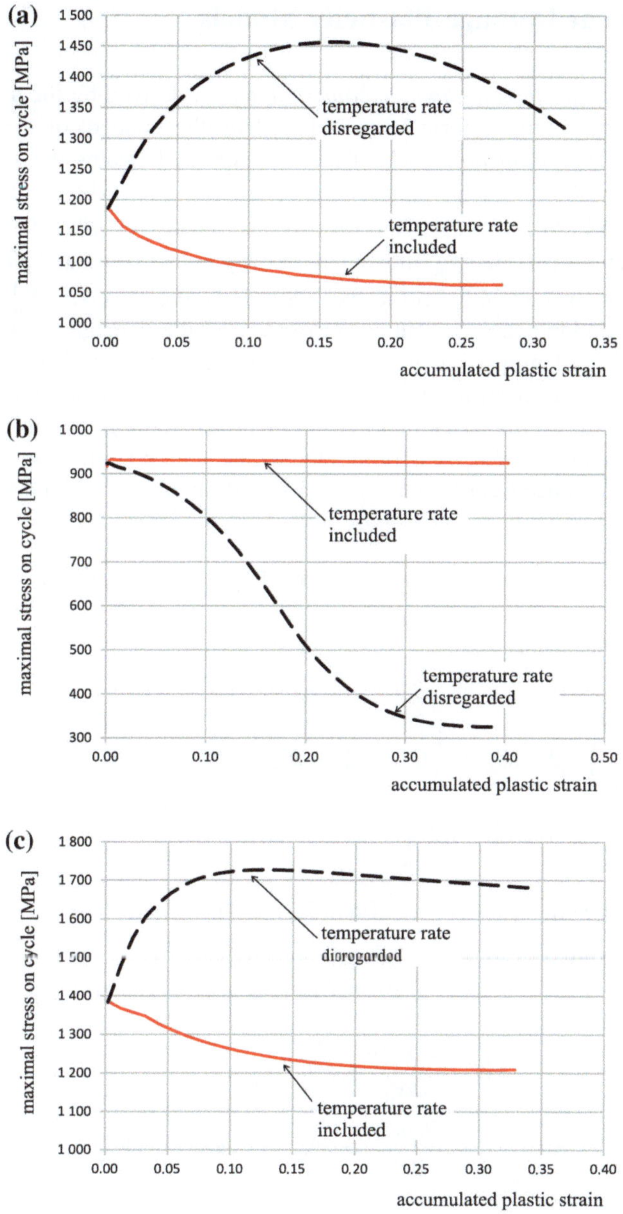

Fig. 7.15 Evolution of maximal stress on cycle versus accumulated plastic strain: **a** test I; **b** test II; **c** test III

7.7 Nonlinear Damage-Plasticity Models

The general *elastic-plastic-damage constitutive law*, derived by the use of the irreversible thermodynamics formalism, is based on the concept of the *state potential* and the *dissipation potentials*. The state potential relates the set of *internal state variables* $V_\alpha = \{D_{ij}; \alpha_{ij}^{\mathrm{p}}, r^{\mathrm{p}}, \alpha_{ij}^{\mathrm{d}}, r^{\mathrm{d}}\}$ to the set of *thermodynamic forces* $J_\alpha = \{-Y_{ij}^{\mathrm{d}}; X_{ij}^{\mathrm{p}}, R^{\mathrm{p}}, X_{ij}^{\mathrm{d}}, R^{\mathrm{d}}\}$. The pairs $(\alpha_{ij}^{\mathrm{p}}, X_{ij}^{\mathrm{p}})$ and $(r^{\mathrm{p}}, R^{\mathrm{p}})$ refer, respectively, to plastic kinematic hardening (movement of the yield surface), and plastic isotropic hardening (dimension of the yield surface). Similarly, the pairs $(\alpha_{ij}^{\mathrm{d}}, X_{ij}^{\mathrm{d}})$ and $(r^{\mathrm{d}}, R^{\mathrm{d}})$ refer to damage kinematic hardening (movement of the damage surface), and damage isotropic hardening (dimension of the damage surface). The additional pair $(D_{ij}, -Y_{ij}^{\mathrm{d}})$ refers to *anisotropic damage variable* and elastic strain energy density release rate as the thermodynamic force conjugated to damage variable. Usually, the *Helmholtz free energy density* ψ is adopted as the state potential, where apart from the damage influenced elastic term $\rho\psi^{\mathrm{e}}$ two additional terms stand for the *plastic hardening* $\rho\psi^{\mathrm{p}}$ and *damage hardening* $\rho\psi^{\mathrm{d}}$ (see Eq. (7.6)):

$$\rho\psi = \rho\psi^{\mathrm{e}}(\varepsilon_{ij}^{\mathrm{e}}, D_{ij}) + \rho\psi^{\mathrm{p}}(\alpha_{ij}^{\mathrm{p}}, r^{\mathrm{p}}) + \rho\psi^{\mathrm{d}}(\alpha_{ij}^{\mathrm{d}}, r^{\mathrm{d}}) \tag{7.69}$$

The specific free energy may be for example assumed in the following nonlinear form [3]:

$$\begin{aligned} \rho\psi = {} & \frac{1}{2}\varepsilon_{ij}^{\mathrm{e}} E_{ijkl}\varepsilon_{kl}^{\mathrm{e}} + \frac{1}{3}C^{\mathrm{p}}\alpha_{ij}^{\mathrm{p}}\alpha_{ij}^{\mathrm{p}} + R_\infty^{\mathrm{p}}\left[r^{\mathrm{p}} + \frac{1}{b^{\mathrm{p}}}\exp(-b^{\mathrm{p}}r^{\mathrm{p}})\right] \\ & + \frac{1}{2}C^{\mathrm{d}}\alpha_{ij}^{\mathrm{d}}\alpha_{ij}^{\mathrm{d}} + R_\infty^{\mathrm{d}}\left[r^{\mathrm{d}} + \frac{1}{b^{\mathrm{d}}}\exp(-b^{\mathrm{d}}r^{\mathrm{d}})\right] \end{aligned} \tag{7.70}$$

Hence, the state equations are (see Eqs. (7.7) and (7.8)):

$$\begin{aligned} \sigma_{ij} &= \frac{\partial(\rho\psi)}{\partial\varepsilon_{ij}^{\mathrm{e}}} = E_{ijkl}(D_{pq})\varepsilon_{kl}^{\mathrm{e}} \\ X_{ij}^{\mathrm{p}} &= \frac{\partial(\rho\psi)}{\partial\alpha_{ij}^{\mathrm{p}}} = \frac{2}{3}C^{\mathrm{p}}\alpha_{ij}^{\mathrm{p}} \\ R^{\mathrm{p}} &= \frac{\partial(\rho\psi)}{\partial r^{\mathrm{p}}} = R_\infty^{\mathrm{p}}[1 - \exp(-b^{\mathrm{p}}r^{\mathrm{p}})] \\ X_{ij}^{\mathrm{d}} &= \frac{\partial(\rho\psi)}{\partial\alpha_{ij}^{\mathrm{d}}} = C^{\mathrm{d}}\alpha_{ij}^{\mathrm{d}} \\ R^{\mathrm{d}} &= \frac{\partial(\rho\psi)}{\partial r^{\mathrm{d}}} = R_\infty^{\mathrm{d}}[1 - \exp(-b^{\mathrm{d}}r^{\mathrm{d}})] \\ Y_{ij}^{\mathrm{d}} &= -\frac{\partial(\rho\psi)}{\partial D_{ij}} \end{aligned} \tag{7.71}$$

In general, each of the two dissipation mechanisms, plasticity and damage, can appear independently, hence two *dissipation surfaces*, for *plasticity* F^{p} and for *damage* F^{d}, may be defined, for example in a following way:

$$
\begin{aligned}
F^{\mathrm{p}} &= f^{\mathrm{p}} + \frac{3}{4}\frac{\gamma^{\mathrm{p}}}{C^{\mathrm{p}}}\widetilde{X}^{\mathrm{p}}_{ij}\widetilde{X}^{\mathrm{p}}_{ij} \\
f^{\mathrm{p}} &= \sqrt{\frac{3}{2}(\widetilde{s}_{ij} - \widetilde{X}'^{\mathrm{p}}_{ij})(\widetilde{s}_{ij} - \widetilde{X}'^{\mathrm{p}}_{ij})} - (R^{\mathrm{p}}_0 + \widetilde{R}^{\mathrm{p}})
\end{aligned}
\tag{7.72}
$$

and

$$
\begin{aligned}
F^{\mathrm{d}} &= f^{\mathrm{d}} + \frac{1}{2}\frac{\gamma^{\mathrm{d}}}{C^{\mathrm{d}}}X^{\mathrm{d}}_{ij}X^{\mathrm{d}}_{ij} \\
f^{\mathrm{d}} &= \sqrt{(Y^{\mathrm{d}}_{ij} - X^{\mathrm{d}}_{ij})(Y^{\mathrm{d}}_{ij} - X^{\mathrm{d}}_{ij})} - (R^{\mathrm{d}}_0 + R^{\mathrm{d}})
\end{aligned}
\tag{7.73}
$$

where symbols R^{p}_0 and R^{d}_0 denote the initial sizes of the yield and damage surfaces. In what is presented above, the loading functions f^{p} and f^{d} stand for the *yield function* and *damage function*, respectively. Loading functions may or may not be equal to the *dissipation potential functions* F^{p} and F^{d}. The cases when $f^{\mathrm{p}} \neq F^{\mathrm{p}}$ and/or $f^{\mathrm{d}} \neq F^{\mathrm{d}}$ refer to the *nonassociated plasticity* and/or *nonassociated damage*. In other words, two potential functions F^{p} and F^{d} serve to define so-called the *generalized normality rules*, for plasticity and damage, as follows (see Eq. (7.20)):

$$
\dot{\varepsilon}^{\mathrm{p}}_{ij} = \dot{\lambda}^{\mathrm{p}}\frac{\partial F^{\mathrm{p}}}{\partial \sigma_{ij}}, \qquad \dot{D}_{ij} = \dot{\lambda}^{\mathrm{d}}\frac{\partial F^{\mathrm{d}}}{\partial Y^{\mathrm{d}}_{ij}}
\tag{7.74}
$$

Two loading functions, f^{p} and f^{d}, serve to determine two dissipation multipliers, $\dot{\lambda}^{\mathrm{p}}$ and $\dot{\lambda}^{\mathrm{d}}$. The multipliers define magnitudes of plastic and damage increments, satisfying current dissipation functions (loading functions) f^{p}_{n+1} and f^{d}_{n+1} (the consistency conditions), as shown in Fig. 7.16. In the case of the associated rules, both potential functions are equal to the dissipation functions, $F^{\mathrm{p}} = f^{\mathrm{p}}$ and $F^{\mathrm{d}} = f^{\mathrm{d}}$, so that the same surfaces are used for the normality rules and the consistency conditions, hence:

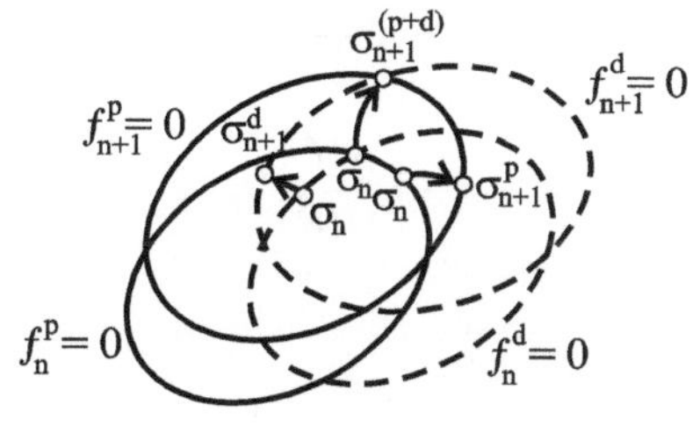

Fig. 7.16 Loading surfaces, after Ganczarski et al. [30]

$$\dot{\varepsilon}^{\mathrm{p}}_{ij} = \dot{\lambda}^{\mathrm{p}} \frac{\partial f^{\mathrm{p}}}{\partial \sigma_{ij}}, \qquad \dot{D}_{ij} = \dot{\lambda}^{\mathrm{d}} \frac{\partial f^{\mathrm{d}}}{\partial Y^{\mathrm{d}}_{ij}} \tag{7.75}$$

and $\dot{f}^{\mathrm{p}} = 0$, $\dot{f}^{\mathrm{d}} = 0$. Note also that both dissipation surfaces, plasticity and damage, can undergo mixed hardening rules (7.72$_2$), (7.73$_2$). Though both mechanisms (plastic slip and/or microcrack growth) can exist separately, the subsequent loading functions f^{p} and f^{d} can be changed on loading steps not only due to plastic strains, but also due to prior damage evolution (if any). This effect is hidden in the symbol tilde ($\widetilde{\ }$) placed over the symbols that refer to the *damage effective variables*, as follows:

$$\widetilde{\sigma}_{ij} = M_{ijkl}\sigma_{kl}, \quad \widetilde{X}^{\mathrm{p}}_{ij} = M_{ijkl}X^{\mathrm{p}}_{kl}, \quad \widetilde{R}^{\mathrm{p}} = \frac{R^{\mathrm{p}}}{1-\sqrt{D_{ij}D_{ij}}} \tag{7.76}$$

Symbols $M_{ijkl}(D_{pq})$ stand for the components of the fourth-order *damage effect tensor*. Chosen representations of this tensor, based on various hypotheses, are given in Sect. 1.2.3.

By the use of the *generalized normality rule* we arrive at the following equations for plasticity and damage:

$$\begin{aligned}
\dot{\varepsilon}^{\mathrm{p}}_{ij} &= \dot{\lambda}^{\mathrm{p}} \frac{\partial F^{\mathrm{p}}}{\partial \sigma_{ij}} = \frac{3}{2}\dot{\lambda}^{\mathrm{p}} \frac{(\widetilde{s}_{kl} - \widetilde{X}'^{\mathrm{p}}_{kl})M_{kilj}}{\sqrt{\frac{3}{2}(\widetilde{s}_{pq} - \widetilde{X}'^{\mathrm{p}}_{pq})(\widetilde{s}_{pq} - \widetilde{X}'^{\mathrm{p}}_{pq})}} \\
\dot{D}_{ij} &= \dot{\lambda}^{\mathrm{d}} \frac{\partial F^{\mathrm{d}}}{\partial Y^{\mathrm{d}}_{ij}} = \dot{\lambda}^{\mathrm{d}} \frac{Y^{\mathrm{d}}_{ij} - X^{\mathrm{d}}_{ij}}{\sqrt{(Y^{\mathrm{d}}_{rs} - X^{\mathrm{d}}_{rs})(Y^{\mathrm{d}}_{rs} - X^{\mathrm{d}}_{rs})}} \\
\dot{\alpha}^{\mathrm{p}}_{ij} &= -\dot{\lambda}^{\mathrm{p}} \frac{\partial F^{\mathrm{p}}}{\partial X^{\mathrm{p}}_{ij}} = -\frac{3}{2}\dot{\lambda}^{\mathrm{p}} M_{ijkl} \left[-\frac{\widetilde{s}_{kl} - \widetilde{X}'^{\mathrm{p}}_{kl}}{\sqrt{\frac{3}{2}(\widetilde{s}_{pq} - \widetilde{X}'^{\mathrm{p}}_{pq})(\widetilde{s}_{pq} - \widetilde{X}'^{\mathrm{p}}_{pq})}} + \frac{\gamma^{\mathrm{p}}}{C^{\mathrm{p}}}\widetilde{X}'^{\mathrm{p}}_{kl} \right] \\
\dot{r}^{\mathrm{p}} &= -\dot{\lambda}^{\mathrm{p}} \frac{\partial F^{\mathrm{p}}}{\partial R^{\mathrm{p}}} = \frac{\dot{\lambda}^{\mathrm{p}}}{1-\sqrt{D_{ij}D_{ij}}} \\
\dot{\alpha}^{\mathrm{d}}_{ij} &= -\dot{\lambda}^{\mathrm{d}} \frac{\partial F^{\mathrm{d}}}{\partial X^{\mathrm{d}}_{ij}} = -\frac{3}{2}\dot{\lambda}^{\mathrm{d}} \left[-\frac{Y^{\mathrm{d}}_{kl} - X^{\mathrm{d}}_{kl}}{\sqrt{(Y^{\mathrm{d}}_{pq} - X^{\mathrm{d}}_{pq})(Y^{\mathrm{d}}_{pq} - X^{\mathrm{d}}_{pq})}} + \frac{\gamma^{\mathrm{d}}}{C^{\mathrm{d}}}X^{\mathrm{d}}_{kl} \right] \\
\dot{r}^{\mathrm{d}} &= -\dot{\lambda}^{\mathrm{d}} \frac{\partial F^{\mathrm{d}}}{\partial R^{\mathrm{d}}} = \dot{\lambda}^{\mathrm{d}}
\end{aligned} \tag{7.77}$$

The thermodynamic force rates $X_{ij}^{\rm p}$, $R^{\rm p}$, and $X_{ij}^{\rm d}$, R^d conjugate of hardening variables $\alpha_{ij}^{\rm p}$, $r^{\rm p}$, and α_{ij}^d, r^d are as follows:

$$\begin{aligned}
\dot{X}_{ij}^{\rm p} &= \frac{2}{3}C^{\rm p}\dot{\varepsilon}_{ij}^{\rm p} - \gamma^{\rm p} M_{ikjl}M_{pkql}X_{pq}^{\rm p}(1-\sqrt{D_{ij}D_{ij}})\dot{r}^{\rm p}\\
\dot{R}^{\rm p} &= b^{\rm p}(R_{\infty}^{\rm p} - R^{\rm p})\dot{r}^{\rm p}\\
\dot{X}_{ij}^{\rm d} &= \left[C^{\rm d}\frac{Y_{ij}^{\rm d}-X_{ij}^{\rm d}}{\sqrt{(Y_{pq}^{\rm d}-X_{pq}^{\rm d})(Y_{pq}^{\rm d}-X_{pq}^{\rm d})}} - \gamma^{\rm d}X_{ij}^{\rm d}\right]\dot{r}^{\rm d}\\
\dot{R}^{\rm d} &= b^{\rm d}(R_{\infty}^{\rm d} - R^{\rm d})\dot{r}^{\rm d}
\end{aligned} \tag{7.78}$$

In what is given above, the plasticity/damage couplings are introduced by the fourth-rank damage effect tensor $M_{ijkl}(D_{pq})$ and the damage equivalent $\sqrt{D_{ij}D_{ij}}$. In the case of nondamage plasticity, the classical Armstrong–Frederick formulas for nonlinear plasticity are recovered (see Eq. (7.34)) whereas both formulas for $\dot{X}_{ij}^{\rm d}$ and $\dot{R}^{\rm d}$ vanish. This also means that the Armstrong and Frederick law may be considered as the intrinsically nonassociated plasticity rule, when the thermodynamic potential-based formulation is used. In order to derive the evolution equations for coupled plasticity (7.75_1) and damage (7.75_2) the postulate of maximum mechanical dissipation may also be used. The nonnegative dissipation function is defined in the form:

$$\sqcap = \sigma_{ij}\dot{\varepsilon}_{ij}^{\rm p} - X_{ij}^{\rm p}\dot{\alpha}_{ij}^{\rm p} - R^{\rm p}\dot{r}^{\rm p} - X_{ij}^{\rm d}\dot{\alpha}_{ij}^{\rm d} - R^{\rm d}\dot{r}^{d} + Y_{ij}^{\rm d}\dot{D}_{ij} \geq 0 \tag{7.79}$$

The dissipation function is subjected to two constraints, $f^{\rm p}=0$ and $f^{\rm d}=0$. Hence, introducing two Lagrange multipliers $\dot{\lambda}^{\rm p}$ and $\dot{\lambda}^{\rm d}$ and maximizing the new functional:

$$\begin{aligned}
\bar{\sqcap} = {} & \sigma_{ij}\dot{\varepsilon}_{ij}^{\rm p} - X_{ij}^{\rm p}\dot{\alpha}_{ij}^{\rm p} - R^{\rm p}\dot{r}^{\rm p} - X_{ij}^{\rm d}\dot{\alpha}_{ij}^{\rm d} - R^{\rm d}\dot{r}^{\rm d} + Y_{ij}^{\rm d}\dot{D}_{ij}\\
& -\dot{\lambda}^{\rm p}f^{\rm p} - \dot{\lambda}^{\rm d}f^{\rm d} \longrightarrow \max
\end{aligned} \tag{7.80}$$

we arrive at the evolution equations for plasticity (7.75_1) and damage (7.75_2) if the implicit formulas for $f^{\rm p}$ and $f^{\rm d}$ are used. For more details, see [3, 26].

The dissipation multipliers $\dot{\lambda}^{\rm p}$ and $\dot{\lambda}^{\rm d}$ obey the following loading/unloading conditions:

$$\begin{aligned}
f^{\rm p} \leq 0 \text{ and } & \begin{cases} \dot{f}^{\rm p} < 0 \text{ and } \dot{\lambda}^{\rm p} = 0 \text{ passive plastic}\\ \dot{f}^{\rm p} = 0 \text{ and } \dot{\lambda}^{\rm p} = 0 \text{ neutral plastic}\\ \dot{f}^{\rm p} = 0 \text{ and } \dot{\lambda}^{\rm p} > 0 \text{ active plastic}\end{cases}\\
f^{\rm d} \leq 0 \text{ and } & \begin{cases} \dot{f}^{\rm d} < 0 \text{ and } \dot{\lambda}^{\rm d} = 0 \text{ passive damage}\\ \dot{f}^{\rm d} = 0 \text{ and } \dot{\lambda}^{\rm d} = 0 \text{ neutral damage}\\ \dot{f}^{\rm d} = 0 \text{ and } \dot{\lambda}^{\rm d} > 0 \text{ active damage}\end{cases}
\end{aligned} \tag{7.81}$$

Both dissipation multipliers are obtained from two consistency conditions (cf. [3]):

$$\begin{aligned}\dot{f}^{\mathrm{p}} &= \frac{\partial f^{\mathrm{p}}}{\partial \sigma_{ij}}\dot{\sigma}_{ij} + \frac{\partial f^{\mathrm{p}}}{\partial X_{ij}^{\mathrm{p}}}\dot{X}_{ij}^{p} + \frac{\partial f^{\mathrm{p}}}{\partial R^{\mathrm{p}}}\dot{R}^{\mathrm{p}} + \frac{\partial f^{\mathrm{p}}}{\partial D_{ij}}\dot{D}_{ij} = 0 \\ \dot{f}^{\mathrm{d}} &= \frac{\partial f^{\mathrm{d}}}{\partial \sigma_{ij}}\dot{\sigma}_{ij} + \frac{\partial f^{\mathrm{d}}}{\partial X_{ij}^{\mathrm{d}}}\dot{X}_{ij}^{\mathrm{d}} + \frac{\partial f^{\mathrm{d}}}{\partial R^{\mathrm{d}}}\dot{R}^{\mathrm{d}} + \frac{\partial f^{\mathrm{d}}}{\partial D_{ij}}\dot{D}_{ij} = 0\end{aligned} \tag{7.82}$$

7.7.1 Conclusions

The description of inelastic behavior of engineering materials requires the mathematical formulations for yield functions, flow rules, and hardening laws, appropriate for the class of materials under consideration, and obeying the mechanical and physical principles. A separate aspect of modeling concerns identifying the constitutive parameters. So far, the influence of damage on most of material characteristics is usually not accounted for in the models due to the existing gap between the formulated constitutive equations and the possibilities to identify the material parameters. However, fast development of computational possibilities allows to simulate numerically even very complex problems. In addition, with the increased attention paid to many innovative materials of complex microstructure, and a deeper understanding of the physical meaning of material characteristics, together with the development of advanced experimental techniques which allow for the determination of structural features such as size and volume fractions of microstructural inhomogeneities in a variety of materials, the identification becomes much more well founded.

References

1. Aboudi, J.: The effect of anisotropic damage evolution on the behavior of ductile and brittle matrix composites. Int. J. Solids Struct. **48**(14–15), 2102–2119 (2011)
2. Abu Al-Rub, R.K., Darabi, M.K.: A thermodynamic framework for constitutive modeling of time-and rate-dependent materials, part I: theory. Int. J. Plast. **34**, 61–92 (2012)
3. Abu Al-Rub, R.K., Voyiadjis, G.Z.: On the coupling of anisotropic damage and plasticity models for ductile materials. Int. J. Solids Struct. **40**, 2611–2643 (2003)
4. Barlat, F., Brem, J.C., Yoon, J.W., Chung, K., Dick, R.E., Lege, D.J., Pourboghrat, F., Choi, S.-H., Chu, E.: Plane stress function for aluminium alloy sheets—part I: theory. Int. J. Plast. **19**, 1297–1319 (2003)
5. Barlat, F., Aretz, H., Yoon, J.W., Karabin, M.E., Brem, J.C., Dick, R.E.: Linear transformation-based anisotropic yield functions. Int. J. Plast. **21**, 1009–1039 (2005)
6. Barlat, F., Gracio, J.J., Lee, M.-G., Rauch, E.F., Vincze, G.: An alternative to kinematic hardening in classical plasticity. Int. J. Plast. **27**, 1309–1327 (2011)
7. Bednarek, Z., Kamocka, R.: The heating rate impact on parameters characteristic of steel behaviour under fire conditions. J. Civ. Eng. Manag. **12**(4), 269–275 (2006)

8. Benallal, A., Bigoni, D.: Effects of temperature and thermo-mechanical couplings on material instabilities and strain localization of inelastic materials. J. Mech. Phys. Solids **52**, 725–753 (2004)
9. Bernhart, G., Moulinier, G., Brucelle, O., Delagnes, D.: High temperature low cycle fatigue behaviour of a martensite forging tool steel. Int. J. Fatigue **21**(2), 179–186 (1999)
10. Blaj, L., Cailletaud, G.: Application of a multimechanism model to the prediction of ratcheting behavior. In: Miannay, D., Costa, P., François, D. (eds.) Advances in Mechanical Behaviour, Plasticity and Damage, SF2M, vol. 2, pp. 1155–1160 (2000)
11. Boudifa, M., Saanouni, K., Chaboche, J.L.: A micromechanical model for inelastic ductile damage prediction in polycrystalline metals for metal forming. Int. J. Mech. Sci. **51**, 453–464 (2009)
12. Cailletaud, G., Saï, K.: Study of plastic/viscoplastic models with various inelastic mechanisms. Int. J. Plast. **11**, 991–1005 (1995)
13. Casey, J.: On elastic-thermo-plastic materials at finite deformations. Int. J. Plast. **14**, 173–191 (1998)
14. Cazacu, O., Barlat, F.: A criterion for description of anisotropy and yield differential effects in pressure-insensitive materials. Int. J. Plast. **20**, 2027–2045 (2004)
15. Cazacu, O., Plunkett, B., Barlat, F.: Orthotropic yield criterion for hexagonal closed packed metals. Int. J. Plast. **22**, 1171–1194 (2006)
16. Chaboche, J.L.: A review of some plasticity and viscoplasticity constitutive theories. Int. J. Plast. **24**, 1642–1693 (2008)
17. Chaboche, J.L., Kruch, S., Maire, J.F., Pottier, T.: Towards a micromechanics based inelastic and damage modeling of composites. Int. J. Plast. **17**, 411–439 (2001)
18. Chaboche, J.L.: Thermodynamic formulation of constitutive equations and application to the viscoplasticity and viscoelasticity of metals and polymers. Int. J. Solids Struct. **34**(18), 2239–2254 (1997)
19. Chaboche, J.L.: Viscoplastic constitutive equations for the description of cyclic and anisotropic behaviour of metals. Bulletin de L'Academie Polonaise des Sciences. Série des Sciences Techniques **XXV**(1), 33–39 (1997)
20. Chaboche, J.L.: Time independent constitutive theories for cyclic plasticity. Int. J. Plast. **2**(2), 149–188 (1986)
21. Chaboche, J.L., Rousselier, G.: On the plastic and viscoplastic constitutive equations, parts I and II. J. Press. Vessel Technol., ASME **105**, 153–164 (1983)
22. Challamel, N., Lanos, C., Casandjian, C.: Strain-based anisotropic damage modeling and unilateral effects. Int. J. Mech. Sci. **47**, 459–473 (2005)
23. Chen, X.F., Chow, C.L.: On damage strain energy release rate Y. Int. J. Damage Mech. **4**, 236–251 (1995)
24. Chow, C.L., Lu, T.J.: An analytical and experimental study of mixed-mode ductile fracture under nonproportional loading. Int. J. Damage Mech. **1**, 191–236 (1992)
25. Egner, H.: Non-isothermal coupled thermo-damage-plasticity. In: Hetnarski, R. (ed.) Encyclopedia of Thermal Stresses, pp. 3356–3368. Springer, Berlin (2014)
26. Egner, H.: On the full coupling between thermo-plasticity and thermo-damage in thermodynamic modeling of dissipative materials. Int. J. Solids Struct. **49**, 279–288 (2012)
27. Egner, H., Egner, W.: Modeling of a tempered martensitic hot work tool steel behavior in the presence of thermo-viscoplastic coupling. Int. J. Plast. **57**, 77–91 (2014)
28. Egner, H., Egner, W.: Modeling of coupled dissipative phenomena in engineering materials. In: Altenbach, H., Kruch, S. (eds.) Advanced Materials Modeling for Structures. Series Advanced Structured Materials, vol. 19, pp. 141–151. Springer, Berlin (2013)
29. Ganczarski, A.: Thermal anisotropy inducing brittle damage. Tech. Mech. **19**, 321–330 (1999)
30. Ganczarski, A.W., Egner, H., Muc, A., Skrzypek, J.J.: Constitutive models for analysis and design of multifunctional technological materials. In: Rustichelli, F., Skrzypek, J.J. (eds.) Innovative Technological Materials. Structural Properties by Neutrons, Synchrotron Radiation and Modelling, pp. 179–220. Springer, New York (2010)

31. Ganczarski, A., Skrzypek, J.J.: A study on coupled thermo-elasto-plastic-damage dissipative phenomena: models and application to some innovative materials. J. Therm. Stress. **32**, 698–751 (2009)
32. Ganczarski, A., Skrzypek, J.: Mechanics of Novel Materials (in Polish). Wydawnictwo Politechniki Krakowskiej, Kraków (2013)
33. Gibbons, C.L., Dunn, J.E.: Investigations of reduced service life of hot work (Cr-Mo) die steel pieces. Ind. Heat. **47**, 6–9 (1980)
34. Golański, G., Mroziński, S.: Low cycle fatigue and cyclic softening behavior of martensitic cast steel. Eng. Fail. Anal. **35**, 692–702 (2013)
35. Golański, G., Mroziński, S.: Fatigue life at 550 °C temperature of aged martensitic cast steel. AASRI Procedia **2**, 249–255 (2013)
36. Halphen, B., Nguyen, Q.S.: Sur les matériaux standards généralisés. Journal de Méchanique **14**, 39–63 (1975)
37. Hartmann, G.: Comparison of the uniaxial behavior of the inelastic constitutive models of Miller and Walker by numerical experiments. Int. J. Plast. **6**, 189–206 (1990)
38. Hershey, A.V.: The plasticity of an isotropic aggregate of anisotropic face centered cubic crystals. J. Appl. Mech. **21**, 241–249 (1954)
39. Houlsby, G.T., Puzrin, A.M.: A thermomechanical framework for constitutive models for rate-independent dissipative materials. Int. J. Plast. **16**, 1017–1047 (2000)
40. Jiang, Y., Zhang, J.: Benchmark experiments and characteristic cyclic plasticity deformation. Int. J. Plast. **24**, 1481–1515 (2008)
41. Khan, A.S., Liu, H.: Strain rate and temperature dependent fracture criteria for isotropic and anisotropic metals. Int. J. Plast. **37**, 1–15 (2012)
42. Khan, A.S., Yu, S., Liu, H.: Deformation enhanced anisotropic responses of Ti-6Al-4V alloy, part II: a stress rate and temperature dependent anisotropic yield criterion. Int. J. Plast. **38**, 14–26 (2012)
43. Kowalsky, U., Ahrens, H., Dinkler, D.: Distorted yield surfaces-modelling by higher order anisotropic hardening tensors. Comput. Mater. Sci. **16**, 81–88 (1999)
44. Krajcinovic, D.: Constitutive theory of damaging materials. ASME J. Appl. Mech. **50**, 355–360 (1983)
45. Kuo, Ch.-M., Lin, Ch.-S.: Static recovery activation energy of pure copper at room temperature. Scripta Materialia **57**, 667–670 (2007)
46. Lebensohn, R.A., Tome, C.N.: A self-consistent anisotropic approach for the simulation of plastic deformation and texture development of polycrystals: application to zirconium alloys. Acta Metall. Mater. **41**, 2611–2624 (1993)
47. Lee, K.D., Krempl, E.: An orthotropic theory of viscoplasticity based on overstress for thermomechanical deformation. Int. J. Solids Struct. **27**, 1445–1459 (1991)
48. Lemaitre, J., Chaboche, J.L.: Mechanics of Solid Materials. Cambridge University Press, London (1990)
49. Litewka, A.: Effective material constants for orthotropically damaged elastic solids. Arch. Mech. Stos. **37**(6), 631–642 (1985)
50. McDowell, D.L.: A nonlinear kinematic hardening theory for cyclic thermoplasticity and thermoviscoplasticity. Int. J. Plast. **8**, 695–728 (1992)
51. Mebarki, N., Delagnes, D., Lamesle, P., Delmas, F., Levaillant, C.: Relationship between microstructure and mechanical properties of a 5%Cr tempered martensitic tool steel. Mater. Sci. Eng. A **387–389**(1–2), 171–175 (2004)
52. Mirzakhani, B., Salehi, M.T., Khoddam, S., Seyedin, S.H., Aboutalebi, M.R.: Investigation of dynamic and static recrystallization behavior during thermomechanical processing in a API-X70 microalloyed steel. J. Mater. Eng. Perform. **18**(8), 1029–1034 (2009)
53. Moreno, V., Jordan, E.H.: Prediction of material thermomechanical response with a unified viscoplastic constitutive model. Int. J. Plast. **2**, 223–245 (1986)
54. Murakami, S., Ohno, N.: A continuum theory of creep and creep damage. In: Ponter, A.R.S., Hayhurst, D.R. (eds.) Creep in Structures, 3rd IUTAM Symposium on Creep in Structures, pp. 422–444. Springer, Berlin (1981)

55. Nixon, M.E., Cazacu, O., Lebensohn, R.A.: Anisotropic response of high-purity α-titanium: experimental characterization and constitutive modeling. Int. J. Plast. **26**, 516–532 (2010)
56. Ohno, N.: Recent topics in constitutive modeling for cyclic plasticity and viscoplasticity. Appl. Mech. Rev. **43**(11), 283–295 (1990)
57. Ohno, N., Takahashi, Y., Kubawara, K.: Constitutive modeling of anisothermal cyclic plasticity of 304 stainless steel. J. Eng. Mater. Technol. **111**, 106–114 (1989)
58. Onsager, L.: Reciprocal relations in irreversible thermodynamics I. Phys. Rev. **37**, 405–426 (1931)
59. Onsager, L.: Reciprocal relations in irreversible thermodynamics II. Phys. Rev. **38**, 2265–2279 (1931)
60. Ottosen, N.S., Ristinmaa, M.: The Mechanics of Constitutive Modeling. Elsevier, Amsterdam (2005)
61. Phillips, A., Tang, J.L.: The effect of loading path on the yield surface at elevated temperatures. Int. J. Solids Struct. **8**, 463–474 (1972)
62. Plunkett, B., Lebensohn, R.A., Cazacu, O., Barlat, F.: Anisotropic yield function of hexagonal materials taking into account texture development and anisotropic hardening. Acta Mater. **54**, 4159–4169 (2006)
63. Plunkett, B., Cazacu, O., Lebensohn, R.A., Barlat, F.: Elastic-viscoplastic modeling of textured metals and validation using the Taylor cylinder impact test. Int. J. Plast. **23**, 1001–1021 (2007)
64. Plunkett, B., Cazacu, O.: Viscoplastic modeling of anisotropic textured metals. In: Cazacu, O. (ed.) Multiscale Modeling of Heterogeneous Materials: From Microstructure to Macro-Scale Properties, pp. 111–126. ISTE/Wiley, New York (2008)
65. Plunkett, B., Cazacu, O., Barlat, F.: Orthotropic yield criteria for description of the anisotropy in tension and compression of sheet metal. Int. J. Plast. **24**, 847–866 (2008)
66. Prager, W.: Non-isothermal plastic deformation. In: Proceedings, Konnickl Nederl. Akademie Van Wetenschappen Te Amsterdam, Series B, vol. 61 (1958)
67. Saanouni, K.: Damage Mechanics in Metal Forming, Advanced Modeling and Numerical Simulation. Wiley, London (2012)
68. Saï, K.: Multi-mechanism models: present state and future trends. Int. J. Plast. **27**, 250–281 (2011)
69. Skrzypek, J.J., Ganczarski, A.: Modeling of damage effect on heat transfer in time-dependent nonhomogeneous solids. J. Therm. Stress. **21**, 205–231 (1998)
70. Skrzypek, J.J., Ganczarski, A.W., Rustichelli, F., Egner, H.: Advanced Materials and Structures for Extreme Operating Conditions. Springer, Berlin (2008)
71. Streilein, T.: Erfassung formativer Verfestigung in viskoplastischen Stoffmodellen, pp. 97–83. Institut für Statik der TU Braunschweig (1997)
72. Swift, H.W.: Plastic instability under plane stress. J. Mech. Phys. Solids **1**, 1–18 (1952)
73. Taleb, L., Cailletaud, G.: An updated version of the multimechanism model for cyclic plasticity. Int. J. Plast. **26**, 859–874 (2010)
74. Velay, V., Bernhart, G., Penazzi, L.: Cyclic behavior modeling of a tempered martensitic hot work tool steel. Int. J. Plast. **22**, 459–496 (2006)
75. Xiao, H., Bruhns, T., Meyers, A.: Thermodynamic laws and consistent Eulerian formulation of finite elastoplasticity with thermal effects. J. Mech. Phys. Solids **55**, 338–365 (2007)
76. Yoon, J.W., Barlat, F., Gracio, J.J., Rauch, E.: Anisotropic strain hardening behavior in simple shear for cube textured aluminum alloy sheets. Int. J. Plast. **21**, 2426–2447 (2005)
77. Yoon, J.W., Barlat, F., Dick, R.E., Karabin, M.E.: Prediction of six or eight ears in a drawn cup based on a new anisotropic yield function. Int. J. Plast. **22**, 174–193 (2006)
78. Yoon, J.W., Lou, Y., Yoon, J., Glazoff, M.V.: Asymmetric yield function based on the stress invariants for pressure sensitive metals. Int. J. Plast. **56**, 184–202 (2014)
79. Yu, D., Chen, X., Yu, W., Chen, G.: Thermo-viscoplastic modeling incorporating dynamic strain aging effect on the uniaxial behavior of Z2CND18.12N stainless steel. Int. J. Plast. **37**, 119–139 (2012)
80. Zhang, Z., Bernhart, G., Delagnes, D.: Cyclic behavior constitutive modeling of a tempered martensitic steel including ageing effect. Int. J. Fatigue **30**, 706–716 (2008)

81. Zhang, Z., Delagnes, D., Bernhart, G.: Ageing effect on cyclic plasticity of a tempered martensitic steel. Int. J. Fatigue **29**(2), 336–346 (2007)
82. Zhang, Z., Qi, Y., Delagnes, D., Bernhart, G.: Microstructure variation and hardness diminution during low cycle fatigue of 55NiCrMoV7 steel. J. Iron Steel Res. **14**(6), 68–73 (2007)
83. Zhang, Z., Delagnes, D., Bernhart, G.: Anisothermal cyclic plasticity modeling of martensitic steels. Int. J. Fatigue **24**, 635–648 (2002)
84. Ziegler, H.: Some extremum principles in irreversible thermodynamics with applications to continuum mechanics. In: Sneddon, I.N., Hill, R. (eds.) Progress is Solid Mechanics, pp. 92–193. North-Holland, Amsterdam (1963)

Index

J.J. Skrzypek and A.W. Ganczarski (eds.), *Mechanics of Anisotropic Materials*,
Engineering Materials, DOI 10.1007/978-3-319-17160-9

I

K

L

M

MIX
Papier aus verantwortungsvollen Quellen
Paper from responsible sources
FSC® C105338
FSC
www.fsc.org

If you have any concerns about our products,
you can contact us on
ProductSafety@springernature.com

In case Publisher is established outside the EU,
the EU authorized representative is:
Springer Nature Customer Service Center GmbH
Europaplatz 3, 69115 Heidelberg, Germany

Printed by Libri Plureos GmbH
in Hamburg, Germany